Foundation Design
Principles and Practices

Donald P. Coduto, PE, GE

Professor of Civil Engineering
California State Polytechnic University, Pomona

Geotechnical Engineer
Yucaipa, California

PRENTICE HALL, Englewood Cliffs, N.J. 07632

Library of Congress Cataloging-in-Publication Data

Coduto, Donald P.
    Foundation Design: principles and practices / Donald P. Coduto
    p. cm.
    Includes bibliological references and index.
    ISBN: 0-13-335381-8
    1. Foundations. I. Title.
TA775.c63   1994
624.1'5--dc20
                             93-22441
                               CIP

Acquisitions editor: *WILLIAM ZOBRIST*
Production editor: *RICHARD DeLORENZO*
Copy editor: *PETER ZURITA*
Cover design: *TERRAPIN GRAPHICS*
Production coordinator: *LINDA BEHRENS*
Editorial assistant: *DANIELLE ROBINSON*
Supplements editor: *ALICE DWORKIN*

©1994 by Prentice-Hall, Inc.
A Simon & Schuster Company
Englewood Cliffs, New Jersey 07632

The author and publisher of this book have used their best efforts in preparing this book. These efforts include the
development, research, and testing of the theories and programs to determine their effectiveness. The author and
publisher make no warranty of any kind, expressed or implied, with regard to these programs or the documentation
contained in this book. The author and publisher shall not be liable in any event for incidental or consequential damages
in connection with, or arising out of, the furnishing, performance, or use of these programs.

Printed in the United States of America

10 9 8 7

ISBN 0-13-335381-8

Prentice-Hall International (UK) Limited, London
Prentice-Hall of Australia Pty. Limited, Sydney
Prentice-Hall Canada Inc., Toronto
Prentice-Hall Hispanoamericana, S.A., Mexico
Prentice-Hall of India Private Limited, New Delhi
Prentice-Hall of Japan, Inc., Tokyo
Simon & Schuster Asia Pte. Ltd., Singapore
Editora Prentice-Hall do Brasil, Ltda., Rio de Janeiro

To CHC and SHC

We will meet one day

Contents

Contents

Preface

Foundation Design: Principles and Practices is primarily intended to be a textbook for undergraduate and graduate-level foundation engineering courses. It also can serve as a reference book for practicing engineers. As the title implies, it is heavily design-oriented, and discusses methods of applying engineering theories, principles, and research to practical design problems.

Foundation engineering draws upon information and techniques from many disciplines, including geotechnical engineering, structural engineering, and construction engineering. Therefore, this book discusses the roles of each discipline, and emphasizes the interrelationships between them. It is written with the assumption that the readers have completed a university-level course in soil mechanics, and have had at least an introduction to structural engineering. Those with more extensive backgrounds in structural engineering will benefit from the additional detail on structural design in Chapters 10 and 18, and the latter half of Chapter 24, whereas others may choose to bypass these chapters.

Although this is not a book on construction, it does cover aspects of foundation construction that impact the design process, and encourages the reader to develop designs that are easily buildable.

Quantitative analyses are an essential part of the design process, and form the core of this book. All formulas have been normalized and thus are usable with any consistent set of units. Each major chapter includes numerical examples to illustrate analysis and design methods, and practice problems for the reader to solve. Approximately half of them use English units of measurement, and the other half use SI units.

The diskette that accompanies this book includes computer programs for IBM or compatible personal computers. These programs relieve much of the tedium of the analyses, and thus help the reader focus on the subject matter. They also encourage parametric studies.

All engineering analyses must be tempered with an understanding of uncertainties. This is especially true in foundation engineering, because our understanding of the true behavior of foundations is often limited, and our ability to quantify the necessary soil properties is even more limited. Therefore, the text frequently discusses the accuracy (or inaccuracy!) of the analyses. The reader will benefit from comparisons of predicted and actual behavior, and from plots of the data used to develop design formulas.

However, analysis is only one part of the design process. Therefore, the book also includes discussions of qualitative issues, construction methods and equipment, economics, creativity, open-ended problems, and other factors. Historical vignettes describe the development of modern foundation engineering, especially those that address the relationships between theory and practice. Studying this history helps us understand how engineers solved problems of the past, and the kinds of techniques we will use in the future to develop new problem solving techniques.

Finally, the book includes extensive bibliographic references, and Appendix A includes lists of recommended resources for further study. These should be useful to those planning to study certain topics in more detail.

Acknowledgments

Many friends, colleagues, and other professionals contributed to this work. Much of the book is the product of their stimulating discussions, constructive reviews, and support. Robert Schneider, Professor Emeritus at Cal Poly University, provided the initial inspiration and encouragement; Richard Handy (Iowa State University), Raymond Moore (University of Kansas), José Pires (University of California, Irvine), Paul Chan (New Jersey Institute of Technology), William and Sandra Houston (Arizona State University), Dean White, Tom Evans, and others reviewed some or all of the manuscript and provided useful suggestions and comments. Parker Miller, M. J. Schiff, and colleagues at Cal Poly provided additional guidance and advice in their specialty areas. Sue Benney and Lore Gravino of the interlibrary loan office at the Cal Poly University Library did an outstanding job of locating countless obscure references. Copy editing by Peter Zurita and coordination by Rick DeLorenzo gave the book its final polish and brought it into production.

A special note of thanks goes to the hundreds of foundation engineering students at Cal Poly University who suffered through the many sets of course notes and draft manuscripts that ultimately became this book. Their constructive comments and suggestions have made this book much more useful. Finally, I appreciate the patience and support from my wife and children during the long process of writing and editing.

I welcome any constructive comments and suggestions from those who use this book.

Donald P. Coduto
Yucaipa, California

Notation and Units of Measurement

There is no universally accepted notation in foundation engineering. However, the notation used in this book, as described in the following table, is generally consistent with popular usage.

Symbol	Description	Typical Units English	Typical Units SI
A	Area	ft^2 or in^2	m^2 or mm^2
A'	Effective area of a spread footing	ft^2 or in^2	m^2 or mm^2
A_c	Cross-sectional area of concrete	in^2	mm^2
A_e	End bearing contact area	ft^2	m^2
A_s	Skin friction contact area	ft^2	m^2
A_s	Cross-sectional area of steel	in^2	mm^2
B	Width of a spread footing or diameter of a deep foundation	ft & in	m
B_b	Base diameter of a deep foundation	ft & in	m
B_r	Reference width	1.0 ft (12 in)	0.30 m (300 mm)
B_s	Shaft diameter of a deep foundation	ft & in	m
B'	Effective width of a spread footing	ft & in	m
b	Unit length of continuous footing or wall	1 ft	1 m
b	Width of a flexural member	in	mm
b_c, b_q, b_γ	Brinch Hansen's base inclination factors	Unitless	Unitless
b_0	Perimeter of critical shear surface	in	mm
C	Compressibility	Unitless	Unitless
C_A	Aging correction factor	Unitless	Unitless
C_B	SPT bore hole diameter factor	Unitless	Unitless
C_c	Compression index	Unitless	Unitless
C_I	Burland and Burbidge's depth of influence correction factor	Unitless	Unitless
C_N	SPT overburden correction factor	Unitless	Unitless
C_{OCR}	Overconsolidation correction factor	Unitless	Unitless
C_P	Grain size correction factor	Unitless	Unitless

$C_{p\phi}$	Evans and Duncan's passive pressure factor	Deg	Deg
C_R	SPT rod correction factor	Unitless	Unitless
C_r	Recompression index	Unitless	Unitless
C_S	SPT sampler correction factor	Unitless	Unitless
C_s	Burland and Burbidge's shape factor	Unitless	Unitless
C_w	Hydroconsolidation coefficient	Unitless	Unitless
$C_{\alpha\varepsilon}$	Coefficient of secondary compression	Unitless	Unitless
$C_{\varepsilon c}$	Compression ratio	Unitless	Unitless
$C_{\varepsilon r}$	Recompression ratio	Unitless	Unitless
C_1	Schmertmann's depth factor	Unitless	Unitless
C_2	Schmertmann's secondary creep factor	Unitless	Unitless
C_3	Schmertmann's shape factor	Unitless	Unitless
c	Cohesion	lb/ft^2	kPa
c	Width of column or wall	in	m
c	Velocity of wave propagation	ft/s	m/s
c	Distance from neutral axis	in	mm
c_{adj}	Adjusted cohesion	lb/ft^2	kPa
c_p	Width of steel base plate	in	mm
c_1, c_2	Plan dimensions of column	in	mm
D	Depth from ground surface to bottom of foundation	ft & in	m
D_{min}	Minimum embedment depth for laterally loaded deep foundations	ft	m
D_r	Relative density	%	%
D_w	Depth from ground surface to groundwater table	ft	m
D_{50}	Grain size at which 50% of the soil is finer	—	mm
d	Effective depth	in	mm
d_b	Nominal diameter of rebar	in	mm
d_c, d_q, d_γ	Brinch Hansen's depth factors	Unitless	Unitless
E	Modulus of elasticity	lb/ft^2	kPa
E	Steel placement factor for rectangular footings	Unitless	Unitless
E_D	Dilatometer modulus	lb/ft^2	kPa
E_m	SPT hammer efficiency	Unitless	Unitless
E_u	Undrained modulus of elasticity	lb/ft^2	kPa
e	Void ratio	Unitless	Unitless
e	Eccentricity	in	mm
e	Base of natural logarithms	2.71828	2.71828
e_B	Eccentricity in the B direction	ft	m

e_L	Eccentricity in the L direction	ft	m
e_{max}	Maximum index void ratio	Unitless	Unitless
e_{min}	Minimum index void ratio	Unitless	Unitless
e_0	Initial void ratio	Unitless	Unitless
F	Factor of safety	Unitless	Unitless
F_a	Allowable normal stress due to axial load	lb/in^2	MPa
F_b	Allowable normal stress in extreme fiber due to flexural load	lb/in^2	MPa
F_v	Allowable shear stress	lb/in^2	MPa
F_y	Yield strength of steel	lb/in^2	MPa
f	Depth from ground surface to maximum moment in deep foundation	ft	m
f_a	Normal stress due to axial load	lb/in^2	MPa
f_b	Normal stress in extreme fiber due to flexural load	lb/in^2	MPa
$f_c{}'$	28-day compressive strength of concrete	lb/in^2	MPa
f_{pc}	Effective prestress stress on gross section	lb/in^2	MPa
f_s	Unit skin friction resistance	lb/ft^2	kPa
f_{sc}	CPT local side friction	t/ft^2	MPa, kg/cm^2
f_v	Shear stress	lb/in^2	MPa
f_y	Ultimate strength of reinforcing steel	lb/in^2	MPa
G_h, G_v	Terzaghi and Peck's earth pressure coefficients (G_h also known as equivalent fluid density)	lb/ft^3	kN/m^3
G_s	Specific gravity of solids	Unitless	Unitless
g_c, g_q, g_γ	Brinch Hansen's ground inclination factors	Unitless	Unitless
H	Thickness of a soil layer	ft	m
h	Pile hammer stroke	in	mm
h	Thickness of soil strata	ft	m
I	Moment of inertia	in^4	mm^4
I_c	Burland and Burbidge's compressibility index	Unitless	Unitless
I_D	Dilatometer material index	Unitless	Unitless
I_r	Rigidity index	Unitless	Unitless
I_z	Schmertmann's strain influence factor	Unitless	Unitless
I_{zp}	Peak value of Schmertmann's strain influence factor	Unitless	Unitless
I_0, I_1	Influence factors	Unitless	Unitless
i_c, i_q, i_γ	Brinch Hansen's load inclination factors	Unitless	Unitless
J	Carter and Kulhawy's correction factor	Unitless	Unitless
j_c	Case method damping constant	Unitless	Unitless
J_s	Smith damping factor	Unitless	Unitless

Symbol	Description	US Units	SI Units
K	Coefficient of lateral earth pressure	Unitless	Unitless
K_a	Coefficient of active earth pressure	Unitless	Unitless
K_D	Dilatometer horizontal stress index	Unitless	Unitless
K_d	Meyerhof's depth factor	Unitless	Unitless
K_h, K_v	Duncan et al.'s earth pressure coefficients	Unitless	Unitless
K_p	Coefficient of passive earth pressure	Unitless	Unitless
K_0	Coefficient of lateral earth pressure at rest	Unitless	Unitless
k	Factor in Boussinesq and Westergaard's stress formulas	Unitless	Unitless
k_c	LCPC cone end bearing factor	Unitless	Unitless
k_s	Modulus of subgrade reaction	lb/in^3	kN/m^3
L	Length of a spread footing	ft & in	m
L'	Effective length of a spread footing	ft & in	m
LL	Liquid limit	Unitless	Unitless
l	Design cantilever distance in spread footing	in	mm
M	Applied moment load	ft-lb	kN-m
M	Constrained modulus	lb/ft^2	kPa
M_c	Evans and Duncan's characteristic moment	ft-lb	kN-m
M_D	Driving moment	ft-lb	kN-m
M_d	Moment dead load	ft-k	kN-m
M_e	Moment earthquake load	ft-k	kN-m
M_h	Moment earth pressure load	ft-k	kN-m
M_l	Moment live load	ft-k	kN-m
M_{max}	Maximum moment	ft-k	kN-m
M_n	Nominal moment load capacity	ft-k	kN-m
M_R	Resisting moment	ft-k	kN-m
M_u	Ultimate moment load	ft-k	kN-m
M_{uc}	Ultimate moment at critical section	ft-k	kN-m
M_w	Moment wind load	ft-k	kN-m
m	Factor in Boussinesq and Westergaard's stress formulas	Unitless	Unitless
m	Evans and Duncan's exponent	Unitless	Unitless
N	Number of piles in a group	Unitless	Unitless
N	SPT blow count	Blows/ft	Blows/0.3 m
N'	SPT blow count corrected for overburden stress	Blows/ft	Blows/0.3 m
N_c, N_q, N_γ	Bearing capacity factors for shallow foundations	Unitless	Unitless
N_c^*, N_q^*, N_γ^*	Bearing capacity factors for deep foundations	Unitless	Unitless

N_{cr}	Carter and Kulhawy's bearing capacity factor	Unitless	Unitless
N_u	Uplift breakout factor	Unitless	Unitless
N_{60}	SPT blow count corrected for field procedures	Blows/ft	Blows/0.3 m
$N_{60}{}'$	SPT blow count corrected for overburden stress and field procedures	Blows/ft	Blows/0.3 m
n	Porosity	%	%
n	Factor in Boussinesq and Westergaard's stress formulas	Unitless	Unitless
n	Evans and Duncan's exponent	Unitless	Unitless
OCR	Overconsolidation ratio	Unitless	Unitless
P	Applied normal load	lb, k	kN
P_a	Net allowable downward load capacity	k	kN
P_a	Normal active earth pressure force	lb	kN
P_{ag}	Net allowable downward load capacity of pile group	k	kN
P_{au}	Net allowable uplift load capacity	k	kN
P_D	Driving force	lb	kN
P_d	Normal dead load	k	kN
P_e	Normal earthquake load	k	kN
P_e	Gross end bearing capacity	k	kN
$P_e{}'$	Net end bearing capacity	k	kN
P_f	Normal load acting on bottom of footing	k	kN
P_g	Normal load on pile group	k	kN
P_h	Normal earth pressure load	k	kN
P_l	Normal live load	k	kN
P_n	Nominal normal load capacity	k	kN
P_{nb}	Nominal bearing strength	lb/in^2	MPa
P_p	Normal passive earth pressure force	lb	kN
P_R	Resisting force	lb	kN
P_r	Reference force	2000 lb	9.0 kN
P_s	Skin friction capacity	k	kN
P_u	Ultimate normal load	k	kN
P_{ub}	Uplift capacity contributed by enlarged base	k	kN
P_w	Normal wind load	k	kN
P_0	Normal at-rest earth pressure force	lb	kN
PI	Plasticity index	Unitless	Unitless
PL	Plastic limit	Unitless	Unitless
p	Lateral soil resistance per unit length of pile	lb/ft	kN/m

Q_c	Compressibility factor	Unitless	Unitless
q	Gross bearing pressure	lb/ft^2	kPa
q	Quake	in	mm
q'	Net bearing pressure	lb/ft^2	kPa
q_a'	Net allowable bearing value	lb/ft^2	kPa
q_a''	Net allowable bearing pressure	lb/ft^2	kPa
q_c	CPT cone resistance	t/ft^2	$MPa, kg/cm^2$
q_{ca}	Equivalent cone bearing resistance at pile tip	lb/ft^2	kPa
q_e'	Net unit end bearing resistance	lb/ft^2	kPa
q'_{equiv}	Equivalent net bearing pressure	lb/ft^2	kPa
q_{er}'	Reduced net unit end bearing resistance	lb/ft^2	kPa
q'_{max}	Maximum net bearing pressure	lb/ft^2	kPa
q'_{min}	Minimum net bearing pressure	lb/ft^2	kPa
q_u'	Net ultimate bearing capacity	lb/ft^2	kPa
q_{uc}	Unconfined compressive strength of rock	lb/in^2	MPa
R	Uplift reduction factor	Unitless	Unitless
R_f	CPT friction ratio	%	%
R_I	Evans and Duncan's moment of inertia ratio	Unitless	Unitless
R_u	Ultimate pile resistance	k	kN
r	Rigidity factor	Unitless	Unitless
S	Degree of saturation	%	%
S	Number of stories	Unitless	Unitless
S	Column spacing	ft	m
S	Slope (inclination) of laterally loaded deep foundation	Deg	Deg
S	Elastic section modulus	in^3	mm^3
S_t	Sensitivity	Unitless	Unitless
s	Shear strength	lb/ft^2	kPa
s	Pile set (penetration) per blow	in	mm
s	Center-to-center spacing of piles	in	mm
s_c, s_q, s_γ	Brinch Hansen's shape factors	Unitless	Unitless
s_u	Undrained shear strength	lb/ft^2	kPa
T	Applied torsion load	ft-lb	kN-m
T	Thickness of spread footing	in	m
T	Torque	ft-lb	kN-m
t	Time or age	yr	yr
t_p	Time to complete primary consolidation	yr	yr
u	Pore water pressure	lb/ft^2	kPa
u_e	Excess pore water pressure	lb/ft^2	kPa
u_h	Hydrostatic pore water pressure	lb/ft^2	kPa

V	Applied shear load	lb, k	kN
V	Volume	ft^3	m^3
V_a	Volume of air	ft^3	m^3
V_a	Shear force acting on wall during active conditions	lb	kN
V_{af}	Allowable shear load on spread footing	k	kN
V_c	Nominal shear load capacity of concrete	k	kN
V_c	Evans and Duncan's characteristic shear load	k	kN
V_d	Shear dead load	k	kN
V_e	Shear earthquake load	k	kN
V_f	Ultimate shear resistance along bottom of footing	k	kN
V_g	Applied shear load on pile group	k	kN
V_h	Shear earth pressure load	k	kN
V_l	Shear live load	k	kN
V_n	Nominal shear load capacity	k	kN
V_p	Shear force acting on wall during passive conditions	lb	kN
V_s	Volume of solids	ft^3	m^3
V_u	Ultimate shear load	k	kN
V_v	Volume of voids	ft^3	m^3
V_w	Shear wind load	k	kN
V_w	Volume of water	ft^3	m^3
W	Weight	lb	kN
W_f	Weight of foundation	lb	kN
W_r	Weight of pile hammer ram	lb	kN
W_s	Weight of solids	lb	kN
W_w	Weight of water	lb	kN
w	Moisture content (water content)	%	%
y	Lateral pile deflection	in	mm
y_t	Lateral deflection at top of pile	in	mm
z	Depth below ground surface	ft	m
z_f	Depth below bottom of footing	ft	m
z_I	Depth of influence below bottom of footing	ft	m
z_i	Depth to imaginary footing	ft	m
z_w	Depth below the groundwater table	ft	m
α	Skin friction adhesion factor	Unitless	Unitless
α	Correlation factor for CPT	Unitless	Unitless
α	Inclination of footing base	Deg	Deg
α	Angular strain	Ratio	Ratio

α	Wetting coefficient	Unitless	Unitless
$\alpha_c{}', \alpha_s{}'$	Nottingham's adhesion factors	Unitless	Unitless
β	Inclination of ground surface	Deg	Deg
β	Duncan and Buchignani's correlation factor	Unitless	Unitless
β	Relative rotation (angular distortion)	Ratio	Ratio
β	Effective stress skin friction coefficient	Unitless	Unitless
β_c	Ratio of long side to short side of column	Unitless	Unitless
γ	Unit weight	lb/ft^3	kN/m^3
γ'	Effective unit weight	lb/ft^3	kN/m^3
γ_b	Buoyant unit weight	lb/ft^3	kN/m^3
γ_d	Dry unit weight	lb/ft^3	kN/m^3
γ_{sat}	Saturated unit weight	lb/ft^3	kN/m^3
γ_w	Unit weight of water	lb/ft^3	kN/m^3
Δ	Change (associated with another parameter)	Varies	Varies
δ	Settlement	in	mm
δ_a	Allowable settlement	in	mm
δ_c	Consolidation settlement	in	mm
δ_D	Differential settlement	in	mm
δ_{Da}	Allowable differential settlement	in	mm
δ_d	Distortion settlement	in	mm
δ_e	Settlement due to elastic compression of pile	in	mm
δ_s	Secondary compression settlement	in	mm
δ_w	Settlement or heave due to wetting (collapsible or expansive soils)	in	mm
ε	Normal strain	Unitless or %	Unitless or %
ε_h	Horizontal normal strain	Unitless or %	Unitless or %
ε_v	Vertical normal strain	Unitless or %	Unitless or %
ε_w	Strain due to wetting (collapsible or expansive soils)	Unitless	Unitless
ε_{50}	Strain at which 50% of the soil strength is mobilized	Unitless	Unitless
η	Group efficiency factor	Unitless	Unitless
θ	Angle	Deg	Deg
θ	Rotation	Ratio	Ratio
θ_a	Allowable rotation	Ratio	Ratio
λ	Frictional capacity coefficient	Unitless	Unitless
λ	Evans and Duncan's stress-strain parameter	Unitless	Unitless
μ	Coefficient of sliding friction	Unitless	Unitless
ν_p	Poisson's ratio	Unitless	Unitless
π	Pi	3.14159	3.14159

ρ	Steel ratio	Unitless	Unitless
ρ_{min}	Minimum steel ratio	Unitless	Unitless
σ	Total stress (in soil)	lb/ft^2	kPa
σ	Normal stress (in structural member)	lb/in^2	MPa
σ'	Effective stress	lb/ft^2	kPa
σ_c'	Preconsolidation stress	lb/ft^2	kPa
σ_D'	Vertical effective stress at a depth D below the ground surface	lb/ft^2	kPa
σ_h	Horizontal total stress or horizontal pressure	lb/ft^2	kPa
σ_h'	Horizontal effective stress	lb/ft^2	kPa
σ_{hs}'	Horizontal swell pressure	lb/ft^2	kPa
σ_m'	Overconsolidation margin	lb/ft^2	kPa
σ_p	Evans and Duncan's representative passive pressure	lb/ft^2	kPa
σ_r	Reference stress	$2000\ lb/ft^2$ $14\ lb/in^2$	100 kPa 0.10 MPa 1.0 kg/cm^2
σ_s	Swell pressure	lb/ft^2	kPa
σ_t	Threshold collapse stress	lb/ft^2	kPa
σ_v	Vertical total stress	lb/ft^2	kPa
σ_v'	Vertical effective stress	lb/ft^2	kPa
σ_{v0}'	Initial vertical effective stress	lb/ft^2	kPa
σ_{vp}'	Initial vertical effective stress at depth of peak strain influence factor	lb/ft^2	kPa
τ	Shear stress	lb/ft^2	kPa
ϕ	Friction angle of a soil	Deg	Deg
ϕ	Strength reduction factor for ultimate strength design of reinforced concrete	Unitless	Unitless
ϕ_a	Friction angle under axisymmetric conditions	Deg	Deg
ϕ_{adj}	Adjusted friction angle	Deg	Deg
ϕ_f	Footing-soil interface friction angle	Deg	Deg
ϕ_{ps}	Friction angle under plane-strain conditions	Deg	Deg
ϕ_s	Soil-pile interface friction angle	Deg	Deg
ϕ_w	Wall-backfill interface friction angle	Deg	Deg
ψ	3-dimensional adjustment coefficient	Unitless	Unitless
ψ	Shear factor for reinforced concrete design	Unitless	Unitless
ω	Tilt	Deg	Deg
ω	DeRuiter and Beringen's soil condition factor	Unitless	Unitless
$-$	"bar" over a variable indicates average value	Varies	Varies

Part A

General Principles

1

Foundations in Civil Engineering

A wise engineer once said "A structure is no stronger than its connections." Although this statement usually invokes images of connections between individual structural members, it also applies to the connections between a structure and the ground that supports it. These are known as its *foundations*. Even the ancient builders knew that the most carefully designed structures can fail if they are not connected to the ground with suitable foundations. The Tower of Pisa in Italy (perhaps the world's most successful foundation "failure") reminds us of this truth.

Although builders have recognized the importance of firm foundations for countless generations, and the history of foundation construction extends for thousands of years, the discipline of *foundation engineering* as we know it today did not begin to develop until the latter part of the nineteenth century.

1.1 THE EMERGENCE OF MODERN FOUNDATION ENGINEERING

Early foundation designs were based solely on precedent, intuition, and common sense. Through trial-and-error, builders developed rules for sizing and building foundations. For example, load-bearing masonry walls built on compact gravel in New York City during the 1880s were supported on spread footings that had a width of 1.5 times that of the wall. Those built on sand or stiff clay were three times the width of the wall (Powell, 1884).

These rules usually produced acceptable results as long as they were applied to structures and soil conditions similar to those encountered in the past. However, the results were often disastrous when the rules were extrapolated to new conditions. This problem became especially troublesome when new methods of building construction began to appear near the end of the nineteenth century. The introduction of steel and reinforced concrete led to a transition away from rigid masonry structures to more flexible frame structures. These new materials also permitted buildings to be taller and heavier than before. Thus, the old rules for foundation design were no longer applicable. These events stimulated the birth of modern foundation engineering. Instead of simply developing new rules, engineers began to investigate the behavior of foundations and develop more rational methods of design. A significant advance came in 1873 when Frederick Baumann, a Chicago architect, published the pamphlet *The Art of Preparing Foundations, with Particular Illustration of the "Method of Isolated Piers" as Followed in Chicago* (Baumann, 1873). He appears to be the first to recommend that the base area of a foundation should be proportional to the applied load and that the loads should act concentrically upon the foundation. He also gave allowable bearing pressures for Chicago soils and specified tolerable limits for total and differential settlements.

Extensive research and experience gained since Baumann's time has furthered our understanding of the behavior of foundations. Thus, modern analysis, design, and construction techniques have made it possible to build foundations with greater capacity and reliability. However, foundation engineering continues to be both an art and a science. Although tests and analyses are important, the foundation engineer also must rely on precedent and engineering judgment.

1.2 THE FOUNDATION ENGINEER

Foundation engineering does not fit completely within any of the traditional subdisciplines within civil engineering. Instead, the foundation engineer must be multidisciplinary and possess a working knowledge of each of the following areas:

- Geotechnical engineering
- Structural engineering
- Construction engineering

All foundations interact with the ground, so the engineering properties of soil and rock strongly influence their design. Thus, the foundation engineer must understand geotechnical engineering. Most foundation engineers also consider themselves to be geotechnical engineers.

A foundation is a structural member, so we must also understand structural engineering. In addition, the foundation supports a structure, so we must understand the sources and nature of structural loads and the structure's tolerance of foundation movements.

Finally, foundations must be built. Although the actual construction is performed by contractors and construction engineers, it is very important for the design engineer to have a thorough understanding of construction methods and equipment to develop a design that can be economically built. A knowledge of construction engineering is also necessary when dealing with problems that develop during construction.

1.3 IGNORANCE AND UNCERTAINTIES

An unknown structural engineer suggested the following definition for structural engineering:

> Structural engineering is the art and science of molding materials we do not fully understand into shapes we cannot precisely analyze to resist forces we cannot accurately predict, all in such a way that the society at large is given no reason to suspect the extent of our ignorance.

We could apply the same definition, even more emphatically, to foundation engineering. In spite of the many advances in foundation engineering theory, there are still many gaps in our understanding. In general, the greatest uncertainties are the result of our limited knowledge of the soil conditions. Although foundation engineers use various investigation and testing techniques in an attempt to define the soil conditions beneath the site of a proposed foundation, even the most thorough investigation program encounters only a small portion of the soils and relies heavily on interpolation and extrapolation.

Limitations in our understanding of the interaction between a foundation and the soil also introduce uncertainties. For example, how does skin friction resistance develop along the surface of a pile? How does the installation of a pile affect the engineering properties of the adjacent soils? These and other questions are the subject of continued research.

It also is difficult to predict the actual service loads that will act on a foundation, especially live loads. Design values, such as those that appear in building codes, are usually conservative.

Because of these and other uncertainties, the wise engineer does not blindly follow the results of tests or analyses. They must be tempered with precedent, common sense, and engineering judgment. It is dangerous to view foundation engineering, or any other type of engineering, as simply a collection of formulas and charts to be followed using some "recipe" for design. This is why it is essential to understand the *behavior* of foundations and the basis and limitations of the analysis methods.

Rationalism and Empiricism

Since we do not fully understand the behavior of foundations, most of our analysis and design methods include a mixture of rational and empirical techniques. Rational techniques are those developed from the principles of physics and engineering science, and are useful ways to describe mechanisms we understand and are able to quantify. Conversely, empirical techniques are based primarily on experimental data and thus are especially helpful when we have a limited understanding of the physical mechanisms.

Methods of analyzing foundation problems often begin as simple rational models with little or no experimental data to validate them, or highly empirical techniques that reflect only the most basic insight into the mechanisms that control the observed behavior. Then, as engineers use these methods we search for experimental data to calibrate the rational methods and discernment to understand the empirical data. These efforts are intended to improve the accuracy of the predictions.

One of the keys to successful foundation engineering is to understand this mix of rationalism and empiricism, the strengths and limitations of each, and how to apply them to practical design problems.

Factors of Safety

In spite of the many uncertainties in foundation analysis and design, the public expects engineers to develop reliable and economical designs in a timely and efficient manner. Therefore, we compensate for these uncertainties by using factors of safety in our designs.

Although it's tempting to think of designs that have a factor of safety greater than some standard value as being "safe" and those with a lower factor of safety as "unsafe," it's better to view these two conditions as having different degrees of reliability or different probabilities of failure. All foundations can fail, but some are more likely to fail than others. The design factor of safety defines the engineer's estimate of the best compromise between cost and reliability. It is based on many factors, including the following:

- Required reliability (i.e., the acceptable probability of failure).
- Consequences of a failure.
- Uncertainties in soil properties and applied loads.
- Construction tolerances (i.e., the potential differences between design and as-built dimensions).
- Ignorance of the true behavior of foundations.
- Cost-benefit ratio of additional conservatism in the design.

Factors of safety in foundations are typically greater than those in the superstructure because of the following:

- Extra weight (another consequence of conservatism) in the superstructure increases the loads on members below, thus compounding the cost increases. However, the foundation is the lowest member in a structure, so additional weight does not affect other members. In fact, additional weight may be a benefit because it increases the foundation's uplift capacity.
- Construction tolerances in foundations are wider than those in the superstructure, so the as-built dimensions are often significantly different than the design dimensions.
- Uncertainties in soil properties introduce significantly more risk.
- Foundation failures can be more costly than failure in the superstructure.

Design values are rarely codified, so engineers develop them using precedent, professional judgment, and perhaps some rudimentary statistical analyses. Typical design values are included throughout this book.

Accuracy of Computations

Those who are new to the field of foundation engineering often make the mistake of expressing the results of computations using too many significant figures. For example, to claim that the predicted settlement of a spread footing is 12.214 mm suggests an accuracy that is well beyond that which is possible using normal exploration and testing methods. Such practices give a false sense of security.

As a general rule, perform most foundation engineering calculations to three significant figures and express the final results and designs to two significant figures. For example, the settlement figure just quoted would best be stated as 12 mm, keeping in mind that the true precision may be on the order of ±25%.

1.4 BUILDING CODES

Building codes govern the design and construction of nearly all foundations. Although foundation engineering is not as codified as some other areas of civil engineering, we must be familiar with the applicable regulations for a particular project.

Many large cities, such as Chicago and New York, have adopted their own building codes. In other areas, regional or national codes are used. Examples of the latter include the *Uniform Building Code*, *The Basic Building Code*, *The Standard Building Code*, and *The National Building Code of Canada*. Even specialized foundations, such as those for offshore oil platforms, are governed by codes and regulations.

Some code requirements are nearly absolute and reflect good practice, whereas others are conservative design parameters that may be superseded by site-specific parameters obtained from an appropriate investigation. However, there is no guarantee that all designs that satisfy code requirements are safe. In some cases, we must exceed these requirements. Therefore, think of codes as guides, not dictators, and certainly not as a substitute for engineering judgment or common sense.

This book does not purport to include an exhaustive summary of building code requirements. However, it does occasionally include some of these requirements to illustrate various concepts or to amplify certain points.

1.5 CLASSIFICATION OF FOUNDATIONS

We will divide foundations into two broad categories: *shallow foundations* and *deep foundations*, as shown in Figure 1.1. Shallow foundations transmit the structural loads to the near-surface soils; deep foundations transmit some or all of the loads to deeper soils. These two categories are discussed in Chapters 5 - 10 and 11 - 18, respectively.

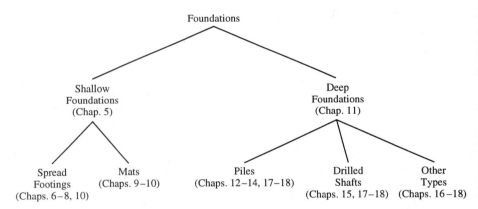

Figure 1.1 Classification of foundations.

1.6 NORMALIZED FORMULAS

Many of the formulas used in foundation engineering analyses are empirical and thus are not always dimensionally consistent. Therefore, those who developed these formulas express them using specific units of measurement. For example, a formula to compute settlement of a footing may require that the load be in tons and the footing width be in feet to produce a settlement in inches.

Such formulas are often a source of confusion and frustration, especially when the user's favorite units do not match those of the original author. Therefore, in this book all such formulas have been normalized by introducing the following *reference parameters*:

B_r = reference width = 1.0 ft = 0.30 m = 12 in = 300 mm

P_r = reference force = 2000 lb = 9.0 kN

σ_r = reference stress = 2000 lb/ft^2 = 100 kPa = 14 lb/in^2 = 0.10 MPa = 1.0 kg/cm^2

Although these are not exact unit conversions,[1] they are sufficiently accurate for empirically based foundation engineering analyses.

When using these normalized formulas, simply use values of B_r, P_r, and σ_r that are dimensionally consistent with the remaining variables in the formula. Thus, every formula in this book may be used with any consistent set of units.

---

[1] For those who wish to use more exact unit conversions:

B_r = reference width = 1.00 ft = 0.305 m = 12.0 in = 305 mm

P_r = reference force = 2000 lb = 8.90 kN

σ_r = reference stress = 2000 lb/ft^2 = 95.7 kPa = 13.9 lb/in^2 = 0.0957 MPa = 0.977 kg/cm^2

2

Performance Requirements

*If a builder builds a house for a man and does not make
its construction firm, and the house which he has built
collapses and causes the death of the owner of the house,
that builder shall be put to death.*

From *The Code of Hammurabi*, Babylon, circa 2000 BC

One of the first steps in any design process is to define the performance requirements. What functions do we expect the final product to accomplish? What are the appropriate design criteria? What constitutes acceptable performance, and what would be unacceptable?

A common misconception, even among some engineers, is that foundations are either perfectly rigid and unyielding, or they are completely incapable of supporting the necessary loads and fail catastrophically. This "it's either black or white" perspective is easy to comprehend, but it is not correct. All engineering products, including foundations, have varying *degrees* of performance that we might think of as various shades of gray. The engineer must determine which shades are acceptable and which are not. Leonards (1982) defined *failure* as "an unacceptable difference between expected and observed performance." For example, consider an engineer who designs a foundation such that it is not expected to settle more than 1 inch when loaded. If it actually settles 1.1 inches, the engineer will probably not consider it to have failed because the difference between the expected and observed performance is small and well within the design factor of safety. However, a settlement of 10 inches would probably be unacceptable and therefore classified as a failure.

2.1 SUPPORTING THE SUPERSTRUCTURE

The most fundamental performance requirement for a foundation is that it support the superstructure. This seemingly simple requirement introduces two important questions:

- What loads are to be supported?
- How far may the foundation settle in response to these loads?

Design Loads

To design a foundation, the engineer must know the type, direction, and magnitude of each applied load. The load types, as shown in Figure 2.1, are:

- *Normal loads*, designated by the variable P
- *Shear loads*, designated by the variable V
- *Moment loads*, designated by the variable M
- *Torsion loads*, designated by the variable T

We will consider normal loads to be those that act parallel to the foundation axis. Usually the axis is vertical, so the normal load becomes the vertical component of the applied load. It may act either downward or upward.

Shear loads are those that act perpendicular to the foundation axis. They may be expressed as two perpendicular components, V_x and V_y. Moment loads also may be expressed using two perpendicular components, M_x and M_y. Torsion loads are usually not significant and may be ignored in most foundation designs.

Each of these loads may include dead load, live load, and special load (i.e., wind, seismic, etc.) components, some of which may vary in magnitude and direction during the life of the structure.

The structural engineer usually determines these design loads as a part of the analysis and design of the superstructure. They are typically a "given" for the foundation engineer, yet we must understand how they are developed and have a feel for the appropriate levels of uncertainty. For example, the structural engineer is able to estimate the actual dead loads with a high degree of certainty by simply adding the weights of the various structural members, so the design dead loads are usually very realistic. However, it is often much more difficult to estimate the actual live loads, so the design values may be very conservative.

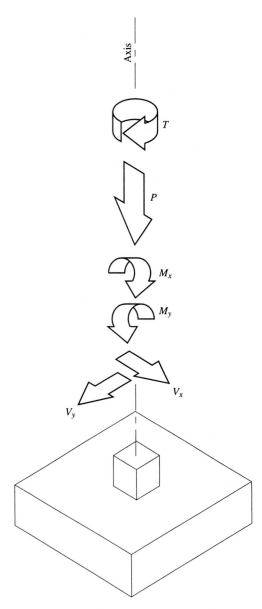

Figure 2.1 Types of applied loads acting on a foundation.

Peck (1967) gave an interesting case history on this subject:

I can recall a discussion of this point in connection with a five-story warehouse for a refinery. The warehouse was built to store quantities of special greases, which were made in batches, in order to be able to fill orders whenever they were received. The containers for the products could be stacked conveniently to heights corresponding to a floor loading of 500 lbs per sq. ft. The structural engineer insisted that such a loading was realistic. Yet, the products were to be handled by fork-lift trucks that require aisles in which to operate. A study indicated that these aisles would always account for at least 20% of the floor area. Furthermore, after the owner reviewed his actual operations, he realized that the economic operation of the warehouse would require that almost half of the storage space would not be loaded at any given time. If a barrel of a given product were replaced as soon as it was sold, there would be no point in having a warehouse to provide a reservoir for the smooth outward flow of products made in batches. The usefulness of the facility was predicated on the idea that the supply of a given product would be allowed to diminish until replenishment was necessary. In the meantime batches of other products would be produced.

In this instance a realistic study, by the owner's operating department, of the purpose for which the warehouse was intended led to such a great reduction in the estimated average long-term loading that the structure could be established upon a raft[1] rather than upon a deep pile foundation. The savings were substantial.

It would be reasonable to expect that certain parts of this warehouse could be loaded to the full 500 lb/ft^2 live load, so the individual floors would probably be designed on this basis. An individual column would probably be designed using a somewhat lower load because it would be less likely that all five floors would be loaded simultaneously to capacity. However, the total force acting between the mat foundation and the soil would include the sum of all the live loads in the building, which would never consist of the full 500 lb/ft^2 acting on all floors simultaneously. Thus, the engineer could safely design this mat using a smaller average live load.

Conversely, if this warehouse were to be supported on spread footings,[2] some of the footings might receive more live load than others, depending on the arrangement of the live loads in the warehouse. Therefore, the design live load would need to be larger.

Some foundations also must resist loads from other sources. For example, foundations for marine structures might be subjected to mooring and impact loads from ship berthings, ice flows, and wave action.

Tolerance of Settlements

Stresses are always accompanied by strains, so the application of a downward load to a foundation always will cause it to settle. The question that faces the foundation engineer

---

[1] A *raft foundation*, also known as a *mat foundation*, consists of a thick reinforced concrete slab that encompasses the entire footprint of the structure. This type of foundation is discussed in Chapters 5 and 9.

[2] *Spread footing* foundations support only a small portion of a structure, such as a single column. They are discussed in Chapters 5 - 8.

is not *if* the foundation will settle, but rather the *amount* of settlement that would be tolerable. This is analogous to the design of beams and the determination of the tolerable deflections.

Foundations can settle in various ways and each affects the performance of the structure differently, as shown in Figure 2.2. The simplest mode, Figure 2.2a, consists of the entire structure settling uniformly. This mode does not distort the structure, so any damage will be related to the interface between the structure and the adjacent ground or adjacent structures. Shearing of utility lines also could be a problem, but this might be accommodated with the use of flexible-extendible connections, as shown in Figure 2.3. For example, an 8 in (203 mm) diameter connector can accommodate differential settlements of up to 15 in (380 mm).

Another possibility is that one side of the structure might settle much more than the opposite side and the portions in between settle proportionately, as shown in Figure 2.2b. This causes the structure to tilt, but it still does not distort. A nominal tilt will not affect the performance of the structure, although it may create aesthetic and public confidence problems. However, if the tilting becomes too great, the structure could be in danger of collapse, especially if it is very tall.

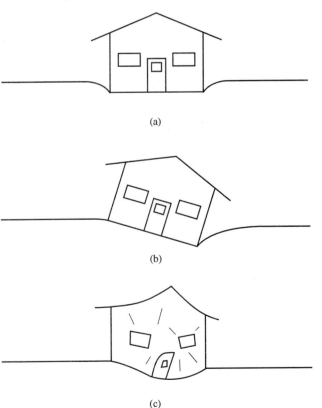

(a)

(b)

(c)

Figure 2.2 Modes of settlement: (a) uniform; (b) tilting with no distortion; and (c) distortion.

Figure 2.3 This flexible-extendible pipe coupler has ball joints at each end and a
telescoping section in the middle. These couplers can be installed where utility lines
enter structures, thus accommodating differential settlement, lateral movement, extension
and compression. (Photo courtesy of EBAA Iron Sales, Inc.).

A much more common mode occurs when the structure settles and distorts, as
shown in Figure 2.2c. This induces additional stresses in the structure that may cause
cracks in walls and floors, jamming of doors and windows, and in extreme cases,
overloading of structural members. An example of a building that suffered this type of
movement is shown in Figure 2.4.

Different types of structures have varying degrees of tolerance to settlements and
distortions. These variations depend on many factors, including the following:

- **The type of construction**—For example, wood-frame buildings with wood siding
  would be much more tolerant than unreinforced brick buildings.
- **The use of the structure**—Even small cracks in a house might be unacceptable,
  whereas much larger cracks in an industrial building might not even be noticed.
- **The presence of sensitive finishes**—Tile or other sensitive finishes are much less
  tolerant of movements.
- **The rigidity of the structure**—If a footing beneath part of a very rigid structure
  settles more than the others, the structure will transfer some of the load away from
  that footing. However, footings beneath flexible structures must settle much more
  before any significant load transfer occurs. Therefore, a rigid structure will have
  less differential settlement than a flexible one.

The structure shown in Figure 2.5 is an example of one that could withstand a large
amount of differential settlement, whereas the one in Figure 2.6 is much less tolerant.

Figure 2.4 This brick building has experienced excessive differential settlements.

Figure 2.5 These buildings have corrugated steel siding and steel framing with no delicate or sensitive finishes. They are examples of structures that could withstand large differential settlements with no loss in serviceability.

Figure 2.6 This unreinforced stone-and-mortar building is very intolerant of differential
settlements. It also is more likely to experience large settlements because of its heavy
weight. Notice the 1.5 in (40 mm) crack in the wall beneath the left window.

Table 2.1 gives a rough indication of the serviceability limits of different types of
structures based on the width of cracks in walls or other structural members.

Burland and Wroth (1974) described the settlement and distortion of buildings using
the parameters shown in Figure 2.7, which are defined as follows:

- The *total settlement*, δ, is the absolute vertical movement of a particular foundation
  from its as-constructed position to its loaded position.
- The *differential settlement*, δ_D, is the difference in the total settlement between two
  foundations.
- The *tilt*, ω, is the rotation of the entire superstructure, or at least a well-defined part
  of it. The Leaning Tower of Pisa is an example of excessive tilt.
- The *rotation*, θ, is the angle between the horizontal and any two foundations or two
  points in a single foundation.
- The *relative rotation* (also known as *angular distortion*), β, is the angle between
  the overall tilt of a structure and the inclination of a specific portion of it.
- The *angular strain*, α, is the change in θ.

- The *relative deflection*, Δ, is the settlement of a point relative to a straight line that connects two reference points a distance S apart. A positive value of Δ indicates a deformation that is concave up, a situation known as *sagging*, whereas a negative value indicates downward concavity, known as *hogging*. The relative deflection is also known as the *relative sag* or *relative hog*.
- The *deflection ratio* (also known as the *sagging ratio* or the *hogging ratio*) is Δ/L_1.

TABLE 2.1 SERVICEABILITY LIMITS

Crack Width (mm)	Degree of Damage			Effect on Structure and Building Use
	Dwelling	Commercial or Public Building	Industrial Building	
≤ 0.1	Insignificant	Insignificant	Insignificant	None
0.1 - 0.3	Very slight	Very slight	Insignificant	None
0.3 - 1	Slight	Slight	Very slight	Aesthetic only
1 - 2	Slight to moderate	Slight to moderate	Very slight	Accelerated weathering to external features
2 - 5	Moderate	Moderate	Slight	Serviceability of the building will be affected, and toward the upper bound, stability may also be at risk
5 - 15	Moderate to severe	Moderate to severe	Moderate	
15 - 25	Severe to very severe	Moderate to severe	Moderate to severe	Increasing risk of structure becoming dangerous
> 25	Very severe to dangerous	Severe to dangerous	Severe to dangerous	

Adapted from Thorburn and Hutchison, 1985.

For some structures, such as bridges, an engineer might be able to determine the tolerable limits of these parameters from a structural analysis. However, the tolerable limits in most structures, especially buildings, will be governed by aesthetic and serviceability requirements, not structural requirements. Unsightly cracks, jamming doors and windows, and other similar problems will develop long before the integrity of the structure is in danger. Unfortunately, it would be very difficult to compute a numerical magnitude for the settlement that would cause these problems. Such a computation would require a complex indeterminate analysis (Wahls, 1981). We would need to consider the type and size of the structure, the engineering properties of both the structural materials and the soils, and the rate and uniformity of settlement. Because the problem is so complex, engineers must rely on empirical correlations between the observed behavior of structures and one or more of Burland and Wroth's parameters.

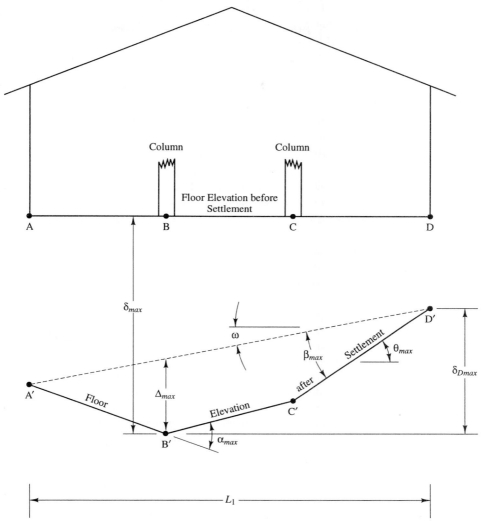

Figure 2.7 Definitions of parameters that describe the movement of building foundations. (Adapted from Burland and Wroth, 1974).

Total Settlement

Some structures have sustained amazingly large total settlements, yet remain in service. For example, many buildings have had little or no ill effects even after settling as much as 10 in (250 mm). Others have experienced some distress, but continue to be used following even greater settlements. Some of the most dramatic examples are located in Mexico City, where buildings have settled more than 7 ft (2 m) and are still in use. Bridges, tanks, and other structures may also be tolerant of very large settlements.

When designing the foundations for a structure, the engineer must determine the allowable total settlement, δ_a. It might be governed by the following:

- **Connections with existing structures**—For example, the floors in a building addition must be at the same elevation as those in the original portion of the building.
- **Shearing or excessive movement of utility lines**—This is especially true of gravity flow lines, such as sewers.
- **Surface drainage**—Excessive settlement might cause rainwater to enter the structure.
- **Access**—Vehicles and pedestrians may need to access the structure, and excessive settlement might impede them.
- **Aesthetics**—Excessive settlement may cause aesthetic problems long before there is any threat to structural integrity or serviceability.
- **The allowable differential settlement**—Greater total settlements tend to generate larger differential settlements. Therefore, the allowable differential settlement may dictate the allowable total settlement.

The word "allowable" means that design values of δ_a must include a factor of safety. The selection of an appropriate factor of safety depends on the precision of the settlement prediction computations and other factors. Typically, buildings on spread footings are designed for $\delta_a = 0.5 - 1.5$ in (10 - 40 mm). Mat foundations, which have more rigidity and thus are less prone to differential settlement, might be designed for $\delta_a = 2 - 4$ in (50 - 100 mm). Highway bridge abutment designs could use $\delta_a = 1.6$ in (40 mm) (after a study by Bozozuk, 1978). However, it's often necessary to place more strict limits on δ_a to keep differential settlements under control.

If the predicted settlement is greater than δ_a, we could consider any or all of the following measures:

- **Use a more elaborate foundation**—For example, we might use piles instead of spread footings, thus reducing the settlement.
- **Improve the properties of the soil**—Many techniques are available to do this; some of them are discussed in Chapter 19.
- **Redesign the structure so it is more tolerant of settlements**—For example, flexible joints could be installed on pipes as was shown in Figure 2.3.

Differential Settlement

Engineers normally design the foundations for a structure such that all of them have the same computed settlement. Thus, in theory, the differential settlement will be zero. Unfortunately, the actual performance of the foundations will not be exactly as predicted, so some differential settlements will inevitably occur. This discrepancy between predicted and observed behavior has many causes, including the following:

- **The soil profile may not be uniform across the site**—This is nearly always true, no matter how uniform it might appear to be.
- **The ratio between the actual load and the design load may be different for each column**—Thus, the column with the lower ratio will settle less than that with the higher ratio.
- **The ratio of dead load to total load may be different for each column**—Therefore, if the settlement computations are based on total load, the actual settlement under dead load only will vary accordingly. For example, a greater percentage of the load on exterior columns is dead load, so they may settle more than interior columns with the same total load.
- **The as-built foundation dimensions may differ from the plan dimensions**—This will cause the actual settlements to be correspondingly different.

Differential settlements are generally more troublesome than total settlements because they distort the structure, as was shown in Figure 2.2c. This causes cracking in walls and other members, jamming in doors and windows, poor aesthetics, and other problems. If allowed to progress to an extreme, differential settlements could threaten the integrity of the structure.

Predicting Differential Settlements

Unfortunately, it is very difficult to predict the magnitude of differential settlements. In addition to evaluating the causes described earlier, the engineer must consider *soil-structure interaction*, as shown in Figure 2.8. The individual foundations are connected to the superstructure, so they do not act independently. Settlement of one foundation may influence the load of an adjacent one. The nature of this interaction depends on the *rigidity* of the structure. For example, a very flexible structure, as shown in Figure 2.8a, could have larger differential settlements because each foundation acts almost independent from the others. In contrast, a more rigid structure, as shown in Figure 2.8b, provides a stiffer connection between the individual foundations. Therefore, if one foundation started to settle excessively, the structure would redistribute some of its load to other foundations, thus reducing the differential settlement. Most structures have a rigidity somewhere between these extremes.

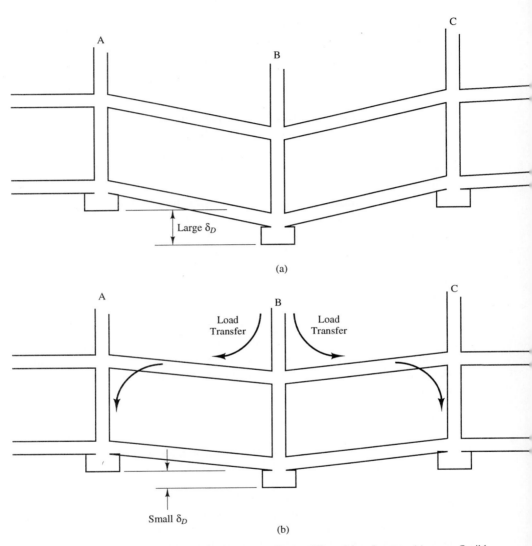

Figure 2.8 Influence of soil-structure interaction on differential settlements: (a) a very flexible structure has little load transfer, and thus could have larger differential settlements; (b) a more rigid structure has greater capacity for load transfer, and thus provides more resistance to excessive differential settlements.

Because of the difficulties in accurately assessing these many factors, engineers use empirical methods to predict differential settlements. These methods are based on observations of the ratio of differential to total settlements in real structures. For example, Bjerrum (1963) compared the total and differential settlements of spread footings on clays and sands. The data for footings on clays are shown in Figure 2.9 and those on sands in Figure 2.10. He considered only data from comparably loaded footings on soils of uniform thickness. In each case, the differential settlement is that which occurred between comparable footings designed for the same total settlement.

For clays, Bjerrum divided the data into two categories: those for relatively rigid structures and those for relatively flexible structures. For his purposes, rigid structures include those with load-bearing brick walls, and flexible structures include those with steel or concrete frames. For structures that have total settlements in the 1 in (25 mm) range, the upper bound δ_D/δ ratio is on the order of 0.8 for flexible structures and 0.5 for rigid structures. As the total settlement increases, which usually corresponds to larger footings mobilizing a larger volume of soil, these ratios reduce to about 0.5 and 0.3, respectively.

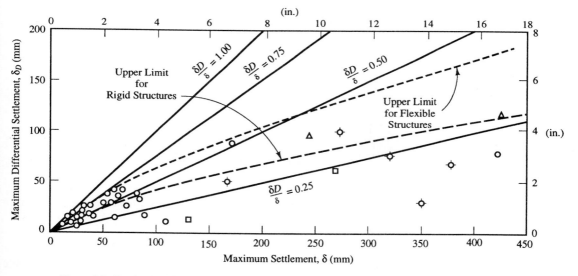

Figure 2.9 Total and differential settlements of spread footings on clays (Adapted from Bjerrum, 1963).

Figure 2.10 indicates that the ratio of differential settlement to total settlement is often much higher in sands than it is in clays. This seems to be due to the greater non-uniformities in natural sand deposits. These differential settlements are higher than had been suggested in the past. The upper bound in this plot approaches a δ_D/δ ratio of 1, which is often a good ratio to use for design. At sites with exceptionally homogeneous sands, such as compacted fills of uniform thickness, we might use a δ_D/δ ratio in the 0.50 to 0.75 range.

Structures supported on deep foundations might have smaller δ_D/δ ratios. However, settlement is generally not a major problem for most deep foundations.

Structures supported on mat foundations are much more rigid than those on spread footings, so the δ_D/δ ratios should be much smaller. We might design them to satisfy the total settlement criteria, and then provide enough flexural rigidity in the mat to control differential settlements. Some mat supported structures might be controlled by tilt criteria.

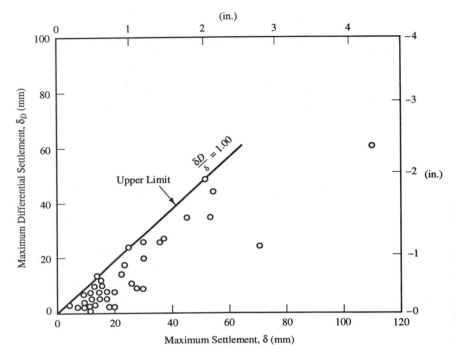

Figure 2.10 Total and differential settlement of spread footings on sands (Adapted from Bjerrum, 1963).

Allowable Differential Settlement

It is convenient to express the allowable differential settlement in terms of the allowable rotation and the column spacing:

$$\delta_{Da} = \theta_a S \qquad\qquad (2.1)$$

Where:

δ_{Da} = allowable differential settlement

θ_a = allowable rotation

S = column spacing (horizontal distance between columns)

The allowable rotation depends on many factors, so we must return to an empirical analysis based on the observed performance of real structures. Comprehensive studies of differential settlements in buildings include Skempton and MacDonald (1956), Polshin and Tokar (1957), and Grant et al. (1974). Skempton and MacDonald's work is based on the observed performance of 98 buildings of various types, 40 of which had evidence of damage due to excessive settlements. Polshin and Tokar reported the results of 25 years of observing the performance of structures in the Soviet Union and reflected Soviet building codes. The study by Grant et al. encompassed data from 95 buildings, 56 of which had damage. Table 2.2 presents a synthesis of this data, expressed in terms of allowable angular distortion. These values include a factor of safety of at least 1.5.

In practice, local custom and precedent often dictate the design values of δ_a and δ_{Da}. Engineers in some areas routinely design structures to accommodate relatively large settlements, and may be willing to accept some long-term maintenance costs (i.e., repairing minor cracking, rebuilding entranceways, etc.) in exchange for reduced construction costs. However, in other areas, even small settlements induce lawsuits, so foundations are designed to meet stricter standards.

TABLE 2.2 ALLOWABLE ROTATION, θ_a

Type of Structure	θ_a
One to two story steel frame warehouse with truss roof and corrugated steel siding	1/200
Steel and reinforced concrete frame buildings	
Without diagonal bracing	1/500
With diagonal bracing	1/600
Overhead traveling crane rails	1/500
Buildings with sensitive finishes	1/1000
Machinery[a]	1/1500
Unreinforced masonry load-bearing walls	
Length/height ≤ 3	1/2500
Length/height ≥ 5	1/1250

[a] Large machines, such as turbines or large punch presses, often have their own foundation, separate from that of the building that houses them.

Example 2.1

A steel-frame building without diagonal bracing has a column spacing of 20 ft. It is to be supported on spread footings founded on a clayey soil. What are the allowable total and differential settlements?

Solution:

1. Compute δ_{Da} based on type of structure and column spacing using Equation 2.1:

$$\begin{aligned}
\delta_{Da} &= \theta_a S \\
&= (1/500)(20) \\
&= 0.04 \text{ ft} = \textbf{0.5 in} \qquad \Leftarrow \textit{Answer}
\end{aligned}$$

2. Use Bjerrum's data (Figure 2.9 and discussion) to compute δ_a:

The total settlement, δ, will probably be on the order of 1 in, and this is a relatively flexible structure. Therefore, the upper limit for $\delta_D/\delta \approx 0.8$ in/1.0 in $= 0.8$ (per Figure 2.9).

$$\delta_a = \delta_{Da}\left(\frac{\delta}{\delta_D}\right) = (0.5)\left(\frac{1}{0.8}\right) = 0.6 \text{ in}$$

Although this structure could accommodate an allowable total settlement of 1.5 inches or more, we must limit it to 0.6 inch to keep the differential settlement within tolerable limits.

$$\delta_a = \textbf{0.6 in} \qquad \Leftarrow \textit{Answer}$$

Tilt

To preserve aesthetics, the tilt, ω, of buildings, chimneys, water towers, silos, and other similar structures should be no more than 1/250 (14 minutes of arc). Greater tilts would be noticeable, especially in taller structures and those that are near other structures. For comparison, the Leaning Tower of Pisa has a tilt of about 1/10.

Rate of Settlement and Sequence of Construction

It also is important to consider the rate of settlement and how it compares with the rate of construction. Foundations in sands settle about as rapidly as the loads are applied, whereas those in saturated clays move much more slowly.

In some structures, much of the load is applied to the foundation before the settlement-sensitive elements are in place. For example, settlements of bridge piers that occur before the deck is placed are far less important than those that occur after. Buildings may not be sensitive to differential settlements until after sensitive finishes, doors, and other architectural items are in place, yet if the foundation is in sand, most of the settlement may have already occurred by then.

However, other structures generate a large portion of their loads after they are completed. For example, the greatest load on a water storage tank is its contents.

Factor of Safety

All of the criteria described in this chapter are all allowable values, which means they already include a factor of safety of at least 1.5 - 2.0 and may be used directly in most design problems. However, we might consider using higher or lower factors of safety (and correspondingly different allowable values) in special cases. For example, a lower factor of safety might be appropriate when the soil conditions are known with a greater degree of certainty, whereas a higher factor might be preferred when there are more unknowns or when the structure is especially sensitive.

2.2 ENDURING HOSTILE ENVIRONMENTS

It often is necessary to build foundations in environments that are less than ideal. The foundation engineer must be aware of potential threats from the environment and be prepared to incorporate appropriate preventive measures in the design. These hostile environments include frozen soils, river bottoms subject to scour, and soils or water that is conducive to chemical, biological, or physical attack.

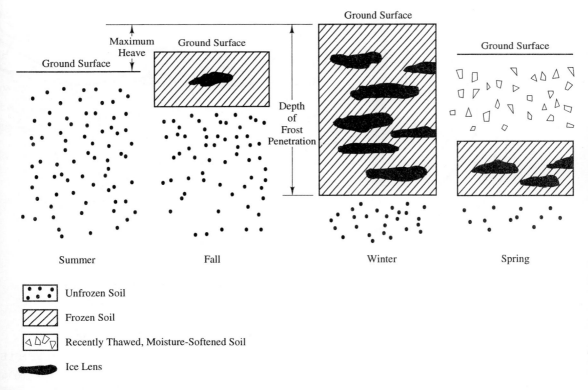

Figure 2.11 Idealized freeze-thaw cycle in temperate climate. During the summer, none of the ground is frozen. During the fall and winter, it progressively freezes from the ground surface down. Then, in the spring, it progressively thaws from the ground surface down.

Frost Heave

In many parts of the world, the air temperature in the winter can fall below the freezing point of water (0 °C) and remain there for extended periods. When this happens, the upper portion of the ground becomes frozen, as shown in Figure 2.11. When the air temperature rises again in the spring, the soil progressively thaws from the ground surface down. Much of the United States, southern Canada, central Europe, and other places with similar climates experience this annual phenomenon.

The greatest depth to which the ground might become frozen at a given locality is known as the *depth of frost penetration*. This distance is part of an interesting thermodynamics problem and is a function of the air temperature and its variation with time, the initial soil temperature, the thermal properties of the soil, and other factors. The deepest penetrations are obtained when very cold air temperatures are maintained for a long duration. Typical depths for the United States are shown in Figure 2.12.

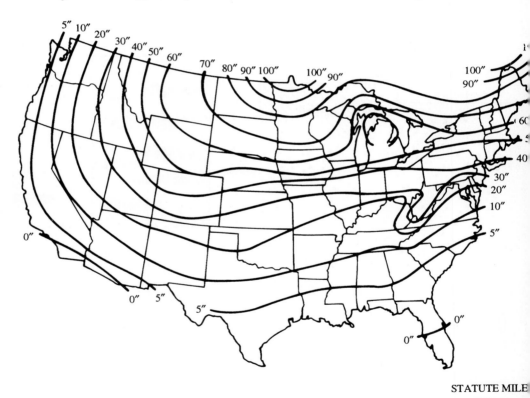

Figure 2.12 Approximate depth of frost penetration in the United States (U.S. Navy, 1982a).

The fact that the ground freezes is of interest to the foundation engineer for three reasons. First, when water freezes, it expands about 9% in volume. If the soil is saturated and has a typical porosity (say, 40%), it will expand about 9% × 40% ≈ 4% in volume when it freezes. In climates comparable to those in the northern United States, this could correspond to surface heaves of as much as 1 - 2 in (25 - 50 mm). Although such heaves are significant, they would probably be fairly uniform and cause relatively little damage.

The second concern is much more insidious and capable of much more damage: If the groundwater table is relatively shallow, capillary action[3] can draw water up to the frozen zone and form ice lenses, as shown in Figure 2.13. This mechanism can move large quantities of water, so it is not unusual for these lenses to produce ground surface heaves of 1 ft (300 mm) or more. This phenomena is known as *frost heave*. Such heaves are likely to be very irregular and create a hummocky ground surface that could cause extensive damage to structures, pavements, and other civil engineering works.

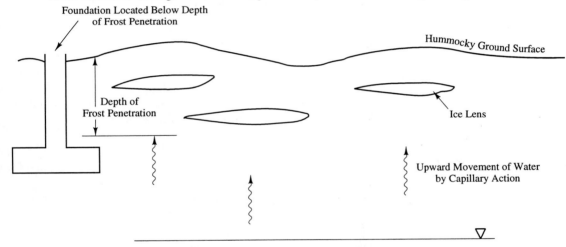

Figure 2.13 Formation of ice lenses. Water is drawn up by capillary action and freezes when it nears the surface. The frozen water forms ice lenses that cause heaving at the ground surface. Foundations placed below the depth of frost penetration are not subject to heaving.

The third concern relates to the soil conditions during the spring thaw. As the ice melts from the ground surface down, it leaves a soil with much more water than was originally present. Because the lower soils will still be frozen for a time, this water temporarily cannot drain away, and the result is a supersaturated soil that is very weak. This condition is often the cause of ruts and potholes in highways and can also effect the performance of shallow foundations and floor slabs. Once all the soil has thawed, the

---

[3] Capillary action is the ability of water to rise through a soil by surface tension. It is the same phenomenon that draws fuel through the wick in a kerosene lantern.

excess water drains down and the soil regains its strength.

The first question the design engineer faces in such circumstances is: "What is the depth of frost penetration at the project site?" For most practical problems, the design depth is based on local experience as documented in building codes. For example, the Chicago Building Code specifies a design frost penetration depth of 42 in (1.1 m). If such information is not available, then maps such as the one in Figure 2.12 may be used. Rarely, if ever, would a rigorous thermodynamic analysis be performed in practice.

Next, the engineer will consider whether ice lenses are likely to form within the frozen zone, thus causing frost heave. This will occur only if both of the following conditions are met:

1. There is a nearby source of water; and
2. The soil is *frost-susceptible.*

To be considered frost-susceptible, a soil must be capable of drawing significant quantities of water up from the groundwater table into the frozen zone. Clean sands and gravels are not frost-susceptible because they are not capable of significant capillary rise. Conversely, clays are capable of raising water through capillary rise, but they have a low permeability and are therefore unable to deliver large quantities of water. Therefore, clays are capable of only limited frost heave. However, intermediate soils, such as silts and fine sands, have both characteristics: They are capable of substantial capillary rise and have a high permeability. Large ice lenses are able to form in these soils, so they are considered to be very frost-susceptible.

The U.S. Army Corps of Engineers has classified frost-susceptible soils into four groups, as shown in Table 2.3. Higher group numbers correspond to greater frost susceptibility and more potential for formation of ice lenses. Clean sands and gravels (i.e., < 3% finer than 0.02 mm) may be considered non frost-susceptible and are not included in this table.

The most common method of protecting foundations from the effects of frost heave is to build them at a depth below the depth of frost penetration, as shown in Figure 2.13. This is usually wise in all soils, whether or not they are frost-susceptible and whether or not the groundwater table is nearby. Even "frost-free" clean sands and gravels will often have silt lenses that are prone to heave, and groundwater conditions can change unexpectedly, thus introducing new sources of water. The small cost of building deeper foundations is a wise investment in such cases. However, foundations supported on bedrock or interior foundations in heated buildings normally do not need to be extended below the depth of frost penetration.

Builders in Canada and Scandinavia often protect buildings with slab-on-grade floors using thermal insulation, as shown in Figure 2.14. This method traps heat stored in the ground during the summer and thus protects against frost heave, even though the foundations are shallower than the normal frost depth. Both heated and nonheated buildings can use this technique (NAHB, 1988 and 1990).

TABLE 2.3 FROST SUSCEPTIBILITY OF VARIOUS SOILS ACCORDING TO THE U.S. ARMY CORPS OF ENGINEERS

Group	Soil Types[a]	USCS Group Symbols[b]
F1 (least susceptible)	Gravels with 3 - 10% finer than 0.02 mm	GW, GP, GW-GM, GP-GM
F2	a. Gravels with 10 - 20% finer than 0.02 mm	GM, GW-GM, GP-GM
	b. Sands with 3 - 15% finer than 0.02 mm	SW, SP, SM, SW-SM, SP-SM
F3	a. Gravels with more than 20% finer than 0.02 mm	GM, GC
	b. Sands, except very fine silty sands, with more than 15% finer than 0.02 mm	SM, SC
	c. Clays with $PI > 12$, except varved clays	CL, CH
F4 (most susceptible)	a. Silts and sandy silts	ML, MH
	b. Fine silty sands with more than 15% finer than 0.02 mm	SM
	c. Lean clays with $PI < 12$	CL, CL-ML
	d. Varved clays and other fine-grained, banded sediments	

Adapted from Johnston (1981).
[a] PI = Plasticity Index (explained in Chapter 3).
[b] See Chapter 3 for an explanation of USCS group symbols.

Alternatively, the natural soils may be excavated to the frost penetration depth and replaced with soils that are known to be frost-free. This may be an attractive alternative for unheated buildings with slab floors to protect both the floor and the foundation from frost heave.

Although frost heave problems are usually due to freezing temperatures from natural causes, it is also possible to freeze the soil artificially. For example, refrigerated buildings such as cold-storage warehouses or indoor ice skating rinks can freeze the soils below and be damaged by frost heave, even in areas where natural frost heave is not a concern (Thorson and Braun, 1975). Placing insulation or air passages between the building and the soil and/or using non frost-susceptible soils usually prevents these problems.

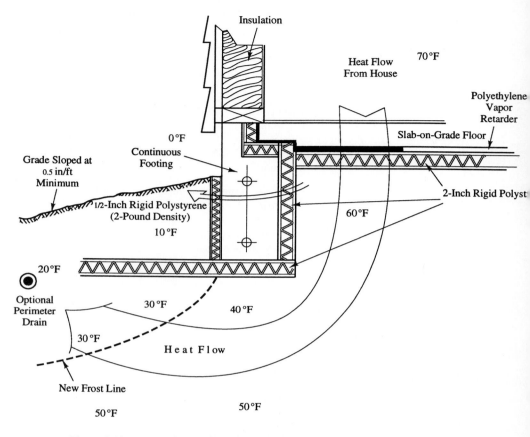

Figure 2.14 Thermal insulation traps heat in the soil, thus protecting a foundation from frost heave (HAHB, 1988, 1990).

A peculiar hazard to keep in mind when foundations or walls extend through frost-susceptible soils is *adfreezing* (CGS, 1985). This is the bonding of soil to a wall or foundation as it freezes. If heaving occurs after the adfreezing, the rising soil will impose a large upward load on the structure, possibly separating structural members. Placing a 0.5 in (10 mm) thick sheet of rigid polystyrene between the foundation and the frozen soil reduces the adfreezing potential.

Permafrost

In areas where the mean annual temperature is less than 0°C, the penetration of freezing in the winter may exceed the penetration of thawing in the summer and the ground can become frozen to a great depth. This creates a zone of permanently frozen soil known as *permafrost*. In the harshest of cold climates, such as Greenland, the frozen ground is continuous, whereas in slightly "milder" climates, such as central Alaska, central Canada, and much of Siberia, the permafrost is discontinuous. Areas of seasonal and continuous permafrost in Canada are shown in Figure 2.15.

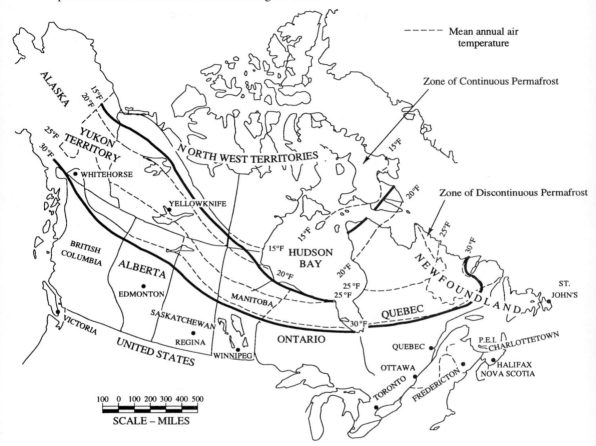

Figure 2.15 Zones of continuous and discontinuous permafrost in Canada (Adapted from Crawford and Johnson, 1971).

In areas where the summer thaws occur, the upper soils can be very wet and weak and probably not capable of supporting any significant loads, while the deeper soils remain permanently frozen. Foundations must penetrate through this seasonal zone and well into the permanently frozen ground below. It is very important that these foundations be designed so that they do not transmit heat to the permafrost, so buildings are typically built with raised floors and a ducting system to maintain subfreezing air temperatures between the floor and the ground surface.

The Alaska Pipeline project is an excellent example of a major engineering work partially supported on permafrost (Luscher et. al, 1975).

Scour

Bridges over rivers often include intermediate piers supported on foundations below the river. The design of these foundations must consider a special river hazard: the *scour* (loss) of the bottom soils during floods. This is also a problem for some waterfront structures.

Scour around the foundations is the most common cause of bridge failure. For example, during the Spring of 1987, there were 17 bridge failures due to scour in the northeastern United States alone (Huber, 1991). The most notable of these was the collapse of the Interstate Route 90 bridge over Schoharie Creek in New York (Murillo, 1987), a failure that killed 10 people.

Scour is part of the natural process that moves river-bottom sediments downstream. It can create large changes in the elevation of the river bottom. For example, Murphy (1908) describes a site on the Colorado River near Yuma, Arizona, where the river bed consists of highly erodible fine silty sands and silts. While passing a flood, the water level at this point rose 14 ft (4.3 m) and the bottom soils scoured to depths of up to 36 ft (11 m)! If a bridge foundation located 35 ft (10.7 m) below the river bottom had been built at this location, it would have been completely undermined by the scour and the bridge would have collapsed.

Scour is often greatest at places where the river is narrowest and constrained by levees or other means. Unfortunately, these are the locations most often selected for bridges. The presence of a bridge pier also creates water flow patterns that intensify the scour. However, methods are available to predict scour depths (Richardson et al., 1991) and engineers can use preventive measures, such as armoring, to prevent scour problems (TRB, 1984).

Durability of Buried Materials

Soil can be a very hostile environment to place engineering materials. Whether they are made of concrete, steel, or wood, structural foundations may be susceptible to chemical and/or biological attack that can adversely affect their integrity.

Corrosion of Steel

Under certain conditions, steel can be the object of extensive corrosion. This can be easily monitored when the steel is above ground, and routine maintenance, such as painting, will usually keep corrosion under control. However, it is impossible to inspect underground steel visually, so it is appropriate to be concerned about its potential for corrosion and long-term integrity.

Owners of underground steel pipelines are especially conscious of corrosion problems. They often engage in extensive corrosion surveys and include appropriate preventive measures in their designs. These procedures are well established and effective, but should they also be used for steel foundations such as H-piles or steel pipe piles?

For corrosion assessment, steel foundations can be divided into two categories: those in marine environments and those in land environments. Both are shown in Figure 2.16.

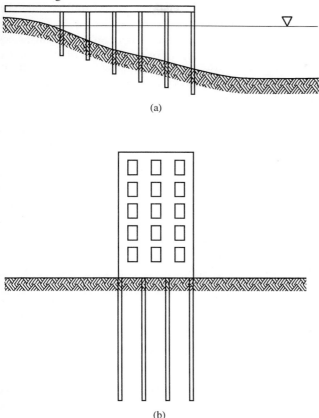

(a)

(b)

Figure 2.16 (a) Marine environments include piers, docks, drilling platforms, and other similar structures where a portion of the foundation is exposed to open water. (b) Land environments include buildings and other structures that are built directly on the ground and the entire foundation is buried.

Steel foundations in marine environments have a significant potential for corrosion, especially those exposed to salt water. Studies of waterfront structures have found that steel is lost at a rate of 0.075 to 0.175 mm/yr (Whitaker, 1976). This corrosion occurs most rapidly in the tidal and splash zones (Dismuke et. al., 1981) and can also be very extensive immediately above the sea floor; then it becomes almost negligible at depths more than about 2 ft (0.5 m) below the sea floor. Such structures may also be prone to abrasion from moving sand, ships, floating debris, and other sources. It is common to protect such foundations with coatings or jackets, at least through the water and splash zones.

However, the situation in land environments is quite different. Based on extensive studies, Romanoff (1962, 1970) observed that no structural failures have been attributed to the corrosion of steel piles in land environments. One likely reason for this excellent performance record is that piles, unlike pipelines, can tolerate extensive corrosion, even to the point of occasionally penetrating through the pile, and remain serviceable.

Romanoff also observed that piles founded in natural soils (as opposed to fills) experienced little or no corrosion, even when the soil could be identified as potentially corrosive. The explanation for this behavior seems to be that natural soils contain very little free oxygen, an essential ingredient for the corrosion process.

However, fills do contain sufficient free oxygen and, under certain circumstances, can be very corrosive environments. Therefore, concern over corrosion of steel piles in land environments can normally be confined to sites where the pile penetrates through fill. Some fills have very little potential for corrosion, whereas others could corrode steel at rates of up to 0.08 mm/yr (Tomlinson, 1987), which means that a typical H-pile section could lose half of its thickness in about 50 years.

Schiff (1982) indicated that corrosion would be most likely in the following soil conditions:

- High moisture content
- Poorly aerated
- Fine grained
- Black or gray color
- Low electrical resistivity
- Low or negative redox potential
- Organic material present
- High chemical content
- Highly acidic
- Sulfides present
- Anaerobic microorganisms present

Areas where the elevation of the groundwater table fluctuates, such as tidal zones, are especially difficult because this scenario continually introduces both water and oxygen to the pile. Contaminated soils, such as sanitary landfills and shorelines near old sewer outfalls, are also more likely to have problems.

One of the most likely places for corrosion on land piles is immediately below a concrete pile cap. Local electrical currents can develop because of the change in materials, with the concrete acting as a cathode and the soil as an anode. Unfortunately, this is also the most critical part of the pile because the stresses are greatest there.

If the foundation engineer suspects that corrosion may be a problem, it is appropriate to retain the services of a corrosion engineer. Detailed assessments of corrosion and the development of preventive designs are beyond the expertise of most foundation engineers.

The corrosion engineer will typically conduct various tests to quantify the corrosion potential of the soil and consider the design life of the foundation to determine whether any preventive measures are necessary. Such measures could include the following:

- Use a different construction material (i.e., concrete, wood).
- Increase the thickness of steel sections by an amount equal to the anticipated deterioration.
- Cover the steel with a protective coating (such as coal tar epoxy) to protect it from the soil. This method is commonly used with underground tanks and pipes, and has also been successfully used with pile foundations. However, consider the possibility that some of the coating may be removed by abrasion when the pile is driven into the ground, especially when sands or gravels are present. Coatings can also be an effective means of combatting corrosion near pile caps, as discussed earlier. In this case, the coating is applied to the portion of the steel that will be encased in the concrete, thus providing the electrical insulation needed to stop or significantly slow the corrosion process.
- Provide a *cathodic protection system*. Such systems consist of applying a DC electrical potential between the foundation (the cathode) and a buried sacrificial metal (the anode). This system causes the corrosion to be concentrated at the anode and protects the cathode (see Figure 2.17). Rectifiers connected to a continuous power source provide the electricity. These systems consume only nominal amounts of electricity. In some cases, it is possible to install a self-energizing system that generates its own current.

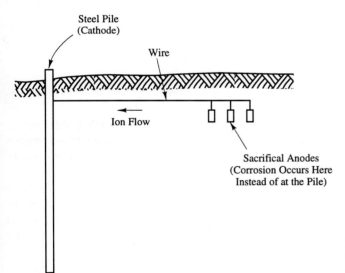

Figure 2.17 Use of a cathodic protection system to protect steel foundations from corrosion.

Sulfate Attack on Concrete

Buried concrete is usually very resistant to corrosion and will remain intact for many years. However, serious degradation can occur in concrete subjected to soils or groundwater that contains high concentrations of sulfates (SO_4). These sulfates can react with the cement to form calcium sulfoaluminate (ettringite) crystals. As these crystals grow and expand, the concrete cracks and disintegrates. In some cases, serious degradation has occurred within 5 to 30 years of construction. Although we do not yet fully understand this process (Mehta, 1983), engineers have developed methods of avoiding these problems.

We can evaluate a soil's potential for sulfate attack by measuring the concentration of sulfates in the soil and/or in the groundwater and comparing them with soils that have had problems with sulfate attack. Soils with some or all of the following properties are most likely to have high sulfate contents:

- Wet
- Fine grained
- Black or gray color
- High organic content
- Highly acidic or highly alkaline

Some fertilizers contain a high concentration of sulfates that may cause problems when building in areas that were formerly used for agricultural purposes. The same is true for some industrial wastes. It is often wise to consult with corrosion experts in such cases. Sea water also has a high concentration: about 2300 ppm.

If the laboratory tests indicate that the soil or groundwater has a high sulfate content, design the buried concrete to resist attack by using one or more of the following methods:

- **Reduce the water:cement ratio**—This reduces the permeability of the concrete, thus retarding the chemical reactions. This is one of the most effective methods of resisting sulfate attack. Suggested maximum ratios are presented in Table 2.4.
- **Increase the cement content**—This also reduces the permeability. Therefore, concrete that will be exposed to problematic soils should have a cement content of at least 6 sacks/yd^3 (564 lb/yd^3 or 335 kg/m^3).
- **Use sulfate-resisting cement**—Type II low-alkali and type V Portland cements are specially formulated for use in moderate and severe sulfate conditions, respectively. Pozzolan additives to a type V cement also help. Type II is easily obtained, but type V may not be readily available in some areas. Table 2.4 gives specific guidelines.
- **Coat the concrete with an asphalt emulsion**—This is an attractive alternative for retaining walls or buried concrete pipes, but not for foundations.

TABLE 2.4 USE OF SULFATE-RESISTING CEMENTS AND LOW WATER:CEMENT RATIOS TO AVOID SULFATE ATTACK OF CONCRETE

Water-Soluble Sulfates in Soil (% by weight)	Sulfates in Water (ppm)	Sulfate Attack Hazard	Cement Type	Maximum Water:Cement Ratio
0.00 - 0.10	0 - 150	Negligible	-	-
0.10 - 0.20	150 - 1500	Moderate	II	0.50
0.20 - 2.00	1500 - 10,000	Severe	V	0.45
> 2.00	> 10,000	Very severe	V plus pozzolan	0.45

Adapted from Kosmatka and Panarese (1988) and PCA (1991).
Used by permission of the Portland Cement Association.

Decay of Timber

The most common use of wood in foundations is timber piles. The life-span of these piles varies depending on their environment. Even untreated timber piles can have a very long life if they are continually submerged below the groundwater table. This was illustrated when a campanile in Venice fell in 1902. The submerged timber piles, which had been driven in AD 900, were found to be in good condition and were used to support the replacement structure (Chellis, 1962). However, when located above the groundwater table, timber can be subject to deterioration from several sources (Chellis, 1961), including:

- **Decay** caused by the growth of fungi. This process requires moisture, oxygen, and favorable temperatures. These conditions are often most prevalent in the uppermost 6 ft (2 m) of the soil. If the wood is continually very dry, then decay will be limited due to the lack of moisture.
- **Insect attack**, including termites, beetles, and marine borers.
- **Fire**, especially in marine structures.

The worst scenario is one in which the piles are subjected to repeated cycles of wetting and drying. Such conditions are likely to be found near the groundwater table because it usually rises and sinks with the seasons and near the water surface in marine applications where splashing and tides will cause cyclic wetting and drying.

To reduce problems of decay, insect attack, and fungi growth, timber piles are usually treated before they are installed. The most common treatment consists of placing them in a pressurized tank filled with creosote or some other preserving chemical. This *pressure treatment* forces some of the creosote into the wood and forms a thick coating on the outside, leaving a product that is almost identical to many telephone poles. When the piles are fully embedded into soil, creosote-treated piles normally have a life at least as long as the design life of the structure.

Timber piles also will lose part of their strength if they are subjected to prolonged high temperatures. Therefore, they should not be used under hot structures such as blast furnaces.

2.3 POSSESSING SUFFICIENT STRUCTURAL INTEGRITY

Foundations are structural members and therefore must have sufficient integrity to carry the design loads safely. The principles and practices of conventional structural design used in the superstructure also apply to foundations and are usually dictated to a large degree by building codes.

The structural design of foundation elements usually includes simplifying assumptions and often uses design strengths that are less than those for comparable materials in the superstructure. These result in designs that are conservative. This approach is justified for the following reasons:

- Foundations are not built with the same degree of precision as the superstructure. For example, spread footings are typically excavated with a backhoe and the sides of the excavation becomes the "formwork" for the concrete, compared to concrete members in the superstructure that are carefully formed with plywood or other materials.
- The structural materials may be damaged when they are installed. For example, cracks and splits may develop in a timber pile during hard driving.
- There is a significant degree of uncertainty in the nature and distribution of the load transfer between foundations and the ground, so the stresses at any point in a foundation are not always known with as much certainty as might be the case in much of the superstructure.
- The consequences of a catastrophic failure are much greater.
- The additional weight brought on by the conservative design is of no consequence. because the foundation is the lowest structural member and therefore does not affect the dead load on any other member. Additional weight in the foundation is actually beneficial in that it increases its uplift resistance.

2.4 BEING ECONOMICAL TO BUILD

Although some degree of conservatism is appropriate in foundation engineering, gross overconservatism is not warranted. An overly conservative design can be very expensive to build, especially with large structures where the foundation is a greater portion of the total project cost. This also is a type of "failure": the failure to produce an economical design.

The nineteenth century engineer Arthur Wellington once said that an engineer's job is that of "doing well with one dollar which any bungler can do with two." We must strive to produce designs that are both safe and cost-effective. Achieving the optimum balance between reliability (safety) and cost is part of good engineering.

A thorough knowledge of construction methods and equipment helps engineers develop economical designs, whereas ignorance of these issues often creates major problems. For example, the design engineer may wish to place a spread footing in a loose sand below the groundwater table. Such a design may be fine on paper, but would be very difficult and expensive to build because of the excavation, shoring, and dewatering problems.

In addition, designs that minimize the required quantity of construction materials does not necessarily minimize the cost. In some cases, designs that use more materials may be easier to build, and thus have a lower overall cost. For example, spread footing foundations are usually made of low-strength concrete, even though it makes them thicker. In this case, the savings in materials and inspection costs are greater than the cost of buying more concrete.

SUMMARY OF MAJOR POINTS

1. The foundation engineer must determine the necessary performance requirements before designing a foundation.
2. A foundation must support normal, shear, moment, and/or torsion loads. The magnitude and direction of these loads may vary during the life of the structure.
3. Design dead loads are usually close to the actual dead loads. However, design live loads are often larger than reality because of the uncertainties in predicting them.
4. All structures will experience both total and differential settlements. These will produce movements and distortions in the structure. We cannot design for no settlement, but we can set upper limits on the tolerable total and differential settlements.
5. Soils that experience freeze-thaw cycles are subject to frost heave. This can produce large movements in foundations unless they are founded below the depth of frost penetration.
6. Permanently frozen ground, known as permafrost, must be protected during and after construction to prevent unwanted ground movements.
7. Scour can undermine foundations beneath riverbeds and other underwater locations. Prevent this by placing the foundation at a suitable depth or by installing protective armor.
8. Steel foundations can be subject to corrosion, especially in marine environments. Preventive measures include using thicker steel sections, applying protective coatings, and installing a cathodic protection system.

9. Concrete foundations in certain soils can be subject to sulfate attack. Control this problem by using dense concrete mixtures, asphalt coatings, and/or sulfate-resisting cements.

10. Wooden foundations are subject to decay, especially if they are exposed to frequent wet-dry cycles. However, wood that is continuously below the groundwater table and not exposed to seawater often has a much longer life. Treatment with creosote or other preservatives is the most common preventive measure.

11. Foundations must have sufficient structural integrity to safely transmit the applied loads to the ground.

12. Because of the uncertainties in the design parameters and the serious consequences of a failure, foundation designs usually are more conservative than those for other structural members. However, gross over-conservatism must be avoided.

QUESTIONS AND PRACTICE PROBLEMS

2.1 Classify the frost susceptibility of the following soils:
   a. Sandy gravel with 3% finer than 0.02 mm.
   b. Well graded sand with 4% finer than 0.02 mm.
   c. Silty sand with 20% finer than 0.02 mm.
   d. Fine silty sand with 35% finer than 0.02 mm.
   e. Sandy silt with 70% finer than 0.02 mm.
   f. Clay with plasticity index = 60.

2.2 A compacted fill is to be placed at a site in North Dakota. The following soils are available for import: Soil 1 - silty sand; Soil 2 - lean clay; Soil 3 - Gravelly coarse sand. Which of these soils would be least likely to have frost heave problems?

2.3 Would it be wise to use slab-on-grade floors for houses built on permafrost? Explain.

2.4 A certain clayey soil contains 0.30% sulfates. Would you anticipate a problem with concrete foundations in this soil? Are any preventive measures necessary? Explain.

2.5 What is the most common cause of failure in bridges?

2.6 A reinforced concrete art gallery is to be built on a site underlain by a clayey soil. This building is to have an unusual architectural design and will include many tile murals on the walls. The column spacing will vary between 5 and 8 meters. Compute the allowable total settlement for the spread footing foundations.

2.7 A single-story building is to be built on a sandy silt in Detroit. How deep must the exterior footings be below the ground surface to avoid problems with frost heave?

2.8 Soon after it was built, a two-story wood-frame house developed several cracks in the interior and exterior walls. These walls are covered with drywall and stucco, respectively.

The width of these cracks ranged from 0.1 to 1.0 mm. The house is now 5 years old and the cracks have become wider and longer. Some of them are now up to 3.0 mm wide. The number of cracks has also increased. Discuss the significance of these cracks and the structural integrity of this house.

2.9 A series of 50 ft long steel piles are to be driven into a natural sandy soil. The groundwater table is at a depth of 35 ft below the ground surface. Would you anticipate a problem with corrosion? What additional data could you gather to make a more informed decision?

2.10 A one-story steel warehouse building similar to that shown in Figure 2.5 is to be built on a natural sandy soil. The roof is to be supported by steel trusses that will span the entire 70 ft width of the building and supported on columns adjacent to the exterior walls. These trusses will be placed 24 ft on center. No interior columns will be present. Compute the allowable total settlement of the footings that support these trusses.

2.11 The grandstands for a small baseball stadium are to be built of structural steel supported on spread footings. The structural engineer plans to use a very wide column spacing (25 m) and no diagonal bracing to provide the best spectator visibility. The soil beneath the foundations is a clay. Compute the allowable total settlement of these foundations.

2.12 The owner of a 100-story building purchased a plumb bob with a very long string. He selected day with no wind, and then gently lowered the plumb bob from his penthouse office window. When it reached the sidewalk, it was 1.0 m from the side of the building. Is this building tilting excessively? Explain.

3

Soil Mechanics

Measure it with a micrometer,
mark it with chalk,
cut it with an axe.

An admonition to maintain a
consistent degree of precision
throughout the analysis, design,
and construction phases of a project.

Engineers classify earth materials into two broad categories: *rock* and *soil*. Although both materials play an important role in foundation engineering, most foundations are supported by soil. In addition, foundations on rock often may be designed much more conservatively because of the rock's greater strength, whereas economics prevents overconservatism when building foundations on soil. Therefore, it is especially important for the foundation engineer to be familiar with soil mechanics.

Users of this book should already have acquired at least a fundamental understanding of the principles of soil mechanics. This chapter reviews these principles and emphasizes those that are most important in foundation engineering analyses. Relevant principles of rock mechanics are included in later chapters within the context of specific applications.

3.1 THE ORIGIN AND CHARACTER OF SOIL DEPOSITS

Most soils are derived from rocks through one or more of the following mechanisms:

- *Residual soils* were formed in-place as a result of physical and chemical decomposition of rock. They have not been transported to another location and often retain much of the character of the parent rock.
- *Colluvial soils* were moved by gravity (such as landslides, falls, flows, or creep).
- *Alluvial soils* were transported by water. Rivers can transport large quantities of alluvial soil and form large alluvial plains.
- *Aeolian soils* were transported by wind. These include windblown dunes, loess, and volcanic dust deposits.
- *Glacial soils* were transported or strongly affected by the action of glaciers.
- *Lacustrine soils* were formed beneath lakes or other bodies of water.

Various weathering processes also will create *clay minerals* in the soil, and the presence of these minerals will have a profound effect on its behavior. Some soils also include organic material derived from the decomposition of plant or animal matter.

By understanding the geology of an area, the foundation engineer is able to understand the origin of the soils and is thus better equipped to evaluate laboratory and field data. This provides some of the basis for making the necessary engineering judgments that are a part of any foundation design.

3.2 THE STRUCTURE OF SOILS

One of the fundamental differences between soil and most other engineering materials is that it is a *particulate material*. This means that it is an assemblage of individual particles rather than being a *continuum* (a single solid mass). The engineering properties of soil, such as strength and compressibility, are dictated primarily by the arrangement of these particles and the interactions between them rather than by their internal properties.

Another important characteristic that differentiates soil from most other materials is that it can contain all three phases of matter: solid, liquid and gas, simultaneously. The solid portion (the particles) includes one or more of the following materials:

- *Rock fragments* such as granite, limestone, and basalt.
- *Rock minerals* such as quartz, feldspar, mica, and gypsum.
- *Clay minerals* such as kaolinite, smectite, and illite.
- *Organic matter* such as decomposed plant materials.
- *Cementing agents* such as calcium carbonate.
- *Miscellaneous materials* such as man-made debris.

Liquids and/or gasses fill the voids between the solid particles. The liquid component is usually water, but it also could contain various chemicals in solution. The latter could come from natural sources, such as calcite leached from limestone, or artificial sources such as gasoline from leaking tanks or pipes. Likewise, the gas component is usually air, but also could consist of other materials, such as methane. For simplicity, we will refer to these components as "water" and "air."

A special exception to this three-phase structure is the case of a saturated soil at a temperature below the freezing point of water. These frozen soils are essentially a completely solid material and require special analysis and design techniques.

Particle Size

Because soil is a particulate material, it is natural to consider the size of these particles and their effect on the behavior of the soil. Several different classification schemes are available, but the one published by ASTM (American Society for Testing and Materials) is the most common system used by geotechnical engineers. This system classifies soil particles, as shown in Table 3.1.

TABLE 3.1 ASTM D2487 PARTICLE SIZE CLASSIFICATION

Sieve Size		Particle Size		Particle
Passes	Retained on	(in)	(mm)	Classification
	12 in	> 12	> 300	Boulder
12 in	3 in	3 - 12	75 - 300	Cobble
3 in	3/4 in	0.75 - 3	19 - 75	Coarse gravel
3/4 in	#4	0.19 - 0.75	4.8 - 19	Fine gravel
#4	#10	0.079 - 0.19	2.0 - 4.8	Coarse sand
#10	#40	0.016 - 0.079	0.42 - 2.0	Medium sand
#40	#200	0.0029 - 0.016	0.075 - 0.42	Fine sand
#200		< 0.0029	< 0.075	Fines (silt & clay)

Copyright ASTM; Used with permission.

Most natural soils contain a wide variety of particle sizes and thus do not fall completely within any of the categories listed in Table 3.1. Thus, the distribution of particle sizes in a particular soil is most easily expressed in the form of a *grain-size distribution curve*, as shown in Figure 3.1. This is a plot of the percentage of the dry soil by weight that is smaller than a certain particle diameter vs. the particle diameter.

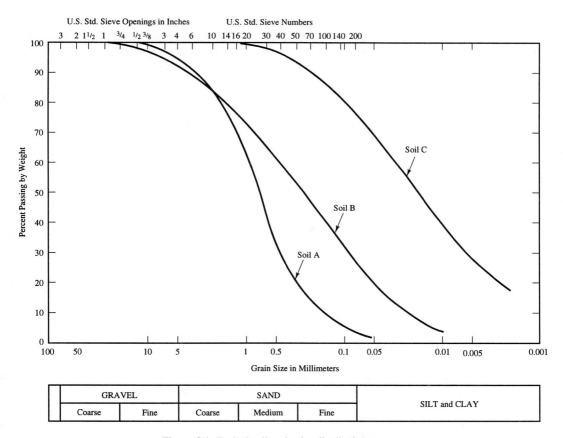

Figure 3.1 Typical soil grain-size distribution curves.

Cohesionless vs. Cohesive Soils

Geotechnical engineers often divide soils into two broad categories: *cohesionless soils* and *cohesive soils*. The distinction depends on whether the individual particles are held together only as a result of gravity or external loads (a cohesionless soil) or if interparticle bonds also appear to be present (a cohesive soil). These characteristics have a significant effect on the behavior of a soil.

Cohesionless soils include gravels, sands, nonplastic silts, and combinations of these materials. The only source of shear strength in these soils is the sliding friction and mechanical interlocking between the particles, which in turn rely on the presence of compressive contact forces. Therefore, they have no strength when unconfined.

Gravels and sands, and to a lesser degree silts, have high coefficients of permeability. Therefore, if they are saturated and volume changes occur, water can rapidly flow into or out of the voids as necessary. This characteristic also has a significant effect on the behavior of cohesionless soils.

Cohesive soils include clays or other soils with a significant clay content as well as cemented soils such as caliche. The interparticle bonding in these soils, which is independent of the normal load, can be the source of a major part of the shear strength. This is why these soils always have some strength, even when unconfined. Cohesive soils also have a much lower coefficient of permeability, which means that water flows in or out of the voids much more slowly. The consequences of this characteristic are discussed in more detail later in this chapter.

In reality, the physical mechanisms that control soil behavior are much more complex, but this simple model is often convenient.

Fissured Clays

Stiff, overconsolidated clays usually contain *fissures* (cracks), and these fissures have a significant effect on the engineering properties of such soils. The shear strength along the fissures will be less than that of the intact soil, so their position, orientation, and spacing influence the behavior of the overall soil mass.

These soils can be especially challenging for the foundation engineer because the spacing between the fissures is typically large compared to the size of most soil samples. Therefore, tests performed on these samples may not represent the larger soil mass.

3.3 WEIGHT-VOLUME RELATIONSHIPS

A knowledge of the relative proportions of solids, water, and air can give important insight into the engineering behavior of a particular soil. A *phase diagram*, as shown in Figure 3.2, describes these proportions.

Geotechnical engineers have developed several standard parameters to define these proportions, and these parameters form much of the basic vocabulary of soil mechanics. Table 3.2 contains the definitions of weight-volume parameters commonly used in foundation engineering, along with typical numerical values. These values are not absolute limits and unusual soils may have properties outside these ranges.

These weight-volume parameters are all related to each other, and many formulas are available to express these relationships. Some of the more useful formulas include the following:

$$e = \frac{w\, G_s}{S} \tag{3.1}$$

$$e = \frac{G_s\, \gamma_w}{\gamma_d} - 1 \tag{3.2}$$

$$\gamma_d = \frac{\gamma}{1 + w} \tag{3.3}$$

$$w = S\left(\frac{\gamma_w}{\gamma_d} - \frac{1}{G_s}\right) \tag{3.4}$$

Convert any parameters expressed as a percentage into decimal form before using them in these formulas.

Geotechnical engineers often use the term "density" instead of "unit weight" for the variable γ. Although density is technically mass/volume (not weight/volume), consider these two terms to be synonymous when used in the context of geotechnical engineering. However, this book uses only the more correct term "unit weight." Table 3.3 presents typical unit weights for various soils.

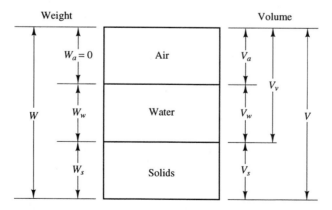

Figure 3.2 A phase diagram describes the relative proportions of solids, water, and air in a soil. The weight of air, W_a, is negligible.

TABLE 3.2 DEFINITIONS AND TYPICAL VALUES OF COMMON SOIL WEIGHT-VOLUME PARAMETERS

Parameter	Symbol	Definition	Typical Range	
			English	SI
Unit weight	γ	$\dfrac{W}{V}$	90 - 130 lb/ft^3	14 - 20 kN/m^3
Dry unit weight	γ_d	$\dfrac{W_s}{V}$	60 - 125 lb/ft^3	9 - 19 kN/m^3
Unit weight of water	γ_w	$\dfrac{W_w}{V_w}$	62.4 lb/ft^3	9.80 kN/m^3
Buoyant unit weight	γ_b	$\gamma_{sat} - \gamma_w$	28 - 68 lb/ft^3	4 - 10 kN/m^3
Degree of saturation	S	$\dfrac{V_w}{V_v} \times 100\%$	2 - 100 %	2 - 100 %
Moisture content	w	$\dfrac{W_w}{W_s} \times 100\%$	3 - 70 %	3 - 70 %
Void ratio	e	$\dfrac{V_v}{V_s}$	0.1 - 1.5	0.1 - 1.5
Porosity	n	$\dfrac{V_v}{V} \times 100\%$	9 - 60 %	9 - 60 %
Specific gravity of solids	G_s	$\dfrac{W_s}{V_s \gamma_w}$	2.6 - 2.8	2.6 - 2.8

γ_{sat} is the unit weight, γ, when $S = 100\%$.

Example 3.1

A 0.320 ft^3 sample of a certain soil has a weight of 38.9 lb, a moisture content of 19.2%, and a specific gravity of solids of 2.67. Find its void ratio and degree of saturation.

Solution: Using Equations 3.1 - 3.4:

$$\gamma = \frac{W}{V} = \frac{38.9}{0.320} = 121.6 \; lb/ft^3$$

$$\gamma_d = \frac{\gamma}{1+w} = \frac{121.6}{1+0.192} = 102.0 \; lb/ft^3$$

$$e = \frac{G_s \gamma_w}{\gamma_d} - 1 = \frac{(2.67)(62.4)}{102.0} - 1 = \textbf{0.633} \qquad \Leftarrow \textit{Answer}$$

$$e = \frac{w \, G_s}{S} \;\rightarrow\; S = \frac{w \, G_s}{e} = \frac{(0.192)(2.67)}{0.633} = \textbf{81.0\%} \qquad \Leftarrow \textit{Answer}$$

TABLE 3.3 TYPICAL UNIT WEIGHTS OF SOIL

Soil Type (See Table 3.4)	Typical Unit Weight, γ			
	Above Groundwater Table		Below Groundwater Table	
	(lb/ft^3)	(kN/m^3)	(lb/ft^3)	(kN/m^3)
GP — Poorly graded gravel	110 - 130	17.5 - 20.5	125 - 140	19.5 - 22.0
GW — Well graded gravel	110 - 140	17.5 - 22.0	125 - 150	19.5 - 23.5
GM — Silty gravel	100 - 130	16.0 - 20.5	125 - 140	19.5 - 22.0
GC — Clayey gravel	100 - 130	16.0 - 20.5	125 - 140	19.5 - 22.0
SP — Poorly graded sand	95 - 125	15.0 - 19.5	120 - 135	19.0 - 21.0
SW — Well graded sand	95 - 135	15.0 - 21.0	120 - 145	19.0 - 23.0
SM — Silty sand	80 - 135	12.5 - 21.0	110 - 140	17.5 - 22.0
SC — Clayey sand	85 - 130	13.5 - 20.5	110 - 135	17.5 - 21.0
ML — Low plasticity silt	75 - 110	11.5 - 17.5	80 - 130	12.5 - 20.5
MH — High plasticity silt	75 - 110	11.5 - 17.5	75 - 130	11.5 - 20.5
CL — Low plasticity clay	80 - 110	12.5 - 17.5	75 - 130	11.5 - 20.5
CH — High plasticity clay	80 - 110	12.5 - 17.5	70 - 125	11.0 - 19.5

Relative Density

The *relative density*, D_r, is a convenient way to express the void ratio of cohesionless soils. It is based on the void ratio of the soil, e, the *minimum index void ratio*, e_{min}, and the *maximum index void ratio*, e_{max}:

$$D_r = \frac{e_{max} - e}{e_{max} - e_{min}} \times 100\% \tag{3.5}$$

The values of e_{min} and e_{max} are determined by conducting standard laboratory tests [ASTM D4254]. The in-situ e could be computed from the unit weight of the soil using Equation 3.2, but accurate measurements of the unit weight of clean sand are difficult or impossible to obtain. Therefore, engineers often obtain D_r from correlations based on in-situ tests, as described in Chapter 4.

If a soil has a relative density of 0%, it is supposedly in its loosest possible condition, while at 100%, it is supposedly in its densest possible condition. Although it is possible for natural soils to have relative densities outside this range, such conditions are very unusual. A relationship between consistency of cohesionless soils and relative density is shown in Table 3.4.

Do not confuse relative density with relative compaction. The latter is based on the Proctor compaction test [ASTM D1557] and is typically used to evaluate compacted fills. Although these two parameters measure similar soil properties, and both are expressed as a percentage, they are not numerically equal.

The relative density applies only to cohesionless soils with less than 15% fines. It can be an excellent indicator of the engineering properties of such soils, and it is therefore an important part of many analysis methods. However, other considerations, such as stress history, mineralogical content, grain-size distribution, and fabric (the configuration of the particles), also affect the engineering properties.

3.4 PLASTICITY

The moisture content, w, is a basic and useful indicator of soil properties, especially in cohesive soils. For example, clays with a low moisture content are stronger and less compressible than those with a high moisture content.

In 1911, the Swedish soil scientist A. Atterberg developed a series of tests to evaluate the relationship between moisture content and soil consistency. In the 1930s Arthur Casagrande adapted these tests for civil engineering purposes and they soon became a routine part of geotechnical engineering. This series includes three separate tests: the *liquid limit test*, the *plastic limit test*, and the *shrinkage limit test*. Together they are known as the *Atterberg limits tests*.

TABLE 3.4 CONSISTENCY OF COHESIONLESS SOILS AT VARIOUS RELATIVE DENSITIES

Relative Density, D_r (%)	Description
0 - 15	Very loose
15 - 35	Loose
35 - 65	Medium dense[a]
65 - 85	Dense
85 - 100	Very dense

a. Lambe and Whitman used the term "medium," but "medium dense" is better because "medium" usually refers to the grain-size distribution.

Adapted from *Soil Mechanics* by Lambe and Whitman, Copyright ©1969 by John Wiley & Sons. Used by permission of John Wiley & Sons, Inc.

Figure 3.3 shows qualitative descriptions of the changes in consistency of a fine-grained soil that occur as its moisture content changes. Dry cohesive soils are hard and brittle, whereas wet cohesive soils are soft and pliable. Although these changes in consistency are gradual, the Atterberg limits tests define the boundaries between the various states in a somewhat arbitrary but standardized way. The test results are expressed in terms of the moisture content with the percent sign dropped:

SL = shrinkage limit
PL = plastic limit
LL = liquid limit

By comparing the moisture content of a soil with its Atterberg limits, an engineer could gain a qualitative sense of its consistency. For example, if a certain soil has a liquid limit of 55, a plastic limit of 20, and a moisture content of 25%, then it would have a consistency comparable to that of a stiff putty (i.e., it is slightly wetter than the plastic limit).

The liquid limit and plastic limit tests are part of many laboratory test programs. They are inexpensive and the results can be quite useful. In contrast, the shrinkage limit has little practical significance for engineers and is rarely measured.

Another useful parameter based on the Atterberg limits is the *plasticity index*:

$$PI = \text{plasticity index}$$
$$= LL - PL \tag{3.6}$$

The plasticity index indicates the range of moisture content at which the soil is in the plastic state. Silty soils have a small PI, which means that adding only a small quantity of water will convert the soil from a semisolid state to a liquid state. Clays have a much higher PI, which means that much more water must be added to obtain the same change in consistency.

The Atterberg limits give the engineer a feel for the soil's behavior, and they form the basis for the Unified Soil Classification System (described in the next section). Engineers also have developed empirical correlations between the Atterberg limits and soil properties such as compressibility and shear strength. Some of these correlations are included in this book.

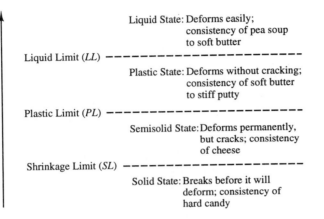

Figure 3.3 The relationship between changes in moisture content and consistency in fine-grained soils. (Adapted and reprinted with permission of MacMillan Publishing Company from *Introductory Soil Mechanics and Foundations: Geotechnical Engineering*, 4th Ed., by George F. Sowers, © 1979 by MacMillan Publishing Company).

3.5 SOIL CLASSIFICATION SYSTEMS

It is very important to have standard methods of classifying soils, and geotechnical engineers have developed a wide variety of soil classification systems. Some of these, such as the AASHTO (American Association of State Highway and Transportation Officials) system or the FAA (Federal Aviation Administration) system, are intended for specific applications (i.e., highways, airports), while others are multipurpose. Geologists, soil scientists, and others also have developed classification systems to suit their needs.

Although no single system is universally accepted, the *Unified Soil Classification System* (USCS) [ASTM D2487] is the most dominant and widely understood method. It is an all-purpose system based on the grain-size distribution of the soil and its Atterberg limits, as shown in Figure 3.4.

Special Soil Classification Terms

In addition to the standard classification systems, engineers often use special terms and names to describe certain soils. These are most likely to be used with soils that have distinctive characteristics. Some of the more common terms are as follows:

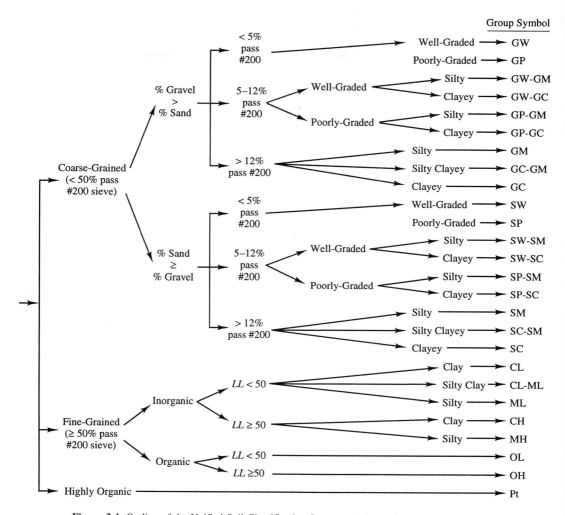

Figure 3.4 Outline of the Unified Soil Classification System. (Adapted from ASTM D2487. Copyright ASTM, used with permission).

- ***Adobe*** — A very plastic silty clay found in the southwestern portion of the United States.
- ***Black cotton*** — A very dark clayey soil found in semitropical areas of India and Southeast Asia.
- ***Caliche*** — A whitish mixture of sand and silt cemented together with calcium carbonate ($CaCO_3$) to form a very hard soil. Could also contain clay and/or gravel. Found in desert areas of Mexico and the southwestern United States where lime-

bearing groundwater has been drawn up by capillary action and evaporated, leaving the mineral deposits behind. Caliche layers range in thickness from less than 1 inch (25 mm) to several feet (≈2 m). This is a type of *calcareous* soil.

- *Diatomaceous earth* — A whitish, fine-grained soil composed of the inorganic remains of microscopic single-celled plants known as *diatoms*. Also known as *diatomite*. Typically has a large amount of water locked into cells, and thus has a high moisture content while appearing to be very dry.
- *Glacial till* — A dense well-graded soil deposited by glaciers. Also known simply as *till*.
- *Gumbo* — A clayey soil found in the southeastern portion of the United States.
- *Hardpan* — Any extremely hard soil, such as caliche.
- *Laterite* — A cemented residual soil found in tropical regions.
- *Loam* — A mixture of silt and sand.
- *Loess* — A very porous aeolian silt that is cemented by a calcareous or clayey binder. Dry and strong in its natural state, but becomes weak and compressible if saturated. Extensive loess deposits exist in the northern plains states of the United States and in many other localities.
- *Varved clay* — A soil consisting of alternating layers of silt (usually lighter in color) and clay (usually darker in color). Each layer is rarely thicker than 1/2 inch (10 mm). Believed to be the result of soils transported by melt water at the end of the Ice Age. Varved clays are very soft and compressible.

3.6 STRESSES IN SOIL

Many of the analyses used in foundation engineering require a knowledge of the stresses in the soil. These stresses are the result of both the force of gravity acting on the soil mass and the action of external loads, such as foundations.

Both normal and shear stresses may be present in a soil. These are represented by the variables σ and τ, respectively. Because the tensile strength of soil is very low, the normal stresses are nearly always compressive, so geotechnical engineers use a sign convention opposite that used by structural engineers: Compressive stresses are positive, while tensile stresses are negative.

When using English units, express stresses in units of lb/ft^2 in the soil and lb/in^2 in structural members. With SI units, use kPa in the soil and MPa in structural members. Many engineers in non-SI metric countries often use units of kg/cm^2 and bars (1 bar = 100 kPa).

Total vs. Effective Stresses

Vertical compressive stresses develop in a soil as a result of its own weight just as pressures develop in a body of water due to its weight. Both are zero at the top and increase with depth at a rate equal to the unit weight of the material. In soils, these are known as *geostatic stresses*.

The compressive stress in a soil may be expressed either in terms of *total stress* or *effective stress*. If only geostatic stresses are present, the vertical total stress at a point in a soil is computed by summing from the ground surface to that point as follows:

$$\sigma_v = \sum \gamma_i h_i \qquad (3.7)$$

Where:

σ_v = vertical total stress at a point in the soil

γ_i = unit weight of soil stratum i

h_i = thickness of soil stratum i

This is analogous to the procedure for computing the pressure in a body of water, except for the additional consideration that the unit weight may vary with depth.

The concept of *effective stress* is slightly more abstract. If the soil is saturated (i.e., all of the voids are filled with water), then all of the pore water is in intimate contact, forming a continuous "column" through the soil. This column of water can and does carry compressive stresses. Therefore, the solid particles carry only a portion of the total stress at a point in a saturated soil; the balance is carried by the pore water. The latter is the *pore water pressure, u*. Under steady-state conditions, compute u as follows:

$$u = \gamma_w z_w \qquad (3.8)$$

Where:

u = pore water pressure at a point in the soil

γ_w = unit weight of water

z_w = depth of that point below the groundwater table

The portion of the total stress carried by the solid particles is called the effective stress, σ':

$$\sigma' = \sigma - u \qquad (3.9)$$

$$\sigma'_v = \sigma_v - u \qquad (3.10)$$

Effective stresses are important because they often control the engineering behavior of soil. For example, the frictional component of the shear strength of a soil depends on the normal forces acting between the particles, which is a function of the effective stress, not the total stress.

An alternative method of computing the vertical effective stress, σ_v', is to use Equation 3.7 with the buoyant unit weight, γ_b, below the groundwater table and the unit weight, γ, above. Both methods produce the same numerical results.

Example 3.2

The soil profile beneath a certain site consists of 5.0 m of silty sand underlain by 13.0 m of clay. The groundwater table is at a depth of 2.8 m below the ground surface. The sand has a unit weight of 19.0 kN/m^3 above the groundwater table and 20.0 kN/m^3 below. The clay has a unit weight of 15.7 kN/m^3. Find the vertical effective stress at a depth of 11.0 m.

Solution:

Using Equations 3.7, 3.8, and 3.10:

$$\sigma_v = \sum \gamma_i h_i = (19.0)(2.8) + (20.0)(2.2) + (15.7)(6.0) = 191 \ kPa$$

$$u = \gamma_w z_w = (9.80)(8.2) = 80 kPa$$

$$\sigma_v' = \sigma_v - u = 191 - 80 = \mathbf{111 \ kPa} \qquad \Leftarrow Answer$$

Horizontal Stress

The shear stress in static water is always equal to zero because water has no shear strength. Therefore, the Mohr's circle must always have a diameter of zero (a Mohr's dot?), as shown in Figure 3.5a and the pressure (compressive stress) at a point must be the same in all directions. However, soil *does* have a shear strength and therefore is able to sustain a Mohr's circle with a diameter greater than zero, as shown in Figure 3.5b. This means the compressive stress in the horizontal direction does not necessarily equal that in the vertical direction. Only by chance will these stresses be equal.

The ratio of the horizontal to vertical effective stresses is the coefficient of lateral earth pressure, K:

$$K = \frac{\sigma_h'}{\sigma_v'} \tag{3.11}$$

The value of K in an undisturbed soil is K_0, the *coefficient of lateral earth pressure at rest.* The value of K_0 can vary between about 0.2 and 6, although it is typically between about 0.35 and 0.7 for normally consolidated soils and between about 0.5 and 3 for overconsolidated soils[1].

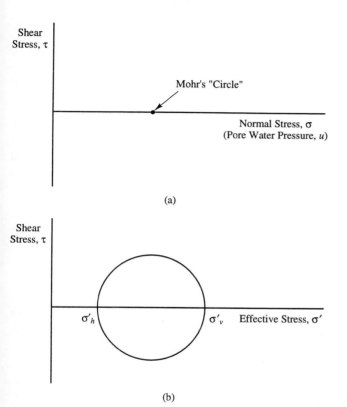

Figure 3.5 (a) Mohr's "circle" at a point in a body of water. Because water has no shear strength, the diameter of the circle must be zero. (b) Mohr's circle at a point in a soil. The shear strength permits a circle with a diameter greater than zero, so the vertical and horizontal stresses are not necessarily equal in magnitude.

The most accurate way to determine K_0 is by measuring $\sigma_h{}'$ in-situ using methods such as the pressuremeter, dilatometer, or stepped blade, and combining it with a computed value of $\sigma_v{}'$. Such methods are not yet commonplace in North America, but are likely to become more popular in the future.

A somewhat less satisfactory but workable method of measuring K_0 is by testing the soil in the laboratory. This can be done using a conventional triaxial compression machine or with the use of specialized equipment, but suffers from unknown sample disturbance, stress history, and aging effects.

The most common method of assessing K_0 is by empirical correlations with other

---

[1] See discussion in Section 3.7 of this chapter for an explanation of the terms "normally consolidated" and "overconsolidated."

soil properties.   Jaky (1944) developed the following relationship for normally consolidated soils[2]:

$$K_0 = 1 - \sin\phi \tag{3.12}$$

Where:

ϕ = drained friction angle

Direct in-situ or laboratory measurements would be more reliable than Equation 3.12. However, when such measurements are not available, this equation provides reasonably good results.

The magnitude of K_0 in an overconsolidated soil is higher than that in a normally consolidated soil and somewhat more difficult to predict. It appears to be primarily a function of the friction angle and the overconsolidation ratio (OCR). Mayne and Kulhawy (1982) developed the following empirical relationship based on laboratory data from 170 soils that ranged from clay to gravel:

$$K_0 = (1 - \sin\phi)\, \text{OCR}^{\sin\phi} \tag{3.13}$$

Where:

OCR = overconsolidation ratio[3]

The coefficient of lateral earth pressure influences many aspects of the engineering behavior of soils. As K increases, the following changes occur (after Schmertmann, 1985):

- Bearing capacity increases
- Slope stability decreases
- Fracture in earth dams becomes less likely
- Lateral pressure acting on retaining walls increases
- Skin friction resistance on deep foundations increases
- Foundation settlements become smaller
- Seismic liquefaction becomes less likely
- Soil improvement becomes more difficult

---

[2] Jaky first published this formula in Hungary during the Second World War, an unusual time to be thinking about lateral earth pressures!

[3] The overconsolidation ratio is defined later in this chapter.

Unfortunately, the current state-of-practice explicitly considers K in only a few analyses, most notably those concerned with skin friction in deep foundations and lateral pressures acting on retaining walls. Therefore, many analyses are less accurate when used in soils with exceptionally high or exceptionally low K values. Therefore, we will probably see more emphasis on this topic in the future, especially as in-situ measurements of K become more common.

Chapter 23 includes a more thorough discussion of horizontal stresses in soil.

Stresses Due to Applied External Loads

External loads, such as those from a foundation, can produce significant stresses in the adjacent soil, and these stresses are superimposed on the geostatic stresses. The magnitude of these stresses depends on the intensity and distribution of the load as well as the proximity of the load to the point where the stress is to be computed. Chapter 7 includes methods of analyzing these stresses.

3.7 COMPRESSIBILITY AND SETTLEMENT

One of the most important soil properties in the context of foundation engineering is its compressibility. This refers to the soil's response to changes in normal stress. For example, if a foundation is built and loaded, the normal stress in the soil below will increase and a corresponding normal strain will occur. This strain will, in turn, produce a corresponding settlement.

Stress-Strain Properties of Soil

Soil is a particulate material, so its stress-strain properties are much more complex than those of other more familiar materials such as steel. These properties depend on the arrangement of these particles, their interaction with each other, their physical and chemical properties, and many other factors. Therefore, the foundation engineer must consider this complex behavior when conducting settlement computations.

Modulus of Elasticity and Poisson's Ratio

Classical engineering mechanics describes the stress-deformation properties of a material in terms of the *modulus of elasticity*, E, (also known as *Young's modulus*) and *Poisson's ratio*, ν_p. If a material is vertically loaded and laterally unconfined ($\sigma_h = 0$), as shown in Figure 3.6a, then these parameters are defined as follows:

$$E = \frac{\sigma_v}{\varepsilon_v} \tag{3.14}$$

$$v_p = \frac{\varepsilon_h}{\varepsilon_v} \tag{3.15}$$

Where:

σ_h = horizontal normal stress

σ_v = vertical normal stress

ε_h = horizontal normal strain

ε_v = vertical normal strain

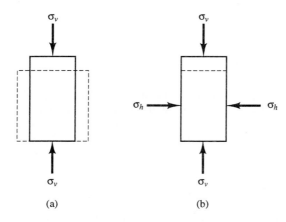

(a)   (b)

Figure 3.6   Stress-strain behavior of a vertically loaded material: (a) laterally unconfined; and (b) subjected to lateral stresses.

Engineers must measure the modulus of elasticity (or some other parameter that describes compressibility) for each new soil, and these measurements form the basis for settlement computations. Methods for doing so are described in Chapter 4 and typical values are listed in Table 3.5. However, engineers rarely measure Poisson's ratio. For practical design problems, simply use the typical values shown in Table 3.6.

Constrained Modulus

Although we could test a soil sample with the boundary conditions shown in Figure 3.6a, such a test would not represent the conditions in the field. This is because an element of soil in the ground is also subjected to lateral stresses, as shown in Figure 3.6b. A general formula for this condition is:

$$\frac{\sigma_v}{\varepsilon_v} = \frac{E}{1 - v_p\,(2\,\sigma_h/\sigma_v)} \tag{3.16}$$

Therefore, as the lateral confinement (i.e., σ_h) increases, the ratio σ_v / ε_v also increases. In other words, a given load on the soil will induce less settlement. The most favorable condition in this context would be the case where no lateral strain is permitted (i.e., $\varepsilon_h = 0$). This is known as the *constrained condition* or a *one-dimensional consolidation condition*. Practical examples of this condition include the soil beneath the center of a fill that extends laterally a great distance and the soil at shallow depths beneath the center of a mat foundation. The consolidation test, which is described in Chapter 4, models this condition in the laboratory.

TABLE 3.5 TYPICAL VALUES OF THE MODULUS OF ELASTICITY FOR SOILS AND ROCKS

Soil or Rock Type and Condition	Modulus of Elasticity, E	
	(lb/ft^2)	(kPa)
Undrained Condition (Also see Equation 7.10)		
Soft clay	30,000 - 200,000	1,500 - 10,000
Medium clay	100,000 - 1,000,000	5,000 - 50,000
Stiff clay	300,000 - 1,500,000	15,000 - 75,000
Drained Condition		
Soft clay	5,000 - 30,000	250 - 1,500
Medium clay	10,000 - 70,000	500 - 3,500
Stiff clay	25,000 - 400,000	1,200 - 20,000
Loose sand	200,000 - 500,000	10,000 - 25,000
Medium dense sand	400,000 - 1,200,000	20,000 - 60,000
Dense sand	1,000,000 - 2,000,000	50,000 - 100,000
Sandstone	1.4×10^8 - 4.0×10^8	7,000,000 - 20,000,000
Granite	5.0×10^8 - 1.0×10^9	25,000,000 - 50,000,000
Steel	4.2×10^9	200,000,000

a. The modulus of elasticity also depends on many other factors, especially effective stress, so these values are only approximate.

b. All soils are considered to be saturated.

c. The concepts of drained and undrained conditions are described later in this chapter.

TABLE 3.6 TYPICAL VALUES OF POISSON'S RATIO FOR SOILS
AND ROCKS

Soil or Rock Type	Poisson's Ratio, v_p
Saturated soil, undrained condition	0.50
Partially saturated clay	0.30 - 0.40
Dense sand, drained condition	0.30 - 0.40
Loose sand, drained condition	0.10 - 0.30
Sandstone	0.25 - 0.30
Granite	0.23 - 0.27

Adapted from Kulhawy, et al., 1983.

The ratio σ_v/ε_v in this condition is known as the *constrained modulus, M*:

$$M = \frac{\sigma_v}{\varepsilon_v} \tag{3.17}$$

The constrained modulus and Young's modulus are related as follows:

$$M = \frac{E(1 - v_p)}{(1 + v_p)(1 - 2v_p)} \tag{3.18}$$

Where:

v_p = Poisson's ratio

For small values of v_p, say, less than 0.25, E and M differ by less than 20%. However, for larger values the difference becomes much greater.

An intermediate condition, where $\varepsilon_x = 0$ and $\varepsilon_y > 0$ (where x and y are two perpendicular horizontal directions) is also possible. This is known as the *plane strain condition* and is found in the field beneath long continuous footings.

Nonlinear and Inelastic Stress-Strain Behavior

Another complication enters when we consider that the stress-strain curve for soil is nonlinear and inelastic, as shown in Figure 3.7. Nonlinear means that stress is not proportional to strain (as compared to the curve for steel, which is nearly straight throughout the normal range of working stresses). Inelastic means some or all of the deformation is plastic (or permanent). Therefore, if the soil is unloaded, it will not rebound to its original volume.

   Because of this behavior, there is no unique value for the modulus of elasticity, even if we consider only one of the lateral constraint conditions described earlier. Sometimes engineers use the *tangent modulus*, which is the slope of the stress-strain curve at a given point. Other times, we use the *secant modulus*, which is the slope of a line that connects the origin with a point on the curve. Both of these modulus values decrease as the normal stress increases. This is part of the reason for the wide ranges in the typical modulus values listed in Table 3.5.

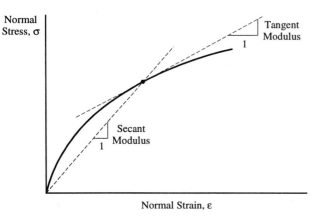

Figure 3.7 Typical stress-strain curve for a soil. Note that this curve is both nonlinear and inelastic.

One-Dimensional Consolidation Settlement

For most soils, the rate and magnitude of settlement are primarily governed by the ability of the soil particles to rearrange into a tighter packing in response to applied normal loads. Karl Terzaghi was the first to address this *consolidation* process. He developed a *theory of consolidation* in the 1920s while teaching in Istanbul and it has become the basis for nearly all consolidation analyses.

   Terzaghi assumed that both the solid particles and the pore water are incompressible. Therefore, the magnitude of the settlement is related directly to the change in volume of the voids, and the rate of settlement in a saturated soil is governed by the rate pore water escapes from the voids.

   The process of consolidation is intimately tied with the buildup and dissipation of *excess pore water pressures* and the corresponding changes in effective stress, as shown in Figure 3.8. When a vertical external load is applied, the total stress immediately rises by an amount $\Delta\sigma_v'$. However, if the soil is saturated, its volume will initially remain constant because both the solid particles and the water are incompressible, and the pore water pressure will rise by an amount equal to $\Delta\sigma_v$ (i.e., $\Delta u = \Delta\sigma_v$). This temporary increase in pore water pressure, Δu, is known as the excess pore water pressure, u_e.

   This increased pore water pressure generates a hydraulic gradient and some of the water gradually moves out of the voids. As the water escapes, consolidation occurs and the volume of the soil decreases. Simultaneously, the excess pore water pressure dissipates and the effective stress increases until u_e eventually becomes equal to zero and σ_v' increases by an amount equal to $\Delta\sigma_v$.

In clayey soils, this process occurs very slowly because their low hydraulic conductivity (coefficient of permeability) impedes the flow of water. However, in sandy soils, the hydraulic conductivity is much higher, the water flows more easily, and consolidation occurs much more rapidly.

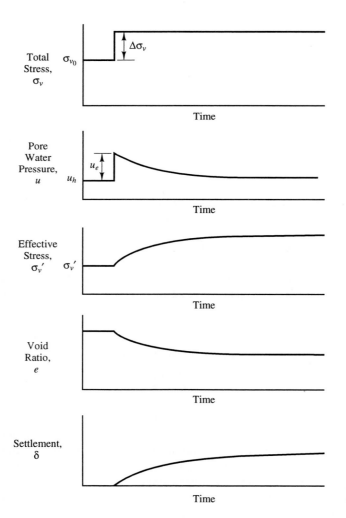

Figure 3.8   Changes in a soil as consolidation occurs.

Engineers have traditionally described the consolidation process using the relationship between the void ratio, e, and the effective stress, σ_v', as shown in the semilogarithmic plot in Figure 3.9. Each soil has a unique plot. With it, we can determine the change in void ratio, Δe, that will occur as a result of a certain change in effective stress, $\Delta\sigma_v'$. This information then produces a prediction of the settlement that will occur in the field.

An alternate method of expressing the consolidation characteristics of a soil is a strain vs. effective stress plot, as shown in Figure 3.10. It has the same shape as Figure 3.9, but is simpler to produce from the laboratory test data and it simplifies the settlement computations. Many also find it easier to visualize the consolidation process by thinking in terms of strain instead of changes in void ratio.

Note how the use of a semi-log plot produces a stress-strain curve that can be idealized as a series of straight lines.

Engineers develop the e-log σ_v' or ε_v-log σ_v' curve for a particular soil by conducting a laboratory *consolidation test* (also known as an *oedometer test*). The test procedure and methods of correcting the test results to compensate for sample disturbance are described in Chapter 4.

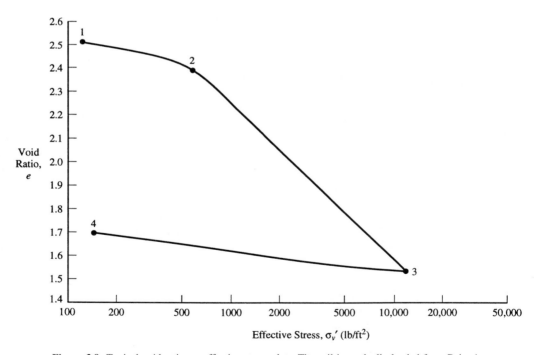

Figure 3.9 Typical void ratio vs. effective stress plot. The soil is gradually loaded from Point 1 to Point 3, then unloaded back to point 4.

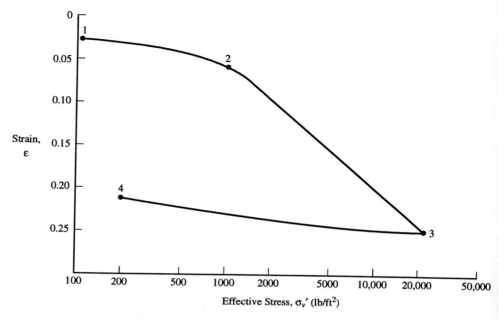

Figure 3.10 Typical strain vs. effective stress plot.

Normally Consolidated vs. Overconsolidated Soils

The change in slope of the consolidation curve at Point 2 in Figures 3.9 and 3.10 represents an important juncture in the consolidation process. The stress at these points is known as the *preconsolidation stress*, σ_c', and represents the greatest vertical effective stress the soil has ever experienced. If the present effective stress in the soil is less than σ_c', the soil is said to be *overconsolidated* (or *preconsolidated*) and the consolidation properties will be as defined by curve 1-2. However, if the present effective stress in the soil is equal to σ_c', the soil is said to be *normally consolidated* and any increase in the effective stress will result in consolidation as defined by curve 2-3. Because the slopes of these curves are different, it is very important to determine whether the soil is normally consolidated or overconsolidated before computing the settlement.

A soil can become overconsolidated for a variety of reasons, including the following:

- Erosion or excavation of the overlying soils.
- Melting of a glacier.
- Rising of the groundwater table.
- Drying of the soil — a process known as *desiccation* (the capillary stresses in a partially saturated soil increases the effective stress).

- Artificial consolidation from soil improvement techniques such as those discussed in Chapter 19.

The degree of overconsolidation may be expressed numerically as the *overconsolidation ratio* (OCR) which is defined as follows:

$$OCR = \frac{\sigma_c'}{\sigma_v'} \qquad (3.19)$$

Where:

OCR = overconsolidation ratio

σ_c' = preconsolidation stress

σ_v' = vertical effective stress

Note that both σ_c' and σ_v' must refer to the conditions at the same point and that the OCR is defined at that point only.

The OCR of a normally consolidated soil is equal to 1; in a lightly overconsolidated soil, it is typically between 1 and 3; a heavily overconsolidated soil may have an OCR as high as 8.

Although the customary way to determine the preconsolidation stress is to measure it directly in a consolidation test, the following empirical relationship (U.S. Navy, 1982a) can be useful when estimating it or when checking test results for reasonableness:

$$\sigma_c' \approx \frac{s_u}{0.11 + 0.0037\,PI} \qquad (3.20)$$

Where:

s_u = undrained shear strength

PI = plasticity index

Adjusting Laboratory Test Results

Even good quality soil samples are adversely influenced by sample disturbance, so the *e*-log σ_v' or ε_v-log σ_v' curves from the laboratory are not exactly the same as those of the soil in the field. Therefore, it is necessary to compensate for these disturbance effects by adjusting the test results.

Casagrande (1936) and Schmertmann (1955) developed the following adjustment procedure:

1. Find the point on the laboratory consolidation curve that has the smallest radius (point A in Figure 3.11).

2. Draw a horizontal line from point A.
3. Draw a line tangent to the laboratory consolidation curve at point A.
4. Bisect the angle made by the lines from steps 2 and 3.
5. Extend the steep portion of the laboratory consolidation curve and locate point B. The effective stress at point B is the preconsolidation stress, σ_c'.
6. Extend a horizontal line from the initial void ratio, e_0 (or from $\varepsilon_v = 0$). The point on this line that corresponds to the vertical effective stress in the ground where the sample was recovered, σ_{v0}', is point C.

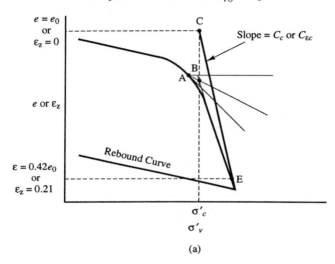

(a)

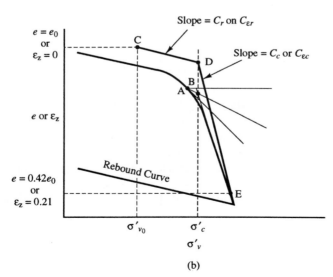

(b)

Figure 3.11 Procedure for adjusting laboratory consolidation test results: (a) for normally consolidated soils, and (b) for overconsolidated soils.

7. Draw a line to the right from point C parallel to the rebound curve until $\sigma_v' = \sigma_c'$. This locates point D, which is directly above point B. If the soil is normally consolidated, points C and D will be at the same spot.

8. Extend a horizontal line from $e = 0.42\,e_0$ (or $\varepsilon_v = 0.21$) until it meets the laboratory consolidation curve. This is point E.

9. Draw a line from point D to point E.

For e-log σ_v' plots, compute the *compression index*, C_c, and the *recompression index*, C_r, as follows:

C_c = compression index = slope of line DE = Δe for each log-cycle (i.e., tenfold) increase in σ_v'. For example, if an increase in effective stress from 1000 lb/ft^2 to 10,000 lb/ft^2 causes the void ratio to decrease from 1.50 to 1.30, then $C_c = 1.50 - 1.30 = 0.20$.

C_r = recompression index = slope of line CD = Δe for each log-cycle increase in σ_v'. Note that this also is the slope of the rebound curve.

For ε_v-log σ_v' plots, compute the *compression ratio*, $C_{\varepsilon c}$, and the *recompression ratio*, $C_{\varepsilon r}$, as follows:

$C_{\varepsilon c}$ = compression ratio = slope of line DE = $\Delta \varepsilon_v$ for each log-cycle (i.e. ten-fold) increase in σ_v'. For example, if an increase in effective stress from 1000 lb/ft^2 to 10,000 lb/ft^2 causes the strain to increase from 0.050 to 0.168, then $C_{\varepsilon c} = 0.168 - 0.050 = 0.118$.

$C_{\varepsilon r}$ = recompression ratio = slope of line CD = $\Delta \varepsilon_v$ for each log-cycle increase in σ_v'. Note that this also is the slope of the rebound curve.

Assessing Soil Compressibility

Once the corrected laboratory test results are available, we need to compute the change in void ratio, Δe, or the strain, ε_v, that would occur when the effective stress in the soil increases from σ_{v0}' to $\sigma_{v0}' + \Delta\sigma_v'$. Although it is possible to do so by graphically interpreting an e-log σ_v' or ε_v-log σ_v' plot, it is more convenient to quantitatively describe the stress-strain behavior using the *compressibility*, C:

$$C = \frac{\varepsilon_v}{\log\left(\dfrac{\sigma_{v0}' + \Delta\sigma_v'}{\sigma_{v0}'}\right)} \tag{3.21}$$

Where:

$\sigma_{v0}{}' =$ initial vertical effective stress

$\Delta\sigma_v{}' =$ change in vertical effective stress

In other words, C is the strain that corresponds to every tenfold increase in vertical effective stress $(d\,\varepsilon_v / d \log \sigma_v{}')$.

Use the following procedure to determine C for a particular soil strata:

1. Determine the preconsolidation stress and the current vertical effective stress at a point in the strata. There are two ways to do this:

    a. If consolidation test results are available, determine the preconsolidation stress, $\sigma_c{}'$, at the sample depth using the Casagrande and Schmertmann procedures described earlier. Then, compute the current vertical effective stress, $\sigma_{v0}{}'$, at the sample depth using Equation 3.10.

    or b. If the overconsolidation ratio, OCR, is known, compute the vertical effective stress, $\sigma_{v0}{}'$, at the midpoint of the strata using Equation 3.10. Then, compute the preconsolidation stress, $\sigma_c{}'$, at the midpoint using Equation 3.19.

2. Compute the *overconsolidation margin*, $\sigma_m{}'$, using:

$$\sigma_m{}' = \sigma_c{}' - \sigma_v{}' \qquad (3.22)$$

3. If $\sigma_m{}' < 2000$ lb/ft^2 (100 kPa) or if the overconsolidation is primarily due to desiccation, compute C using either of the following formulas:

$$C = \frac{C_c}{1 + e_0} \qquad (3.23)$$

$$C = C_{\varepsilon c} \qquad (3.24)$$

Where:

$e_0 =$ initial void ratio

4. If $\sigma_m{}' \geq 2000$ lb/ft^2 (100 kPa) and the overconsolidation is not primarily due to desiccation, compute C using either of the following formulas:

$$C = \frac{C_r}{1 + e_0} \qquad\qquad (3.25)$$

$$C = C_{\varepsilon r} \qquad\qquad (3.26)$$

This procedure treats slightly overconsolidated soils as if they are normally consolidated because of the following:

- The change in effective stress, $\Delta\sigma'$, due to the applied external load may increase the effective stress in the soil to a value greater than σ_c'.
- Soil creep effects may make these soils behave as if they were normally consolidated.
- Sample disturbance effects make σ_c' difficult to measure.

In addition, soils overconsolidated by desiccation appear to have a high OCR when dry. However, when they are wetted, this apparent OCR dissipates (Stark and Duncan, 1991). The soil has little or no memory of this apparent preconsolidation, especially at higher normal stresses. Therefore, desiccated soils that might become wet should be evaluated as if they were normally consolidated.

The compressibility of the soil may be classified using Table 3.7.

TABLE 3.7 CLASSIFICATION OF SOIL COMPRESSIBILITY

Compressibility, C	Classification
0 - 0.05	Very slightly compressible
0.05 - 0.10	Slightly compressible
0.10 - 0.20	Moderately compressible
0.20 - 0.35	Highly compressible
> 0.35	Very highly compressible

Empirical correlations that relate compressibility with other soil properties (Azzouz et al., 1976; Al-Khafaji and Andersland, 1992) also are available. These are useful for preliminary analyses and for checking laboratory test results.

For saturated normally consolidated clays, Kulhawy and Mayne (1990) suggested:

$$C_c \approx \frac{PI}{74} \tag{3.27}$$

They also found that the compressibility of overconsolidated clays is about 20% of that for the same clay in a normally consolidated condition. Thus, for overconsolidated clays:

$$C_r \approx \frac{PI}{370} \tag{3.28}$$

Unfortunately, correlations between compressibility and Atterberg limits have a large scatter of data, especially with overconsolidated soils (Sivakugan, 1990). Therefore, Equations 3.27 and 3.28 have potential errors of about ±40% and ±70%, respectively.

Lambe and Whitman (1969) found a somewhat better correlation between C and the natural moisture content for saturated, normally consolidated cohesive soils, as shown in Figure 3.12.

The compressibility index, C_c, for loose uniform sands with $\sigma_v' = 2000 - 8000$ lb/ft^2 is typically between 0.05 and 0.06. The same soil in a dense condition will typically have C_c between 0.02 and 0.03 (U.S. Navy, 1982a). These correspond to C values of about 0.03 - 0.04 and 0.01 - 0.02, respectively.

The compressibility, C, is related to the constrained modulus, M, as follows:

$$M = \frac{2.30 \, \sigma_v'}{C} \tag{3.29}$$

Computing Consolidation Settlement

The consolidation settlement caused by an applied load is:

$$\delta_c = \int \varepsilon_v \, dz = \int C \log \left(\frac{\sigma_{v0}' + \Delta\sigma_v'}{\sigma_{v0}'} \right) dz \tag{3.30}$$

Where:
  δ_c = consolidation settlement at the ground surface
  ε_v = vertical strain
  σ_{v0}' = initial vertical effective stress (before load is placed)
  $\Delta\sigma_v'$ = increase in vertical effective stress due to the applied load
  z = depth below the ground surface

For practical analyses, engineers evaluate the integral in Equation 3.30 by dividing the soil profile into n finite layers, computing δ_c for each layer, and summing:

$$\delta_c = \sum_{i=1}^{n} C H_i \log \left(\frac{\sigma'_{v0} + \Delta\sigma'_v}{\sigma'_{v0}} \right) \tag{3.31}$$

Where:

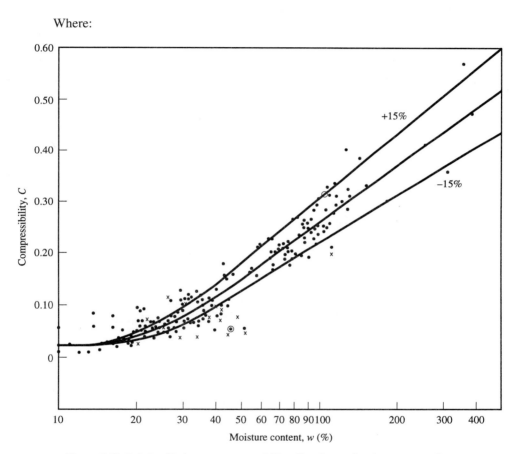

Figure 3.12 Relationship between compressibility, C, and natural moisture content for saturated, normally consolidated cohesive soils (Adapted from *Soil Mechanics* by Lambe and Whitman, Copyright ©1969 by John Wiley & Sons. Used by permission of John Wiley & Sons, Inc.).

Compute σ'_{v0} and $\Delta\sigma'_v$ at the midpoint of each layer. Because these parameters vary with depth, often in a nonlinear fashion, the accuracy of the analysis will improve as the number of layers increases. When performing hand computations, most engineers use only a couple of layers so the computational effort does not become excessive. However, computer-based solutions can easily use many more layers.

Example 3.3

A 3.0 m thick fill that is very long and wide is to be placed on the soil profile shown in the figure below. The results of a consolidation test performed on a sample of the clay are: $C_{\varepsilon c} = 0.16$, $C_{\varepsilon r} = 0.05$, $\sigma_c' = 65$ kPa. The dense sand and dense sand and gravel strata may be assumed to be incompressible in comparison to the clay. Find the settlement that will occur as a result of the weight of this fill.

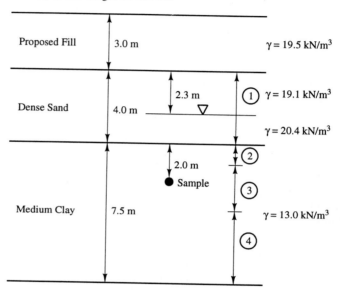

Proposed Fill — 3.0 m — $\gamma = 19.5$ kN/m³

Dense Sand — 4.0 m — 2.3 m — ① $\gamma = 19.1$ kN/m³

$\gamma = 20.4$ kN/m³

② 2.0 m ● Sample ③

Medium Clay — 7.5 m — $\gamma = 13.0$ kN/m³

④

Dense Sand and Gravel

Solution:

1. Determine whether the clay is normally consolidated or overconsolidated:

@ sample: $\sigma_{v0}' = (19.1)(2.3) + (20.4)(1.7) + (13.0)(2.0) - (9.8)(3.7) = 68$ kPa

$\sigma_m' = 65 - 68 = -3$ kPa   ∴ Clay is normally consolidated; $C = C_{\varepsilon c} = 0.16$

2. Divide into layers and compute settlement:

$\Delta\sigma_v' = (19.5)(3) = 58.5$ kPa for all layers because the width of fill is large compared to depth to the compressible soils

Compute the settlement using Equation 3.31:

Layer No.	Upper Depth (m)	Lower Depth (m)	H_i (m)	σ_{v0}' (kPa)	C	δ_{ci} (m)
1	0.0	4.0	4.0	38.2	0	0
2	4.0	5.0	1.0	63.6	0.16	0.045
3	5.0	7.5	2.5	69.1	0.16	0.107
4	7.5	11.5	4.0	79.6	0.16	0.153
					$\delta_c =$	0.305

$$\delta_c = 300 \text{ mm} \qquad \Leftarrow Answer$$

This is a large settlement that will probably require special design and construction measures.

Secondary Compression Settlement

Another source of settlement in some soils is *secondary compression*. It is the result of creep, viscous behavior of the clay-water system, the compression and decomposition of organic matter, and other physical and chemical processes. Secondary compression settlement is a function of time, but it is independent of $\Delta\sigma_v'$ as shown in the following formula:

$$\delta_s = C_{\alpha\varepsilon} H \log\left(\frac{t}{t_p}\right) \qquad (3.32)$$

Where:
δ_s = secondary compression settlement
$C_{\alpha\varepsilon}$ = coefficient of secondary compression[4]
H = thickness of compressible strata
t = time after application of load
t_p = time required to complete primary consolidation

The coefficient of secondary compression, $C_{\alpha\varepsilon}$, is the change in void ratio due to secondary compression that occurs over one log-cycle of time $(d\,\varepsilon_v / d \log t)$. It may be measured in a laboratory consolidation test.

---

[4] The coefficient of secondary compression is sometimes expressed in terms of the change in void ratio $(de/d \log t)$ and other times in terms of strain $(d\varepsilon_v/d \log t)$. These two are not numerically equal, so this is often a point of confusion. We will use only the latter definition.

For overconsolidated clays, $C_{\alpha\varepsilon}$ is typically less than 0.001, which means that secondary compression settlement will be negligible. For normally consolidated clays, it might be between 0.001 and 0.01, as shown in Figure 3.13.   Secondary compression

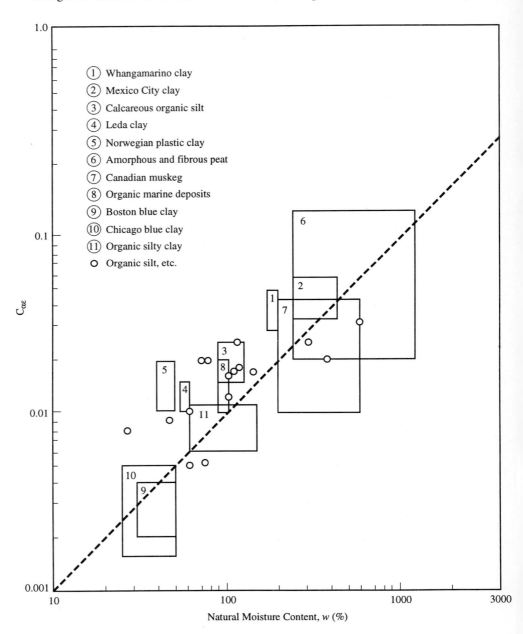

Figure 3.13 Coefficient of secondary compression for saturated, normally consolidated soils (Adapted from Mesri, 1973; Holtz and Kovacs, 1981; Used with permission).

settlements in these soils might be important. However, organic soils pose the greatest concern because their coefficient might be as high as 0.15, thus potentially creating large settlements.

If secondary compression is a concern, add it to the computed consolidation settlement to find the total settlement, δ:

$$\delta = \delta_c + \delta_s \qquad (3.33)$$

Spread footing foundations and other small loaded areas also are influenced by a third type of settlement, distortion settlement, as discussed in Chapter 7.

3.8 STRENGTH

When geotechnical engineers refer to the strength of a soil, we are nearly always thinking about shear strength. This is because the tensile strength of soil is very small (almost universally assumed to be zero) and because applied compressive loads will eventually cause the soil to fail in shear, not compression. This behavior is once again due to the particulate nature of soil. It fails when the stresses between the particles are such that they roll or slide past one another, as shown in Figure 3.14.

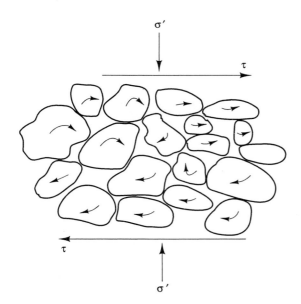

Figure 3.14 Microscopic view of soil failing in shear. The individual particles slide and roll past each other.

Shear strength is another important soil property in the context of foundation engineering because the design of the foundations must be such that the stresses induced in the soil do not cause it to fail in shear.

We can consider the sources of shear strength (i.e., the forces that resist sliding or rolling of the particles) to be divided into two categories:

- **Frictional strength**: This is similar to classic sliding friction problems from basic physics. The force that resists sliding is equal to the normal force multiplied by the coefficient of friction, μ.

  The shear strength due to sliding friction is directly proportional to the applied normal force acting between the particles. In other words, this component of the shear strength is directly proportional to the effective stress, σ'.

- **Cohesive strength**: In some soils, such as overconsolidated clays, many partially saturated soils, and cemented soils, the individual particles are bonded together. This can be viewed as another source of shear strength that is independent of the normal force.

In general, the shear strength of a soil varies with the effective stress, as shown in Figure 3.15. The cohesive strength is represented by the y-intercept and is known as the *cohesion*, c. The frictional strength is seen in the slope of the line, and expressed as the *angle of internal friction* (or *friction angle*), ϕ. Note that $\phi = \tan^{-1}\mu$, where μ reflects both the surface roughness of the particles and their mechanical interlocking.

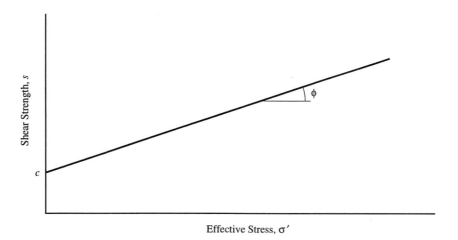

Figure 3.15 A plot of shear strength vs. effective stress can usually be taken to be a straight line. The y-intercept is the cohesion, c, and the angle from the horizontal is the friction angle, ϕ.

Based on the relationship in Figure 3.16, the shear strength, s, of a soil may be expressed using the *Mohr-Coulomb Formula*:

$$s = c + \sigma' \tan\phi \qquad (3.34)$$

The values of c and ϕ depend on many factors, including:

- Soil type
- Moisture content
- Rate of loading
- Drainage conditions
- Stress history
- Shear strain
- Lateral strain
- Intermediate principal stress

The Mohr-Coulomb Formula implicitly considers most of these factors when we use undisturbed soil samples and simulate the field conditions in the laboratory. However, the lateral strain and intermediate principal stress are difficult to model in conventional laboratory tests and are therefore often not accurately reflected in the c and ϕ values. Geotechnical engineers normally consider two such stress conditions:

- The *axisymmetric* (or *transversely isotropic*) *stress condition*, where the intermediate principal stress, σ_2, equals the minor principal stress, σ_3. This would be very close to the conditions present in the field beneath the tip of a pile foundation, as shown in Figure 3.16a.
- The *plane strain condition*, where the lateral strain is zero, such as would occur beneath a very long spread footing, as shown in Figure 3.16b.

The shear strength of soils, especially sands, is higher in the plane strain condition. Ladd et al. (1977) reported that ϕ_{ps} is 4° to 9° larger than ϕ_a in dense sands and 2° to 4° larger in loose sands. Lade and Lee (1976) suggested the following conservative relationships for finding the plane strain strength from the axisymmetric strength:

$$\phi_{ps} = 1.5\,\phi_a - 17° \qquad \text{For } \phi_a > 34° \qquad (3.35)$$

$$\phi_{ps} = \phi_a \qquad \text{For } \phi_a \leq 34° \qquad (3.36)$$

Where:

ϕ_a = friction angle under axisymmetric conditions

ϕ_{ps} = friction angle under plane strain conditions

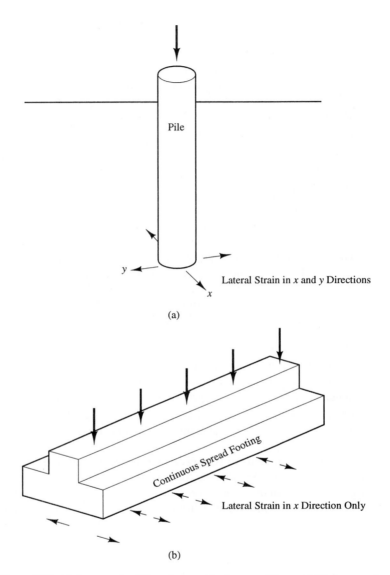

Figure 3.16 (a) An example of the axisymmetric stress condition; and (b) An example of
the plane strain stress condition.

Methods are available to express the shear strength of a soil in such a way that the
intermediate principal stress and strain, as well as other factors, are considered more fully
(Griffiths, 1990). Although these methods may be used in lieu of the Mohr-Coulomb
Formula, thus far they have been used only as research tools. The Mohr-Coulomb

Formula, adjusted as necessary when plane strain conditions govern, is simple and conservative and will probably continue to dominate for some time.

Undrained vs. Drained Strength

A frequently misunderstood yet important aspect of soil strength is the issue of *drainage*. Foundations are usually designed for one of two conditions: the *undrained condition* or the *drained condition*, so we must know which applies to a given set of circumstances.

Consider the foundation shown in Figure 3.17 and presume that it is founded on a saturated sand. Initially, the applied load, P, is zero and the magnitudes of σ_v, u, σ_v', and s at a point in the soil beneath the footing are as shown on the left side of the diagrams in Figure 3.18.

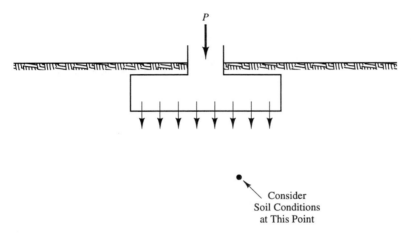

P

Consider
Soil Conditions
at This Point

Figure 3.17 A typical foundation.

When a load is applied to the footing, the total stress in the soil increases by an amount $\Delta\sigma_v$. This causes the soil-water matrix to compress and some of the water is squeezed out. Because sands have a high hydraulic conductivity (coefficient of permeability), this water is able to move quickly and easily. The potential drainage rate is at least as great as the loading rate. Thus, the vertical effective stress, σ_v', (computed using Equation 3.10) increases as P and σ_v increase.

This is known as the *drained condition*. It is important because the shear strength, s (computed using Equation 3.34), increases as P increases, and we can rely on this increased strength to resist the shear stresses created by P. Thus, a shear failure is less likely to occur.

Now, consider the same footing founded on a saturated clay. When the load is applied, the total stress in the soil increases by an amount $\Delta\sigma_v$ as before. However, clays have a much lower hydraulic conductivity, so the drainage rate is slow compared to the loading rate. This is the *undrained condition*.

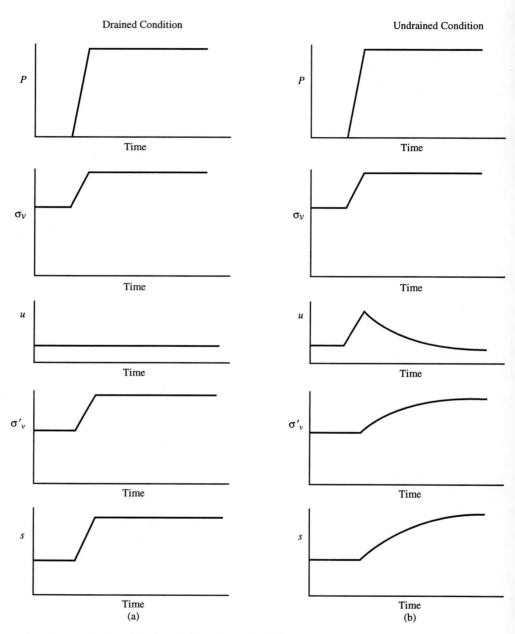

Figure 3.18 Variation of applied load, P; total vertical stress, σ_v; pore water pressure, u; effective vertical stress, σ_v'; and shear strength, s with time: (a) saturated sand (drained conditions); and (b) saturated clay (undrained conditions).

The pore water is essentially incompressible in comparison to the assemblage of solid particles, so $\Delta\sigma_v$ is initially carried entirely by the pore water. Therefore, the pore water pressure, u, increases by an increment u_e (the excess pore water pressure). Since $u_e = \Delta\sigma_v$, the vertical effective stress, σ_v', is initially unchanged as demonstrated below:

Before application of load:

$$\sigma_v' = \sigma_v - u$$

Immediately after application of load:

$$\sigma_v' = (\sigma_v + \Delta\sigma_v) - (u + u_e)$$
$$= (\sigma_v + \Delta\sigma_v) - (u + \Delta\sigma_v)$$
$$= \sigma_v - u$$

The excess pore water pressure also induces a hydraulic gradient in the soil, so some of the pore water will eventually flow away from the stressed zone. This allows the soil to consolidate, and gradually $\Delta\sigma_v$ is transferred to the solid particles. Thus, the excess pore water pressure slowly dissipates and eventually the effective stress, σ_v', increases by an amount $\Delta\sigma_v$, as shown in Figure 3.18b. According to Equation 3.34, the shear strength also increases as σ_v' increases.

Notice how the shear strength increases much more slowly than it does when drained conditions prevail. There is a period where P has reached its maximum value, but s is still low. Thus, there is a greater potential for shear failure in the soil. For design purposes, we would need to use the lower value of s, as shown in Figure 3.18.

Most foundation engineering analyses consider sands to be drained and clays to be undrained because normal rates of loading on foundations are slower than the rate of drainage in sands and faster than that in clays. However, when loaded very rapidly (such as during an earthquake), sands can be undrained. Likewise, very slow loading on clays produces drained conditions.

Measuring and Expressing the Drained Strength

When drained conditions prevail, an increase in σ will quickly cause corresponding increases in σ' and the shear strength, s, as shown in Figure 3.18a. Therefore, s may be computed using Equation 3.34 with the post-loading value of σ'. The shear strength parameters, c and ϕ, also must reflect drained conditions and may be determined using any of a variety of methods as discussed in Chapter 4.

For most cohesionless soils, it is best to neglect any cohesive strength that may appear in the laboratory or field tests and rely only on the frictional strength. Typical values of the coefficient of friction, ϕ, are shown in Figure 3.19.

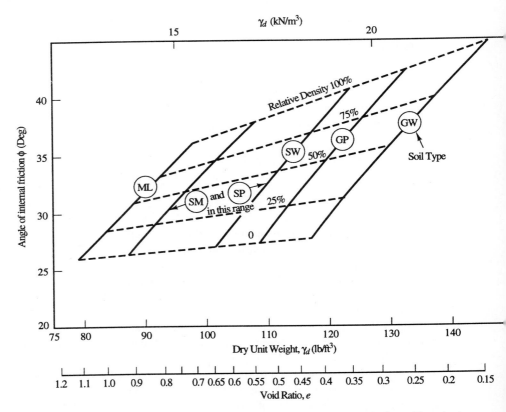

Figure 3.19 Typical drained φ values for cohesionless soils without plastic fines. The cohesion of these soils can be considered to be equal to zero (Adapted from U.S. Navy, 1982a).

Although most analyses with cohesive soils use the undrained strength, some use the drained strength. If the soil is saturated and normally consolidated, use a drained cohesion, c, of zero. The drained φ correlates approximately with the plasticity index (*PI*) as follows (after Mitchell, 1993; Kulhawy and Mayne, 1990):

$$\phi \approx \sin^{-1}(0.8 - 0.094 \ln PI) \tag{3.37}$$

Measuring and Expressing the Undrained Strength

When dealing with undrained conditions, the effective stress does not increase immediately following the application of a load, so the shear strength will not increase as it does under drained loading conditions. In addition, if excessive shearing were to occur in the soil, additional positive excess pore water pressures may develop. Therefore, it is most convenient to express the shear strength in terms of s_u (the subscript denotes "undrained") instead of c and ϕ. The values of s_u measured in the lab or in the field will implicitly consider any excess pore water pressures as long as the strength test also is conducted under undrained conditions. These test procedures are discussed in Chapter 4.

For analyses that require values of c and ϕ, we can simply set $c = s_u$ and $\phi = 0$. Thus, the undrained condition is often called a "$\phi = 0$" condition. However, the statement "$\phi = 0$" is actually a misnomer, because the soil still has frictional strength; it simply does not gain *additional* frictional strength when an incremental σ is added. The s_u value implicitly contains both the cohesive strength (if any) and the frictional strength with appropriate compensation for excess pore water pressures.

A relationship between undrained strength and the consistency of cohesive soils is presented in Table 3.8. This information could be used to generate initial estimates of s_u or to check the results of lab or field tests.

Although the undrained shear strength is a useful parameter, it is not truly a material property. The magnitude of s_u for a given soil depends on strain rate, stress state, and many other factors (Kulhawy and Mayne, 1990). Thus, the various methods of measuring it often produce different results.

Assessing Drainage Conditions for Design Purposes

Both drained and undrained conditions can occur in real soils, and partially drained conditions are also quite common. The question that faces the foundation engineer is: Which strength to use for design? In the context of foundations subject to static loads, the following guidelines are usually appropriate:

- **Clean sands and gravels**: The hydraulic conductivity of these soils is quite high compared to the rate at which the soil is being loaded. Therefore, assume that drained conditions will prevail.

- **Saturated clays**: The hydraulic conductivity of these soils is very low and very little of the pore water will escape while the load is being applied. Therefore, assume that undrained conditions will prevail immediately after the load is applied and drained conditions will prevail at some later time.

- **Intermediate soils (i.e., mixtures of sand, clay, and silt)**: These soils have a moderate hydraulic conductivity and most likely will be partially drained when subjected to loads from a foundation. Because the stress conditions under a foundation are such that the soil compresses and positive excess pore water pressures develop, it is conservative to assume undrained conditions prevail immediately after the load is applied. However, this is an area where engineering judgment is required.

TABLE 3.8 RELATIONSHIP BETWEEN CONSISTENCY OF COHESIVE SOILS AND UNDRAINED SHEAR STRENGTH

Consistency	Undrained Shear Strength, s_u		Visual Identification
	(lb/ft^2)	(kPa)	
Very soft	< 250	< 12	Thumb can penetrate more than 1 in (25 mm)
Soft	250 - 500	12 - 25	Thumb can penetrate about 1 in (25 mm)
Medium	500 - 1000	25 - 50	Penetrated with thumb with moderate effort
Stiff	1000 - 2000	50 - 100	Thumb will indent soil about 1/4 in (8 mm)
Very stiff	2000 - 4000	100 - 200	Thumb will not indent, but readily indented with thumbnail
Hard	> 4000	> 200	Indented by thumbnail with difficulty or cannot indent with thumbnail

Adapted from Terzaghi & Peck, 1967 (used with Permission) and ASTM D2488-90 (Copyright ASTM; used with permission).

Partially Saturated Soils

The shear strength of a partially saturated soil is higher than if it was saturated. However, do not rely on this additional strength because of the possibility that the soil may become wetted sometime in the future. Therefore, it is usually best to soak all soil samples in the laboratory before testing them.

When conducting drained tests, this additional strength will be manifested as cohesion, so be cautious about using cohesive strength except in overconsolidated clays or in soils cemented with non water-soluble agents. When conducting undrained tests, strength gains due to partial saturation are more insidious because they increase the measured s_u. Therefore, it is appropriate to determine the degree of saturation of laboratory test samples after they have been soaked to verify that they have been wetted sufficiently to match the worst-case condition that might appear in the field. However, this does not necessarily correspond to 100% saturation.

Soils that are overconsolidated due to desiccation will appear to have a large cohesive strength. However, after several cycles of wetting, much of this cohesion dissipates (Stark and Duncan, 1991). Therefore, do not rely on cohesive strength in desiccated soils.

Variation of Strength with Depth in Cohesive Soils

The undrained shear strength in a uniform deposit of cohesive soil will increase with depth because the soils at greater depths have been consolidated under a correspondingly greater effective stress. If the soil is normally consolidated, the ratio s_u/σ_v' will be constant with depth. For overconsolidated soils, s_u/σ_c' will be constant, where σ_c' is the preconsolidation stress.

Jamiolkowski et al. (1985) expressed this relationship for soils with a plasticity index of less than 60 as follows:

$$\frac{s_u}{\sigma_v'} = (0.23 \pm 0.04)\,\text{OCR}^{0.8} \tag{3.38}$$

The coefficient in this formula probably trends toward the higher end of the stipulated range in soils with a high plasticity index, and toward the lower end in those with a low plasticity index. We also could use this approximate relationship to estimate the overconsolidation ratio (OCR) of a soil.

Most deposits of supposedly normally consolidated soil actually have an overconsolidated *crust* near the ground surface. Desiccation due to fluctuations in the groundwater table can create such crusts. We can detect them by examining the s_u/σ_v' ratio, as shown in Figure 3.20.

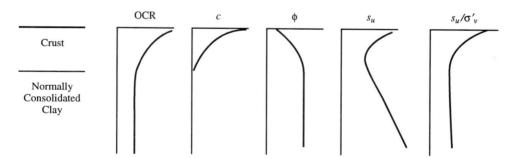

Figure 3.20 Typical crust near the ground surface in a supposedly normally consolidated cohesive soil.

Sensitivity

Most clayey soils will have a lower shear strength if they are remolded than they would have at the same unit weight and moisture content in an undisturbed condition. This strength loss is known as *sensitivity*, S_t:

$$S_t = \frac{\text{undisturbed } s_u}{\text{remolded } s_u} \tag{3.39}$$

A clay of low sensitivity might have $S_t = 1$ to 2, whereas a highly sensitive clay might have an $S_t \approx 8$. There are some extremely sensitive clays (for example, Leda clay in eastern Canada or many of the clays in Scandinavia) that have sensitivities greater than 50. These are known as *quick clays*. In extreme cases, engineers have observed sensitivities greater than 1000.

Sensitivity is important when the construction process might disturb the soil. For example, driving piles into a sensitive clay will remold it and cause it to lose some of its strength.

SUMMARY OF MAJOR POINTS

1. The foundation engineer must have a reasonably complete understanding of the soil conditions at the project site before he or she is able to proceed with the analysis and design of a foundation.

2. Soil is a particulate material and the voids between the particles are filled with gasses and/or liquids (usually air and/or water). The engineering behavior of soil depends primarily on the interaction of these particles with each other and with the pore fluids.

3. The plastic limit, liquid limit, and plasticity index are quantitative ways of expressing the relationship between consistency and moisture content for soils with a significant fines content.

4. The Unified Soil Classification System (USCS) is the most common method of classifying soils for foundation engineering projects. This method assigns a group symbol and a group name to the soil.

5. The stresses at a point in a soil are the result of geostatic stresses (those due to the weight of the soil above) and applied stresses (those due to external loads).

6. The compressive stress at a point in a soil can be viewed two ways: The total stress and the effective stress. The total stress, σ, is the actual compressive stress at that point, while the effective stress, σ' is that portion of the total stress carried by the soil particles. The difference between σ and σ' is the pore water pressure, u, which is the compressive stress (pressure) in the pore water at that point.

7. The horizontal effective stress at a point in a soil usually does not equal the vertical effective stress. The ratio of these two stresses is the coefficient of lateral earth pressure, K.

8. The compressibility of a soil depends on many factors, including soil type and stress conditions. The stress-strain curve for soil is nonlinear.

9. Soil compressibility may be expressed using the modulus of elasticity, E, the constrained modulus, M, or the compressibility, C.

10. A normally consolidated soil is one that has never experienced a higher vertical effective stress. In contrast, an overconsolidated soil is one that has had a higher vertical effective stress sometime in the past.

11. The shear strength of a soil has two components: frictional strength and cohesive strength. These are represented by the parameters ϕ and c, respectively.

12. When water is able to flow easily in or out of the voids in a soil, no excess pore water pressures will develop and the soil is said to be in a drained condition. Conversely, when little or no flow is able to occur, large excess pore water pressures might develop and the soil is said to be in the undrained condition. Select the proper strength (drained or undrained) depending on the conditions that will be present in the field.

QUESTIONS AND PRACTICE PROBLEMS

3.1 Explain the difference between moisture content and degree of saturation.

3.2 A certain saturated sand ($S = 100\%$) has a moisture content of 25.1% and a specific gravity of solids of 2.68. It also has a maximum index void ratio of 0.84 and a minimum index void ratio of 0.33. Compute its relative density and classify its consistency.

3.3 Consider a soil that is being placed as a fill and compacted using a sheepsfoot roller (a piece of construction equipment). Will the action of the roller change the void ratio of the soil? Explain.

3.4 A sample of soil has a volume of 0.45 ft^3 and a weight of 53.3 lb. After being dried in an oven, it has a weight of 45.1 lb. It has a specific gravity of solids of 2.70. Compute its moisture content and degree of saturation before it was placed in the oven.

3.5 A site is underlain by a soil that has a unit weight of 18.7 kN/m^3 above the groundwater table and 19.9 kN/m^3 below. The groundwater table is located at a depth of 3.5 m below the ground surface. Compute the total vertical stress, pore water pressure, and effective vertical stress at the following depths below the ground surface:

   a. 2.2 m
   b. 4.0 m
   c. 6.0 m

3.6 Estimate the friction angle of the following soils:

   a. Silty sand with a dry unit weight of 110 lb/ft^3.
   b. Poorly-graded gravel with a relative density of 70%.
   c. Very dense well-graded sand.

3.7 What is a typical range of values for the constrained modulus, M, in sandstone?

3.8 The soil profile at a certain site is as follows:

Depth (ft)	Unit Weight (lb/ft^3)	Cohesion (lb/ft^2)	Friction Angle (deg)
0 - 12	119	1000	0
12 - 20	126	200	20
20 - 32	129	0	32

The groundwater table is at a depth of 15 ft.

Develop plots of pore water pressure, total stress, effective stress, and shear strength vs. depth. All four of these plots should be superimposed on the same diagram with the parameters on the horizontal axis (increasing to the right) and depth on the vertical axis (increasing downward).

Hint: Because the cohesion and friction angle suddenly change at the strata boundaries, the shear strength also may change suddenly at these depths.

3.9 Repeat Problem 3.8 using the following data:

Depth (m)	Unit Weight (kN/m^3)	Cohesion (kPa)	Friction Angle (deg)
0 - 5	18.5	50.0	0
5 - 12	20.0	8.4	21
12 - 20	20.5	0.0	35

The groundwater table is at a depth of 7 m.

3.10 Explain the difference between drained strength and undrained strength.

3.11 Determine the compression index, recompression index, and preconsolidation stress for the soil described in Figure 3.9. Use $e_0 = 2.55$ and assume the soil is normally consolidated.

3.12 Determine the compression ratio, recompression ratio and preconsolidation stress for the soil described in Figure 3.10. The in-situ vertical effective stress at the sample location is 400 lb/ft^2.

3.13 A 2 m thick fill is to be placed on the soil shown in the profile below. Once it is compacted, this fill will have a unit weight of 19.5 kN/m^3. Compute the ultimate settlement due to consolidation of the underlying clay. You may presume that the dense sand and the bedrock are incompressible in comparison to the clay.

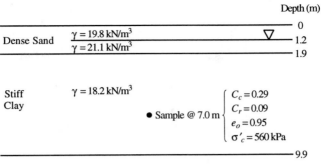

Depth (m)

Dense Sand $\gamma = 19.8 \, \text{kN/m}^3$ — 0
 — 1.2
 $\gamma = 21.1 \, \text{kN/m}^3$ — 1.9

Stiff
Clay $\gamma = 18.2 \, \text{kN/m}^3$

● Sample @ 7.0 m $\begin{cases} C_c = 0.29 \\ C_r = 0.09 \\ e_o = 0.95 \\ \sigma'_c = 560 \, \text{kPa} \end{cases}$

 — 9.9

Bedrock

3.14 A 9 ft thick fill is to be placed on the soil shown in the profile below. Once it is compacted, this fill will have a unit weight of 122 lb/ft³. Compute the ultimate settlement due to consolidation of the underlying clay. You may presume that the dense sand and the bedrock are incompressible in comparison to the clay.

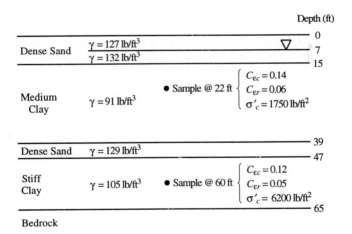

Depth (ft)

Dense Sand $\gamma = 127 \, \text{lb/ft}^3$ — 0
 — 7
 $\gamma = 132 \, \text{lb/ft}^3$ — 15

Medium
Clay $\gamma = 91 \, \text{lb/ft}^3$

● Sample @ 22 ft $\begin{cases} C_{\varepsilon c} = 0.14 \\ C_{\varepsilon r} = 0.06 \\ \sigma'_c = 1750 \, \text{lb/ft}^2 \end{cases}$

 — 39

Dense Sand $\gamma = 129 \, \text{lb/ft}^3$ — 47

Stiff
Clay $\gamma = 105 \, \text{lb/ft}^3$ ● Sample @ 60 ft $\begin{cases} C_{\varepsilon c} = 0.12 \\ C_{\varepsilon r} = 0.05 \\ \sigma'_c = 6200 \, \text{lb/ft}^2 \end{cases}$

 — 65

Bedrock

4

Site Characterization

Make a large number of trial holes to find the different strata in order to be sure that an apparently good soil does not overlay a clay, a sandy soil, or some other soil which can be compressed under a load. If the trial holes cannot be made, then the earth may be beaten with a wooden rafter six or eight feet long; if the sound is dry and light, and the soil offers resistance, then the earth is firm, but a heavy sound and poor resistance mean a worthless foundation.

Paraphrased from *L'Architecture Pratique*, a 1691 book of practical design and construction guidelines by the French engineer Bullet (after Heyman, 1972; Reprinted with permission of Cambridge University Press)

One of the fundamental differences between the practices of structural engineering and geotechnical engineering is the way each determines the engineering properties of the materials with which they work. For practical design problems, structural engineers normally find the necessary material properties by referring to handbooks. For example, if one wishes to use A36 steel, its engineering properties (strength, modulus of elasticity, etc.) are well known and can be found from a variety of sources. It is not necessary to measure the strength of A36 steel each time we use it in a design (although routine strength tests may be performed later as a quality control function). Conversely, the geotechnical engineer works with soil and rock, both of which are natural materials with unknown engineering properties. Therefore, we must identify and test the materials at each new site before conducting any analyses.

Modern soil investigation and testing techniques have progressed far beyond Bullet's method of beating the earth with wooden rafters. A variety of techniques are available, as discussed in this chapter, yet this continues to be the single largest source of uncertainties in foundation engineering. Our ability to perform analyses far exceeds our ability to determine the appropriate soil properties to input into these analyses. Therefore, it is very important for the foundation engineer to be familiar with the available techniques, know when to use them, and understand the degree of precision (or lack of precision!) associated with them.

For purposes of this discussion, we will divide these techniques into three categories:

- **Site investigation** includes methods of defining the soil profile and other relevant data and recovering soil samples.
- **Laboratory testing** includes testing the soil samples in order to determine relevant engineering properties.
- **In-situ testing** includes methods of testing the soils in-place, thus avoiding the difficulties associated with recovering samples.

4.1 SITE INVESTIGATION

The objectives of the site investigation phase include:

- Determining the locations and thicknesses of the soil strata.
- Determining the location of the groundwater table as well as any other groundwater-related characteristics.
- Recovering soil samples.
- Defining special problems and concerns.

Typically, we accomplish these goals using a combination of literature searches and on-site exploration techniques.

Background Literature Search

Before conducting any new exploration at a project site, gather whatever information is already available, both for the proposed structure and the subsurface conditions at the site. Important information about the structure would include:

- Its location and dimensions.
- The type of construction, column loads, column spacing, and allowable settlements.
- Its intended use.
- The finish floor elevation.
- The number and depth of any basements.
- The depth and extent of any proposed grading.
- Local building code requirements.

The literature search also should include an effort to obtain at least a preliminary idea of the subsurface conditions. It would be very difficult to plan an exploration program with no such knowledge. Fortunately, many methods and resources are often available to gain a preliminary understanding of the local soil conditions. These may include one or more of the following:

- Determining the geologic history of the site, including assessments of anticipated rock and soil types, the proximity of faults, and other geologic features.
- Gathering copies of boring logs and laboratory test results from previous investigations on this or other nearby sites.
- Reviewing soil maps developed for agricultural purposes.
- Reviewing old and new aerial photographs and topographic maps (may reveal previous development or grading at the site).
- Reviewing water well logs (helps establish historic groundwater levels).
- Locating underground improvements, such as utility lines, both onsite and immediately offsite.
- Locating foundations of adjacent structures, especially those that might be impacted by the proposed construction.

At some sites, this type of information may be plentiful, whereas at others it may be scarce or nonexistent.

Field Reconnaissance

Along with the background literature search, the foundation engineer should visit the site and perform a field reconnaissance. Often such visits will reveal obvious concerns that may not be evident from the literature search or the logs of the exploratory borings.

The field reconnaissance would include obtaining answers to such questions as the following:

- Is there any evidence of previous development on the site?
- Is there any evidence of previous grading on the site?
- Is there any evidence of landslides or other stability problems?

- Are nearby structures performing satisfactorily?
- What are the surface drainage conditions?
- What types of soil and/or rock are exposed at the ground surface?
- Will access problems limit the types of subsurface exploration techniques that can be used?
- Might the proposed construction affect existing improvements? (For example, a fragile old building adjacent to the site might be damaged by vibrations from pile driving.)
- Do any offsite conditions affect the proposed development? (For example, potential flooding, mudflows, rockfalls, etc.)

Subsurface Exploration and Sampling

The heart of the site investigation phase consists of exploring the subsurface conditions and sampling the soils. These efforts provide most of the basis for developing a design soil profile. A variety of techniques are available to accomplish these goals.

Exploratory Borings

The most common method of exploring the subsurface conditions is to drill a series of vertical holes in the ground. These are known as *borings* or *exploratory borings* and are typically 3 to 24 in (75 - 600 mm) in diameter and 10 to 100 ft (3 - 30 m) deep. They can be drilled with hand augers or with portable power equipment, but they are most commonly drilled using a truck-mounted rig, as shown in Figure 4.1.

Figure 4.1 A truck-mounted drill rig.

A wide variety of drilling equipment and techniques are available to accommodate the various subsurface conditions that might be encountered. Sometimes it is possible to drill an open hole using a *flight auger* or a *bucket auger*, as shown in Figure 4.2. However, if the soil is subject to *caving* (i.e., the sides of the boring fall in) or *squeezing* (the soil moves inward, reducing the diameter of the boring), then it will be necessary to provide some type of lateral support during drilling. Caving is likely to be encountered in clean sands, especially below the groundwater table, while squeezing is likely in soft saturated clays.

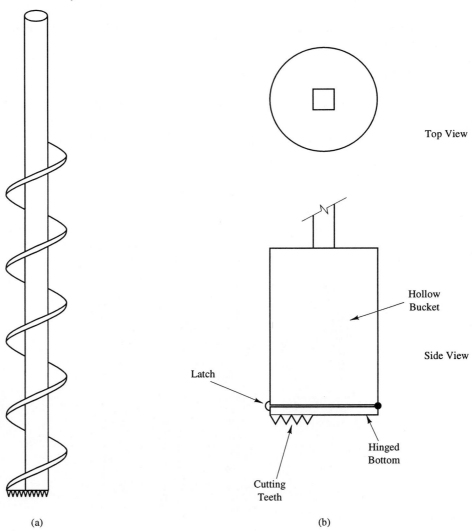

(a)   (b)

Figure 4.2 (a) Flight auger; and (b) Bucket auger

One method of dealing with caving or squeezing soils is to use *casing*, as shown in Figure 4.3a. This method involves temporarily lining some or all of the boring with a steel pipe. Alternatively, we could use a *hollow-stem auger*, as shown in Figure 4.3b. The driller screws each of these augers into the ground and obtains soil samples by lowering sampling tools through a hollow core. When the boring is completed, the augers are removed. Finally, we could use a *rotary wash boring*, as shown in Figure 4.3c. These borings are filled with a bentonite slurry (a combination of bentonite clay and water) to provide hydrostatic pressure on the sides of the boring and thus prevent caving.

Drilling through rock, especially hard rock, requires different methods and equipment. Engineers often use *coring*, which recovers intact cylindrical specimens of the rock.

The appropriate number, location, and depth of these borings depend on a number of factors, including the following:

- Is the soil profile erratic, or is it consistent across the site?
- Will the structure most likely need a shallow foundation or a deep foundation?
- How critical is the structure (i.e., what would be the consequences of a foundation failure)?
- How large is the structure?

We will not know the final answer to some of these questions (such as the type of foundation) until the exploration and testing programs are completed. However, we must have at least an approximate idea of the final design in mind so the exploration program may be planned accordingly.

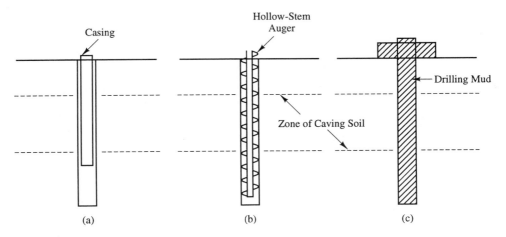

Figure 4.3 Methods of dealing with caving or squeezing soils: (a) casing; (b) hollow-stem auger; and (c) rotary wash boring.

Sowers (1979) suggested boring spacing guidelines for projects on typical soil profiles. These guidelines are shown in Table 4.1. He also suggests that these spacings might be doubled at sites with very uniform soil conditions or halved at sites with very erratic soil conditions. The *Basic Building Code* (BOCA, 1990) requires a closer spacing: at least one boring for every 2500 ft^2 (230 m^2) of built-over area.

TABLE 4.1 GUIDELINES FOR SPACING EXPLORATORY BORINGS

	Approximate Spacing of Borings	
Type of Structure	(ft)	(m)
Multistory buildings	50 - 150	15 - 45
One-story manufacturing plants	100 - 300	30 - 90

Reprinted with permission of MacMillan Publishing Company from *Introductory Soil Mechanics and Foundations: Geotechnical Engineering*, 4th Ed., by George F. Sowers, Copyright © 1979 by MacMillan Publishing Company.

Borings should extend to a depth such that at the change in effective stress due to the new construction is no more than 10% of the initial effective stress. For buildings on spread footings, Sowers (1979) suggests this criteria will be met if borings are drilled to the depths specified in Table 4.2. If fill is present, the borings must extend through it and into the natural ground below.

TABLE 4.2 GUIDELINES FOR CHOOSING DEPTH OF EXPLORATORY BORINGS FOR BUILDINGS WITH SHALLOW FOUNDATIONS

	Minimum Depth of Borings[a]	
Type of Building	(ft)	(m)
Narrow and light	$10 \, S^{0.7}$	$3 \, S^{0.7}$
Wide and heavy	$20 \, S^{0.7}$	$6 \, S^{0.7}$

Reprinted with permission of MacMillan Publishing Company from *Introductory Soil Mechanics and Foundations: Geotechnical Engineering*, 4th Ed., by George F. Sowers, Copyright © 1979 by MacMillan Publishing Company.

[a] S is the number of stories in the building.

For heavy structures, at least some of the borings should be carried to bedrock, if possible. If a deep foundation is anticipated, the borings should extend well below the anticipated tip elevations.

On large projects, the drilling program might be divided into two phases: a preliminary phase to determine the general soil profile, and a final phase that is planned based on the results of the preliminary borings.

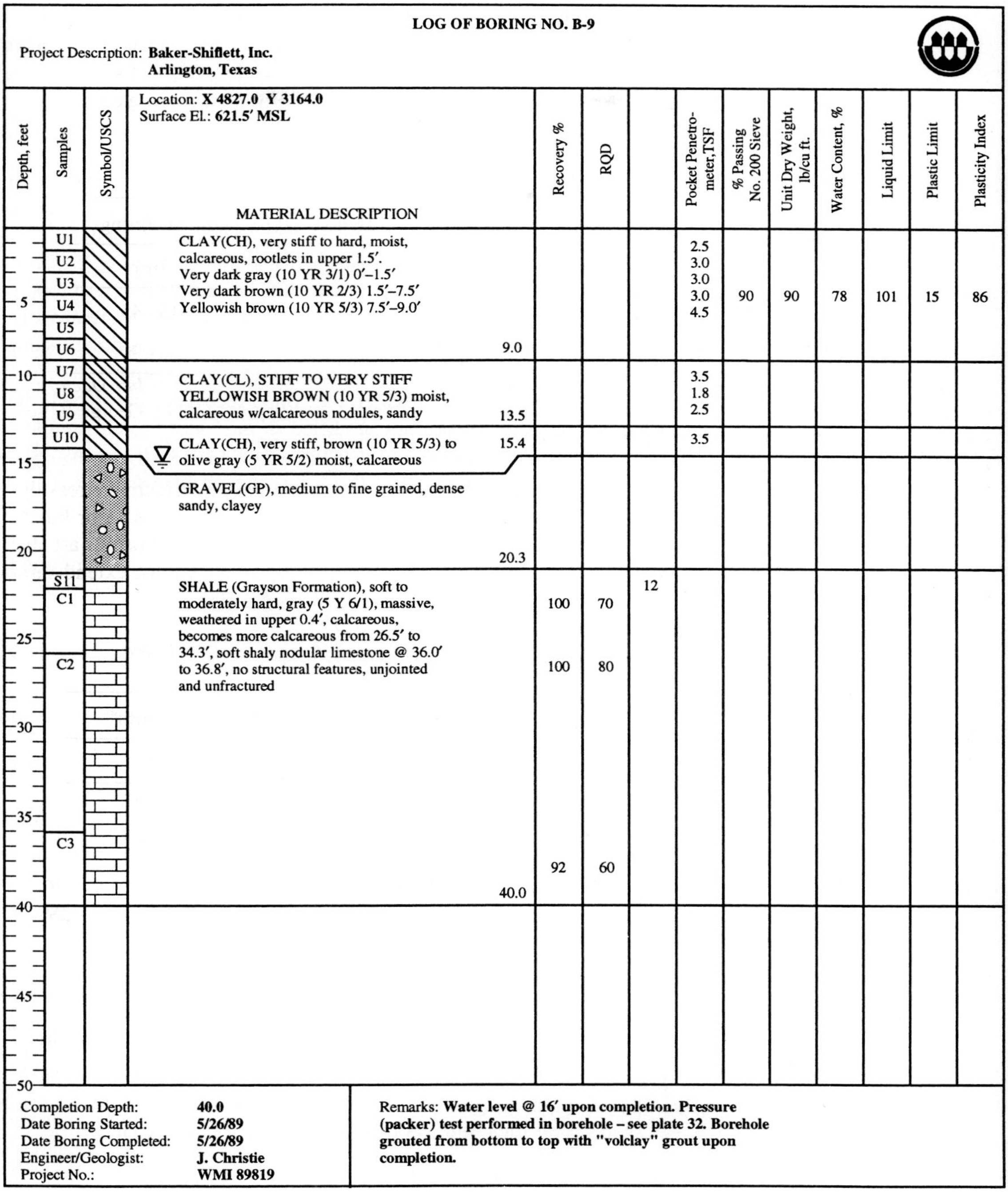

LOG OF BORING NO. B-9

Project Description: **Baker-Shiflett, Inc.**
Arlington, Texas

Location: **X 4827.0 Y 3164.0**
Surface El.: **621.5′ MSL**

Depth, feet	Samples	Symbol/USCS	MATERIAL DESCRIPTION	Recovery %	RQD	Pocket Penetrometer, TSF	% Passing No. 200 Sieve	Unit Dry Weight, lb/cu ft.	Water Content, %	Liquid Limit	Plastic Limit	Plasticity Index
	U1		CLAY(CH), very stiff to hard, moist, calcareous, rootlets in upper 1.5′. Very dark gray (10 YR 3/1) 0′–1.5′ Very dark brown (10 YR 2/3) 1.5′–7.5′ Yellowish brown (10 YR 5/3) 7.5′–9.0′			2.5						
	U2					3.0						
	U3					3.0						
5	U4					3.0	90	90	78	101	15	86
	U5					4.5						
	U6		9.0									
10	U7		CLAY(CL), STIFF TO VERY STIFF YELLOWISH BROWN (10 YR 5/3) moist, calcareous w/calcareous nodules, sandy			3.5						
	U8					1.8						
	U9		13.5			2.5						
	U10		CLAY(CH), very stiff, brown (10 YR 5/3) to olive gray (5 YR 5/2) moist, calcareous 15.4			3.5						
15			GRAVEL(GP), medium to fine grained, dense sandy, clayey									
20			20.3									
	S11		SHALE (Grayson Formation), soft to moderately hard, gray (5 Y 6/1), massive, weathered in upper 0.4′, calcareous, becomes more calcareous from 26.5′ to 34.3′, soft shaly nodular limestone @ 36.0′ to 36.8′, no structural features, unjointed and unfractured						12			
	C1			100	70							
25												
	C2			100	80							
30												
35												
	C3			92	60							
40			40.0									
45												
50												

Completion Depth:	**40.0**	Remarks: **Water level @ 16′ upon completion. Pressure (packer) test performed in borehole – see plate 32. Borehole grouted from bottom to top with "volclay" grout upon completion.**
Date Boring Started:	**5/26/89**	
Date Boring Completed:	**5/26/89**	
Engineer/Geologist:	**J. Christie**	
Project No.:	**WMI 89819**	

Figure 4.4 A typical boring log (Courtesy of Geotechnical Computer Applications).

The conditions encountered in an exploratory boring are normally presented in the form of a *boring log*, as shown in Figure 4.4. These logs also indicate the sample locations and might include some of the laboratory test results.

Soil Sampling Techniques

One of the primary purposes of drilling the exploratory borings is to obtain representative soil samples. We use these samples to determine the soil profile and to perform laboratory tests.

There are two categories of samples: disturbed and undisturbed. A *disturbed sample* (sometimes called a *bulk* sample) is one in which there is no attempt to retain the in-place structure of the soil. The driller might obtain such a sample by removing the cuttings off the bottom of a flight auger and placing them in a bag. Disturbed samples are suitable for many purposes, such as classification and Proctor compaction tests.

A truly *undisturbed sample* is one in which the soil is recovered completely intact and its in-place structure and stresses are not modified in any way. Such samples are desirable for laboratory tests that depend on the structure of the soil, such as consolidation tests and shear strength tests. Unfortunately, the following problems make it almost impossible to obtain a truly undisturbed soil sample:

- Shearing and compressing the soil during the process of inserting the sampling tool
- Relieving the sample of its in-situ stresses
- Possible drying and desiccation
- Vibrating the sample during recovery and transport

Some soils are more subject to disturbance than others. For example, techniques are available to obtain good quality samples of medium clays, while clean sands are almost impossible to sample without extensive disturbance. However, even the best techniques produce samples that are best described as "relatively undisturbed."

A variety of sampling tools is available. Some of these tools are shown in Figure 4.5. Those with thin walls produce the least disturbance, but they may not have the integrity needed to penetrate hard soils.

Monitoring Groundwater Conditions

The position and movements of the groundwater table are very important factors in foundation design. Therefore, subsurface investigations must include an assessment of groundwater conditions. This is often done by installing an *observation well* in the completed boring to monitor groundwater conditions. Such wells typically consist of slotted or perforated PVC pipe, as shown in Figure 4.6. Once the groundwater level has stabilized, we can locate it by lowering a probe into the observation well.

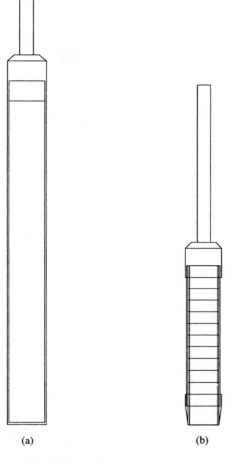

(a)   (b)

Figure 4.5 Common soil sampling tools: (a) Shelby tube samplers have thin walls to reduce sample disturbance; and (b) Ring-lined barrel samplers have thicker walls to withstand harder driving. Both are typically 2.5 - 4 in. (60 - 100 mm) in diameter.

Exploratory Trenches

Sometimes it is only necessary to explore the upper 10 ft (3 m) of soil. This might be the case for lightweight structures on sites where the soil conditions are good, or on sites with old shallow fills of questionable quality. Additional shallow investigations might also be necessary to supplement a program of exploratory borings. In such cases, it can be very helpful to dig *exploratory trenches* (also known as *test pits*) using a backhoe. They provide more information than a boring of comparable depth (because more of the soil is exposed) and are often less expensive. Disturbed samples can easily be recovered with a shovel, and undisturbed samples can be obtained using hand-held sampling equipment. The log from a typical exploratory trench is shown in Figure 4.7.

Two special precautions are in order when using exploratory trenches: First, these trenches must be adequately shored before anyone enters them. Many individuals (including one of the author's former colleagues) have been killed by neglecting to enforce this basic safety measure. Second, these trenches must be properly backfilled to avoid creating an artificial soft zone that might affect future construction.

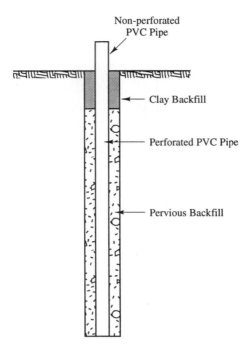

Figure 4.6 A typical observation well.

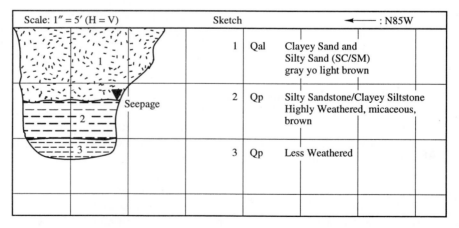

Figure 4.7 Log from an exploratory trench (Courtesy of Converse Consultants Inland Empire).

Geophysical Methods

Direct subsurface exploration methods, such as borings and trenches, often can be supplemented by various indirect methods. Many of these methods fall within the domain of the art and science of *geophysics*.

Most geophysical methods were originally developed for use in prospecting for mineral deposits (Dobrin, 1988). Some of them have found practical use in various aspects of geotechnical engineering, including foundation engineering. Although there are a relatively few projects in which such methods are of use, they can be extremely helpful in those cases.

Perhaps the most common geophysical method in the context of foundation engineering is the technique of seismic refraction, as shown in Figure 4.8. This method consists of generating shock waves in the ground, allowing them to travel through soil and bedrock, then using sensors known as *geophones* to measure the waves as they return to the ground surface. It can produce fair to good estimates of the depth to bedrock across a site.

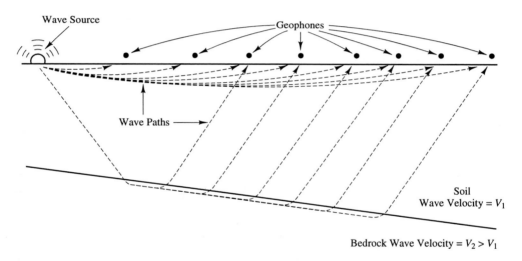

Figure 4.8 Use of seismic refraction to measure the depth to a hard layer, such as bedrock.

4.2 LABORATORY TESTING

Soil samples obtained from the field are normally brought to a soil mechanics laboratory for further classification and testing. The purpose of the testing program is to determine the appropriate engineering properties of the soil.

Classification, Weight-Volume, and Index Tests

Several routine tests are usually performed on many of the samples to ascertain the general characteristics of the soil profile. These include:

- Moisture content
- Unit weight (density)
- Atterberg limits (plastic limit, liquid limit)
- Grain-size distribution

These tests are inexpensive and can provide a large quantity of valuable information.

Shear Strength

The analysis and design of foundations rely heavily on shear strength information. Therefore, one of the principal goals of the site investigation and soil testing program is to develop design soil strength parameters.

Several different laboratory tests are commonly used to measure the shear strength of soils. Each has its advantages and disadvantages, and no one test is suitable for all circumstances. When selecting a test method, we must consider many factors, including the following:

- Soil type
- Initial moisture content and need, if any, to saturate the sample
- Required drainage conditions (drained or undrained)

Direct Shear Test

The French engineer Alexandre Collin may have been the first to measure the shear strength of a soil (Head, 1982). His tests, conducted in 1846, were similar to the modern direct shear test. The test as we now know it [ASTM D3080] was perfected by several individuals during the first half of the twentieth century.

The apparatus, shown in Figure 4.9, typically accepts a 2.5 to 3.0 inch (60 - 75 mm) diameter cylindrical sample and subjects it to a certain effective stress. The shear stress is then slowly increased until the soil fails along the surface shown in the figure. The test is normally repeated on new samples of the same soil until three sets of effective stress and shear strength measurements are obtained. A plot of this data, such as that shown in Figure 3.14, will produce values of the cohesion, c, and friction angle, ϕ.

The direct shear test has the advantage of being simple and inexpensive and it is an appropriate method when we need the drained strength of cohesionless soils. However, other tests are usually more appropriate when attempting to measure the undrained strength or when working with cohesive soils.

Unconfined Compression Test

The unconfined compression test [ASTM D2166], shown in Figure 4.10, uses a tall, cylindrical sample of cohesive soil subjected to an axial load. This load is applied quickly (i.e., only a couple of minutes to failure) to maintain undrained conditions.

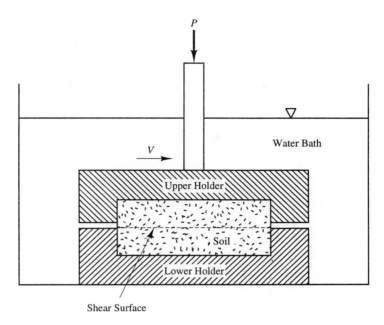

Figure 4.9 Direct shear test apparatus.

At the beginning of the test, this load and the stresses in the soil are both equal to zero. As the load increases, the stresses in the soil increase, as shown by the Mohr's circles in Figure 4.11, until the soil fails. The soil appears to fail in compression, and the test results are often expressed in terms of the compressive strength[1]; it actually fails in shear on diagonal planes, as shown in Figure 4.11. The undrained shear strength, s_u, is then:

$$s_u = \frac{P_f}{2A_f} \tag{4.1}$$

Where:

P_f = axial load at failure

A_f = cross-sectional area at failure

This test is inexpensive and in common use. It does not force failure to occur along a predetermined surface, and thus it reflects the presence of weak zones or planes. It

---

[1] When reviewing unconfined compression test results, be sure to note whether they are expressed in terms of "compressive" strength or shear strength. In this test, the shear strength is equal to half of the compressive strength.

usually provides conservative (low) results because the horizontal stress is zero rather than what is was in the field and because of sample disturbance. Tests of fissured clays are often misleading unless a large sample is used because the fissures in small samples are rarely representative of those in the field.

Figure 4.10 Unconfined compression test apparatus.

Triaxial Compression Test

The triaxial compression test [ASTM D2850] may be thought of as an extension of the unconfined compression test. The cylindrical soil sample is now located inside a pressurized chamber that supplies the desired lateral stress, as shown in Figure 4.12. Although the apparatus required to perform this test is much more complex, it also allows more flexibility and greater control over the test. It can measure either the drained or undrained strength of nearly any type of soil. In addition, unsaturated soils can be effectively saturated before testing.

The three most common types of triaxial compression tests are as follows:

- The ***unconsolidated-undrained (UU) test*** (also known as a *quick* or *Q* test): Horizontal and vertical stresses, usually equal to the vertical stress that was present in the field, are applied to the sample. No consolidation is permitted, and the soil is sheared under undrained conditions. The result is expressed as an s_u value.

- The ***consolidated-drained (CD) test*** (also known as a *slow* or *S* test): Horizontal and vertical stresses, usually equal to or greater than the vertical stress that was present in the field, are applied and the soil is allowed to consolidate. Then, it is sheared under drained conditions. Typically, three of these tests are performed at different confining stresses to find the drained values of *c* and ϕ.

- The ***consolidated-undrained test*** (also known as a *rapid* or *R* test): The initial stresses are applied as with the CD test and the soil is allowed to consolidate. However, the shearing occurs under undrained conditions. The results could be expressed as a value of s_u, but it is also possible to obtain the drained *c* and ϕ by measuring pore water pressures during the test and computing the effective stresses.

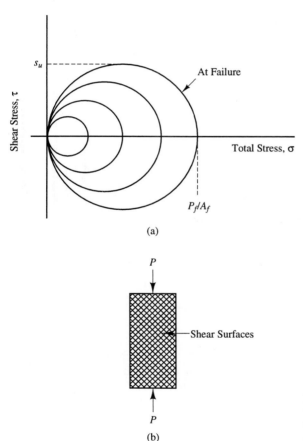

Figure 4.11 (a) Mohr's circles at various times during an unconfined compression test; (b) The soil fails in shear along diagonal planes.

Simple Hand-Operated Devices

Small portable testing devices are also available to conduct preliminary strength tests on soil samples. The *pocket penetrometer* and *TorVane*, both of which are shown in Figure 4.13, are among the most common. The pocket penetrometer is a calibrated probe that is pressed into the soil. The force required to penetrate a fixed distance correlates with the undrained shear strength. The TorVane is a small vane shear test. Either test may be used for preliminary evaluations, but they should be followed by more conventional testing.

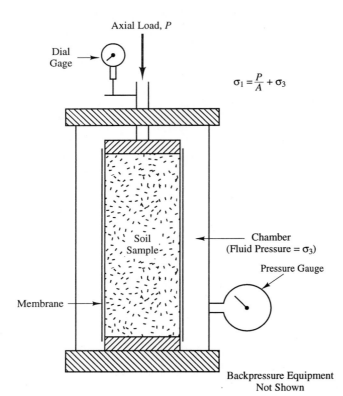

$$\sigma_1 = \frac{P}{A} + \sigma_3$$

Backpressure Equipment Not Shown

Figure 4.12 Triaxial compression test apparatus.

Consolidation

If cohesive soils have been recovered, we will be interested in assessing their compressibility, so one or more *consolidation tests* (also known as *oedometer tests*) [ASTM D2435 or D4186] will usually be in order. Karl Terzaghi conceived the idea of a consolidation test in the 1920s as a natural extension of his theory of consolidation, and the first test apparatus, known as a *consolidometer* (or *oedometer*), appeared in the early 1930s. This is now one of the most common soil tests.

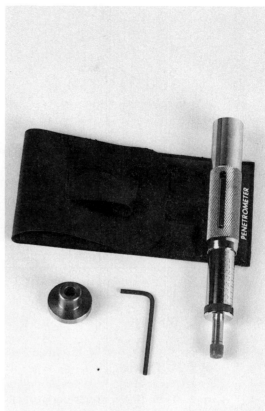

Figure 4.13 Simple hand-operated testing devices: (a) Pocket penetrometer; and (b) TorVane. (Photos courtesy of ELE/Soiltest, Inc.)

A consolidometer, as shown in Figure 4.14, contains a cylindrical soil sample that is confined laterally and set between two porous stones. It also includes provisions for applying normal loads and measuring the resulting strains. The test consists of applying a series of normal loads, allowing the soil to consolidate under each of these loads, and measuring the corresponding strain. The test results may be expressed either in the form of a void ratio vs. log effective stress plot or a normal strain vs. log effective stress plot, as discussed in Chapter 3.

The consolidation test is very sensitive to sample disturbance and therefore should be conducted only on good to excellent quality samples. As discussed earlier, the tools required to obtain suitable samples of most cohesive soils are readily available, but it is very difficult or impossible to obtain suitable samples of cohesionless soils. Therefore, consolidation tests are best performed only on clays and silts. In-situ tests provide a better measure of the compressibility of clean sands.

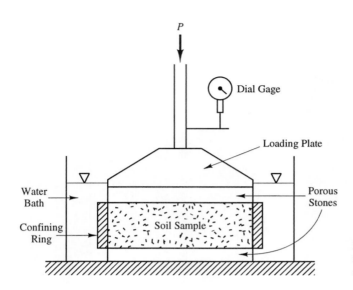

Figure 4.14 A consolidometer used to measure the consolidation properties of a soil.

4.3 IN-SITU TESTING

Laboratory tests on "undisturbed" soil samples are no better than the quality of the sample. Depending on the type of test, the effects of sample disturbance can be significant, especially in sands. Fortunately, we can often circumvent these problems by using *in-situ* (in-place) testing methods. These entail bringing the test equipment to the field and testing the soils in-place.

In addition to bypassing sample disturbance problems, in-situ tests have the following advantages:

- They are usually less expensive, so a greater number of tests can be performed, thus characterizing the soil in more detail.
- The test results are available immediately.

However, they also have disadvantages, including:

- Often no sample is obtained, thus making soil classification more difficult.
- The engineer has less control over confining stresses and drainage.

In most cases, we must use empirical correlations and calibrations to convert in-situ test results to appropriate engineering properties for design. Many such methods have been published, and some of them are included in this book. Most of these correlations were developed for clays of low to moderate plasticity or for quartz sands, and thus may not be appropriate for special soils such as very soft clays, organic soils, sensitive clays, fissured clays, cemented soils, calcareous sands, micaceous sands, collapsible soils, and frozen soils.

Some in-situ test methods have been in common use for several decades, while others are relative newcomers. Many of these tests will probably continue to become more common in engineering practice.

Standard Penetration Test (SPT)

One of the most common in-situ tests is the standard penetration test, or SPT. This test was originally developed in the late 1920s and has been used most extensively in North and South America, the United Kingdom, and Japan. Because of this long record of experience, the test is well established in engineering practice. Unfortunately, it is also plagued by many problems that affect its accuracy and reproducibility and is slowly being replaced by other test methods, especially on larger and more critical projects.

Test Procedure

The test procedure was not standardized until 1958 when ASTM standard D1586 first appeared. It is essentially as follows[2]:

1. Drill a 2.5 to 8 in (60 - 200 mm) diameter exploratory boring to the depth of the first test.
2. Insert the SPT sampler (also known as a *split-spoon sampler*) into the boring. The shape and dimensions of this sampler are shown in Figure 4.15. It is connected via steel rods to a 140 lb (63.5 kg) hammer, as shown in Figure 4.16.
3. Using either a rope and cathead arrangement or an automatic tripping mechanism, raise the hammer a distance of 30 in (760 mm) and allow it to fall. This energy drives the sampler into the bottom of the boring. Repeat this process until the sampler has penetrated a distance of 18 in (450 mm), recording the number of hammer blows required for each 6 in (150 mm) interval. Stop the test if more than 50 blows are required for any of the intervals, or if more than 100 total blows are required. Either of these events is known as *refusal* and is so noted on the boring log.
4. Compute the N value by summing the blow counts for the last 12 in (300 mm) of penetration. The blow count for the first 6 in (150 mm) is retained for reference purposes, but not used to compute N because the bottom of the boring is likely to be disturbed by the drilling process and may be covered with loose soil that fell from the sides of the boring. Note that the N value is the same regardless of whether the engineer is using English or SI units.

---

[2] See the ASTM D1586 standard for the complete procedure.

5. Remove the SPT sampler; remove and save the soil sample.

6. Drill the boring to the depth of the next test and repeat steps 2 through 6 as required.

Thus, *N* values may be obtained at intervals no closer than 18 in (450 mm).

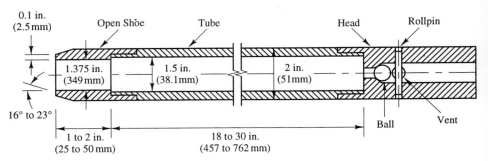

Figure 4.15 The SPT sampler (Adapted from ASTM D1586; Copyright ASTM, reprinted with permission).

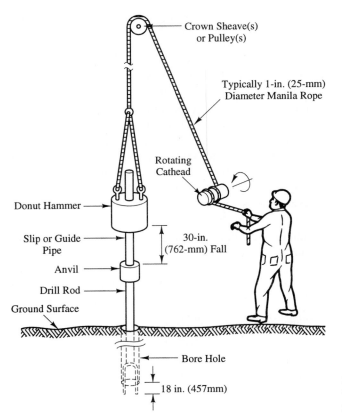

Figure 4.16 The SPT sampler in place in the boring with hammer, rope and cathead in place (Adapted from Kovacs et al., 1981).

Unfortunately, the procedure used in the field varies, partially due to changes in the standard, but primarily as a result of variations in the test procedure and poor workmanship. The test results are sensitive to these variations, so the N value is not as repeatable as we would like. The principal variants are as follows:

- Method of drilling
- How well the bottom of the hole is cleaned before the test
- Presence or lack of drilling mud
- Diameter of the drill hole
- Location of the hammer (surface type or down-hole type)
- Type of hammer, especially whether it has a manual or automatic tripping mechanism
- Number of turns of the rope around the cathead
- Actual hammer drop height (manual types are often as much as 25 percent in error)
- Mass of the anvil that the hammer strikes
- Friction in rope guides and pulleys
- Wear in the sampler drive shoe
- Straightness of the drill rods
- Presence or absence of liners inside the sampler (this seemingly small detail can alter the test results by 10 - 30%)
- Rate at which the blows are applied

As a result of these variables, the test results will vary depending on the crew and equipment. These variations, as well as other aspects of the test, were the subject of increased scrutiny during the 1970s and 1980s along with efforts to further standardize the "standard" penetration test (DeMello, 1971; Nixon, 1982). Based on these studies, Seed et al. (1985) recommended the following additional criteria be met when conducting standard penetration tests:

- Use the rotary wash method to create a boring that has a diameter between 4 and 5 in (200 - 250 mm). The drill bit should provide an upward deflection of the drilling mud (tricone or baffled drag bit).
- If the sampler is made to accommodate liners, then these liners should be used so the inside diameter is 1.38 in (35 mm).
- Use A or AW size drill rods for depths less than 50 ft (15 m) and N or NW size for greater depths.
- Use a hammer that has an efficiency of 60%.
- Apply the hammer blows at a rate of 30 to 40 per minute.

Fortunately, automatic hammers are becoming more popular. They are much more consistent than hand-operated hammers, and thus improve the reliability of the test.

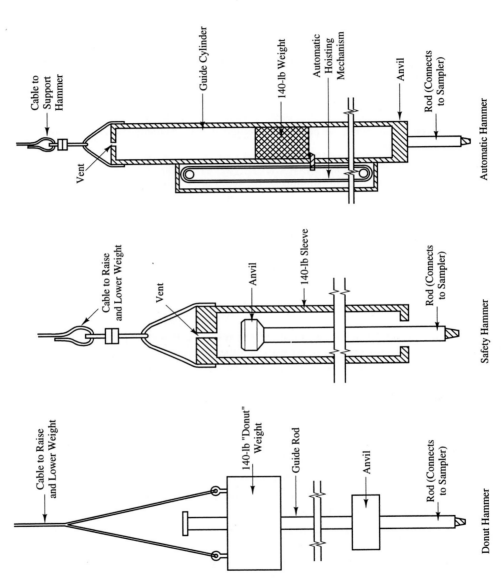

Figure 4.17 Types of SPT Hammers.

TABLE 4.3 SPT HAMMER EFFICIENCIES

Country	Hammer Type	Hammer Release Mechanism	Hammer Efficiency E_m
Argentina	Donut	Cathead	0.45
Brazil	Pin Weight	Hand Dropped	0.72
China	Automatic	Trip	0.60
	Donut	Hand dropped	0.55
	Donut	Cathead	0.50
Colombia	Donut	Cathead	0.50
Japan	Donut	Tombi trigger	0.78 - 0.85
	Donut	Cathead 2 turns + special release	0.65 - 0.67
UK	Automatic	Trip	0.73
USA	Safety	2 turns on cathead	0.55 - 0.60
	Donut	2 turns on cathead	0.45
Venezuela	Donut	Cathead	0.43

Adapted from Clayton (1990).

TABLE 4.4 BOREHOLE, SAMPLER, AND ROD CORRECTION FACTORS

Factor	Equipment Variables	Value
Borehole diameter factor, C_B	2.5 - 4.5 in (65 - 115 mm)	1.00
	6 in (150 mm)	1.05
	8 in (200 mm)	1.15
Sampling method factor, C_S	Standard sampler	1.00
	Sampler without liner (not recommended)	1.20
Rod length factor, C_R	10 - 13 ft (3 - 4 m)	0.75
	13 - 20 ft (4 - 6 m)	0.85
	20 - 30 ft (6 - 10 m)	0.95
	> 30 ft (> 10 m)	1.00

Adapted from Skempton (1986).

Although much has been said about the disadvantages of the SPT, it does have at least three important advantages over other in-situ test methods: First, it obtains a sample of the soil being tested. This permits direct soil classification. Most of the other methods do not include sample recovery, so soil classification must be based on conventional sampling from nearby borings and on correlations between the test results and soil type. Second, it is very fast and inexpensive because it is performed in borings that would have been drilled anyway. Finally, nearly all drill rigs used for soil exploration are equipped to perform this test, whereas other in-situ tests require specialized equipment that may not be readily available.

Corrections to Test Data

We can improve the raw SPT data by applying certain correction factors. The variations in testing procedures may be at least partially compensated by converting the measured N to N_{60} as follows (Skempton, 1986):

$$N_{60} = \frac{E_m\, C_B\, C_S\, C_R\, N}{0.60} \qquad (4.2)$$

Where:

N_{60} = SPT N value corrected for field procedures

E_m = hammer efficiency (from Table 4.3)

C_B = borehole diameter correction (from Table 4.4)

C_S = sampler correction (from Table 4.4)

C_R = rod length correction (from Table 4.4)

N = measured SPT N value

Many different hammer designs are in common use, none of which is 100% efficient. Some common hammer designs are shown in Figure 4.17, and typical hammer efficiencies are listed in Table 4.3. Many of the SPT-based design correlations were developed using hammers that had an efficiency of about 60%, so Equation 4.2 corrects the results from other hammers to that which would have been obtained if a 60% efficient hammer was used.

The SPT data also may be adjusted using an *overburden correction* that compensates for the effects of effective stress. Deep tests in a uniform soil deposit will have higher N values than shallow tests in the same soil, so the overburden correction adjusts the measured N values to what they would have been if the vertical effective stress, σ_v', was 2000 lb/ft^2 (100 kPa). The corrected value, N_{60}', is:

$$N_{60}' = C_N N_{60} \qquad (4.3)$$

Skempton's (1986) functions for the overburden correction factor, C_N, are:

For normally consolidated fine sands:

$$C_N = \frac{2}{1 + \sigma_v'/\sigma_r} \qquad (4.4)$$

For normally consolidated coarse sands:

$$C_N = \frac{3}{2 + \sigma_v'/\sigma_r} \qquad (4.5)$$

For overconsolidated sands:

$$C_N = \frac{1.7}{0.7 + \sigma_v'/\sigma_r} \qquad (4.6)$$

Where:

N_{60}' = SPT N value corrected for field procedures and overburden stress

σ_r = reference stress = 2000 lb/ft^2 = 100 kPa

σ_v' = vertical effective stress at the test location

N_{60} = SPT N value corrected for field procedures

The use of correction factors is often a confusing issue. Corrections for field procedures are always appropriate, but the overburden correction may or may not be appropriate depending on the procedures used by those who developed the analysis method under consideration.

Uses of SPT Data

The SPT N value, as well as many other tests, is only an *index* of soil behavior. It does not directly measure any of the conventional engineering properties of soil and is useful only when appropriate correlations are available. Many such correlations exist, all of which were obtained empirically.

Unfortunately, most of these correlations are very approximate, especially those based on fairly old data that were obtained when test procedures and equipment were different from those now used. In addition, because of the many uncertainties in the SPT results, all of these correlations have a wide margin of error.

Be especially cautious when using correlations between SPT results and engineering properties of clays because these functions are especially crude. In general, the SPT should be used only in sandy soils.

Correlation with Relative Density

Early correlations (i.e., Terzaghi and Peck, 1967) gave direct relationships between the relative density of sandy soils, D_r, and the SPT N value. However, later research has shown that overburden stress, overconsolidation ratio, particle size, and other factors are also important. Figure 4.18 gives a relationship for D_r as a function of N_{60} and the vertical effective stress, σ_v' at the test location. Equation 4.7 gives a more sophisticated

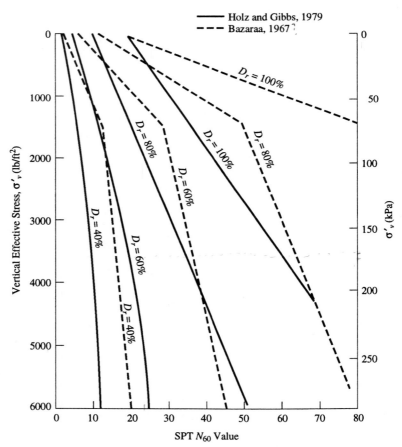

Figure 4.18 Relative density, D_r, determined from SPT N_{60} and the vertical effective stress, σ_v', at the test location (Adapted from USBR, 1974; Bazaraa, 1967).

relationship that also considers the overconsolidation ratio and grain-size distribution (Kulhawy and Mayne, 1990).

The relative density forms the basis for some methods of foundation analysis. It also may be used to classify the consistency of sandy soils using the system described in Table 3.4.

$$D_r = \sqrt{\frac{N'_{60}}{C_P C_A C_{OCR}}} \tag{4.7}$$

$$C_P = 60 + 25 \log D_{50} \tag{4.8}$$

$$C_A = 1.2 + 0.05 \log\left(\frac{t}{100}\right) \tag{4.9}$$

$$C_{OCR} = OCR^{0.18} \tag{4.10}$$

Where:

D_r = relative density (in decimal form)

N'_{60} = SPT N value corrected for field procedures and overburden stress

C_P = grain-size correction factor

C_A = aging correction factor

C_{OCR} = overconsolidation correction factor

D_{50} = grain size at which 50% of the soil is finer (mm)

t = age of soil (time since deposition) (years)

OCR = overconsolidation ratio

Correlation with Shear Strength

DeMello (1971) suggested a correlation between SPT results and the friction angle of uncemented sands, ϕ, as shown in Figure 4.19. This correlation should be used only at depths greater than about 7 ft (2 m).

Correlation with Compressibility

Many correlations have been proposed to relate SPT N values to the modulus of elasticity, E, or the constrained modulus, M. Unfortunately, the various methods produce dramatically different results (Mitchell and Gardner, 1975), so it is difficult to determine which, if any, are correct. Kulhawy and Mayne (1990) suggested the following approximate relationships:

For sands with fines:

$$E \approx 5 \, \sigma_r \, N_{60} \tag{4.11}$$

For clean, normally consolidated sands:

$$E \approx 10 \, \sigma_r \, N_{60} \tag{4.12}$$

For clean, overconsolidated sands:

$$E \approx 15 \, \sigma_r \, N_{60} \tag{4.13}$$

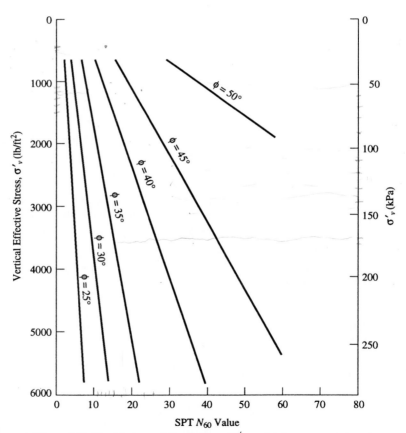

Figure 4.19 Empirical correlation between N'_{60} and ϕ for uncemented sands (Adapted from DeMello, 1971).

Cone Penetration Test (CPT)

The *cone penetration test* [ASTM D3441], or CPT, is another common in-situ test (Schmertmann, 1978; De Ruiter, 1981; Meigh, 1987; Robertson and Campanella, 1989; Briaud and Miran, 1991). Most of the early development of this test occurred in western Europe in the 1930s and again in the 1950s. Although many different styles and configurations have been used, the current standard grew out of work performed in the Netherlands, so it is sometimes called the *Dutch Cone*. The CPT has been used extensively in Europe for many years and is becoming increasingly popular in North America and elsewhere.

Two types of cones are commonly used: the *mechanical cone* and the *electric cone*, as shown in Figure 4.20. Both have two parts, a 35.7 mm diameter cone-shaped tip with a 60° apex angle and a 35.7 mm diameter × 133.7 mm long cylindrical sleeve. A hydraulic ram pushes this assembly into the ground and instruments measure the resistance to penetration. The *cone resistance*, q_c, is the total force acting on the cone

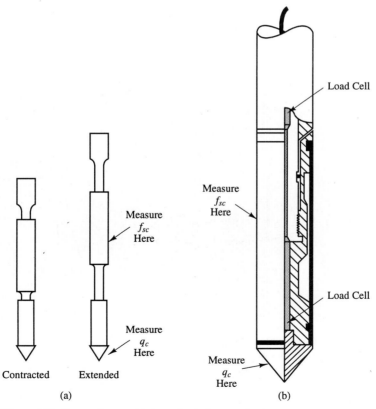

Figure 4.20 Types of cones: (a) A mechanical cone (also known as a Begemann Cone); and (b) An electric cone (also known as a Fugro Cone).

divided by the projected area of the cone (10 cm^2); and the *local side friction, f_{sc}*, is the total frictional force acting on the friction sleeve divided by its surface area (150 cm^2). It is common to express the side friction in terms of the *friction ratio, R_f*, which is equal to $f_{sc}/q_c \times 100\%$.

The operation of the two types of cones differs in that the mechanical cone is advanced in stages and measures q_c and f_{sc} at intervals of about 20 cm, whereas the electric cone includes built-in strain gages and is able to measure q_c and f_{sc} continuously with depth. In either case, the CPT defines the soil profile with much greater resolution than does the SPT.

CPT rigs are often mounted in large three-axle trucks such as the one in Figure 4.21. These are typically capable of producing maximum thrusts of 10 - 20 tons (100 - 200 kN). Smaller, trailer-mounted or truck-mounted rigs also are available.

The CPT has been the object of extensive research and development (Robertson and Campanella, 1983) and thus is becoming increasingly useful to the practicing engineer. Some of this research effort is now being conducted using cones equipped with pore pressure transducers in order to measure the excess pore water pressures that develop while conducting the test. These are known as *piezocones*, and the enhanced procedure is known as a CPTU test. These devices promise to be especially useful in saturated clays.

A typical plot of CPT results is shown in Figure 4.22.

Figure 4.21 A truck-mounted CPT rig. A hydraulic ram, located inside the truck, pushes the cone into the ground, using the weight of the truck as a reaction.

Although the CPT has many advantages over the SPT, there are at least three important disadvantages:

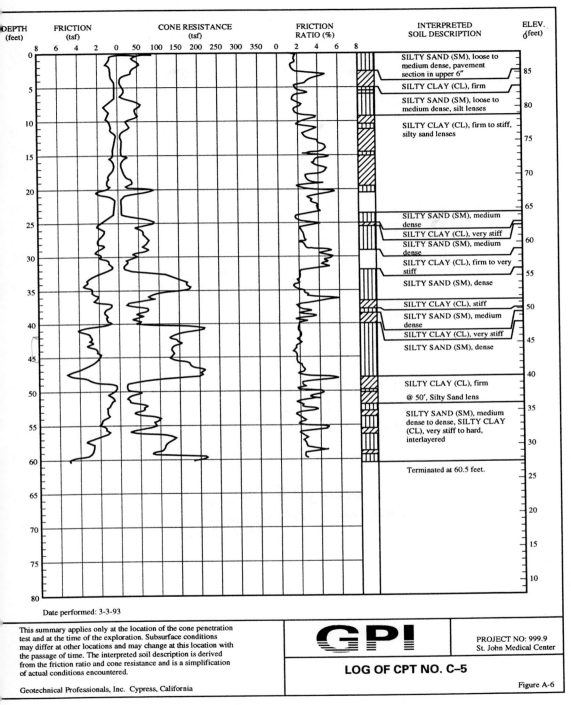

Figure 4.22 Typical CPT test results (Courtesy of Geotechnical Computer Applications).

- No soil sample is recovered, so there is no opportunity to inspect the soils.
- The test is unreliable or unusable in soils with significant gravel content.
- Although the cost per foot of penetration is less than that for borings, it is necessary to mobilize a special rig to perform the CPT.

Uses of CPT Data

The CPT is an especially useful and inexpensive way to evaluate soil profiles. Since it retrieves data continuously with depth (with electric cones) or at very close intervals (with mechanical cones), the CPT is able to detect fine changes in the stratigraphy. Therefore, engineers often use the CPT in the first phase of subsurface investigation, saving boring and sampling for the second phase.

The CPT also can be used to assess the engineering properties of the soil through the use of empirical correlations. Those correlations intended for use in cohesionless soils are generally more accurate and most commonly used. They are often less accurate in cohesive soils because of the presence of excess pore water pressures and other factors. However, the piezocone may overcome this problem.

Correlations are also available to directly relate CPT results with foundation behavior. These are especially useful in piles as discussed in Chapter 12.

Correlation with Soil Classification

Because the CPT does not recover any soil samples, it is not a substitute for conventional exploratory borings. However, it is possible to obtain an approximate soil classification using the correlation shown in Figure 4.23.

Correlation with Relative Density

Kulhawy and Mayne (1990) developed the following approximate relationship between CPT results and the relative density of cohesionless soils:

$$D_r = \sqrt{\left(\frac{1}{305\, Q_c\, \mathrm{OCR}^{0.18}}\right)\left(\frac{q_c}{\sigma_r}\right)\sqrt{\frac{\sigma_r}{\sigma_v'}}} \qquad (4.14)$$

Where:

Q_c = compressibility factor

       = 0.91 for highly compressible sands

       = 1.00 for moderately compressible sands

       = 1.09 for slightly compressible sands

OCR = overconsolidation ratio

q_c = cone resistance

$\sigma_v{}' = $ vertical effective stress

$\sigma_r = $ reference stress $= 2000 \text{ lb/ft}^2 = 100 \text{ kPa} = 1 \text{ bar} = 1 \text{ kg/cm}^2$

For purposes of solving this formula, a sand with a high fines content or a high mica content is "highly compressible," whereas a pure quartz sand would be "slightly compressible."

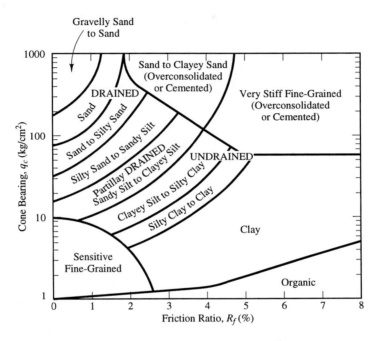

Figure 4.23 Classification of soil based on CPT test results (Adapted from Robertson and Campanella, 1983).

Correlation with Shear Strength

The CPT results also have been correlated with shear strength parameters, especially in sands. Figure 4.24 presents Robertson and Campanella's 1983 correlation for uncemented, normally consolidated quartz sands. For overconsolidated sands, subtract $1 - 2°$ from the friction angle obtained from this figure.

Correlation with Compressibility

The relationship between CPT results and the constrained modulus, M, for cohesionless soils may be expressed as follows:

$$M = \alpha q_c \tag{4.15}$$

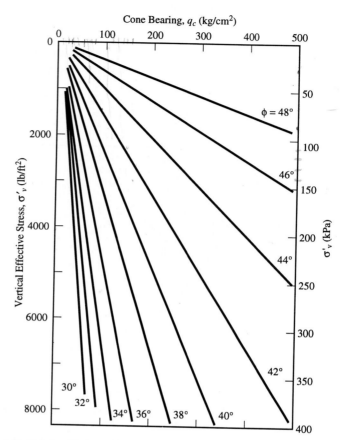

Figure 4.24 Relationship between CPT results, overburden stress and friction angle for uncemented, normally consolidated quartz sands (Adapted from Robertson and Campanella, 1983).

The coefficient α depends on the mineralogy and grain-size distribution of the soil, its relative density, and its overconsolidation ratio among other factors. Jamiolkowski et al. (1988) studied this relationship for Ticino Sand (a medium to coarse quartz sand with about 5% mica). The results of their study are shown in Figure 4.25. This also seems to be at least approximately correct for other sands.

Correlation with SPT N Value

Since the SPT and CPT are the two of the most common in-situ tests, it often is useful to convert results from one to the other. The ratio q_c/N_{60} as a function of the mean grain size, D_{50}, is shown in Figure 4.26. Note that N_{60} does not include an overburden correction.

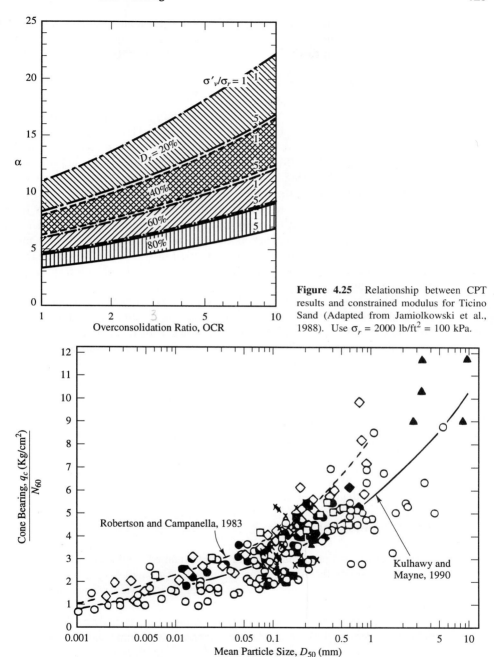

Figure 4.25 Relationship between CPT results and constrained modulus for Ticino Sand (Adapted from Jamiolkowski et al., 1988). Use $\sigma_r = 2000$ lb/ft^2 = 100 kPa.

Figure 4.26 Correlation between q_c/N_{60} and the mean grain size, D_{50} (Adapted from Kulhawy and Mayne, 1990). Copyright ©1990 Electric Power Research Institute, reprinted with permission.

Be cautious about converting CPT data to equivalent N values, and then using SPT-based analysis methods. This technique compounds the uncertainties because it uses two correlations—one to convert to N, and then another to compute the desired quantity.

Vane Shear Test (VST)

The Swedish engineer John Olsson developed the vane shear test (VST) in the 1920s to test the sensitive Scandinavian marine clays in-situ. The VST has grown in popularity, especially since the Second World War, and is now used around the world.

This test [ASTM D2573] consists of inserting a metal vane into the soil, as shown in Figure 4.27, and rotating it until the soil fails in shear. The undrained shear strength may be determined from the torque at failure, the vane dimensions, and other factors. The vane can be advanced to greater depths by simply pushing it deeper (especially in softer soils) or the test can be performed below the bottom of a boring and repeated as the boring is advanced. However, because the vane must be thin to minimize soil disturbance, it is only strong enough to be used in soft to medium cohesive soils. The test is performed rapidly (about 1 minute to failure) and therefore measures only the undrained strength.

There are no universally accepted test procedures or vane dimensions, both of which affect the relationship between torque and shear strength. However, for typical procedures and typical vanes, such as that shown in Figure 4.27 with a 2:1 height:diameter ratio, used in soils with moderate plasticity, we can use the following relationship (Chandler, 1988):

$$s_u = \frac{0.86\,T}{\pi\,d^3} \tag{4.16}$$

Where:

s_u = undrained shear strength

T = torque at failure

d = diameter of vane

Pressuremeter Test (PMT)

In 1954, a young French engineering student named Louis Ménard began to develop a new type of in-situ test: the pressuremeter test. Although Kögler had done some limited work on a similar test some 20 years earlier, it was Ménard who made it a practical reality.

The pressuremeter is a cylindrical balloon that is inserted into the ground and inflated, as shown in Figure 4.28. Measurements of volume and pressure can be used to evaluate the in-situ stress, compressibility, and strength of the adjacent soil and thus the behavior of a foundation (Baguelin et al., 1978; Briaud, 1992).

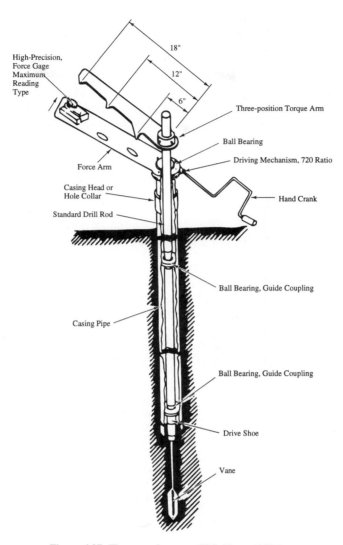

Figure 4.27 The vane shear test (U.S. Navy, 1982a).

The PMT may be performed in a carefully drilled boring or the test equipment can be combined with a small auger to create a self-boring pressuremeter. The latter design provides less soil disturbance and more intimate contact between the pressuremeter and the soil.

The PMT produces much more direct measurements of soil compressibility and lateral stresses than do the SPT and CPT. Thus, in theory, it should form a better basis for settlement analyses, and possibly for pile capacity analyses. However, the PMT is a difficult test to perform and is limited by the availability of the equipment and personnel trained to use it.

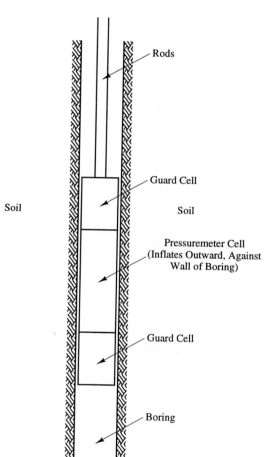

Figure 4.28 Schematic of the pressuremeter test.

Although the PMT is widely used in France and Germany, it is used only occasionally in other parts of the world. However, it may become more popular in the future.

Dilatometer Test (DMT)

The dilatometer (Marchetti, 1980; Schmertmann, 1986b, 1988a, and 1988b), which is one of the newest in-situ test devices, was developed during the late 1970s in Italy by Silvano Marchetti. It is also known as a *flat dilatometer* or a *Marchetti dilatometer* and consists of a 95 mm wide, 15 mm thick metal blade with a thin, flat, circular, expandable steel membrane on one side, as shown in Figure 4.29.

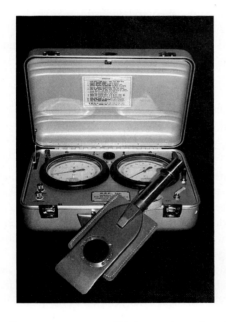

Figure 4.29 The Marchetti dilatometer along with its control unit and nitrogen gas bottle.   (Photo courtesy GPE, Inc., Gainesville, FL).

The dilatometer test (DMT) is conducted as follows (Schmertmann, 1986a):

1. Press the dilatometer into the soil to the desired depth using a CPT rig or some other suitable device.
2. Apply nitrogen gas pressure to the membrane to press it outward. Record the pressure required to move the center of the membrane 0.05 mm into the soil (the *A* pressure) and that required to move its center 1.10 mm into the soil (the *B* pressure).
3. Depressurize the membrane and record the pressure acting on the membrane when it returns to its original position. This is the *C* pressure and is a measure of the pore water pressure in the soil.
4. Advance the dilatometer 150 to 300 mm deeper into the ground and repeat the test. Continue until reaching the desired depth.

Each of these test sequences typically requires 1 to 2 minutes to complete, so a typical *sounding* (a complete series of DMT tests between the ground surface and the desired depth) may require about 2 hours. In contrast, a comparable CPT sounding might be completed in about 30 minutes.

The primary benefit of the DMT is that it measures the lateral stress condition and compressibility of the soil. These are determined from the *A*, *B*, and *C* pressures and certain equipment calibration factors and expressed as the *DMT indices:*

I_D = material index (a normalized modulus)
K_D = horizontal stress index (a normalized lateral stress)
E_D = dilatometer modulus (theoretical elastic modulus)

Researchers have developed correlations between these indices and certain engineering properties of the soil (Schmertmann, 1988b; Kulhawy and Mayne, 1990), including:

- Classification
- Coefficient of lateral earth pressure, K_0
- Overconsolidation ratio, OCR
- Modulus of elasticity, *E*, or constrained modulus, *M*

The CPT and DMT are complementary tests (Schmertmann, 1988b). The cone is a good way to evaluate soil strength, whereas the dilatometer assesses compressibility and in-situ stresses. These three kinds of information form the basis for most foundation engineering analyses. In addition, the dilatometer blade is most easily pressed into the ground using a conventional CPT rig, so it is a simple matter to conduct both CPT and DMT tests while mobilizing only a minimum of equipment.

The dilatometer test is a relative newcomer, and thus has not yet become a common engineering tool. Engineers have had only limited experience with it and the analysis and design methods based on DMT results are not yet well developed. However, its relatively low cost, versatility, and compatibility with the CPT suggest that it may enjoy widespread use in the future.

Becker Penetration Test

Soils that contain a large percentage of gravel and those that contain cobbles or boulders create problems for most in-situ test methods. Often, the in-situ test device is not able to penetrate through such soils (it meets *refusal*) or the results are not representative because the particles are about the same size as the test device. Frequently, even conventional drilling equipment cannot penetrate through these soils.

One method of penetrating through these very large-grained soils is to use a *Becker hammer drill*. This device, developed in Canada, uses a small diesel pile-driving hammer and percussion action to drive a 5.5 to 9.0 inch (135 - 230 mm) diameter double-wall steel casing into the ground. The cuttings are sent to the top by blowing air through the casing. This technique has been used successfully on very dense and coarse soils.

The Becker hammer drill also can be used to assess the penetration resistance of these soils using the *Becker penetration test*, which is monitoring the hammer blow-count. The number of blows required to advance the casing 1 ft (300 mm) is the Becker blowcount, N_B. Several correlations are available to convert it to an equivalent SPT N value (Harder and Seed, 1986). One of these correlation methods also considers the bounce chamber pressure in the diesel hammer.

Comparison of In-Situ Test Methods

Each of the in-situ test methods has its strengths and weaknesses. Table 4.5 compares some of the important attributes of the tests described in this chapter.

4.4 SYNTHESIS OF FIELD AND LABORATORY DATA

Investigation and testing programs often generate large amounts of information that can be difficult to sort through and synthesize. Real soil profiles are nearly always very complex, so the borings will not correlate and the test results will often vary significantly. Therefore, we must develop a simplified soil profile before proceeding with the analysis. In many cases, this simplified profile is best defined in terms of a one-dimensional function of soil type and engineering properties vs. depth; an idealized boring log. However, when the soil profile varies significantly across the site, one or more vertical cross-sections may be in order.

The development of these simplified profiles requires a great deal of engineering judgment along with interpolation and extrapolation of the data. It is important to have a feel for the approximate magnitude of the many uncertainties in this process and reflect them in an appropriate degree of conservatism. This judgment comes primarily with experience combined with a thorough understanding of the field and laboratory methodologies.

4.5 ECONOMICS OF INVESTIGATION AND TESTING

The site investigation and soil testing phase of foundation engineering is the single largest source of uncertainties. No matter how extensive it is, there is always some doubt whether the borings accurately portray the subsurface conditions, whether the samples are representative, and whether the tests are correctly measuring the soil properties. Engineers attempt to compensate for these uncertainties by applying factors of safety in our analyses. Unfortunately, this solution also increases construction costs.

In an effort to reduce the necessary level of conservatism in the foundation design, the engineer may choose a more extensive investigation and testing program to better define the soils. The additional costs of such efforts will, to a point, result in decreased

TABLE 4.5 ASSESSMENT OF IN-SITU TEST METHODS

	Standard Penetration Test	Cone Penetration Test	Vane Shear Test	Pressuremeter Test	Dilatometer Test	Becker Penetration Test
Simplicity and Durability of Apparatus	Simple; rugged	Complex; rugged	Simple; rugged	Complex; delicate	Complex; moderately rugged	Simple, rugged
Ease of Testing	Easy	Easy	Easy	Complex	Easy	Easy
Continuous Profile or Point Values	Point	Continuous	Point	Point	Point	Continuous
Basis for Interpretation	Empirical	Empirical; theory	Direct measurement	Empirical; theory	Empirical; theory	Empirical
Suitable Soils	All except gravels	All except gravels	Soft clays	All	All except gravels	Sands through boulders
Equipment Availability and Use in Practice	Universally available; used routinely	Generally available; used routinely	Generally available; used routinely	Difficult to locate; used on special projects	Difficult to locate; used on special projects	Difficult to locate; used on special projects
Potential for Future Development	Limited	Great	Uncertain	Great	Great	Uncertain

Modified from Mitchell, 1978; Used by permission of ASCE.

construction costs, as shown in Figure 4.30.   However, at some point, this becomes a matter of diminishing returns, and eventually the incremental cost of additional investigation and testing does not produce an equal or larger reduction in construction costs. The minimum on this curve represents the optimal level of effort.

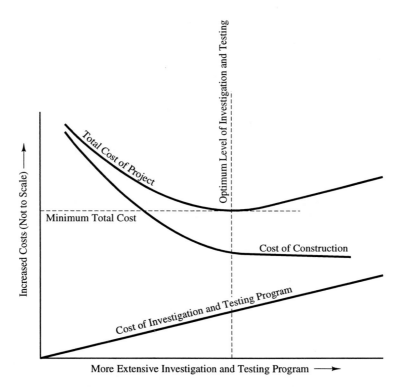

Figure 4.30 Cost effectiveness of more extensive investigation and testing programs.

An example of this type of economic consideration is a decision regarding the value of conducting full-scale pile load tests (see Chapter 10). On large projects, the potential savings in construction costs will often justify one or more tests, whereas on small projects, a conservative design developed without the benefit of load tests may result in a lower total cost.

We also must decide whether to conduct a large number of moderately precise tests (such as the SPT) or a smaller number of more precise but expensive tests (such as the PMT). Handy (1980) suggested the most cost-effective test is the one with a variability consistent with the variability of the soil profile. Thus, a few precise tests might be appropriate in a uniform soil deposit, but more data points, even if they are less precise, are more valuable in an erratic deposit.

SUMMARY OF MAJOR POINTS

1. Soil and rock are natural materials. Therefore, their engineering properties vary from site to site and must be determined individually for each project. This process can be divided into three parts: site investigation, laboratory testing, and in-situ testing.

2. Site investigations typically include conducting literature searches, performing a site reconnaissance, drilling exploratory borings, and obtaining soil samples. They also might include digging backhoe trenches, geophysical exploration, and other methods.

3. Usually, we bring soil samples to a laboratory to conduct standard laboratory tests. These tests help to classify the soil, determine its strength, and assess its compressibility. Other tests might also be conducted.

4. In-situ tests are those conducted in the ground. These techniques are especially useful in soils that are difficult to sample, such as clean sands. In-situ methods include the standard penetration test (SPT), the cone penetration test (CPT), the vane shear test (VST), the pressuremeter test (PMT), the dilatometer test (DMT), and the Becker penetration test.

5. An investigation and testing program will typically generate large amounts of data, even though only a small portion of the soil is actually tested. The engineer must synthesize this data into a simplified form to be used in analyses and design.

6. Investigation and testing always involves uncertainties and risks. These can be reduced, but not eliminated, by drilling more borings, retrieving more samples, and conducting more tests. However, there are economic limits to such endeavors, so we must determine what amount of work is most cost effective.

QUESTIONS AND PRACTICE PROBLEMS

4.1 Describe a scenario that would require a very extensive site investigation and laboratory testing program (i.e., one in which a large number of borings and many laboratory and/or in-situ tests would be necessary).

4.2 Which laboratory and in-situ tests would be appropriate for finding the undrained strength of a cohesive soil?

4.3 Which laboratory and in-situ tests would be appropriate for finding the drained strength of a cohesionless soil?

4.4 How would you go about determining the location of the groundwater table in the design soil profile. Recall that this is not necessarily the same as the groundwater table that was present when the borings were made.

4.5 Which laboratory tests require undisturbed soil samples? Why? Which do not? Why?

4.6 You anticipate the foundation for a proposed highway bridge will consist of end-bearing piles driven through the overlying soils to bedrock. However, you are not certain that the bedrock is shallow enough to make such a foundation economically feasible. You wish to obtain at least an approximate idea of the depth to bedrock before you begin drilling the exploratory borings. How would you go about finding this information?

4.7 A two-story reinforced concrete building is to be built on a vacant parcel of land. This building will be 100 ft wide and 200 ft long. Based on information from other borings on adjacent properties, you are reasonably certain that the soils below a depth of 5 to 8 feet (1.5 to 2.5 m) are strong and relatively incompressible. However, the upper soils are questionable because several uncompacted fills have been found in the neighborhood. Not only are these uncompacted fills loose, they have often contained various debris such as wood, rocks, and miscellaneous trash. However, none of these deleterious materials is present at the ground surface at this site.

Plan a site investigation program for this project and present your plan in the form of written instructions to your field crew. This plan should include specific instructions regarding what to do, where to do it, and any special instructions. You should presume that the field crew is experienced in soil investigation work, but is completely unfamiliar with this site.

4.8 Discuss the advantages of the cone penetration test over the standard penetration test.

4.9 The following data were obtained from direct shear tests on a series of 60 mm diameter samples of a certain soil:

Test No.	Effective Stress at Failure (kPa)	Shear Strength (kPa)
1	75.0	51.2
2	150.0	82.7
3	225.0	110.1

Find the values of c and ϕ.

4.10 A standard penetration test was performed in a 150 mm diameter boring at a depth of 9.5 m below the ground surface. The driller used a UK style automatic trip hammer and a standard SPT sampler. The actual blow count, N, was 19. The soil is a normally consolidated fine sand with a unit weight of 18.0 kN/m^3. The groundwater table is at a depth of 15 m.

Compute the following:

   a. N_{60}
   b. N_{60}'
   c. D_r
   d. Consistency (based on Table 3.4)
   e. ϕ

4.11 Using the cone penetration test data in Figure 4.22, a unit weight of 115 lb/ft^3, and an overconsolidation ratio of 3, compute the following for the soil between depths of 42 and 45 ft. Use a groundwater depth of 15 ft below the ground surface.

   a. Soil classification
   b. D_r (assume the soil has some fines, but no mica)
   c. Consistency (based on Table 3.4)
   d. ϕ
   e. M
   f. N_{60} (use an estimated D_{50} of 0.20 mm)

Part B

Shallow Foundation Analysis and Design

5

Shallow Foundations

> *The most important thing is*
> *to keep the most important thing*
> *the most important thing*

Shallow foundations are those that transmit structural loads to the near-surface soils. These include *spread footing foundations* and *mat foundations*.

5.1 SPREAD FOOTING FOUNDATIONS

A spread footing (also known as a *footer* or simply a *footing*) is an enlargement at the bottom of a column or bearing wall that spreads the applied structural loads over a sufficiently large soil area. Typically, each column and each bearing wall has its own spread footing, so each structure may include dozens of individual footings.

Spread footings are by far the most common type of foundation, primarily because of their low cost and ease of construction. They are most often used in small- to medium-size structures on sites with moderate to good soil conditions.

Spread footings may be built in different shapes and sizes to accommodate individual needs, as shown in Figure 5.1. These include the following:

- *Square spread footings* (or simply *square footings*) have plan dimensions of $B \times B$. The depth from the ground surface to the bottom of the footing is D and the thickness is T. Square footings usually support a single centrally-located column.

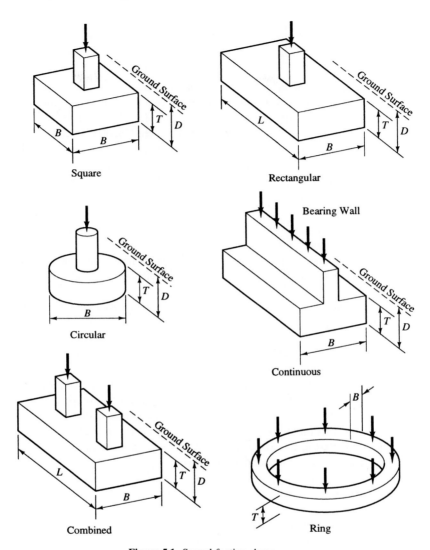

Figure 5.1 Spread footing shapes.

- *Rectangular spread footings* have plan dimensions of $B \times L$, where L is the longest dimension. These are useful when obstructions prevent construction of a square footing with a sufficiently large base area and when large moment loads are present.
- *Circular spread footings* are round in plan view. These are most frequently used as foundations for light standards, flagpoles, and power transmission lines. If these foundations extend to a large depth (i.e., D/B greater than about 3), they may behave more like a deep foundation (see Chapter 15).

    Large cylindrical above-ground storage tanks also will behave as a circular foundation. Although the walls may be supported on an annular-shaped continuous footing (a *ring footing*) and the roof may be supported by columns founded on square footings, the contents of the tank will be distributed evenly across the tank floor.
- *Continuous spread footings* (also known as *wall footings* or *strip footings*) are used to support bearing walls.
- *Combined footings* are those that support more than one column. These are useful when columns are located too close together for each to have its own footing.

Sometimes it is necessary to build spread footings very close to a property line, another structure, or some other place where no construction may occur beyond one or more of the exterior walls. This circumstance is shown in Figure 5.2. Because such a footing cannot be centered beneath the column, the load is eccentric. This can cause the footing to rotate and thus produce undesirable moments and displacements in the column.

One solution to this problem is to use a *strap footing* (also known as a *cantilever footing*), which consists of an eccentrically loaded footing under the exterior column connected to the first interior column using a *grade beam*. This arrangement, which is similar to a combined footing, provides the necessary moment in the exterior footing to counter the eccentric load. Sometimes grade beams connect all of the spread footings to provide a more rigid foundation system.

Materials

Before the mid nineteenth century, almost all spread footings were made of masonry, as shown in Figure 5.3. *Dimension-stone footings* were built of stones cut and dressed to specific sizes and fit together with minimal gaps, while *rubble-stone footings* were built from random size material joined with mortar (Peck et al., 1974). These footings had very little tensile strength, so builders had to use large height-to-width ratios to keep the flexural stresses tolerably small and thus avoid tensile failures.

Although masonry footings were satisfactory for small structures, they became large and heavy when used in heavier structures, often encroaching into the basement. For example, the masonry footings beneath the nine-story Home Insurance Building in Chicago (built in 1885) had a combined weight equal to that of one of the stories (Peck, 1948). As these larger structures became more common, it was necessary to develop footings that were shorter and lighter, yet still had the same base dimensions. This required structural materials that could sustain flexural stresses.

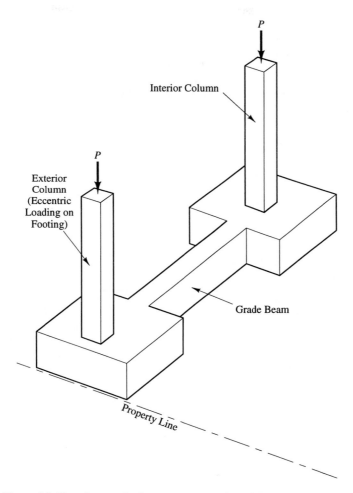

Figure 5.2 Use of a strap footing to support exterior columns when construction cannot occur beyond the wall.

The *steel grillage*, as used in the 10-story Montauk Block Building in Chicago in 1882, may have been the first spread footings designed to resist flexure. They included several layers of railroad tracks, as shown in Figure 5.4. The flexural strength of the steel permitted construction of a short and lightweight footing. Steel grillage footings, modified to use I-beams instead of railroad tracks, soon became the dominant design. They prevailed until the advent of reinforced concrete in the early twentieth century.

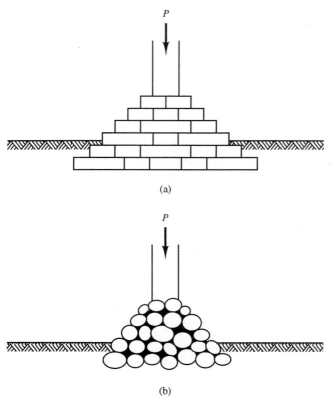

(a)

(b)

Figure 5.3 (a) Dimension-stone footing; and (b) Rubble-stone footings.

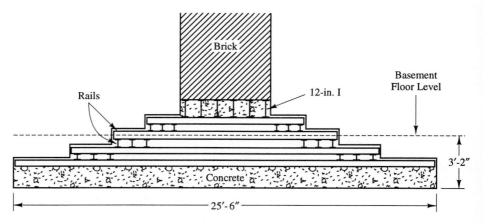

Figure 5.4 Steel grillage footing made from railroad tracks, Montauk Block Building, Chicago, 1882. The concrete that surrounded the steel was for corrosion protection only (Peck, 1948).

Figure 5.5 shows a typical reinforced concrete footing. These are very strong, economical, durable, and easy to build. Reinforced concrete footings also can be very thin and thus do not require large excavations and do not intrude into basements. Thus, nearly all spread footings are now made of reinforced concrete.

Construction Methods

Contractors usually use a *backhoe* to excavate spread footings, as shown in Figure 5.6. Typically, some hand work is also necessary to produce a clean excavation. Once the excavation is open, it is important to check the exposed soils to verify that they are comparable to those used in the design. Inspectors often check the firmness of these soils using a 0.5 in (12 mm) diameter steel probe. If the soil type is not as anticipated, or if it is not sufficiently firm, it may be necessary to revise the design accordingly.

Shallow wood forms are placed above the excavation, as shown in Figure 5.7a, so the top of the footing is at the proper elevation and has the proper width. Most soils have sufficient strength to stand vertically until it is time to pour the concrete. This procedure of pouring the concrete directly against the soil is known as pouring a *neat footing*.

If the soil will not stand vertically, such as with clean sands or gravels, it is necessary to make a larger excavation and build a full-depth wooden form, as shown in Figure 5.7b. This is known as a *formed footing*.

With either method, the contractor then places the reinforcing steel, pours the concrete, and then removes the wooden forms.

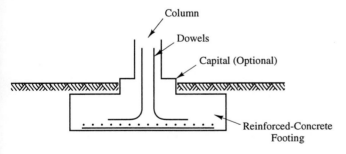

Reinforced-Concrete Footing

Figure 5.5 Reinforced concrete spread footing.

5.2 MAT FOUNDATIONS

As the applied loads become larger or the soil becomes weaker and more compressible, the size of spread footings must correspondingly increase. If the footings become so large that they cover more than about half of the building footprint, then some other type of foundation will probably be more economical. The alternatives are a mat foundation or some type of deep foundation.

Figure 5.6 A backhoe making an excavation for a spread footing.

A mat foundation (also known as a *raft foundation*) is a very large spread footing that supports most or all of the structure, as shown in Figure 5.8. This design has the following advantages over individual spread footings:

- Spreads the structural load over a larger area, thus reducing the bearing pressure.
- Provides much more structural rigidity and thus reduces the potential for excessive differential settlements.
- Is easier to waterproof.
- Has a greater weight and thus is able to resist greater uplift loads.
- Distributes lateral loads into the soil more evenly and efficiently.

Chapter 9 describes mat foundations in more detail.

5.3 SHALLOW VS. DEEP FOUNDATIONS

Shallow foundations often cannot support large structures or those on poor soils. Then the engineer must consider a deep foundation that transfers some or all of the structural loads to deeper soil strata. Deep foundations are also appropriate when the upper soils are subject to scour, such as beneath a bridge pier.

Soils usually become better with depth, so most deep foundations can support large loads. However, they are also more expensive to build. Chapters 11 to 18 describe deep foundations in more detail.

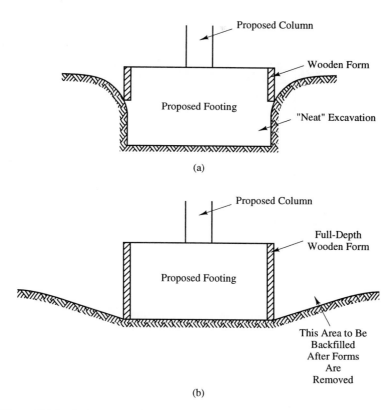

(a)

(b)

Figure 5.7 Methods of forming footings: (a) neat excavation with nominal wooden form at the top; and (b) Formed footing with full-depth wooden form.

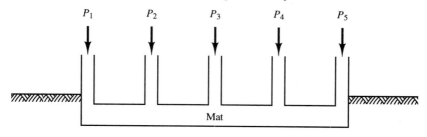

Figure 5.8 A mat foundation.

6

Spread Footings –
Bearing Capacity

The problems of soil mechanics may be divided into two principal
groups — the stability problems and the elasticity problems.
Karl Terzaghi, 1943

The construction and loading of a spread footing generates both normal and shear stresses in the ground. This introduces two concerns:

1. Will the shear stresses exceed the shear strength of the soil, resulting in a *bearing capacity failure*? This is what Terzaghi calls a stability problem.
2. Will the normal strains induced by the increase in normal stress cause the footing to settle excessively? Terzaghi would call this an elasticity problem.

Geotechnical engineers address these concerns by conducting site investigations and soil tests as described in Chapter 4 and using the results in engineering analyses. This chapter describes analysis techniques for evaluating bearing capacity, Chapter 7 describes those for settlement, and Chapter 8 describes methods of synthesizing the results.

6.1 BEARING PRESSURE

The most fundamental parameter that defines the interface between a footing and the soil that supports it is the *bearing pressure*. This is the contact force per unit area along the bottom of the footing. Engineers recognized the importance of bearing pressure during the nineteenth century, thus forming the basis for later developments in bearing capacity and settlement theories.

Distribution of Bearing Pressure

Although the integral of the bearing pressure across the base area of the footing must be equal to the force acting between the footing and the soil, this pressure is not necessarily distributed evenly. Theoretical studies and field measurements (Schultze, 1961; Dempsey and Li, 1989; and others) indicate that the actual distribution depends on several factors, including the following:

- Eccentricity, if any, of the applied load
- Magnitude of the applied moment, if any
- Structural rigidity of the footing
- Stress-strain properties of the soil
- Roughness of the bottom of the footing

Figure 6.1 shows the distribution of bearing pressure along the base of spread footings subjected to concentric vertical loads. Perfectly flexible footings bend as necessary to maintain a uniform bearing pressure, as shown in Figures 6.1a and 6.1b, whereas perfectly rigid footings settle uniformly but have variations in the bearing pressure, as shown in Figures 6.1c and 6.1d.

Real footings are close to being perfectly rigid, so the bearing pressure distribution is not uniform. However, bearing capacity and settlement analyses based on such a distribution would be very complex, so it is customary to assume that the pressure beneath concentric vertical loads is uniform across the base of the footing, as shown in Figure 6.1e. The error introduced by this simplification is not significant.

The analysis becomes slightly more complex if the footing is subjected to eccentric and/or moment loads. In these cases, the bearing pressure is biased toward one side, as shown in Figure 6.2a. Once again, it is customary to simplify the pressure distribution, as shown in Figure 6.2b.

When analyzing spread footings, it is customary and reasonable to neglect any sliding friction along the sides of the footing and assume that the entire load is transmitted to the bottom. This is an important analytical difference between footings and deep foundations and will be explored in more detail in Chapters 11 to 16.

Gross vs. Net Bearing Pressure

Geotechnical engineers have two ways of defining bearing pressure: *gross bearing pressure* and *net bearing pressure*. Unfortunately, it is not always clear which is being used and this has often been a point of confusion.

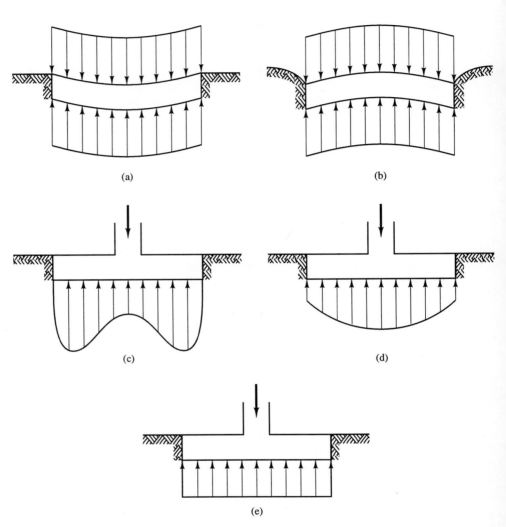

Figure 6.1 Distribution of bearing pressure along the base of a spread footings subjected to concentric vertical loads: (a) flexible footing on clay, (b) flexible footing on sand, (c) rigid footing on clay, (d) rigid footing on sand, and (e) simplified distribution (after Taylor, 1948)

The gross bearing pressure, q, is the actual contact pressure between the bottom of a spread footing and the soil below. For concentric vertical loads, the formula for q is:

$$q = \frac{P + W_f}{A} \tag{6.1}$$

Where:

   P = applied normal load

   W_f = weight of footing (includes weight of any soil above the footing)

   A = base area of footing

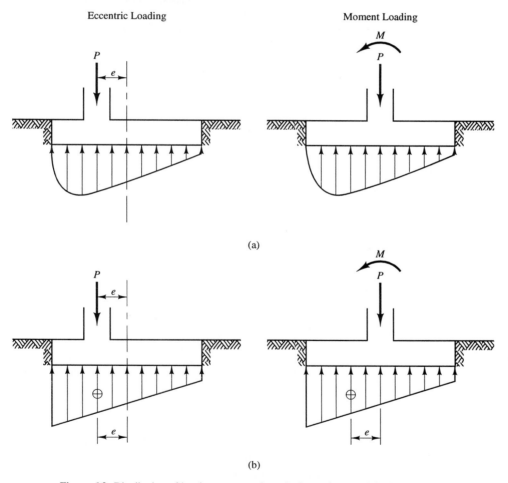

Figure 6.2 Distribution of bearing pressure along the base of a spread footing subjected to eccentric and/or moment loads: (a) actual distribution, and (b) simplified distribution.

Conversely, the net bearing pressure (or *effective bearing pressure*), q', is the difference between the gross bearing pressure and the effective stress, σ_0', that was present in the soil immediately below the proposed footing before it was built:

$$q' = q - \sigma_0'$$

$$= \frac{P + W_f}{A} - \sigma_0' \tag{6.2}$$

In other words, q is a measure of the stress in the soil immediately below the completed footing, and q' is the portion of this stress that is due to the construction of the footing and the application of the structural loads. The latter is more useful for foundation analysis and design, so we will use it whenever possible.

If we make the simplifying assumption that the weight of the footing equals the weight of the soil excavated to build the footing, then:

$$\sigma_0' = \frac{W_f}{A} \tag{6.3}$$

Therefore:

$$q' = \frac{P}{A} \tag{6.4}$$

For continuous footings, express the applied normal load as the force per unit length, P/b, perhaps using units of k/ft or kN/m. Equations 6.1 and 6.4, respectively, then become:

$$q = \frac{P/b + W_f/b}{B} \tag{6.5}$$

$$q' = \frac{P/b}{B} \tag{6.6}$$

Where:

b = unit length of continuous footing (usually 1 ft or 1 m)

P/b = normal load per unit length of footing

W_f/b = weight of footing per unit length

Example 6.1

A 5 ft square footing supports a column load of 100 k. The bottom of the footing is at a depth of 2 ft below the ground surface. Compute the gross and net bearing pressures.

Solution:

Use 150 lb/ft^3 for the unit weight of concrete.

$$W_f = (5 \text{ ft})(5 \text{ ft})(2 \text{ ft})(150 \text{ lb/ft}^3) = 7500 \text{ lb}$$

Note: We have assumed that the concrete extends to the ground surface. Some footings are covered with soil, in which case W_f should equal the weight of the footing plus the soil cover.

$$A = (5 \text{ ft})(5 \text{ ft}) = 25 \text{ ft}^2$$

Using Equation 6.1:

$$q = \frac{P + W_f}{A} = \frac{100,000 \text{ lb} + 7500 \text{ lb}}{25 \text{ ft}^2} = \textbf{4300 lb/ft}^2 \qquad \Leftarrow \textit{Answer}$$

Using Equation 6.4:

$$q' = \frac{P}{A} = \frac{100,000 \text{ lb}}{25 \text{ ft}^2} = \textbf{4000 lb/ft}^2 \qquad \Leftarrow \textit{Answer}$$

Example 6.2

A 0.70 m wide continuous footing supports a wall load of 110 kN/m. The bottom of this footing is at a depth of 0.50 m below the adjacent ground surface and the soil has a unit weight of 17.5 kN/m^3. Compute the gross and net bearing pressures.

Solution:

Use 23.6 kN/m^3 for the unit weight of concrete.

$$W_f/b = (0.70 \text{ m})(0.50 \text{ m})(23.6 \text{ kN/m}^3) = 8 \text{ kN/m}$$

Using Equation 6.5:

$$q = \frac{P/b + W_f/b}{B} = \frac{110 \text{ kN/m} + 8 \text{ kN/m}}{0.70 \, m} = \textbf{169 kPa} \qquad \Leftarrow \textit{Answer}$$

Using Equation 6.6:

$$q' = \frac{P/b}{B} = \frac{110 \text{ kN/m}}{0.70 \text{ m}} = \textbf{157 kPa} \qquad \Leftarrow Answer$$

Footings with Eccentric or Moment Loads

Some footings must carry eccentric or moment loads. These loads may be permanent, such as with retaining walls, or they may be temporary, such as with wind or seismic loads. When such loads are present, the bearing pressure is higher at one end of the footing than the other, as shown in Figure 6.3.

We may express either of these loading conditions in terms of the eccentricity, e, of the resultant force that acts on the base of the footing, as shown in Figure 6.3. If the applied load is eccentric, then the resultant at the base acts immediately below the applied load and both have the same eccentricity. For moment loads on square or rectangular footings, e is:

$$e = \frac{M}{P} \tag{6.7}$$

Where:

M = applied moment

P = applied normal load

If $e \leq B/6$ and the eccentricity or applied moment is in the plane of the B dimension only, the minimum and maximum net bearing pressures on a square, circular or rectangular footings are:

$$q'_{min} = \frac{P}{A}\left(1 - \frac{6e}{B}\right) \tag{6.8}$$

$$q'_{max} = \frac{P}{A}\left(1 + \frac{6e}{B}\right) \tag{6.9}$$

Where:

q'_{min} = minimum net bearing pressure

q'_{max} = maximum net bearing pressure

A = base area of footing

If $e < B/6$, the bearing pressure distribution has a trapezoidal shape, as shown in Figure 6.3a. If $e = B/6$ (i.e., the resultant force acts at the third-point of the footing),

q'_{min} equals 0 and the bearing pressure distribution is triangular as shown in Figure 6.3b. Either of these conditions is normally acceptable because compressive stresses are present along the entire base of the footing.

However, if $e > B/6$, the resultant force at the base is outside the third-point and the pressure distribution is as shown in Figure 6.3c. There can be no tension between the footing and the soil, so the heel of the footing lifts off the soil. In addition, the high bearing pressure at the toe may cause a large settlement there. The net result is an excessive tilting of the footing, which is not desirable. Therefore, it is accepted practice to design footings so that $e \leq B/6$.

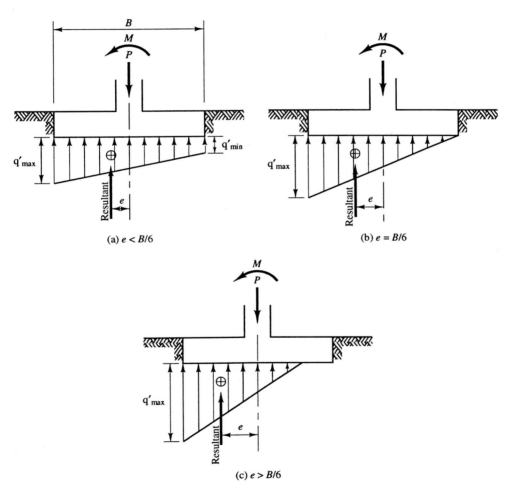

Figure 6.3 Distribution of bearing pressure beneath footings with various eccentricities: (a) $e < B/6$, (b) $e = B/6$, and (c) $e > B/6$.

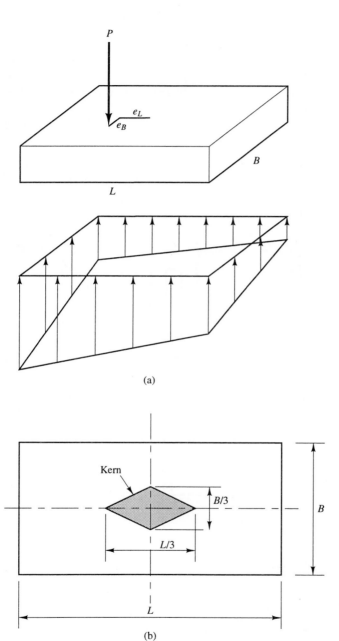

(a)

(b)

Figure 6.4 (a) Pressure distribution beneath spread footing with vertical load that is eccentric in both the B and L directions. (b) To maintain $q' \geq 0$ along the entire base of the footing, the resultant force must be located within this diamond-shaped kern .

The maximum net bearing pressure, q'_{max}, when $B/6 < e < B/2$ is:

$$q'_{max} = \frac{2P}{3A(0.5 - e/B)} \tag{6.10}$$

For continuous footings, use Equations 6.7 - 6.10, but substitute P/b and M/b for P and M, respectively, and substitute B for A.

If the resultant load acting on the base is eccentric in both the B and L directions, it must fall within the diamond-shaped kern shown in Figure 6.4 for the contact pressure to be compressive along the entire base of the footing. It falls within the kern if the following statement is true:

$$\frac{6\,e_B}{B} + \frac{6\,e_L}{L} \le 1.0 \tag{6.11}$$

If Equation 6.11 is true, the magnitude of q' at the four corners is:

$$q' = \frac{P}{A}\left(1 \pm \frac{6\,e_B}{B} \pm \frac{6\,e_L}{L}\right) \tag{6.12}$$

Where:
e_B = eccentricity in the B direction
e_L = eccentricity in the L direction

Example 6.3

A 5.00 ft wide continuous footing is subjected to a concentric vertical load of 12.0 k/ft and a moment load of 8.0 ft-k/ft acting laterally across the footing, as shown in the figure. Determine whether the resultant force on the base of the footing acts within the middle third and compute the maximum and minimum net bearing pressures.

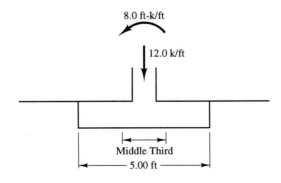

Solution:

Using a modification of Equation 6.7:

$$e = \frac{M/b}{P/b} = \frac{8.0 \text{ ft-k/ft}}{12.0 \text{ k/ft}} = 0.667 \text{ ft}$$

$$\frac{B}{6} = \frac{5 \text{ ft}}{6} = 0.833 \text{ ft}$$

$e < B/6$;   therefore **the resultant is in the middle third**   ⇐ *Answer*

Using modifications of Equations 6.8 and 6.9:

$$q'_{min} = \frac{P/b}{B}\left(1 - \frac{6e}{B}\right) = \frac{12,000}{5.0}\left(1 - \frac{(6)(0.667)}{5.0}\right) = 480 \text{ lb/ft}^2 \quad ⇐ Answer$$

$$q'_{max} = \frac{P/b}{B}\left(1 + \frac{6e}{B}\right) = \frac{12,000}{5.0}\left(1 + \frac{(6)(0.667)}{5.0}\right) = 4320 \text{ lb/ft}^2 \quad ⇐ Answer$$

When designing combined footings, try to arrange the footing dimensions and column locations so the resultant of the applied loads acts through the centroid of the footing. This produces a uniform bearing pressure distribution. Some combined footing designs accomplish this by using a trapezoidal shaped footing (as seen in plan view) with the more lightly loaded column on the narrow side of the trapezoid.

6.2 PRESUMPTIVE BEARING PRESSURES

One of the first developments in modern foundation engineering occurred in the 1870s when engineers recognized the importance of bearing pressures and developed *allowable bearing pressures*. This was an attempt to control problems with excessive settlements that were common at the time (Terzaghi and Peck, 1967). Theories of bearing capacity and settlement were not yet available, so these allowable bearing pressures were based on experience and were generally expressed as a function of the soil type. We now call these *presumptive bearing pressures* (also known as *prescriptive bearing pressures*) because the engineer must presume a design value based on very little information. Once the allowable pressure was determined, the engineer would simply size the footings such that the bearing pressure did not exceed the allowable value.

The concept of presumptive bearing pressures was a natural development from structural engineering. It seemed reasonable to tabulate allowable bearing pressures for various types of soil, as had been done for steel, wood, and other engineering materials. As a result, by the 1920's most building codes included tables of allowable bearing pressures and engineers routinely used them to design spread footings.

Although the use of the allowable bearing pressures listed in building codes usually produced satisfactory designs, some structures still had problems with excessive settlement. Engineers attributed these problems to improper soil classification that resulted in the use of an incorrect allowable bearing pressure. However, these problems also initiated some of the early research in the new discipline of soil mechanics.

TABLE 6.1 PRESUMPTIVE BEARING PRESSURES FROM VARIOUS BUILDING CODES[a]

Soil or Rock Classification	Allowable Bearing Pressure, lb/ft^2 (kPa)		
	Uniform Code[b] (ICBO, 1991a)	Basic Code (BOCA, 1990)	Canadian Code (NRCC, 1990)
Massive crystalline bedrock	4,000 - 12,000 (200 - 600)	12,000 (600)	40,000 - 200,000 (2,000 - 10,000)
Sedimentary or foliated rock	2,000 - 6,000 (100 - 300)	6,000 (300)	10,000 - 80,000 (500 - 4000)
Sandy gravel or gravel	2,000 - 6,000 (100 - 300)	5,000 (250)	4,000 - 12,000 (200 - 600)
Sand, silty sand, clayey sand, silty gravel, or clayey gravel	1,500 - 4,500 (75 - 225)	3,000 (150)	2,000 - 8,000 (100 - 400)
Clay, sandy clay, silty clay, or clayey silt	1,000 - 3,000 (50 - 150)	2,000 (100)	1,000 - 12,000 (50 - 600)

ICBO values reproduced from the 1991 edition of the *Uniform Building Code*, © 1991, with permission of the publisher, the International Conference of Building Officials.

[a] The values in this table are for illustrative purposes only and are not a complete description of the code provisions. Portions of the table include the author's interpretations to classify the presumptive bearing values into uniform soil groups. Refer to the individual codes for more details.

[b] The Uniform Building Code values in soil are intended to provide a factor of safety of at least 3 against a bearing capacity failure, and a total settlement of no more than 0.5 in (12 mm) (ICBO, 1991c).

Our understanding of the behavior of soils and foundations is much better now than it was then, and we are able to develop more reliable allowable bearing pressures based on soil tests and analyses. These newer methods, described in the remainder of this chapter, are especially useful for heavily loaded footings. Nevertheless, presumptive bearing pressure tables continue to appear in many building codes and are still useful in some circumstances. Usually, these values are very conservative and may be used only for small structures, say, those with column loads less than about 50 k (200 kN), that are at sites known to have good soils. When limited to such circumstances, they are appropriate. However, the use of presumptive bearing pressures is inappropriate for larger structures or those on marginal or poor soils.

Table 6.1 shows presumptive bearing pressures from three widely used building codes. Notice how the values for a given soil often vary considerably from one code to another. Some of these discrepancies are due to differing degrees of conservatism, and others reflect the specific subsurface conditions in the area where the code is used.

6.3 PLATE LOAD TESTS

The problems with presumptive bearing values caused engineers to develop a new method: the *plate load test*. This approach consisted of making an excavation to the depth of the proposed footings, temporarily placing a 1 ft (305 mm) square plate on the base of the excavation and loading it to obtain in-situ load-settlement data. Some engineers carried the test to failure, and then designed the footing using a bearing pressure equal to half the pressure at failure. Others used the bearing pressure that corresponded to some specific settlement of the plate, such as 0.5 in or 1.0 cm.

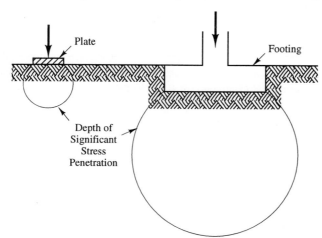

Figure 6.5 The stresses induced by a plate load test do not penetrate very deep into the soil, so its load-settlement behavior is not the same as that of a full-size footing.

Although this may seem to be a reasonable approach, experience has proven otherwise. Its most fundamental flaw is that the depth of influence (about $2B$ to $4B$) is much shallower than that of the real footing, as shown in Figure 6.5, so the test reflects only the properties of the near-surface soils. This can introduce large errors, and several complete foundation failures occurred in spite of the use of plate load tests (Terzaghi and Peck, 1967).

Although some attempts have been made to account for the size differences between the plate and the footing, the test still reflects only the properties of the nearby soils and can still be very misleading. For example, D'Appolonia et al. (1968) conducted a series of plate load tests in northern Indiana and found that, even after adjusting the test results for scale effects, the plate load tests underestimated the actual settlements by an average of a factor of 2. The test for a certain 12 ft wide footing at the site was in error by a factor of 3.2.

Fortunately, the development of bearing capacity and settlement theories and test methods to obtain the needed soil properties have eliminated the need for plate load tests. The remainder of this chapter concentrates on these methods.

6.4 BEARING CAPACITY FAILURES

The magnitudes of the shear stresses in the soil beneath a footing depend largely on the net bearing pressure and the size of the footing. If the bearing pressure is large enough, or the footing is small enough, these shear stresses may exceed the shear strength of the soil, resulting in a *bearing capacity failure*.

Researchers have identified three types of bearing capacity failures: a *general shear failure*, a *local shear failure*, and a *punching shear failure*, as shown in Figure 6.6. A typical load-displacement curve for each is shown in Figure 6.7.

The general shear failure is the most common mode. It occurs in soils that are relatively incompressible and reasonably strong, or in saturated, normally consolidated clays that are loaded rapidly enough that the undrained strength prevails. The failure surface is well defined and failure occurs quite suddenly, as illustrated by the load-displacement curve. A clearly formed bulge appears on the ground surface adjacent to the foundation. Although these bulges may appear on both sides of the footing, ultimate failure occurs on one side only, accompanied by tilting of the footing.

The opposite extreme is the punching shear failure. It occurs in very loose sands, in a thin crust of strong soil underlain by a very weak soil, or in weak clays loaded under slow, drained conditions. The high compressibility of such soil profiles causes large settlements and poorly defined vertical shear surfaces. Little or no bulging occurs at the ground surface and failure develops gradually, as illustrated by the ever-increasing load-settlement curve.

The local shear failure is an intermediate case. The shear surfaces are well defined under the footing, and then become vague near the ground surface. A small bulge may occur, but considerable settlement, perhaps on the order of half the footing width, is necessary before a clear shear surface forms near the ground. Even then, a sudden failure does not occur, as happens in the general shear case. The footing just continues to sink ever deeper into the ground.

Vesić (1973) reported the results of load tests of model footings in sand, as shown in Figure 6.8. Although these test results apply only to Vesić's sand and cannot necessarily be generalized to other soils, it does give a general relationship between the mode of failure, relative density, and the D/B ratio. It shows how spread footings (small D/B) on sands can fail in any of the three modes, depending on the relative density. However, deep foundations (large D/B) are always governed by punching shear.

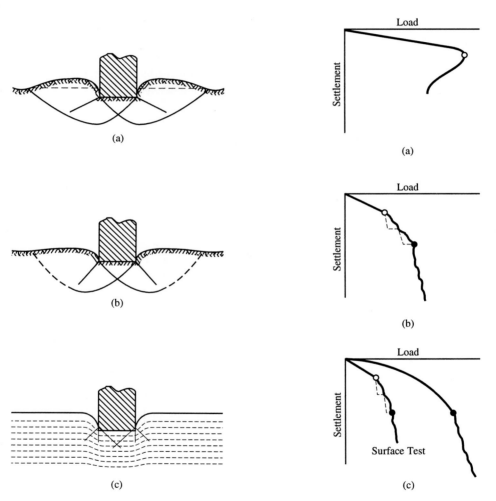

Figure 6.6 Modes of bearing capacity failures beneath spread footings: (a) general shear failure, (b) local shear failure, and (c) punching shear failure (Adapted from Vesić, 1963).

Figure 6.7 Typical load-displacement curves for different modes of bearing capacity failure: (a) general shear failure, (b) local shear failure, and (c) punching shear failure (Adapted from Vesić, 1963).

Complete quantitative criteria have yet to be developed to determine which of these three modes of failure will govern in any given circumstance, but the following guidelines may be helpful:

- Footings in clays are governed by the general shear case.
- Footings in dense sands are governed by the general shear case. In this context, a dense sand is one with a relative density, D_r, greater than about 67%.
- Footings on loose to medium dense sands ($30\% < D_r < 67\%$) are probably governed by local shear.
- Footings on very loose sand ($D_r < 30\%$) are probably governed by punching shear.

For most practical design problems, it is only necessary to check the general shear case, and then conduct settlement analyses to verify that the footing will not settle excessively. These settlement analyses implicitly protect against local and punching shear failures.

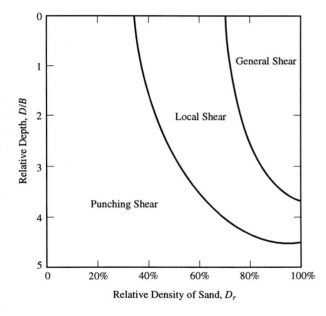

Figure 6.8 Modes of failure of model footings in Chattahoochee Sand (Adapted from Vesić, 1963 and 1973).

6.5 BEARING CAPACITY ANALYSES — GENERAL SHEAR CASE

Methods of Analyzing Bearing Capacity

To analyze spread footings for bearing capacity failures and design in a way to avoid such failures, we must understand the relationship between bearing capacity, load, footing dimensions, and soil properties. Various researchers have studied these relationships using a variety of techniques, including:

- Theoretical stress analyses
- Assessments of the performance of real foundations
- Full-scale load tests
- Load tests on model footings
- Finite element method (FEM) analyses

   Model footing tests have been used quite extensively, mostly because the cost of these tests is far below that for full-scale tests. Unfortunately, model tests have their limitations, especially when conducted in sands, because of uncertainties in applying the proper scaling factors. However, the advent of centrifuge model tests has partially overcome this problem.

   In spite of extensive work in this area, no general solution has yet been found that completely satisfies the laws of statics. However, a variety of methods have been proposed, all of which include simplifying assumptions, so we are now able to analyze the bearing capacity of spread footings with an accuracy that is more than sufficient for nearly all practical problems.

   Although Rankine briefly addressed this subject as early as 1857, most modern methods of bearing capacity analysis have their root in Prandtl's studies of the punching resistance of metals (Prandtl, 1920). He considered the ability of very thick masses of metal (i.e., not sheet metal) to resist concentrated loads. If we apply soil engineering terminology to Prandtl's work, we would say that he considered only purely cohesive material ($\phi = 0$) with no unit weight ($\gamma = 0$). With these assumptions, he defined the shape of the shear zones and developed a method for determining the force required to punch through the metal.

   The shear strength of soils is more complex than that of metals and includes both cohesive and frictional components, as described in Chapter 3:

$$s = c + \sigma' \tan \phi$$

Therefore, to extend Prandtl's work to soils, we must consider both c and ϕ.

Terzaghi's Bearing Capacity Formulas

Various methods of computing bearing capacity of soils were advanced in the first half of the twentieth century, but the first one to achieve widespread acceptance was that of Terzaghi (1943). His method includes the following assumptions:

- The depth of the footing is less than or equal to its width ($D \leq B$).
- The bottom of the footing is sufficiently rough that no sliding occurs between the footing and the soil.

- The soil beneath the footing is a homogeneous semi-infinite mass (i.e., the soil extends for a great distance below the footing and the soil properties are uniform throughout).
- The shear strength of the soil is described by the formula $s = c + \sigma' \tan \phi$.
- The general shear mode of failure governs.
- No consolidation of the soil occurs (i.e., settlement of the footing is due only to the shearing and lateral movement of the soil).
- The footing is very rigid in comparison to the soil.

These assumptions are generally reasonable and conservative for analyses of general shear failure, although in some cases, it is difficult to model an erratic or stratified soil deposit with equivalent "homogeneous" soil parameters.

Terzaghi considered three zones in the soil, as shown in Figure 6.9. Immediately beneath the footing is a *wedge zone* that remains intact and moves downward with the footing. Next, a *radial shear zone* extends from each side of the wedge, where he took the shape of the shear planes to be logarithmic spirals. Finally, the outer portion is the *linear shear zone* in which the soil shears along planar surfaces.

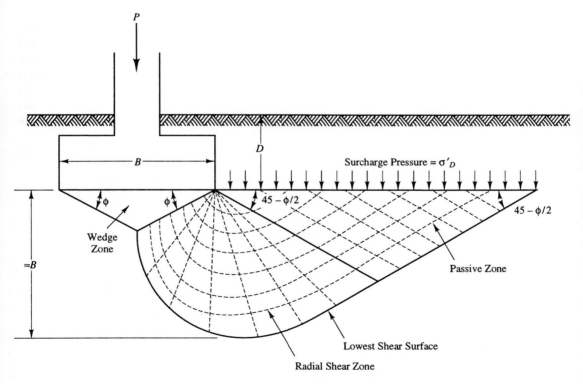

Figure 6.9 Physical model for Terzaghi bearing capacity formulas.

Terzaghi terminated the shear zones at a level even with the bottom of the footing. This means that he considered the soil from the ground surface to a depth D to be only a surcharge: It contributes weight (and therefore adds to the effective stress and strength in the soil below), but has no shear resistance. This assumption is conservative, and is part of the reason for limiting the method to relatively shallow footings ($D < B$).

Terzaghi developed his theory for continuous footings. This is the simplest case because it is a two-dimensional problem. He then extended it to square and round footings by adding empirical coefficients. These formulas, written in terms of net pressures,[1] are as follows:

For square footings:

$$q_u' = 1.3cN_c + \sigma_D'(N_q - 1) + 0.4\gamma B N_\gamma \qquad (6.13)$$

For continuous footings:

$$q_u' = cN_c + \sigma_D'(N_q - 1) + 0.5\gamma B N_\gamma \qquad (6.14)$$

For circular footings:

$$q_u' = 1.3cN_c + \sigma_D'(N_q - 1) + 0.3\gamma B N_\gamma \qquad (6.15)$$

Where:

q_u' = net ultimate bearing capacity

c = soil cohesion (use $c = s_u$ when analyzing undrained conditions)

σ_D' = effective stress at depth D below the ground surface
    ($\sigma_D' = \gamma D$ if depth to groundwater table is greater than D)

γ = soil unit weight

D = depth of footing below ground surface

B = width (or diameter) of footing

N_c, N_q, N_γ = bearing capacity factors $= f(\phi)$
    (use $\phi = 0$ when analyzing undrained conditions)

---

[1] Terzaghi, as well as most researchers, wrote his formulas in terms of gross pressure. However, for the sake of consistency with the rest of this chapter, they have been rewritten in terms of net pressure.

TABLE 6.2 BEARING CAPACITY FACTORS

ϕ (deg)	Terzaghi			Brinch Hansen		
	N_c	N_q	N_γ	N_c	N_q	N_γ
0	5.7	1.0	0.0	5.1	1.0	0.0
1	6.0	1.1	0.1	5.4	1.1	0.0
2	6.3	1.2	0.1	5.6	1.2	0.0
3	6.6	1.3	0.2	5.9	1.3	0.0
4	7.0	1.5	0.3	6.2	1.4	0.0
5	7.3	1.6	0.4	6.5	1.6	0.1
6	7.7	1.8	0.5	6.8	1.7	0.1
7	8.2	2.0	0.6	7.2	1.9	0.2
8	8.6	2.2	0.7	7.5	2.1	0.2
9	9.1	2.4	0.9	7.9	2.3	0.3
10	9.6	2.7	1.0	8.3	2.5	0.4
11	10.2	3.0	1.2	8.8	2.7	0.5
12	10.8	3.3	1.4	9.3	3.0	0.6
13	11.4	3.6	1.6	9.8	3.3	0.8
14	12.1	4.0	1.9	10.4	3.6	1.0
15	12.9	4.4	2.2	11.0	3.9	1.2
16	13.7	4.9	2.5	11.6	4.3	1.4
17	14.6	5.5	2.9	12.3	4.8	1.7
18	15.5	6.0	3.3	13.1	5.3	2.1
19	16.6	6.7	3.8	13.9	5.8	2.5
20	17.7	7.4	4.4	14.8	6.4	2.9
21	18.9	8.3	5.1	15.8	7.1	3.5
22	20.3	9.2	5.9	16.9	7.8	4.1
23	21.7	10.2	6.8	18.0	8.7	4.9
24	23.4	11.4	7.9	19.3	9.6	5.7
25	25.1	12.7	9.2	20.7	10.7	6.8
26	27.1	14.2	10.7	22.3	11.9	7.9
27	29.2	15.9	12.5	23.9	13.2	9.3
28	31.6	17.8	14.6	25.8	14.7	10.9
29	34.2	20.0	17.1	27.9	16.4	12.8
30	37.2	22.5	20.1	30.1	18.4	15.1
31	40.4	25.3	23.7	32.7	20.6	17.7
32	44.0	28.5	28.0	35.5	23.2	20.8
33	48.1	32.2	33.3	38.6	26.1	24.4
34	52.6	36.5	39.6	42.2	29.4	28.8
35	57.8	41.4	47.3	46.1	33.3	33.9
36	63.5	47.2	56.7	50.6	37.8	40.1
37	70.1	53.8	68.1	55.6	42.9	47.4
38	77.5	61.5	82.3	61.4	48.9	56.2
39	86.0	70.6	99.8	67.9	56.0	66.8
40	95.7	81.3	121.5	75.3	64.2	79.5

Terzaghi's bearing capacity factors are:

$$N_q = \frac{a_\theta^2}{2\cos^2(45 + \phi/2)} \tag{6.16}$$

$$a_\theta = e^{\pi(0.75 - \phi/360)\tan\phi} \tag{6.17}$$

$$N_c = 5.7 \qquad\qquad \text{for } \phi = 0 \tag{6.18}$$

$$N_c = \frac{N_q - 1}{\tan\phi} \qquad\qquad \text{for } \phi > 0 \tag{6.19}$$

$$N_\gamma = \frac{\tan\phi}{2}\left(\frac{K_{p\gamma}}{\cos^2\phi} - 1\right) \tag{6.20}$$

These bearing capacity factors are also presented in tabular form in Table 6.2.

Terzaghi did not clearly describe the mathematical function for $K_{p\gamma}$. However, he did provide plots of N_c, N_q, and N_γ as a function of ϕ, so this shortcoming was of little practical concern until the age of computers and the desire to program these functions. Therefore, instead of Equation 6.20, we may use the following formula to compute N_γ:

$$N_\gamma \approx \frac{2(N_q + 1)\tan\phi}{1 + 0.4\sin(4\phi)} \tag{6.21}$$

The author developed Equation 6.21 by fitting a curve to match Terzaghi's. It produces N_γ values within about 10 percent of Terzaghi's values. Alternatively, Kumbhojkar (1993) provides a more precise, but more complex, formula for N_γ.

Example 6.4

A 3 ft 3 in square footing is supported by a soil with the following properties: $c = 150$ lb/ft², $\phi = 30°$, and $\gamma = 121$ lb/ft³. The bottom of this footing is at a depth of 2 ft 0 in below the ground surface. The groundwater table is at a great depth. Compute the net ultimate bearing capacity and the ultimate column load.

Solution:

For $\phi = 30°$: $N_c = 37.2$, $N_q = 22.5$, $N_\gamma = 20.1$ (from Table 6.2)

The groundwater table is deep, therefore:

$$\sigma_D' = \gamma D$$

Using Equation 6.13:

$$
\begin{aligned}
q_u' &= 1.3 c N_c + \gamma D (N_q - 1) + 0.4 \gamma B N_\gamma \\
&= (1.3)(150)(37.2) + (121)(2)(22.5 - 1) + (0.4)(121)(3.25)(20.1) \\
&= 7200 + 5200 + 3200 \\
&= \mathbf{15{,}600 \ lb/ft^2} \qquad\qquad\qquad \Leftarrow\!\textit{Answer}
\end{aligned}
$$

$$
\begin{aligned}
P_u &= q_u' B^2 \\
&= (15{,}600)(3.25)^2 \\
&= 165{,}000 \ lb \\
&= \mathbf{165 \ k} \qquad\qquad \Leftarrow\!\textit{Answer}
\end{aligned}
$$

According to this analysis, a column load of 165 k would cause a bearing capacity failure beneath this footing. Note that about half of this capacity comes from the first term in the bearing capacity formula and is therefore dependent on the cohesion of the soil. Since the cohesive strength is rather tenuous, it is prudent to use conservative values of c in bearing capacity analyses. In contrast, the frictional strength is comparatively "reliable" and does not need to be interpreted as conservatively.

Example 6.5

A continuous footing is 0.7 m wide and 0.4 m deep. The soil beneath this footing is an undrained clay with $s_u = 120$ kPa. Compute the net ultimate bearing capacity and the ultimate wall load.

Solution:

This clay is in the undrained condition, so its shear strength is defined by s_u. Since the friction angle under undrained conditions is zero, the Mohr-Coulomb strength formula, $s = c + \sigma' \tan \phi$, reduces to $s_u = c_u$ (the u subscript denotes undrained conditions). Therefore, we may use Terzaghi's formula with $c = 120$ kPa and $\phi = 0$.

For $\phi = 0$: $N_c = 5.7$, $N_q = 1$, $N_\gamma = 0$ (from Table 6.2)

Using Equation 6.14:

$$q_u' = cN_c + \sigma_D'(N_q - 1) + 0.5\gamma B N_\gamma$$
$$= (120)(5.7) + \sigma_D'(1-1) + 0.5\gamma B(0)$$
$$= 684 \text{ kPa} \qquad \Leftarrow \textit{Answer}$$

$$P_u/b = q_u'(B)$$
$$= (684)(0.7)$$
$$= 480 \text{ kN/m} \qquad \Leftarrow \textit{Answer}$$

Terzaghi's method is still often used, primarily because it is simple and familiar. However, it does not consider special cases, such as rectangular footings, inclined loads, or footings with large depth:width ratios. Many others have extended Terzaghi's work; some have developed more general solutions; others have concentrated on special cases. Most of them also are more accurate.

Brinch Hansen's Bearing Capacity Formulas

The topic of bearing capacity has spawned large amounts of research and numerous methods of analysis. Skempton (1951), Meyerhof (1953), Brinch Hansen (1961b), DeBeer and Ladanyi (1961), Meyerhof (1963), and many others have contributed.[2] The formulas developed in Denmark by Brinch Hansen (1970) reflect the theoretical and experimental findings from these and other sources and is an excellent alternative to Terzaghi. It produces more accurate bearing values and it applies to a much broader range of loading and geometry conditions. The primary disadvantage is its added complexity.

Brinch Hansen retained Terzaghi's basic format and added the following additional factors[3]:

s_c, s_q, s_γ = shape factors
d_c, d_q, d_γ = depth factors
i_c, i_q, i_γ = load inclination factors

---

[2] See Vesić (1973) for a bibliography.

[3] Brinch Hansen was not completely clear regarding the formulas for some of these factors. Those presented here are the author's interpretation.

b_c, b_q, b_γ = base inclination factors

g_c, g_q, g_γ = ground inclination factors

For soils with $\phi > 0$, the basic formula written in terms of net bearing pressure is:

$$q_u' = cN_c s_c d_c i_c b_c g_c + \sigma_D' (N_q s_q d_q i_q b_q g_q - 1) + 0.5\gamma B N_\gamma s_\gamma d_\gamma i_\gamma b_\gamma g_\gamma \quad (6.22)$$

For clays that fail in undrained conditions ($\phi = 0$ soils), Brinch Hansen felt it was more appropriate to use additive constants instead of factors, giving:

$$q_u' = 5.14 s_u (1 + s_c^a + d_c^a - i_c^a - b_c^a - g_c^a) - \sigma_D' \quad (6.23)$$

Terzaghi's formulas consider only vertical loads acting on a footing with a horizontal base with a level ground surface, whereas Brinch Hansen's inclination factors allow any or all of these to vary. The notation for these factors is shown in Figure 6.10.

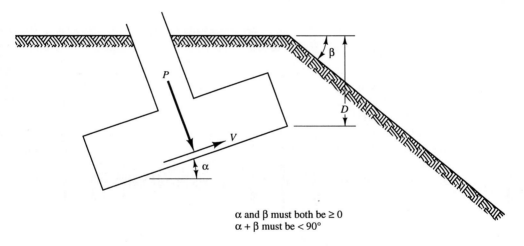

α and β must both be ≥ 0
α + β must be < 90°

Figure 6.10 Notation for Brinch Hansen's load inclination, base inclination and ground inclination factors. Note that θ is measured about the base of the footing and is zero when the load acts perpendicular to the base. All angles are expressed in degrees.

Shape Factors

Brinch Hansen also considered a broader range of footing shapes and defined them in his *s* factors.

There are three sets of shape factors: one to be used when the load is perpendicular to the base of the footing, one for loads inclined in the plane of the *B* dimension, and one for loads inclined in the plane of the *L* dimension. Usually, the first set is used, but if the load is inclined, square and rectangular footings tend to behave more like continuous footings and the shape factors must be adjusted accordingly. These factors are listed in Table 6.3.

TABLE 6.3 SHAPE FACTORS FOR BRINCH HANSEN'S BEARING CAPACITY FORMULAS

Shape Factor	Direction of Load		
	Perpendicular to Footing Base	Inclined in the Plane of the B Dimension	Inclined in the Plane of the L Dimension
s_c	$1 + 0.2\,(B/L)$	$1 + 0.2\,(B/L)\,i_{cB}$	$1 + 0.2\,(L/B)\,i_{cL}$
$s_c{}^a$	$0.2\,(B/L)$	$0.2\,(B/L)\,i_{cB}$	$0.2\,(L/B)\,i_{cL}$
s_q	$1 + (B/L)\sin\phi$	$1 + (B\,i_{qB}/L)\sin\phi$	$1 + (L\,i_{qL}/B)\sin\phi$
s_γ	$1 - 0.4\,(B/L)$	$1 - 0.4\,(Bi_{\gamma B}/Li_{\gamma L}) \geq 0.6$	$1 - 0.4\,(Li_{\gamma L}/Bi_{\gamma B}) \geq 0.6$

Exercise special care when using Brinch Hansen's formulas with rectangular footings that have the load inclined in the plane of the L dimension. In such cases, it is not immediately clear whether the bearing capacity failure would occur in the plane of the B dimension or in the plane of the L dimension, so check both possibilities. When checking for failure in the B direction, use the factors with the B subscript; for failure in the L direction, use those with the L subscript. Whichever gives the lowest bearing capacity is the one that controls.

For continuous footings ($B/L \rightarrow \infty$), s_c, s_q, and s_γ become equal to 1 and $s_c{}^a$ becomes equal to 0, regardless of the direction of the load. This means that the s factors can be ignored when analyzing continuous footings.

Depth Factors

Unlike Terzaghi, Brinch Hansen has no limitations on the depth of the footing. This method could even be used for deep foundations. The depth of the footing is considered in the following depth factors:

$$d_c = 1 + 0.4\,k \tag{6.24}$$

$$d_c{}^a = 0.4\,k \tag{6.25}$$

$$d_q = 1 + 2k \tan\phi \ (1 - \sin\phi)^2 \tag{6.26}$$

$$d_\gamma = 1 \tag{6.27}$$

For relatively shallow footings ($D/B \le 1$), use $k = D/B$. For deeper footings, use $k = \tan^{-1}(D/B)$ with the $\tan^{-1}$ term expressed in radians. Note that this will cause a discontinuity at $D/B = 1$.

Load Inclination Factors

The load inclination factors are for loads that do not act perpendicular to the base of the footing, but still act through its centroid (eccentric loads are discussed later in this chapter). The variable P refers to the component of the load that acts perpendicular to the bottom of the footing, and V refers to the component that acts parallel to the bottom. The load inclination factors are:

$$i_c = \sqrt{1 - \frac{V}{Ac}} \qquad \text{for } V/Ac < 1 \tag{6.28}$$

$$i_c = 0 \qquad \text{for } V/Ac \ge 1 \tag{6.29}$$

$$i_c^a = 0.5 - 0.5\sqrt{1 - \frac{V}{A s_u}} \qquad \text{for } V/Ac \le 1 \tag{6.30}$$

$$i_q = \left[1 - \frac{0.5V}{P + Ac \ \cot\phi}\right]^5 \ge 0 \tag{6.31}$$

$$i_\gamma = \left[1 - \frac{0.7V}{P + Ac \ \cot\phi}\right]^5 \ge 0 \tag{6.32}$$

Where:

A = base area of footing
c = cohesion
s_u = undrained shear strength

If the load acts perpendicular to the base of the footing, the i factors in may be neglected.

Base Inclination Factors

The vast majority of footings are built with horizontal bases. However, if the applied load is inclined at a large angle from the vertical, it may be better to incline the base of the footing to the same angle so the applied load acts perpendicular to the base.

The inclined base factors are:

$$b_c = 1 - \frac{\alpha}{147} \tag{6.33}$$

$$b_c{}^a = \frac{\alpha}{147} \tag{6.34}$$

$$b_q = e^{-0.0349\,\alpha\,\tan\phi} \tag{6.35}$$

$$b_\gamma = e^{-0.0349\,\alpha\,\tan\phi} \tag{6.36}$$

If the base of the footing is level, which is the usual case, all of the b factors may be ignored.

Ground Inclination Factors

Footings located near the top of a slope will have a lower bearing capacity than those on level ground. Brinch Hansen's ground inclination factors, presented below, account for this. However, there are also other matters to consider when placing footings near slopes. These are discussed in Chapter 8.

$$g_c = 1 - \frac{\beta}{147} \tag{6.37}$$

$$g_c{}^a = \frac{\beta}{147} \tag{6.38}$$

$$g_q = g_\gamma = \left[1 - 0.5 \tan\beta\right]^5 \tag{6.39}$$

If the ground surface is level ($\beta = 0$), the g factors may be ignored.

Bearing Capacity Factors

Brinch Hansen recommended the use of Prandtl's formulas for computing N_c and N_q:

$$N_q = e^{\pi \tan\phi} \tan^2(45 + \phi/2) \tag{6.40}$$

$$N_c = \frac{N_q - 1}{\tan\phi} \qquad\qquad \text{for } \phi > 0 \quad (6.41)$$

$$N_c = 5.14 \qquad\qquad \text{for } \phi = 0 \quad (6.42)$$

Most other authorities also accept these formulas or others that produce very similar results. However, there is much more disagreement regarding the proper value of N_γ. Relatively small changes in the geometry of the failure surface below the footing can create significant differences in N_γ, especially in soils with high friction angles. Brinch Hansen recommended the following formula:

$$N_\gamma = 1.5 (N_q - 1) \tan\phi \tag{6.43}$$

Brinch Hansen's bearing capacity factors are also presented in tabular form in Table 6.2.

6.6 GROUNDWATER EFFECTS

Regardless of the method used to compute the ultimate bearing capacity, we also must consider the possible influence of a shallow groundwater table. Although it is customary to perform bearing capacity analyses using the saturated strength of the soil (to obtain worst-case values of c and ϕ), a shallow groundwater table will further diminish the strength of the soil. This additional loss is the result of pore water pressures and the corresponding reduction in effective stress.

When exploring the subsurface conditions, determine the current location of the groundwater table and worst-case (highest) location that might reasonably be expected during the life of the proposed structure. The depth from the ground surface to the groundwater table is D_w. If $D_w < B + D$, the groundwater table will affect the bearing capacity because the effective stress in the shear zone is reduced (Meyerhof, 1955). In soils with $\phi > 0$, this reduction causes a diminished soil strength.

Groundwater conditions may be divided into three cases, as described below and shown in Figure 6.11:

- Case 1: $D_w \leq D$
- Case 2: $D < D_w < D + B$
- Case 3: $D + B \leq D_w$

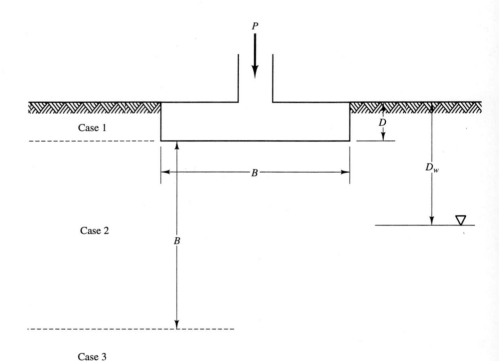

Figure 6.11 Three groundwater cases for bearing capacity analyses.

A simple, yet sufficiently accurate, method of accounting for a shallow groundwater table (Vesić, 1973) is to replace the unit weight, γ, in the third term of the bearing capacity formulas with the *effective unit weight*, γ'. This is an extension of the buoyant unit weight method of computing effective stresses discussed in Chapter 3.

The effective unit weight is the value that, when multiplied by the appropriate soil thickness will give the effective stress. It varies between the buoyant unit weight, γ_b, and the unit weight, γ, depending on the position of the groundwater table. Compute it as follows:

For case 1:

$$\gamma' = \gamma_b = \gamma_{sat} - \gamma_w \tag{6.44}$$

For case 2:

$$\gamma' = \gamma - \gamma_w \left(1 - \left(\frac{D_w - D}{B} \right) \right) \tag{6.45}$$

For case 3 (no groundwater correction is necessary):

$$\gamma' = \gamma \tag{6.46}$$

In case 1, the second term in the bearing capacity formulas also is affected, but the appropriate correction will be implicit in the computation of σ_D'.

Example 6.6

Compute the net ultimate bearing capacity of a 1.0 m square footing founded at a depth of 0.5 m below the ground surface. The groundwater table is at a depth of 0.8 m below the ground surface and the soil is a clayey sand with $c = 5.0$ kPa, $\phi = 21°$, and $\gamma = 18.5$ kN/m³.

Solution:

Determine groundwater case:

$D_w = 0.8$ m; $D = 0.5$ m; $B = 1.0$ m $D < D_w < D + B$ ∴ Case 2

Using Equation 6.45:

$$\gamma' = \gamma - \gamma_w \left(1 - \left(\frac{D_w - D}{B} \right) \right) = 18.5 - 9.8 \left(1 - \left(\frac{0.8 - 0.5}{1.0} \right) \right) = 11.6 \text{ kN/m}^3$$

Use Terzaghi's method (Equation 6.13)

For $\phi = 21°$: $N_c = 18.9$, $N_q = 8.3$, $N_\gamma = 5.1$ (from Table 6.2)

$$\begin{aligned}
\sigma_D' &= \sigma - u \\
&= \gamma D - 0 \\
&= (18.5)(0.5) - 0 \\
&= 9.2 \text{ kPa}
\end{aligned}$$

$$q_u^{/} = 1.3cN_c + \sigma_D^{/}(N_q-1) + 0.4\gamma BN_\gamma$$
$$= (1.3)(5.0)(18.9) + (9.2)(8.3-1) + (0.4)(11.6)(1.0)(5.1)$$
$$= 122.8 + 67.2 + 23.7$$
$$= \mathbf{214\ kPa} \qquad \Leftarrow\textbf{\textit{Answer}}$$

6.7 SELECTION OF SOIL STRENGTH PARAMETERS

Saturated Strength

Engineers nearly always use the saturated strength of the soil when performing bearing capacity analyses, even if the soil is not saturated in the field. It is not unusual for naturally dry soils to become wet; so it is prudent to design for the worst-case condition. For example, the soil could become saturated if a water pipe accidentally broke, if surface drainage was poor and rainwater infiltrated into the soil, or from a variety of other sources.

Plane Strain Conditions

If a bearing capacity were to occur beneath a continuous footing, the soil would fail in a plane strain condition. The soil beneath footings of other shapes would be somewhere between the axisymmetric and plane strain conditions. However, the fundamental bearing capacity formulas are based on continuous footings. Formulas for other shapes were derived from the continuous footing using empirical adjustment factors. Therefore, in theory, we should use the plane strain strength when conducting bearing capacity analyses, regardless of the footing shape. However, in practice, engineers rarely consider the differences between plane strain and axisymmetric strengths.

Drained vs. Undrained Strength

Footings located on saturated cohesive soils generate positive excess pore water pressures, so the most likely time for a bearing capacity failure is immediately after the load is applied. These excess pore water pressures dissipate with time, so the soil gradually becomes stronger and the factor of safety increases. Therefore, conduct bearing capacity analyses on saturated clays using the undrained strength.

For footings on cohesionless soils, any excess pore water pressures are very small and dissipate very rapidly. Therefore, evaluate such footings using the drained strength.

Intermediate soils are likely to be partially drained, and engineers have varying opinions on how to handle them. The more conservative approach is to use the undrained strength, but many engineers use design strengths somewhere between the drained and undrained strength.

Collapse of the Fargo Grain Elevator

One of the most dramatic bearing capacity failures was the Fargo Grain Elevator collapse of 1955. This grain elevator, shown in Figure 6.12, was built near Fargo, North Dakota, in 1954. It was a reinforced concrete structure composed of 20 cylindrical bins and other appurtenant structures, all supported on a 52 ft (15.8 m) wide, 218 ft (66.4 m) long, 2 ft 4 in (0.71 m) thick mat foundation.

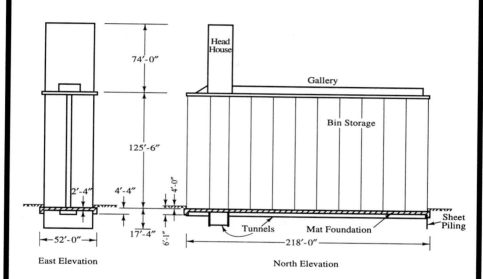

Figure 6.12 Elevation views of the elevator (Nordlund and Deere, 1970; Reprinted by permission of ASCE).

The average net bearing pressure, q', due to the weight of the empty structure was 1590 lb/ft^2 (76.1 kPa). When the bins began to be filled with grain in April 1955, q' began to rise, as shown in Figure 6.13. In this type of structure, the live load (i.e. the grain) is much larger than the dead load; so by mid-June, the average net bearing pressure had tripled and reached 4750 lb/ft^2 (227 kPa). Unfortunately, as the bearing pressure rose, the elevator began to settle at an accelerating rate, as shown in Figure 6.14.

Early on the morning of June 12, 1955, the elevator collapsed and was completely destroyed. This failure was accompanied by the formation of a 6 ft (2 m) bulge, as shown in Figure 6.15.

No geotechnical investigation had been performed prior to the construction of the elevator, but Nordlund and Deere (1970) conducted an extensive after-the-fact investigation. They found that the soils were primarily saturated clays with s_u=600-1000 lb/ft² (30-50 kPa). Bearing capacity analyses based on this data indicated a net ultimate bearing capacity of 4110 to 6520 lb/ft² (197-312 kPa) which compared well with the q' at failure of 4750 lb/ft² (average) and 5210 lb/ft² (maximum).

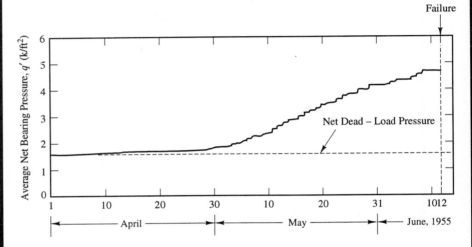

Figure 6.13 Rate of loading (Nordlund and Deere, 1970; Reprinted by permission of ASCE).

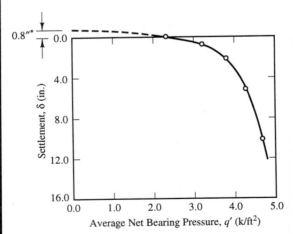

*Probable Settlement Before Installation of Elevation Benchmarks

Figure 6.14 Settlement at centroid of mat (Nordlund and Deere, 1970; Reprinted by permission of ASCE).

In other words, this was a classic bearing capacity failure that easily could have been predicted in advance.

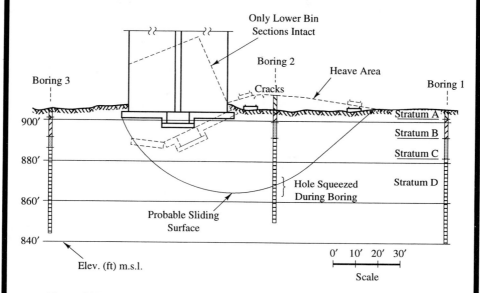

Figure 6.15 Cross-section of collapsed elevator (Nordlund and Deere, 1970; Reprinted by permission of ASCE)

Although bearing capacity failures of this size are unusual, this failure was not without precedent. A very similar failure occurred in 1913 at a grain elevator near Winnipeg, Manitoba, approximately 200 miles (320 km) north of Fargo (Peck and Bryant, 1953; White, 1953). This elevator rotated to an inclination of 27° from the vertical when the soil below experienced a bearing capacity failure at an average q' of 4680 lb/ft^2 (224 kPa). The soil profiles at the two sites are very similar, as are the average q' values at failure. This is a classic example of engineers failing to learn from the mistakes of others.

Unsaturated soils are more complex and thus more difficult to analyze. Usually, it is best to saturate them in the lab and design the foundation as if they are saturated in the field. Thus, use the undrained strength for cohesive soils and the drained strength for cohesionless soils.

6.8 FACTOR OF SAFETY AND ALLOWABLE BEARING CAPACITIES

Geotechnical engineers use a working stress approach for most strength related analyses. Therefore, we divide the net ultimate bearing capacity by a factor of safety to obtain the *net allowable bearing capacity, q_a'*:

$$q_a' = \frac{q_u'}{F} \tag{6.47}$$

Then, we design the footing so that the net bearing pressure is no greater than the net allowable bearing capacity (q' must be $\leq q_a'$).

Building codes do not specify design factors of safety. Therefore, engineers can use their own discretion and professional judgment when selecting F. Items to consider when selecting a design factor of safety include the following:

- The type of soil (use higher values for clays, lower for sands).
- The level of uncertainty in defining the soil profile and design shear strength parameters.
- The importance of the structure and the consequences of a failure.
- The likelihood of the design load ever actually occurring (see the discussion in Chapter 2).

The actual factor of safety is probably much greater than the design factor of safety, because of the following:

- The shear strength data are normally interpreted conservatively, so the design values of c and ϕ implicitly contain another factor of safety.
- The service loads are probably less than the design loads.
- Settlement, not bearing capacity, often controls the final design, so the footing will likely be larger than that required to satisfy bearing capacity criteria.
- Spread footings are commonly built somewhat larger than the plan dimensions.

Vesić (1975) suggested the design factors of safety listed in Table 6.4.

TABLE 6.4   GUIDELINES FOR SELECTING MINIMUM FACTOR OF SAFETY FOR SPREAD FOOTINGS

Category	Typical Structures	Characteristics of the Category	Design Factor of Safety	
			Thorough and Complete Soil Exploration	Limited Soil Exploration
A	Railway bridges, warehouses, blast furnaces, hydraulic, retaining walls, silos	Maximum design load likely to occur often; consequences of failure disastrous	3.0	4.0
B	Highway bridges, light industrial and public buildings	Maximum design loads may occur occasionally, consequences of failure serious	2.5	3.5
C	Apartment and office buildings	Maximum design load unlikely to occur	2.0	3.0

Adapted from Vesić, 1975.

1. For temporary structures, reduce these factors to 75% of the above values. However, the design safety factor should never be less than 2.0.
2. For exceptionally tall structures, such as chimneys and towers, or generally whenever progressive bearing capacity failure is possible, increase these factors by 20 to 50%.
3. Consider the possibility of flooding of the foundation soil and/or removal of existing overburden by scour or excavation.
4. Check both the short-term (end-of-construction) and long-term stability unless one of these two conditions clearly controls.
5. Conduct an independent settlement analysis because it may govern the design.

Example 6.7

A proposed public building includes a bearing wall that will carry a dead load of 70 kN/m and a live load of 50 kN/m. This wall will be supported on a 0.3 m deep continuous footing. Based on a moderately thorough soil investigation, we have determined that the soil beneath this footing is a clay with $s_u = 125$ kPa and $\gamma = 17.3$ kN/m^3. The groundwater table is at a depth of 5 m below the ground surface. Determine the required footing width, B.

Solution:

$$P/b = 70 + 50 = 120 \text{ kN/m}$$

Use Brinch Hansen's formula (Equation 6.23). The depth factor, d_c^a, is a function of B, which has not yet been determined. Therefore, we must develop an estimate of B. Assuming q_a' is on the order of 4000 lb/ft² (200 kPa), B is probably about (120 kN/m)/200 kPa = 0.6 m. Also note that s_c^a is based on a footing length of ∞, not 1 m.

From Table 6.3:

$$s_c^a = 0.2\,(B/L) = 0.2\,(B/\infty) = 0$$

Using Equation 6.33:

$$D/B \approx 0.3/0.6 = 0.5 \leq 1 \qquad \therefore\ k = D/B \approx 0.5$$

$$d_c^a = 0.4\,k = (0.4)(0.5) = 0.2$$

$$\sigma_D' = (17.3\ \text{kN/m}^3)(0.3\ \text{m}) = 5.2\ \text{kPa}$$

Using Equation 6.23:

$$\begin{aligned}
q_u' &= 5.14\,s_u\,(1 + s_c^a + d_c^a - 0 - 0 - 0) - \sigma_D' \\
&= 5.14\,(125)\,(1 + 0 + 0.2 - 0 - 0 - 0) - 5 \\
&= 766\ \text{kPa}
\end{aligned}$$

Use $F = 3$ (per Table 6.4).

Using Equation 6.47:

$$q_a' = \frac{q_u'}{F} = \frac{766}{3} = 255\ \text{kPa}$$

$$B = \frac{P/b}{q_a'} = \frac{120\ \text{kN/m}}{255\ \text{kPa}} = \mathbf{0.5\ m} \qquad\qquad \Leftarrow \textit{Answer}$$

Since the computed B of 0.5 m is close to the estimated width of 0.6 m, there is no need to recompute the value of d_c^a.

6.9 FOOTINGS WITH ECCENTRIC OR MOMENT LOADS

When evaluating the bearing capacity of footings with eccentric or moment loads, we must select which bearing pressure to compare with the allowable bearing capacity. Meyerhof (1963), Brinch Hansen (1970), and others have proposed the following approach:

1. Determine the eccentricity in the B and/or L directions (e_B, e_L).
2. Compute the effective footing dimensions:

$$B' = B - 2e_B \qquad (6.48)$$

$$L' = L - 2e_L \qquad (6.49)$$

This produces an equivalent footing with an area $A' = B' \times L'$ as shown in Figure 6.16.

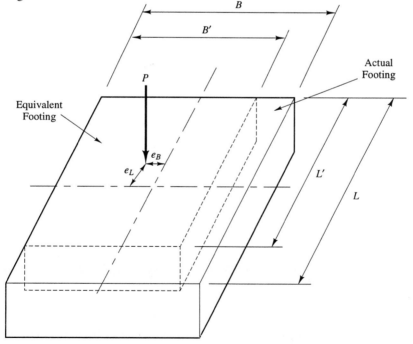

Figure 6.16 Equivalent footing for evaluating the bearing capacity of footings with eccentric or applied moment loads. Note that the equivalent footing has no eccentricity.

3. Compute the allowable bearing capacity using the dimensions of the equivalent footing in Brinch Hansen's or another appropriate formula.

4. Design the footing so the equivalent net bearing pressure, q'_{equiv}, does not exceed the allowable bearing capacity:

$$q'_{equiv} = \frac{P}{A'} \leq q'_a \qquad (6.50)$$

Example 6.8

A 5 ft square footing supports a vertical load of 75 k and a moment load of 50 ft-k. The allowable bearing capacity of the soil is 3500 lb/ft^2. Is this design satisfactory?

Solution:

Using Equation 6.7:

$$e = \frac{M}{P} = \frac{50 \text{ ft-k}}{75 \text{ k}} = 0.67 \text{ ft}$$

Using Equation 6.48:

$$B' = B - 2e_B = 5 - (2)(0.67) = 3.66 \text{ ft}$$

Using Equation 6.50:

$$q'_{equiv} = \frac{P}{A'} = \frac{75000 \text{ lb}}{(3.66 \text{ ft})(5 \text{ ft})} = 4100 \text{ lb/ft}^2$$

Since $q'_{equiv} > q'_a$ (4100 > 3500), this design is not satisfactory. A larger B is required.

6.10 BEARING CAPACITY ANALYSES — LOCAL AND PUNCHING SHEAR CASES

As discussed earlier, engineers rarely need to compute the local or punching shear bearing capacities because settlement analyses implicitly protect against this type of failure. In addition, a complete analysis would be more complex because of the following:

• These modes of failure do not have well defined shear surfaces, such as those shown in Figure 6.9, and are therefore more difficult to evaluate.

- The soil can no longer be considered incompressible (Ismael and Vesić, 1981).
- The failure is not catastrophic (refer to Figure 6.8), so the failure load is more difficult to define.
- Scale effects make it difficult to properly interpret model footing tests.

Terzaghi (1943) suggested a simplified way to compute the local shear bearing capacity using the general shear formulas with appropriately reduced values of c and ϕ:

$$c_{adj} = 0.67\, c \tag{6.51}$$

$$\phi_{adj} = \tan^{-1}(0.67 \tan\phi) \tag{6.52}$$

Vesić (1975) expanded upon this concept and developed the following adjustment formula for sands with a relative density, D_r, less than 67%:

$$\phi_{adj} = \tan^{-1}[(0.67 + D_r - 0.75 D_r^2)\tan\phi] \tag{6.53}$$

Where:

c_{adj} = adjusted cohesion
ϕ_{adj} = adjusted friction angle
D_r = relative density of sand $(0 \le D_r \le 67\%)$

Although Vesić's formula was confirmed with a few model footing tests, both methods are flawed because the failure mode is not being modeled correctly. However, local or punching shear will normally only govern the final design with shallow, narrow footings on loose sands, so an approximate analysis is acceptable. The low cost of such footings does not justify a more extensive analysis, especially if it would require additional testing.

An important exception to this conclusion is the case of a footing supported by a thin crust of strong soil underlain by very weak soil. This would likely be governed by punching shear and would justify a custom analysis.

6.11 BEARING CAPACITY OF ROCK

In comparison to foundations on soil, those on bedrock usually present few difficulties for the designer (Peck, 1976). The greatest problems often involve difficulties in construction, such as excavation problems and proper removal of weathered or disturbed material to provide good contact between the footing and the bedrock.

The allowable bearing pressure on rock may be determined in at least four ways (Kulhawy and Goodman, 1980):

- Presumptive values found in building codes
- Empirical rules
- Rational methods based on bearing capacity and settlement analyses
- Full-scale load tests

When supported on good quality rock, spread footings are normally able to support moderately large loads with very little settlement. Engineers usually design them using presumptive bearing pressures, preferably those developed for the local geologic conditions. Typical values are listed in Table 6.5.

TABLE 6.5 TYPICAL ALLOWABLE BEARING PRESSURES FOR FOUNDATIONS ON BEDROCK

Rock Type	Rock Consistency	Allowable Bearing Pressure	
		(lb/ft^2)	(kPa)
Massive crystalline igneous and metamorphic rock: Granite, diorite, basalt, gneiss, thoroughly cemented conglomerate	Hard and sound (minor cracks OK)	120,000 - 200,000	6000 - 10,000
Foliated metamorphic rock: Slate, schist	Medium hard, sound (minor cracks OK)	60,000 - 80,000	3000 - 4000
Sedimentary rock: Hard cemented shales, siltstone, sandstone, limestone without cavities	Medium hard, sound	30,000 - 50,000	1500 - 2500
Weathered or broken bedrock of any kind; compaction shale or other argillaceous rock in sound condition	Soft	16,000 - 24,000	800 - 1200

Adapted from U.S. Navy, 1982b.

Nearly all bedrocks contain *discontinuities* such as joints, seams, faults, or bedding planes. The intact rock between these discontinuities may be very strong, but the discontinuities are much weaker. Therefore, the spacing, orientation and strength of the discontinuities will strongly influence the strength of the rock mass.

Carter and Kulhawy (1988) offered a semi-empirical approach that considers the discontinuities:

$$q_u^{/} = J c N_{cr} \tag{6.54}$$

Where:

$q_u^{/}$ = net ultimate bearing capacity

J = correction factor (from Figure 6.17)

c = cohesive strength of the rock mass

ϕ = frictional strength of the rock mass

N_{cr} = bearing capacity factor (from Figure 6.18)

H = vertical spacing of discontinuities

S = horizontal spacing of discontinuities

B = width of footing

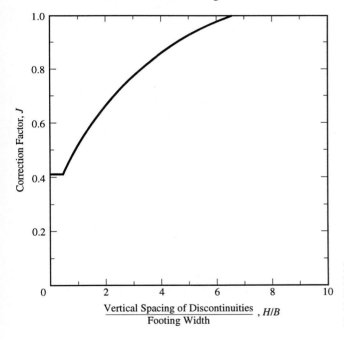

Figure 6.17 Correction factor, J, for Equation 6.54 (Adapted from Carter and Kulhawy, 1988. Copyright ©1988 Electric Power Research Institute, reprinted with permission).

Values of c and ϕ measured in the laboratory do not consider the effect of the discontinuities. Therefore, they must be adjusted before being used in the analysis. Do this by multiplying the laboratory ϕ by a reduction factor of 0.50 to 0.75 and c by a reduction factor of α_E. The value of α_E depends on the rock quality designation (RQD), which is determined by obtaining a core sample of the rock, summing the lengths of all pieces in the sample that are longer than 4 in (10 cm) and dividing by the total length of the core. Use $\alpha_E = 0.1$ for RQD $\leq 70\%$, linearly increasing to 0.6 at RQD $= 100\%$.

If the rock mass is very strong, the strength of the concrete may governs the bearing capacity of spread footings. Therefore, do not use an allowable bearing value, $q_a{'}$, greater than one-third of the compressive strength of the concrete ($0.33\,f_c{'}$).

When working with bedrock, be aware of certain special problems. For example, soluble rocks, including limestone, gypsum, and rock salt, may have underground cavities that might collapse, causing *sinkholes* to form at the ground surface. These have caused extensive damage to buildings in Florida and elsewhere.

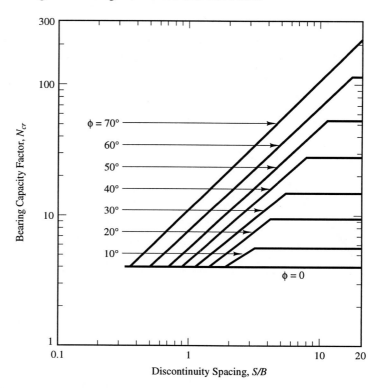

Figure 6.18 Bearing capacity factor, N_{cr}, for Equation 6.54 (Adapted from Carter and Kulhawy, 1988. Copyright ©1988 Electric Power Research Institute, reprinted with permission).

SUMMARY OF MAJOR POINTS

1. The contact force per unit area along the base of a spread footing is known as the *bearing pressure*. This may be expressed in two ways: the *gross bearing pressure*, q, is the actual contact pressure, while the *net bearing pressure*, $q\,{'}$, is the increase in pressure at that elevation due to the construction of the footing and application of the structural loads.

2. The bearing pressure beneath footings with concentric vertical loads only is taken to be uniformly distributed. For an eccentric load or an applied moment, it is taken to vary linearly across the bottom of the footing. Footings with eccentric loads should be designed such that the resultant force at the base acts within the middle one-third of the footing.

3. A *bearing capacity failure* occurs when the soil beneath the footing fails in shear. There are three types of bearing capacity failures: general shear, local shear, and punching shear.

4. A variety of formulas have been developed to compute the *allowable bearing capacity*, $q_a{'}$, including Terzaghi's formulas and Brinch Hansen's formulas. Terzaghi's are most appropriate for quick hand calculations, whereas Brinch Hansen's are more useful when greater precision is needed or special loading or geometry conditions must be considered.

5. Shallow groundwater tables reduce the effective stress in the near-surface soils and can therefore adversely affect bearing capacity. Adjustment factors are available to account for this effect.

COMPUTER SOFTWARE - PROGRAM FTGBC

A computer program that computes the bearing capacity of spread footings, FTGBC, is included with this book. This program uses Terzaghi's method, Brinch Hansen's method and the method proposed by Meyerhof (1963).[4] The user can easily change the dimensions of the footing or the soil properties, thus permitting parametric studies to be performed more easily.

To access this program, install the software in accordance with the instructions in Appendix C, and then type **FTGBC** at the DOS prompt. A brief introductory screen will appear, followed by the worksheet. Then, follow this procedure:

1. Select the units of measurement by pressing **ALT-U** and selecting ENGLISH or SI from the menu. English units are the default.

2. Select the shape of the footing by pressing **ALT-P**. The program will consider square, rectangular, continuous, or circular footings. The default shape is square.

3. Use the arrow keys to move the cursor and input the title, footing geometry, and soil properties. All of the **XXXXX** indicators should be replaced with appropriate data, as shown in Figure 6.19.

    Exception: If the shear load, V, is zero, then it is not necessary to input the normal load, P.

---

[4] This is Meyerhof's method for bearing capacity analysis, which is not the same as Meyerhof's method of computing settlement, as discussed in Chapter 7.

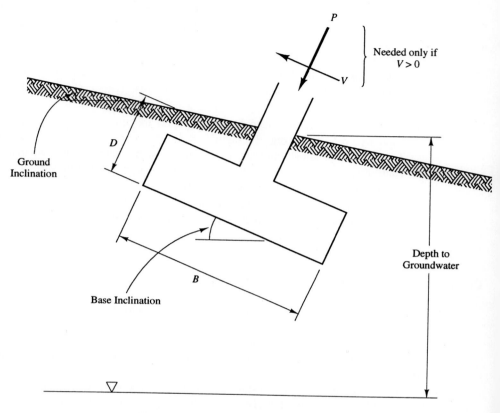

Figure 6.19 Input data for the FTGBC program.

4. Review the output shown in the right portion of the worksheet. Notice that both the allowable bearing capacities and the allowable loads are given.

5. If a printout is needed, press **ALT-M** to access the menu and select one of the following:

> **PRINT** - Prints the worksheet on a printer.
> **PRINT TO DISK** - Prints the worksheet to a disk file.

> If you select **PRINT TO DISK**, the program will ask for a disk file name. You must select a new file name; the program will not write over an old file.

6. To exit to DOS, select **QUIT** from the menu.

Example 6.9

Use FTGBC to solve Example 6.7.

Output:

```
          FTGBC   Version 1.0   (c) 1994 by Prentice Hall, Inc.
                Bearing Capacity Analysis of Spread Footings

Title:          Rework Problem 6.7

Unit System:    SI                    Date: 05-May-93   Time: 11:58 PM
(Press ALT-U to set unit system)
                                      *************************************
Footing Shape:  Continuous            *              RESULTS             *
(Press ALT-S to set footing shape)    *                                  *
                                      *ALLOWABLE BEARING CAPACITY (kPa)   *
                                      *                                  *
Footing Width        =    0.50  m     *                          Brinch  *
Footing Depth        =    0.30  m     *         Terzaghi  Meyerhof Hansen *
Base Inclination     =       0  deg   * Gross      240      242      266  *
Ground Inclination   =       0  deg   * Net        238      240      264  *
Soil Cohesion        =     125  kPa   *                                  *
Soil Friction Angle  =       0  deg   *ALLOWABLE WALL LOAD (Kn/m)         *
Soil Unit Weight     =    17.3  kN/cu m*                                 *
Depth to Groundwater =     5.0  m     *                          Brinch  *
Factor of Safety     =    3.00        *         Terzaghi  Meyerhof Hansen *
                                      *          119.0    119.9    131.9  *
Applied Loads (Needed only if shear>0) *************************************
    Normal           P =  XXXXXX  kN
    Shear            V =  XXXXXX  kN
```

QUESTIONS AND PRACTICE PROBLEMS

6.1   Explain the difference between gross bearing pressure and net bearing pressure.

6.2   A 400 kN vertical downward column load acts at the centroid of a 1.5 m square footing. The bottom of this footing is 0.4 m below the ground surface. Compute the gross and net bearing pressures.

6.3   A bearing wall carries a dead load of 5.0 k/ft and a live load of 3.0 k/ft. It is supported on a 3 ft wide, 2 ft deep continuous footing. Compute the gross and net bearing pressures.

6.4   A 5 ft square, 2 ft deep spread footing is subjected to a concentric vertical load of 60 k and an overturning moment of 30 ft-k. The overturning moment acts parallel to one of the sides of the footing. Determine whether the resultant force acts within the middle third of the footing, compute the minimum and maximum net bearing pressures and show the distribution of bearing pressure in a sketch.

6.5   Consider the footing and loads in Problem 6.4, except that the overturning moment now acts at a 45° angle from the side of the footing (i.e. it acts diagonally across the top of the footing). Determine whether the resultant force acts within the kern, compute the net bearing pressure at each corner of the footing and show the pressure distribution in a sketch similar to Figure 6.4.

6.6   List the three types of bearing capacity failures and explain the differences between them.

6.7   A 1.2 m square, 0.4 m deep spread footing is underlain by a soil with the following properties: $\gamma = 19.2$ kN/m^3, $c = 5$ kPa, $\phi = 30°$. The groundwater table is at a great depth.

    a.   Use hand computations to compute its net allowable bearing value and allowable column load using Terzaghi's formulas and a factor of safety of 2.5.

    b.   Repeat part a using Brinch Hansen's formulas.

    c.   Repeat parts a and b using program FTGBC.

6.8   A column carrying a vertical downward dead load and live load of 150 k and 120 k, respectively, is to be supported on a square spread footing. The soil beneath this footing is an undrained clay with $s_u = 3000$ lb/ft^2. Compute the width B required to obtain a factor of safety of 3 against a bearing capacity failure.

6.9   A 120 ft diameter cylindrical tank is located on the ground surface. This tank is to be filled with water. The soil below is an undrained clay with $s_u = 1000$ lb/ft^2. The empty tank weighs 1,100,000 lb. Compute the maximum allowable depth of the water in the tank that will maintain a factor of safety of 3.0 against a bearing capacity failure.

6.10   The footing designed in Problem 6.7a has been built. Then, sometime after construction, the groundwater table rose to a depth of 0.5 m below the ground surface. Using hand computations with Terzaghi's method, compute the new factor of safety. Confirm your computations using program FTGBC.

6.11   Conduct a bearing capacity analysis on the Fargo Grain Elevator (see sidebar) and back-calculate the average undrained shear strength of the soil. The groundwater table is at a depth of 6 ft below the ground surface. Soil strata A and B have unit weights of 110 lb/ft^3; stratum D has 95 lb/ft^3. The unit weight of stratum C is unknown. Assume that the load on the foundation acted through the centroid of the mat.

6.12   A certain column carries a vertical downward load of 1200 kN and a horizontal shear load of 700 kN. It is to be supported on a 1 m deep, square footing. The soil beneath this footing has the following properties: $\gamma = 20.5$ kN/m^3, $c = 5$ kPa, $\phi = 36°$. The groundwater table is at a depth of 1.5 m below the ground surface. Using program FTGBC, compute the footing width required for a factor of safety of 3.5, and then check the answer by conducting hand computations.

Advanced Questions and Practice Problems

6.13   Three columns, A, B, and C, are collinear, 500 mm in diameter, and 2.0 m on-center. They have vertical downward loads of 1000, 550, and 700 kN, respectively, and are to be supported on a single, 1.0 m deep rectangular combined footing. The soil beneath this proposed footing has the following properties: $\gamma = 19.5$ kN/m^3, $c = 10$ kPa, and $\phi = 31°$.

The groundwater table is at a depth of 25 m below the ground surface.

a. Determine the minimum footing length, L, and the placement of the columns on the footing that will place the resultant load at the centroid of the footing. The footing must extend at least 500 mm beyond the edges of columns A and C.

b. Using the results from part a, determine the minimum footing width, B, that will maintain a factor of safety of 2.5 against a bearing capacity failure. Show the final design in a sketch.

Hint: Assume a value for B, compute the allowable bearing capacity, then solve for B. Repeat this process until the computed B is approximately equal to the assumed B.

6.14 Two columns, A and B, are to be built 6 ft 0 in apart (measured from their centerlines). Column A has a vertical downward dead load and live loads of 90 k and 80 k, respectively. Column B has corresponding loads of 250 k and 175 k. The dead loads are always present, but the live loads may or may not be present at various times during the life of the structure. It is also possible that the live load would be present on one column, but not the other.

These two columns are to be supported on a 4 ft 0 in deep rectangular spread footing founded on a soil with the following parameters: unit weight = 122 lb/ft^3, friction angle = 37°, and cohesion = 100 lb/ft^2. The groundwater table is at a very great depth.

a. The location of the resultant of the loads from columns A and B depends on the amount of live load acting on each at any particular time. Considering all of the possible loading conditions, how close could it be to column A? To column B?

b. Using the results of part a, determine the minimum footing length, L, and the location of the columns on the footing necessary to keep the resultant force within the middle third of the footing under all possible loading conditions. The footing does not need to be symmetrical. The footing must extend at least 24 in beyond the centerline of each column.

c. Determine the minimum required footing width, B, to maintain a factor of safety of at least 2.5 against a bearing capacity failure under all possible loading conditions.

d. If the B computed in part c is less than the L computed in part b, then use a rectangular footing with dimensions $B \times L$. If not, then redesign using a square footing. Show your final design in a sketch.

7

Spread Footings – Settlement

Men cease to build where the foundation sinkes.
John Webster: *The Duchess of Malfi* (1624), Act III, Scene V

By the 1950s, engineers were performing bearing capacity analyses as a part of many routine design projects. However, during that period, many engineers seemed to have the misconception that any footing designed with an adequate factor of safety against a bearing capacity failure would not settle excessively. Although settlement analysis methods were available, Hough (1959) observed that these analyses, if conducted at all, were considered to be secondary. Fortunately, Hough and others emphasized that bearing capacity and settlement do not go hand-in-hand and that independent settlement analyses also need to be performed. We now know that settlement frequently controls the design of spread footings, especially when B is large, and that the bearing capacity analysis is, in fact, often secondary.

Although this chapter concentrates on settlements caused by the application of structural loads on the footing, other sources of settlement also may be important. These include the following:

- Settlements due to the weight of a recently placed fill.
- Settlements due to a falling groundwater table.
- Settlements caused by underground mining or tunneling.

- Settlements caused by the formation of sinkholes.
- Lateral movements resulting from nearby excavations that indirectly cause settlements.

Methods of settlement analysis developed along two parallel paths: one for footings on cohesive soils and the other for footings on cohesionless soils. Although the basic mechanisms controlling settlement are similar for both, some important differences dictate the use of different analysis methods. These differences include the following:

- The permeability of cohesive soils is very low, so consolidation settlements require more time, with undrained conditions prevailing during the early stages. In contrast, cohesionless soils usually settle as quickly as the load is applied and drained conditions prevail.
- High quality undisturbed samples of cohesive soils can be obtained fairly inexpensively, which means that laboratory tests can be performed on representative samples. However, it is very difficult or impossible to obtain high quality samples of most cohesionless soils. Even when they are remolded in the laboratory to the same unit weight as in the field, the test results are not accurate. Therefore, these soils are best tested in-situ.
- Most cohesionless soils are much less compressible, so the magnitude of the settlements are generally smaller.

7.1 STRESSES IN SOIL BENEATH A SPREAD FOOTING

To perform a settlement analysis, we must know how the footing load changes the vertical effective stress in the soil. This change in vertical effective stress, $\Delta\sigma_v'$, causes strains in the soil that, in turn, cause the footing to settle.

The most commonly used methods of computing $\Delta\sigma_v'$ are based on the theory of elasticity. In the context of soil mechanics, this means that the soil is presumed to have a linear stress-strain curve, which is far from the truth. However, for purposes of computing the stress distribution, this assumption is acceptable as long as the soil is reasonably homogeneous and the strains are not excessive.

Note that the use of elastic theory to compute *strains* is an entirely different matter. Most methods of computing strains (and therefore settlements) implicitly or explicitly consider the nonlinearity of the stress-strain curve.

Boussinesq's Method

The classic solution for stresses in an elastic material is the one developed by Boussinesq (1885). He wrote it for point loads, but it can be integrated over any shape to produce a formula for computing stresses beneath perfectly flexible loaded areas. In the context of spread footings on soil, the stress at a point below a perfectly flexible loaded rectangle or square is as follows (after Newmark, 1935):

If $m^2 + n^2 + 1 < m^2 n^2$:

$$\Delta \sigma_v' = \frac{q'}{k\pi} \left[\frac{2mn\sqrt{m^2+n^2+1}}{m^2+n^2+1+m^2n^2} \frac{m^2+n^2+2}{m^2+n^2+1} + \pi - \sin^{-1}\left(\frac{2mn\sqrt{m^2+n^2+1}}{m^2+n^2+1+m^2n^2} \right) \right] \quad (7.1)$$

Otherwise:

$$\Delta \sigma_v' = \frac{q'}{k\pi} \left[\frac{2mn\sqrt{m^2+n^2+1}}{m^2+n^2+1+m^2n^2} \frac{m^2+n^2+2}{m^2+n^2+1} + \sin^{-1}\left(\frac{2mn\sqrt{m^2+n^2+1}}{m^2+n^2+1+m^2n^2} \right) \right] \quad (7.2)$$

Where:

$\Delta \sigma_v'$ = change in vertical effective stress at a point below a perfectly flexible loaded rectangle

q' = net bearing pressure

B = footing width

L = footing length (use $L = B$ for square footings)

z_f = depth from bottom of footing to point

k, m, n = factors from Table 7.1

Note: The $\sin^{-1}$ terms must be expressed in radians.

It also is possible to integrate Boussinesq's equation over circular footings. However, for most practical purposes, it is sufficient to evaluate circular footings as if they were square footings with the same area.

For continuous footings, $n \rightarrow \infty$ and Equations 7.1 and 7.2 reduce to the following:

If $m \geq 1$:

$$\Delta \sigma_v' = \frac{q'}{k\pi} \left[\frac{2m}{1+m^2} + \pi - \sin^{-1}\left(\frac{2m}{1+m^2} \right) \right] \quad (7.3)$$

Otherwise:

$$\Delta \sigma_v' = \frac{q'}{k\pi} \left[\frac{2m}{1+m^2} + \sin^{-1}\left(\frac{2m}{1+m^2} \right) \right] \quad (7.4)$$

TABLE 7.1 k, m, AND n FACTORS FOR EQUATIONS 7.1 - 7.5

Footing Shape	Horizontal Location Beneath Footing	k	m	n
Square	Center	1	$0.5\ B/z_f$	$0.5\ B/z_f$
Square	Midpoint of edge	2	$0.5\ B/z_f$	B/z_f
Square	Corner	4	B/z_f	B/z_f
Rectangular	Center	1	$0.5\ B/z_f$	$0.5\ L/z_f$
Rectangular	Midpoint of short edge	2	$0.5\ B/z_f$	L/z_f
Rectangular	Midpoint of long edge	2	B/z_f	$0.5\ L/z_f$
Rectangular	Corner	4	B/z_f	L/z_f
Continuous	Centerline	1	$0.5\ B/z_f$	
Continuous	Edge	2	B/z_f	

Example 7.1

A 1.0 m square footing supports a column load of 150 kN. The bottom of this footing is 0.3 m below the ground surface, the groundwater table is at a great depth, and the unit weight of the soil is 19.0 kN/m^3. Compute the initial vertical effective stress, σ_{v0}', and the change in vertical effective stress, $\Delta\sigma_v'$, at a point 1.5 m below the center of the bottom of this footing.

Solution:

Using Equations 3.7 and 3.10:

$$\sigma_{v0}' = \gamma h - u = (19.0)(0.3 + 1.5) - 0 = \mathbf{34\ kPa} \qquad \Leftarrow Answer$$

From Table 7.1:

$$k = 1$$

$$m = n = 0.5\ (1/1.5) = 0.333$$

$$m^2 + n^2 + 1 = (0.333)^2 + (0.333)^2 + 1 = 1.22$$

$$m^2 n^2 = (0.333)^2 (0.333)^2 = 0.01$$

$$m^2 + n^2 + 1 \geq m^2 n^2 \qquad \therefore \text{ Use Equation 7.2}$$

$$q' = \frac{P}{A} = \frac{150}{1.0} = 150 \text{ kPa}$$

$$\frac{2mn\sqrt{m^2+n^2+1}}{m^2+n^2+1+m^2n^2} = \frac{2(0.333)(0.333)\sqrt{0.333^2+0.333^2+1}}{0.333^2+0.333^2+1+0.333^20.333^2} = 0.199$$

$$\Delta\sigma_v' = \frac{q'}{k\,\pi} \left[\frac{2mn\sqrt{m^2+n^2+1}}{m^2+n^2+1+m^2n^2} \frac{m^2+n^2+2}{m^2+n^2+1} + \sin^{-1}\left(\frac{2mn\sqrt{m^2+n^2+1}}{m^2+n^2+1+m^2n^2} \right) \right]$$

$$= \frac{150}{1\,\pi} \left[0.199 \frac{0.333^2+0.333^2+2}{0.333^2+0.333^2+1} + \sin^{-1}(0.199) \right]$$

= 27 kPa $\Leftarrow$ *Answer*

Boussinesq's method also can compute $\Delta\sigma_v'$ at other locations, both beneath and beyond the footing. Newmark (1935) did this by defining rectangular loaded areas, as shown in Figure 7.1, and computing the stress by superposition.

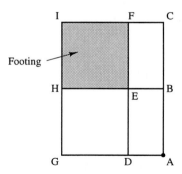

To Compute Stress at Point A Due to Load from Footing EFHI:
$(\Delta\sigma_v')_A = (\Delta\sigma_v')_{ACGI} - (\Delta\sigma_v')_{ACDF} - (\Delta\sigma_v')_{ABGH} + (\Delta\sigma_v')_{ABDE}$

Figure 7.1   Using Newmark's form of Boussinesq's equation to find $\Delta\sigma'$ at any point.

Example 7.2

Another footing, identical to the one described in Example 7.1, is located adjacent to it. These two footings are separated by a distance of 0.2 m. Compute the change in vertical effective stress, $\Delta\sigma_v'$, at a point 1.5 m below the center of the bottom of the first footing that results from the loading on the second footing.

Solution:

Use two imaginary footings, A and B, as shown in the figure below. Using the principle of superposition, the $\Delta\sigma_v'$ due to the load on the real footing is that due to footing A minus that due to footing B.

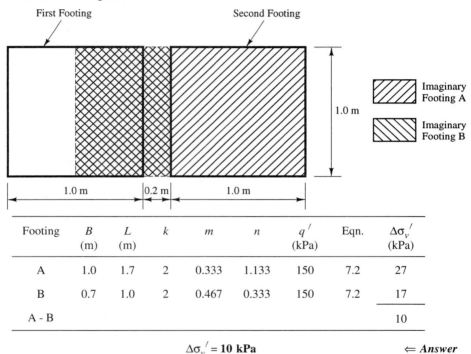

Footing	B (m)	L (m)	k	m	n	q' (kPa)	Eqn.	$\Delta\sigma_v'$ (kPa)
A	1.0	1.7	2	0.333	1.133	150	7.2	27
B	0.7	1.0	2	0.467	0.333	150	7.2	17
A - B								10

$$\Delta\sigma_v' = 10 \text{ kPa} \qquad\qquad \Leftarrow Answer$$

It often is convenient to express the results of Equations 7.1 - 7.4 as *pressure bulbs*. These are contours of equal $\Delta\sigma_v'/q'$ as shown in Figure 7.2. Note how the stresses from a continuous footing penetrate much deeper than those from a square footing.

Westergaard's Method

Boussinesq's method presumes that the material, in this case soil, is isotropic, which means the modulus of elasticity and other relevant parameters are constant in all directions. However, many real soils have distinct horizontal layers that tend to alter the

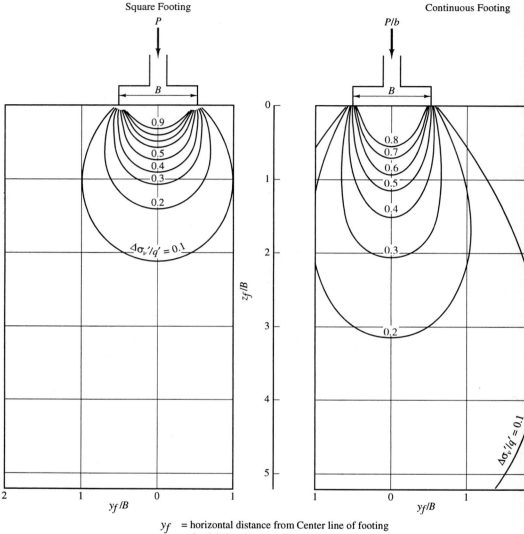

Figure 7.2 Pressure bulbs based on Newmark's solution of Boussinesq's method for square and continuous footings.

y_f = horizontal distance from Center line of footing
z_f = depth below bottom of footing
$\Delta\sigma_v'$ = change in vertical effective stress
q' = net bearing pressure

stress distribution. Some believe Westergaard's (1938) solution is a more accurate solution. He specifically had soil in mind (Boussinesq's method was a general solution for all materials) and considered it to be a soft elastic material reinforced by closely spaced horizontal flexible but unstretchable sheets. The combined thickness of these sheets is considered to be negligible. This approach generally produces higher values of $\Delta\sigma_v'$.

Westergaard developed his formula for a point load and left it to others to integrate it over flexible loaded areas. Taylor's (1948) solution for rectangular areas may be expanded into the following form:

$$\Delta\sigma_v' = \frac{2\,q'}{k\,\pi}\,\cot^{-1}\sqrt{\left(\frac{1-2v_p}{2-2v_p}\right)\left(\frac{1}{m^2}+\frac{1}{n^2}\right)+\left(\frac{1-2v_p}{2-2v_p}\right)^2\left(\frac{1}{m^2n^2}\right)} \qquad (7.5)$$

Where:

v_p = Poisson's ratio $(v_p < 0.5)$
The $\cot^{-1}$ term is expressed in radians

Simplified Method

The Boussinesq and Westergaard formulas can be placed into a computer program, such as the FTGSETT program included with this book. Various chart solutions also are available (for example, Newmark 1942). However, to obtain a quick approximation of $\Delta\sigma_v'$ without the use of a computer, the charts, or Figure 7.2, the simplified method is appropriate.

The author developed this method by experimentally fitting a simple formula to the Boussinesq function for $\Delta\sigma_v'$ at a depth z_f below the center of the bottom of a square, rectangular, or continuous footing. The resulting formulas are as follows:

For square or rectangular footings:

$$\Delta\sigma_v' = \left(\frac{1.7\,P}{(B+z_f)\,(L+z_f)}\right) - 0.05\,q' \qquad 0 \le \Delta\sigma_v' \le q' \qquad (7.6)$$

For circular footings, use Equation 7.6 with a B that represents the width of a square with an equivalent area.

For continuous footings:

$$\Delta\sigma_v' = \frac{1.4\,P/b}{B+1.3z_f} - 0.05\,q' \qquad 0 \le \Delta\sigma_v' \le q' \qquad (7.7)$$

Where:

$\Delta\sigma_v'$ = change in vertical effective stress at a point beneath the center of the footing

P = column load

P/b = wall load per unit length

B = footing width

L = footing length

z_f = depth from the bottom of the footing to the point

q' = net bearing pressure

The difference between these formulas and Boussinesq's formulas is less than 5% of q'.

Which Method to Use?

When solving practical problems, consider the following guidelines to select an appropriate method of computing stresses:

- If the soil beneath the footing is reasonably homogeneous, Boussinesq's method is probably most appropriate.
- If the soil beneath the footing is highly anisotropic, or if is has distinct horizontal stratification, then Westergaard's method may be preferable.
- For quick, informal calculations, especially when the appropriate charts or computational facilities are not available, then the simplified method will provide acceptable results.

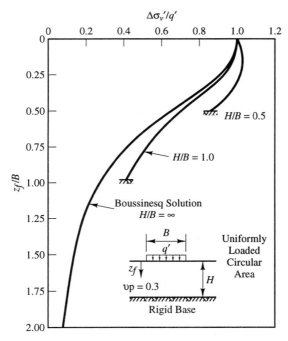

Figure 7.3 Stress distribution in a compressible soil of thickness h that is underlain by an infinitely rigid layer (Adapted from Foott and Koutsoftas, 1984).

Stresses in Stratified Soils

Each of these methods of computing $\Delta\sigma_v{}'$ is based on the assumption that the soil is homogeneous to a great depth below the footing. However, if this is not the case, the stress distribution will be altered.

If the soil strata extend only to a depth h, as shown in Figure 7.3, and are underlain by a very rigid stratum, such as hard rock, the load spreading will be inhibited and $\Delta\sigma_v{}'/q'$ will be as much as 20% larger. The opposite occurs if a stiff stratum is underlain by a softer stratum, as shown in Figure 7.4.

Usually, engineers do not explicitly consider these effects, but we must be mindful of them to properly interpret settlement analyses. Poulos and Davis (1974) provide methods for computing these stresses in situations where an explicit analysis is warranted. Alternatively, a finite element analysis could be used.

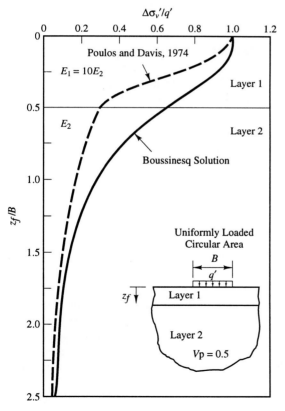

Figure 7.4 Stress distribution in a profile consisting of a stiff soil underlain by a softer soil (Adapted from Foott and Koutsoftas, 1984).

7.2 SETTLEMENT OF FOOTINGS ON COHESIVE SOILS

Chapter 3 described how to conduct a settlement analysis when confined conditions (no lateral strain) prevail. However, this method does not strictly apply to settlement analyses of spread footings because the boundary conditions are different. The soil can strain laterally when subjected to a vertical load over a limited area, such as a footing. Therefore, we must modify the analysis.

One method of analysis would be to measure the soil compressibility in the laboratory using boundary conditions comparable to those in the field. For example, we could simulate the conditions beneath a square or circular footing by testing a sample in a triaxial compression machine using a confining stress comparable to that present in the field (Davis and Poulos, 1968). Plane strain conditions, such as those present beneath a continuous footing, would be more difficult to model in the lab and would require special custom-built equipment.

An alternative method is to use standard consolidation test results (which model the confined compression case) and modify the computed settlement to account for lateral strains. Although this is not as accurate as the first method, it is suitable for nearly all practical problems. Therefore, nearly all settlement analyses use this method.

We will divide the settlement of a footing, δ, into three components:

- *Distortion settlement*, δ_d, reflects the lateral displacement of the soil beneath the footing.
- *Consolidation settlement*, δ_c (also known as *primary consolidation* settlement), reflects the change in volume of the soil that results from changes in the effective stress.
- *Secondary compression settlement*, δ_s, reflects the reduction of the soil volume at a constant effective stress and is the result of decomposition of organic material and other physical and chemical processes.

The total settlement, δ, is the sum of these:

$$\delta = \delta_d + \delta_c + \delta_s \tag{7.8}$$

Distortion Settlement

The distortion settlement (sometimes called *immediate settlement*, *initial settlement*, or *undrained settlement*) is that due to lateral spreading of the soil, as shown in Figure 7.5. This occurs under undrained conditions and therefore involves no change in volume. This portion of the settlement probably occurs as rapidly as the load is applied.

Based on elastic theory, the distortion settlement beneath the center of a perfectly flexible footing on clay is:

$$\delta_d = \frac{q'B}{E_u} I_0 I_1 \tag{7.9}$$

Where:

δ_d = distortion settlement

q' = net bearing pressure

B = footing width

I_0, I_1 = influence factors

E_u = undrained modulus of elasticity of soil

Soil does not have linear stress-strain properties, so the modulus, E_u, must represent an equivalent linear material.

Janbu, Bjerrum, and Kjaernsli (1956) first proposed this formula. Since then, Christian and Carrier (1978) revised the procedure and Taylor and Matyas (1983) shed additional light on its theoretical basis. The updated functions are shown in Figure 7.6.

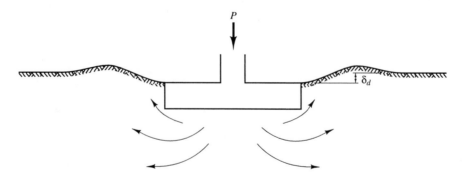

Figure 7.5 Lateral spreading of soil at a constant volume beneath a footing: the cause of distortion settlement.

Some have written Equation 7.9 with a $(1 - v_p^2)$ term in the numerator, where v_p is Poisson's ratio. Although that is the correct form based on elastic theory, the I_0 and I_1 factors presented in Figure 7.6 implicitly include a Poisson's ratio of 0.5. Therefore, in this context, it is appropriate to write the equation without the $(1 - v_p^2)$ term.

When using this method, we cannot always model continuous footings using $L = \infty$. Figure 7.6 suggests that if both H/B and L/B are infinite, I_1 also becomes infinite. To avoid this difficulty, limit I_1 to a maximum value of 2.5, which corresponds to an L/B of about 50.

The undrained modulus of elasticity, E_u, is the most difficult factor to assess. One method of measuring it is to apply incremental loads on an undisturbed sample in a triaxial compression machine and measure the corresponding deformations. Unfortunately, this method tends to underestimate the modulus, sometimes by a large margin (Simons, 1987). It appears that measurements of the modulus are exceptionally sensitive to sample disturbance and the test results can be in error by as much as a factor of 3. Although careful sampling and special laboratory test techniques can reduce this error, direct laboratory testing is generally not a reliable method of measuring the modulus of elasticity.

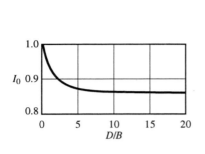

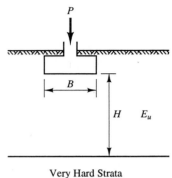

Very Hard Strata

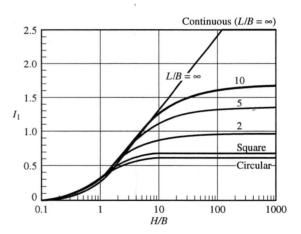

Figure 7.6 Influence factors I_0 and I_1 for use in Equation 7.9 (Adapted from Christian and Carrier, 1978; used with permission). The recommended function for continuous footings is the author's interpretation. H is the distance from the bottom of the footing to some very hard strata, such as bedrock. Usually, the H/B ratio is very large.

The usual response to sample disturbance problems is to conduct in-situ tests. The pressuremeter and dilatometer tests are especially appropriate for this application. These tests measure the modulus in a semidirect fashion and provide reasonably good values as long as the soil is reasonably isotropic (both tests measure the horizontal modulus, whereas the footing responds to the vertical modulus).

The pressuremeter and dilatometer tests are not yet widely used, so engineers must usually rely on other methods. The most common approach is to use empirical correlations between E_u and the undrained shear strength, s_u. The Duncan and

Buchignani (1976) correlation is shown in Figure 7.7 and Equation 7.10.

$$E_u = \beta\, s_u \tag{7.10}$$

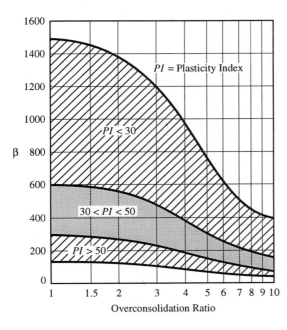

Figure 7.7 Correlation between the undrained modulus of elasticity for clay and the undrained shear strength (Duncan and Buchignani, 1976).

The broad bands in Duncan and Buchignani's correlation reflect its very approximate nature and the level of uncertainty involved. Select a suitably conservative value of β to account for this uncertainty.

If the soil has a large organic content, the modulus may be smaller than suggested by Figure 7.7 and the distortion settlement will be correspondingly higher. For example, Foott and Ladd (1981) reported $\beta \approx 100$ for normally consolidated Taylor River Peat at shear stress levels comparable to those that might be found beneath a spread footing.

If the computed distortion settlement is large, it may be necessary to obtain a more precise assessment of E_u, perhaps using pressuremeter or dilatometer tests. In that case, a more sophisticated analysis, such as that proposed by D'Appolonia, Poulos, and Ladd (1971) may be justified.

Example 7.3

The square footing shown in the sketch below is founded on a clayey soil that extends to a great depth (more than 10B). Compute the distortion settlement of this footing.

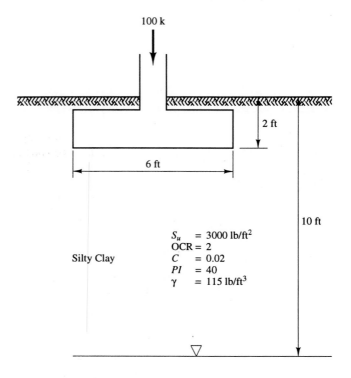

100 k

2 ft

6 ft

10 ft

S_u = 3000 lb/ft^2
OCR = 2
C = 0.02
PI = 40
γ = 115 lb/ft^3

Silty Clay

Solution:

$$\beta = 350 \quad \text{(conservative value from Figure 7.7)}$$

Using Equation 7.10:

$$E_u = \beta\, s_u = (350)(3000) = 1{,}000{,}000 \text{ lb/ft}^2$$

$$D/B = 2/6 = 0.3 \qquad \therefore\ I_0 = 0.98$$

$$L/B = 1, \quad H/B = \infty \qquad \therefore\ I_1 = 0.7$$

Using Equations 6.4 and 7.9:

$$q' = \frac{P}{A} = \frac{100{,}000 \text{ lb}}{6^2 \text{ ft}^2} = 2780 \text{ lb/ft}^2$$

$$\delta_d = \frac{q'\,B}{E_u} I_0 I_1 = \frac{(2780)(6)}{1{,}000{,}000}(0.98)(0.7) = 0.01 \text{ ft} = \textbf{0.1 in} \qquad \Leftarrow \textbf{\textit{Answer}}$$

Consolidation Settlement

The second component of settlement is that due to the primary consolidation. This is the type of settlement discussed in Chapter 3 in the context of loads created by large loaded areas, such as fills. The same general mechanism governs consolidation beneath small loaded areas, such as a footings, with the following additional considerations:

- The change in vertical effective stress, $\Delta\sigma_v'$, is no longer constant with depth. It is largest immediately below the footing and becomes progressively smaller with depth as discussed earlier.
- The consolidation is no longer one-dimensional because the soil may consolidate both vertically and horizontally.

The formula for computing the consolidation settlement beneath a footing is:

$$\delta_c = r \int \psi \, C \log\left(\frac{\sigma_{v0}' + \Delta\sigma_v'}{\sigma_{v0}'}\right) dz \tag{7.11}$$

Where:

δ_c = consolidation settlement

r = rigidity factor

ψ = 3-dimensional adjustment coefficient

C = compressibility

σ_{v0}' = initial vertical effective stress (i.e., before footing load is applied)

$\Delta\sigma_v'$ = increase in vertical effective stress due to footing load

z = depth

Rather than integrating this function, it is customary to divide the soil into n layers, compute the settlement of each layer, and sum them. Thus, Equation 7.11 becomes:

$$\delta_c = r \sum_{i=1}^{n} \psi \, C \, H_i \log\left(\frac{\sigma_{v0}' + \Delta\sigma_v'}{\sigma_{v0}'}\right) \tag{7.12}$$

where H_i is the thickness of layer i. Usually, about three layers provide sufficient accuracy, but more layers may be necessary if the soil is stratified or if additional accuracy is required. When using three layers, choose their thicknesses approximately as shown in Table 7.2. When solving Equation 7.12, compute σ_{v0}' and $\Delta\sigma_v'$ at the midpoint of each layer.

TABLE 7.2 APPROXIMATE THICKNESSES OF SOIL LAYERS FOR MANUAL COMPUTATION OF CONSOLIDATION SETTLEMENT

Layer Number	Approximate Layer Thickness	
	Square Footing	Continuous Footing
1	$B/2$	B
2	B	$2B$
3	$2B$	$4B$

1. Adjust the number and thickness of the layers to account for changes in soil properties. Locate each layer entirely within one soil stratum.
2. For rectangular footings, use layer thicknesses between those given for square and continuous footings.
3. Use somewhat thicker layers (perhaps up to 1.5 times the thicknesses shown) if the groundwater table is very shallow.
4. For quick, but less precise, analyses, use a single layer with a thickness of about $3B$ (square footings) or $6B$ (continuous footings).

The conventional Terzaghi one-dimensional consolidation analysis, as described in Chapter 3, often overestimates the settlement of spread footings. Reasons for this include:

- Sample disturbance
- Lateral strains in the soil
- The stress path in the laboratory consolidation test is not the same as that in the soil beneath a loaded footing (Simons, 1987)

These are the reasons for the adjustment coefficient, ψ, in Equations 7.1 and 7.12. Some engineers have developed values of ψ based on the observed behavior of footings. These values are typically between 0.50 and 0.75 and can be very useful when working in soils similar to those used to develop the factor. When locally derived values are not available, consider those in Table 7.3.

TABLE 7.3 TYPICAL ψ FACTORS

Soil Type	Typical OCR	ψ
Very sensitive clays	1.0	1.0 - 1.2
Normally consolidated clays and silts	1.0 - 1.2	0.7 - 1.0
Overconsolidated clays and silts	1.2 - 5	0.4 - 0.7
Heavily overconsolidated clays and silts	> 5	0.3 - 0.6

Adapted from Skempton and Bjerrum (1957).

The rigidity factor, r, accounts for the difference between the settlement of a perfectly flexible loaded area (as assumed by the consolidation analysis) and that beneath a rigid loaded area such as a spread footing.

Davis and Poulos (1968) suggested the following approximate relationships:

For circular footings:	$\delta \approx 0.50 \, (\delta_{center} + \delta_{edge})$
For rectangular footings:	$\delta \approx 0.33 \, (2\delta_{center} + \delta_{corner})$
For continuous footings:	$\delta \approx 0.50 \, (\delta_{center} + \delta_{edge})$

Where δ_{center}, δ_{edge} and δ_{corner} are the settlements at the center, edge and corner of a perfectly flexible loaded area, and δ is the settlement of a perfectly rigid loaded area.

Based on Boussinesq's stress distribution, $\delta_{edge} / \delta_{center}$ ranges between 0.5 and 1.0, whereas $\delta_{corner} / \delta_{center}$ ranges between 0.25 and 1.0 depending on the variation of compressibility with depth. By combining this with Davis and Poulos' relationships, r must be between 0.75 and 1.00. For normal spread footings, use an r value of 0.85, as shown in Table 7.4.

Example 7.4

Using the data from Example 7.3, compute the consolidation settlement and the total settlement. Assume that secondary compression settlement is negligible.

Solution:

Because there is only one soil type present, use the layer thicknesses from Table 7.2. Compute $\Delta\sigma_v{}'$ using simplified method (Equation 7.6) and δ_{ci} using Equation 7.12.

Layer No.	Depth (ft)	H_i (ft)	At Midpoint of Layer				ψ	C	δ_{ci}/r (ft)
			Depth (ft)	z_f (ft)	$\sigma_{v0}{}'$ (lb/ft^2)	$\Delta\sigma_v{}'$ (lb/ft^2)			
1	2 - 5	3.0	3.5	1.5	402	2780	0.6	0.02	0.032
2	5 - 11	6.0	8.0	6.0	920	1042	0.6	0.02	0.024
3	11 - 23	12.0	17.0	15.0	1518	246	0.6	0.02	0.009
							$\Sigma =$		0.065 ft
							r		$\times$ 0.85
							$\delta_c =$		0.055 ft
									= 0.7 in

$$\delta = \delta_d + \delta_c + \delta_s = 0.1 + 0.7 + 0 = \textbf{0.8 in} \qquad \Leftarrow Answer$$

TABLE 7.4 RIGIDITY FACTORS

Flexibility of Loaded Area	Rigidity Factor, r
Perfectly flexible	1.00
Perfectly rigid (use for footings)	0.85

Secondary Compression Settlement

Secondary compression is the continued straining of a soil after the excess pore water pressures that drive primary consolidation have dissipated. Usually, the secondary compression settlement beneath spread footings is relatively small and may be neglected. However, case histories have been reported where the secondary compression was significant. It is generally a concern only in soils with a high organic content or extremely plastic clays.

Use Equation 3.32 to compute the magnitude of the secondary compression settlement, δ_s.

Rate of Settlement

If the clay is saturated, it is safe to assume the distortion settlement occurs as rapidly as the load is applied. The consolidation settlement will occur over some period, depending on the drainage rate.

Terzaghi's theory of consolidation includes a methodology for computing the rate of consolidation settlement in saturated soils. It is controlled by the rate water is able to squeeze out of the pores and drain away. However, because the soil beneath a footing is able to drain in three dimensions, not one as assumed in Terzaghi's theory, the water will drain away more quickly, so consolidation settlement also will occur more quickly. Davis and Poulos (1968) observed this behavior when they reviewed 14 case histories. In four of these cases, the rate was very much faster than predicted, and in another four cases, the rate was somewhat faster. In the remaining six cases, the rate was very close to or slightly slower than predicted, but this was attributed to the drainage conditions being close to one-dimensional. They also presented a method of accounting for this effect.

Rate estimates become more complex for some partially saturated soils, as discussed in Chapters 20 and 21.

Computer Software - Program FTGSETT

A computer program that computes the settlement of spread footings on cohesive soils, FTGSETT, is included with this book. It uses the analysis techniques described earlier, and can consider square, rectangular, circular and continuous footings. The user can easily change the footing dimensions, loads, and soil properties, thus reducing the tedium of hand computations and making parametric studies much easier to perform.

To access this program, install the software in accordance with the instructions in Appendix C, and then type **FTGSETT** at the DOS prompt. A brief introductory screen will appear, followed by the worksheet. Then, follow this procedure:

1. Choose the footing shape (square, rectangular, continuous, or circular) by pressing **ALT-S** and selecting from the menu. Square is the default.

2. Choose the units of measurement (English or SI) by pressing **ALT-U** and selecting from the menu. English units are the default.

3. Choose the method of computing $\Delta\sigma_v{}'$ (Boussinesq or simplified) by pressing **ALT-D**. The default is Boussinesq.

4. Use the arrow keys to move the cursor and input the title, footing geometry, and soil properties. Replace all of the **XXXXX** indicators with the appropriate data.

5. Hold the down-arrow key until the lower portion of the worksheet scrolls onto the screen. Notice that the computer has divided the soil into many very thin layers. The depths in the first two columns are measured from the ground surface. Move the curser to the first layer and input the unit weight, γ, and press the Enter key.

6. Leave the cursor at the unit weight for the first layer and copy it to the lower layers by pressing **ALT-M** and selecting **COPY** from the menu. This will copy it down to the bottom of the worksheet. If the unit weight changes with depth, move the cursor to the shallowest layer with the new unit weight, type in the value, and then use the copy function as before. Thus, it is possible to input any distribution of unit weight with depth.

7. Move the cursor to the first layer and input the compressibility, C. Use the copy function to create the desired distribution of compressibility with depth.

8. Review the output using the arrow keys or the Home key. The computed distortion, consolidation, and total settlements are shown in the upper right portion of the worksheet.

9. If a printout is needed, press **ALT-M** to access the menu and select one of the following:

> **PRINT** - Prints the worksheet on a printer
> **PRINT TO DISK** - Prints the worksheet to a disk file

> If you select **PRINT TO DISK**, the program will ask for a disk file name. You must select a new file name; the program will not write over an old file.

10. To erase the user input data, press **ALT-M** and select **REFRESH** from the menu.

11. To exit to DOS, press **ALT-M** and select **QUIT** from the menu.

Example 7.5

Solve Examples 7.3 and 7.4 using program FTGSETT.

Partial Output (The full output extends to a depth of 56 ft):

```
          FTGSETT   Version 1.0    (c) 1994 by Prentice Hall, Inc.
          Settlement Analysis of Spread Footings on Cohesive Soils

Title: Example 7.5

Footing Shape:        Square        *******************************
(Press ALT-S to change)            *                               *
                                   *            RESULTS            *
Units of Measurement: English      *                               *
(Press ALT-U to change)            *SETTLEMENT                     *
                                   *                               *
Stress Distribution: Boussinesq    *  Distortion      0.14  in    *
(Press ALT-D to change)            *  Consolidation   0.69  in    *
                                   *                  -----------  *
   Footing Width    =     6.00  ft *  Total            0.83  in    *
   Footing Depth    =     2.00  ft *                               *
   Applied Load     =      100  k  *                               *
   Soil Modulus     =  1000000 psf *BEARING PRESSURE               *
   Groundwater Depth =   10.00  ft *                               *
   3-D Adjus. Coeff. =    0.600    *   Gross   q  = 3008  psf     *
                                   *   Net     q' = 2778  psf     *
                                   *******************************

Date:                 07-May-93
Time:                 04:58 PM
```

Layer Depth (ft) Top	Bottom	Unit Weight (pcf)	Compressi-bility C	Initial Eff. Stress at Midpoint (psf)	Change in Eff. Stress at Midpoint (psf)	Consol Settle-ment (ft)	Strain (%)
0.00	2.00	115.0	0.020	115	0	0.000	0.00
2.00	2.50	115.0	0.020	259	2777	0.005	1.09
2.50	3.00	115.0	0.020	316	2748	0.005	1.01
3.00	3.50	115.0	0.020	374	2655	0.005	0.93
3.50	4.00	115.0	0.020	431	2496	0.004	0.85
4.00	4.50	115.0	0.020	489	2289	0.004	0.77
4.50	5.00	115.0	0.020	546	2061	0.003	0.69
5.00	5.50	115.0	0.020	604	1835	0.003	0.62
5.50	6.00	115.0	0.020	661	1623	0.003	0.55
6.00	6.50	115.0	0.020	719	1432	0.002	0.49
6.50	7.00	115.0	0.020	776	1264	0.002	0.43
7.00	7.50	115.0	0.020	834	1117	0.002	0.38
7.50	8.00	115.0	0.020	891	990	0.002	0.33
8.00	9.00	115.0	0.020	978	832	0.003	0.27
9.00	10.00	115.0	0.020	1093	669	0.002	0.21
10.00	11.00	115.0	0.020	1176	547	0.002	0.17
11.00	12.00	115.0	0.020	1229	453	0.001	0.14
12.00	13.00	115.0	0.020	1282	381	0.001	0.12
13.00	14.00	115.0	0.020	1334	324	0.001	0.10
14.00	15.00	115.0	0.020	1387	279	0.001	0.08
15.00	16.00	115.0	0.020	1439	242	0.001	0.07

Factor of Safety

Notice that the settlement analysis does not include a factor of safety in the same way that a bearing capacity analysis does. These analyses produce an estimate (although possibly a conservative one) of the real settlement of the footing. The factor of safety is implicit within the tolerable settlements presented in Chapter 2.

7.3 SETTLEMENT OF FOOTINGS ON COHESIONLESS SOILS

The design of spread footings on cohesionless soils is nearly always governed by settlement, not bearing capacity. The only likely exceptions to this rule would be very narrow and shallow footings, especially when a shallow groundwater table is present. Therefore, the primary emphasis is on the analysis of settlement.

Techniques for computing settlement in cohesionless soils are nearly always based on in-situ tests to avoid the sample disturbance problems. The test results are then combined with empirical analysis methods.

Analyses Based on the Standard Penetration Test

Although the standard penetration test does not directly measure the stress-strain properties of soils, it does provide a fair correlation with compressibility. Dense, relatively incompressible sands will have high N values, while loose, compressible sands will have low N values. However, this correlation can only be found experimentally.

Many methods have been proposed to compute settlements of sands based on the SPT (Jeyapalan and Boehm, 1986). Only two are presented here: the Modified Meyerhof's Method and Burland and Burbidge's Method.

Modified Meyerhof's Method

Meyerhof's Method, introduced in 1956, has been one of the more popular ways to compute settlement based on standard penetration test data. Since Meyerhof had very little test data, he intentionally made this method very conservative. Therefore, it nearly always overpredicts the settlement, often by a factor of 3 or more. Therefore, Meyerhof (1965) suggested adjusting his original formulas by a factor of 1.5. Let us call this revised procedure the *Modified Meyerhof's Method*. The following formulas include this adjustment factor:

For $B \leq 4$ ft (1.2 m):

$$\frac{\delta}{B_r} = \frac{0.44 \, q' / \sigma_r}{\bar{N}_{60} \, K_d} \tag{7.13}$$

For $B > 4$ ft (1.2 m):

$$\frac{\delta}{B_r} = \frac{0.68\, q'/\sigma_r}{\overline{N}_{60}\, K_d} \left(\frac{B}{B + B_r}\right)^2 \tag{7.14}$$

Where:

δ = settlement

B_r = reference width = 1 ft = 0.3 m = 12 in = 300 mm

q' = net bearing pressure

σ_r = reference stress = 2000 lb/ft^2 = 100 kPa

$\overline{N}_{60}$ = average SPT N_{60} value between the bottom of the footing and a depth $2B$ below the bottom

B = footing width

K_d = depth factor = $1 + 0.33\, D/B \le 1.33$

Meyerhof considered these formulas to be valid for footings of all shapes. Although this is a simplification of reality, it may be sufficiently accurate for normal design purposes.

Do not correct the field N_{60} values for overburden, but do adjust them using Equation 7.15 when the soil is a dense silty sand below the groundwater table and $N_{60} > 15$.

$$N_{60\,\text{adjusted}} = 15 + 0.5\,(N_{60\,\text{field}} - 15) \tag{7.15}$$

Meyerhof suggested that groundwater table effects would be implicitly incorporated into the SPT results. However, consider adjusting the measured N_{60} values if the sand was dry during testing but may become saturated later.

Even with the 1.5 adjustment factor, the Modified Meyerhof Method still tends to be conservative. It overestimates the settlement about 75 percent of the time.

Example 7.6

A 250 k column load is to be supported on a square spread footing founded at a depth of 3 ft below the ground surface. The maximum allowable total settlement is 0.75 in. The soil is a normally consolidated silty sand and the groundwater table is at a depth of 15 ft below the ground surface. The SPT N_{60} values are:

Depth (ft)	4	7	10	13	16	20	25
N_{60}	15	13	19	23	27	32	30

Determine the required footing width.

Solution:

Estimate $q_a' = 4000$ lb/ft^2 and use it to obtain an approximate value of B:

$$B = \sqrt{\frac{P}{q_a'}} \approx \sqrt{\frac{250,000}{4000}} = 7.9 \text{ ft}$$

Therefore, the zone of soil between the bottom of the footing and a depth $2B$ below the bottom is represented by the N_{60} values at 4, 7, 10, 13, and 16 ft. The value at 16 ft must be adjusted using Equation 7.15, so it changes from 27 to 21. Thus:

$$\bar{N}_{60} = \frac{15 + 13 + 19 + 23 + 21}{5} = 18$$

The depth factor is approximately:

$$K_d = 1 + 0.33 \ D/B \approx 1 + 0.33 \ (3/7.9) = 1.12$$

Compute the required footing width using Equation 7.14:

$$\frac{\delta}{B_r} = \frac{0.68 \ q'/\sigma_r}{\bar{N}_{60} \ K_d} \left(\frac{B}{B + B_r}\right)^2$$

$$= \frac{0.68 \ (P/B^2)/\sigma_r}{\bar{N}_{60} \ K_d} \left(\frac{B}{B + B_r}\right)^2$$

$$\frac{0.75}{12} = \frac{0.68 \ (250,000/B^2)/2000}{(18)(1.12)} \left(\frac{B}{B + 1}\right)^2$$

$$0.0148 \ B^2 = \left(\frac{B}{B + 1}\right)^2$$

$$B = \textbf{7 ft 3 in} \ \text{(solved to nearest 3 in)} \qquad \Leftarrow \textit{Answer}$$

In this case, there is no need to go back and revise K_d.

Burland and Burbidge's Method

Burland and Burbidge (1985) presented another empirical method of using SPT data to compute the settlement of spread footings. They developed this method from a database of more than 200 records of measured settlements on sands and gravels. Since it is based on such a large database, this method is more precise and less conservative than the Modified Meyerhof Method. However, it still suffers from the uncertainties of the standard penetration test.

Their procedure is as follows:

1. Obtain a series of SPT N_{60} values at various depths below the proposed footing. Do not apply an overburden correction. However, if the soil is a fine sand or a silty sand below the groundwater table and it has an $N_{60} > 15$, adjust N_{60} using Equation 7.15.

   If the soil is a gravel or sandy gravel, Burland and Burbidge recommend multiplying N_{60} by an adjustment factor of 1.25. However, be very cautious when attempting to use SPT based methods in gravels because the N values are usually very unreliable.

2. Compute the depth of influence below the bottom of the footing. If the adjusted SPT N_{60} values are constant or increasing with depth, then the depth of influence is:

$$\frac{z_I}{B_r} = 1.4 \left(\frac{B}{B_r} \right)^{0.75} \tag{7.16}$$

   Where:

   z_I = depth of influence below bottom of footing

   B = footing width

   B_r = reference width = 1 ft = 0.3 m = 12 in = 300 mm

   If the SPT N_{60} values consistently decrease with depth, use $z_I = 2B$ or the depth to the bottom of the soft layer, whichever is less.

3. Compute the average of the adjusted N_{60} values between the bottom of the footing and the depth of influence. This average value is $\overline{N}_{60}$.

4. Compute the compressibility index, I_c:

   For normally consolidated soils:

$$I_c = \frac{1.71}{(\overline{N}_{60})^{1.4}} \tag{7.17}$$

   For overconsolidated soils:

$$I_c = \frac{0.57}{(\overline{N}_{60})^{1.4}} \tag{7.18}$$

5. Compute the depth of influence correction factor:

$$C_I = \frac{H}{z_I}\left(2 - \frac{H}{z_I}\right) \le 1 \qquad (7.19)$$

Where:

C_I = depth of influence correction factor

H = depth from bottom of footing to bottom of compressible soil

This factor is of concern only when a loose soil is underlain by a much denser soil, and the interface between these layers is at a depth of less than z_I below the bottom of the footing.

6. Compute the shape factor:

$$C_s = \left[\frac{1.25\ L/B}{(L/B) + 0.25}\right]^2 \qquad (7.20)$$

Where:

C_s = shape factor

B = width of foundation

L = length of foundation

Note that $C_s = 1$ for square and circular footings and 1.56 for continuous footings.

7. Compute the settlement:

For normally consolidated soils:

$$\frac{\delta}{B_r} = 0.14\ C_s\ C_I\ I_c\left(\frac{B}{B_r}\right)^{0.7}\left(\frac{q'}{\sigma_r}\right) \qquad (7.21)$$

For overconsolidated soils with $q' \le \sigma_c'$:

$$\frac{\delta}{B_r} = 0.047\ C_s\ C_I\ I_c\left(\frac{B}{B_r}\right)^{0.7}\left(\frac{q'}{\sigma_r}\right) \qquad (7.22)$$

For overconsolidated soils with $q' > \sigma_c'$:

$$\frac{\delta}{B_r} = 0.14\ C_s\ C_I\ I_c\left(\frac{B}{B_r}\right)^{0.7}\left(\frac{q' - 0.67\ \sigma_c'}{\sigma_r}\right) \qquad (7.23)$$

Where:

    δ = settlement

    q' = net bearing pressure

    B_r = reference width = 1 ft = 0.3 m = 12 in = 300 mm

    σ_r = reference stress = 2000 lb/ft^2 = 100 kPa

    σ_c' = preconsolidation stress

It is very difficult to assess σ_c' in cohesionless soils, and to determine whether they are normally consolidated or overconsolidated. Therefore, Equations 7.22 and 7.23 should be used with caution.

    Reapplying this method to the data from which it was obtained suggests that the 95% confidence interval corresponds to about ±50% of the computed settlement. Thus, it appears to be slightly more precise than the Modified Meyerhof's method. However, Burland and Burbidge's method is intended to produce a best estimate of the true settlement, so it should overpredict the settlement as often as it underpredicts. This is fundamentally different from the Modified Meyerhof's method which is intended to be close to an upper-bound solution.

Long-Term Performance

Very few footings have been monitored for extended periods, but the little data that are available indicate that some secondary settlement continues to occur, even after the primary settlement (as predicted by the earlier formulas) is complete. This secondary settlement may be due to the presence of silt or clay layers within the zone of influence. However, a series of long-term measurements made on structures in Poland (Bolenski, 1973) indicates that footings with loads that remain fairly constant (such as buildings) have less secondary settlement than those with varying loads (such as storage tanks). This suggests that repeated cycles of loading may influence the sand, especially if these cycles are large, thus creating secondary settlements.

    Burland and Burbidge suggest that under relatively constant or slightly fluctuating loads, the settlement of footings on sand 30 years after construction might be as much as 1.5 times the immediate post-construction settlement. For footings subjected to heavily fluctuating loads, it might be as much as 2.5 times the immediate post-construction settlement.

Example 7.7

Rework Example 7.6 using Burland and Burbidge's method and compare the results.

Solution:

Estimate B = 7.9 ft (from Example 7.6).

Using Equation 7.16:

$$\frac{z_I}{B_r} = 1.4 \left(\frac{B}{B_r}\right)^{0.75}$$

$$\frac{z_I}{1} = 1.4 \left(\frac{7.9}{1}\right)^{0.75}$$

$$z_I = 6.6 \text{ ft}$$

This is much less than the 15.8 ft depth of influence used in Example 7.6.

In this case, only the N_{60} values at 4 and 7 ft lie within the depth of influence. (Most real design problems would have SPT data from more than one boring, so we would be able to use correspondingly more test data).

$$\bar{N}_{60} = \frac{15 + 13}{2} = 14$$

Using Equation 7.17:

$$I_c = \frac{1.71}{(\bar{N}_{60})^{1.4}} = \frac{1.71}{14^{1.4}} = 0.0425$$

$$C_I = 1$$

$$C_s = 1$$

Using Equation 7.21:

$$\frac{\delta}{B_r} = 0.14 \, C_s \, C_I \, I_c \left(\frac{B}{B_r}\right)^{0.7} \left(\frac{q'}{\sigma_r}\right)$$

$$= 0.14 \, C_s \, C_I \, I_c \left(\frac{B}{B_r}\right)^{0.7} \left(\frac{P/B^2}{\sigma_r}\right)$$

$$\frac{0.75}{12} = (0.14)(1)(1)(0.0425) \left(\frac{B}{1}\right)^{0.7} \left(\frac{250,000/B^2}{2000}\right)$$

$$B = \mathbf{6\,ft\,9\,in} \quad (\text{solved to nearest 3 in}) \quad \Leftarrow Answer$$

Even though we used a smaller $\overline{N}_{60}$ than in Example 7.6 (due to the smaller effective depth), the computed B was smaller. This is because Burland and Burbidge's method produces a best estimate of the settlement, whereas the Modified Meyerhof's method is close to an upper-bound solution.

Schmertmann's Method

Schmertmann's method (Schmertmann, 1970, 1978; and Schmertmann et al., 1978) is a popular and useful technique for computing the settlement of footings on cohesionless soils. This method is more precise than either the Modified Meyerhof or Burland and Burbidge methods because:

- It is based on cone penetration test data, which has much greater resolution and precision than the standard penetration test. (It's also possible to use this method with other data, including the SPT).
- It allows the engineer to divide the soil into layers and assign a different modulus to each layer, while the other methods require a single N_{60} value. This is especially useful in complex soil profiles.
- This method considers the relative importance of each layer through the use of strain influence factors. These factors were developed from model load tests and finite element analyses.

The Schmertmann procedure is as follows:

1. Perform appropriate in-situ tests to determine the modulus, E, of the soil and its variation with depth. This is most frequently done using the cone penetration test (CPT). Correlations between E and q_c are presented in Table 7.5.

   As mentioned earlier, it is very difficult to determine whether cohesionless soils are normally consolidated or overconsolidated. Therefore, the E/q_c ratio for overconsolidated soils in Table 7.5 should be used cautiously.

TABLE 7.5 CORRELATIONS BETWEEN E AND q_c

Soil Type	E/q_c
Young, normally consolidated silica sands (age < 100 years)	2.5
Aged, normally consolidated silica sands (age > 3000 years)	3.5
Overconsolidated silica sands	6.0

Adapted from Schmertmann et al. (1978), and Robertson and Campanella (1989).

2. Consider the soil from the base of the footing to the depth of influence ($2B$ for square footings, $4B$ for continuous footings) below the base or to an incompressible strata, whichever is shallower. Divide this zone into layers and assign a

representative E value to each layer. The required number of layers and the thickness of each layer depend on the variations in the E vs. depth profile. Typically 5 to 10 layers are appropriate.

3. The strain influence factor varies with depth, as shown in Figure 7.8. To compute the magnitude of this factor for each layer, it is first necessary to determine the peak value, I_{zp}:

$$I_{zp} = 0.5 + 0.1 \sqrt{\frac{q'}{\sigma'_{vp}}} \qquad (7.24)$$

Where:

I_{zp} = peak strain influence factor

q' = net bearing pressure

σ'_{vp} = initial vertical effective stress at depth of peak strain influence factor (depth = $D + B/2$ for square and circular footings, $D + B$ for continuous footings)

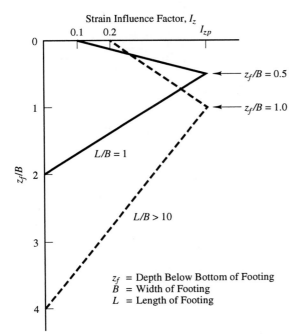

z_f = Depth Below Bottom of Footing
B = Width of Footing
L = Length of Footing

Figure 7.8 Distribution of strain influence factor with depth under square and continuous footings (Adapted from Schmertmann, 1978; Used by permission of ASCE).

4. Compute the strain influence factor at the midpoint of each layer:

Square and circular footings:

For $z_f = 0$ to $B/2$: $\qquad\qquad$ $I_z = 0.1 + (z_f/B)\,(2I_{zp} - 0.2)$ $\qquad$ (7.25)

For $z_f = B/2$ to $2B$: $\qquad\qquad$ $I_z = 0.667\,I_{zp}\,(2 - z_f/B)$ $\qquad$ (7.26)

Continuous footings ($L/B \ge 10$):

For $z_f = 0$ to B: $\qquad\qquad\quad$ $I_z = 0.2 + (z_f/B)\,(I_{zp} - 0.2)$ $\qquad$ (7.27)

For $z_f = B$ to $4B$: $\qquad\qquad\;$ $I_z = 0.333\,I_{zp}\,(4 - z_f/B)$ $\qquad$ (7.28)

Rectangular footings ($1 < L/B < 10$):

$$I_z = I_{zs} + 0.111\,(I_{zc} - I_{zs})\,(L/B - 1) \qquad (7.29)$$

Where:

z_f = depth from bottom of footing to midpoint of layer

I_z = strain influence factor

$I_{zc} = I_z$ for a continuous footing

I_{zp} = peak I_z

$I_{zs} = I_z$ for a square footing

5. Compute the settlement:

$$\delta = C_1\,C_2\,C_3\,q'\sum_{i=1}^{n}\frac{I_{zi}\,H_i}{E_i} \qquad (7.30)$$

$$C_1 = 1 - 0.5\left(\frac{\sigma'_D}{q'}\right) \qquad (7.31)$$

$$C_2 = 1 + 0.2\,\log\left(\frac{t}{0.1}\right) \qquad (7.32)$$

$$C_3 = 1.03 - 0.03\,L/B \quad \ge 0.73 \qquad (7.33)$$

Where:

δ = settlement of footing

C_1 = depth factor

C_2 = secondary creep factor

C_3 = shape factor

$\sigma_D{}'$ = effective stress at a depth D below the ground surface

n = number of soil layers

$I_{zi} = I_z$ at midpoint of soil layer i

H_i = thickness of soil layer i

E_i = modulus of soil layer i

t = time since application of load (yr) $(t \geq 0.1$ yr$)$

B = footing width

L = footing length

These formulas may be used with any consistent set of units, except that t must be expressed in years.

The greatest change in vertical effective stress, $\Delta\sigma_v{}'$, occurs in the soil immediately below the footing, as shown in Figure 7.2. However, model studies and finite element analyses indicate the strain influence factor reaches a peak value at a depth of $B/2$ to B below the bottom of the footing. This behavior may be due to the sliding friction between the bottom of the footing and the soil, and the corresponding lateral restraint on the soil.

Using Other Test Methods

Although Schmertmann's method was developed for use with cone penetration test (CPT) data, it also may be used with modulus data from other test methods, including the standard penetration test (SPT), the dilatometer test (DMT), and the pressuremeter test (PMT). The latter two tests are especially promising because they measure the deformation properties more directly. Leonards and Frost (1988) proposed a method of combining CPT and DMT data to assess both compressibility and overconsolidation and use the results in a modified version of Schmertmann's method.

It is also possible to convert standard penetration test (SPT) results to equivalent CPT q_c values using Figure 4.26, and then use these values to compute the modulus.

Example 7.8

The results of a CPT sounding performed at McDonald's Farm near Vancouver, British Columbia, are shown in Figure 7.9. The soils at this site consist of young, normally consolidated sands with some interbedded silts. The groundwater table is at a depth of 2.0 m below the ground surface.

A 375 kN/m load is to be supported on a 2.5 m × 30 m footing to be founded at a depth of 2.0 m in this soil. Use Schmertmann's method to compute the settlement of this footing soon after construction and the settlement 50 years after construction.

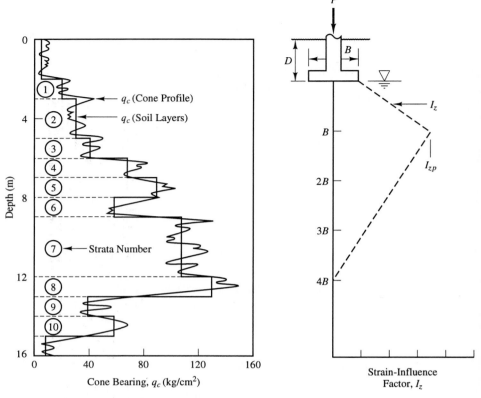

Figure 7.9 CPT results at McDonald's Farm (Adapted from Robertson and Campanella, 1988, and Brown, 1984).

Solution:

Depth of influence $= D + 4B = 2.0 + 4(2.5) = 12.0$ m

Layer No.	Depth (m)	q_c (kg/cm^2)	E (kPa)
1	2.0 - 3.0	20	5,000
2	3.0 - 5.0	30	7,500
3	5.0 - 6.0	41	10,000
4	6.0 - 7.0	71	17,500
5	7.0 - 8.0	92	22,500
6	8.0 - 9.0	61	15,000
7	9.0 - 12.0	112	27,500

Using Equation 6.6:

$$q' = \frac{P/b}{B} = \frac{375 \text{ kN/m}}{2.5 \text{ m}} = 150 \text{ kPa}$$

Use $\gamma = 17$ kN/m³ above groundwater table and 20 kN/m³ below (from Table 3.3).

$$\sigma'_{vp}(at\, z = D + B) = \Sigma\gamma h - u = (17)(2) + (20)(2.5) - (9.8)(2.5) = 59.5 \text{ kPa}$$

Using Equations 7.24 - 7.33:

$$I_{zp} = 0.5 + 0.1\sqrt{\frac{q'}{\sigma'_{vp}}} = 0.5 + 0.1\sqrt{\frac{150}{59.5}} = 0.659$$

Layer No.	E (kPa)	z_f (m)	I_{zi}	H_i (m)	$I_{zi}\,H_i/E$
1	5,000	0.5	0.292	1.0	5.84×10^{-5}
2	7,500	2.0	0.567	2.0	15.12×10^{-5}
3	10,000	3.5	0.571	1.0	5.71×10^{-5}
4	17,500	4.5	0.483	1.0	2.76×10^{-5}
5	22,500	5.5	0.395	1.0	1.76×10^{-5}
6	15,000	6.5	0.308	1.0	2.05×10^{-5}
7	27,500	8.5	0.132	3.0	1.44×10^{-5}
				$\Sigma =$	34.68×10^{-5}

$$\sigma'_D = \gamma D = (17)(2) = 34 \text{ kPa}$$

$$C_1 = 1 - 0.5\left(\frac{\sigma'_D}{q'}\right) = 1 - 0.5\left(\frac{34}{150}\right) = 0.887$$

$$C_3 = 1.03 - 0.03\, L/B \geq 0.73$$
$$= 1.03 - 0.03\,(30/2.5)$$
$$= 0.67$$
$$Use\ C_3 = 0.73$$

At $t = 0.1$ yr:

$$C_2 = 1$$

$$\delta = C_1\, C_2\, C_3\, q' \sum_{i=1}^{n} \frac{I_z H_i}{E}$$

$$= (0.887)(1)(0.73)(150)(34.68 \times 10^{-5})$$
$$= 0.034 \text{ m}$$
$$= \textbf{34 mm} \qquad\qquad \Leftarrow \textit{Answer}$$

At $t = 50$ yr:

$$C_2 = 1 + 0.2 \log\!\left(\frac{t}{0.1}\right) = 1 + 0.2 \log\!\left(\frac{50}{0.1}\right) = 1.54$$

$$\delta = 34(1.54) = \textbf{52 mm} \qquad\qquad \Leftarrow \textit{Answer}$$

Computer Software - Program SCHMERT

A computer program, SCHMERT, that computes the settlement of spread footings on cohesionless soils using Schmertmann's method is included with this book. It can consider square, rectangular, circular and continuous footings. The user can easily change the footing dimensions, loads, and soil properties, thus reducing the tedium of hand computations and making parametric studies much easier to perform.

To access this program, install the software in accordance with the instructions in Appendix C, then type **SCHMERT** at the DOS prompt. A brief introductory screen will appear, followed by the worksheet. Then, follow this procedure:

1. Choose the units of measurement (English or SI) by pressing **ALT-U** and selecting from the menu. English units are the default.

2. Choose the footing shape (square, rectangular, continuous, or circular) by pressing **ALT-S** and selecting from the menu. Square is the default.

3. Use the arrow keys to move the cursor and input the title, footing geometry, and soil properties. Replace all of the **XXXXX** indicators with the appropriate data.

4. Hold the down-arrow key until the lower portion of the worksheet scrolls onto the screen. Notice that the computer has divided the soil into many very thin layers. The depths in the first two columns are measured from the ground surface. Move the curser to the first layer and input the modulus, E, and press the Enter key.

5. Leave the cursor at the modulus value for the first layer and copy it to the lower layers by pressing **ALT-M** and selecting **COPY** from the menu. This will copy it down to the bottom of the worksheet. If the modulus changes with depth, move the cursor to the shallowest layer with the new modulus, type in the value, and then use the copy function as before. Thus, it is possible to input any distribution of modulus with depth.

6. Review the output using the arrow keys or the Home key. The computed settlement is shown in the upper right portion of the worksheet.

7. If a printout is needed, press **ALT-M** to access the menu and select one of the following:

PRINT - Prints the worksheet on a printer
PRINT TO DISK - Prints the worksheet to a disk file

If you select **PRINT TO DISK**, the program will ask for a disk file name. You must select a new file name; the program will not write over an old file.

8. To erase the user input data, press **ALT-M** and select **REFRESH** from the menu.

9. To exit to DOS, press **ALT-M** and select **QUIT** from the menu.

Example 7.9

Solve Example 7.8 using program SCHMERT.

Partial Output:

```
     SCHMERT - Version 1.0            (c) 1994 by Prentice Hall, Inc.
       Settlement of Spread Footings on Cohesionless Soils
                   Using Schmertmann's Method

Title:   Example 7.9

Units of Measurement:  SI            Date:      07-May-93
        (Press ALT-U to change)      Time:      06:21 PM

Footing Shape:         Rectangular
        (Press ALT-S to change)      *****************************
                                     *                           *
                                     *          RESULTS          *
                                     *                           *
Footing Length     =    30.00  m     *Bearing Pressure           *
Footing Width      =     2.50  m     * Gross   q =       184 kPa *
Footing Depth      =     2.00  m     * Net     q'=       150 kPa *
Applied Load       =    11250  kN    *                           *
Elapsed Time       =    50.00  yr    *Settlement =     51.16 mm  *
Soil Unit Weight                     *                           *
   Above Groundwater =    17.0  kN/m3 *****************************
   Below Groundwater =    20.0  kN/m3
Depth to Groundwater =     2.0  m
```

:Depth Interval :		Soil	:Influence:		:	:
: Upper :	Lower :	Modulus :	Factor	:Compression:	Strain	:
: (m) :	(m) :	(kPa) :	Iz	: (m) :	(%)	:
: 0.0 :	2.0 :		:	:	:	
: 2.0 :	2.1 :	5000 :	0.209 :	0.00063 :	0.63	:
: 2.1 :	2.2 :	5000 :	0.228 :	0.00068 :	0.68	:
: 2.2 :	2.3 :	5000 :	0.246 :	0.00074 :	0.74	:
: 2.3 :	2.4 :	5000 :	0.264 :	0.00079 :	0.79	:
: 2.4 :	2.5 :	5000 :	0.283 :	0.00084 :	0.84	:
: 2.5 :	2.6 :	5000 :	0.301 :	0.00090 :	0.90	:
: 2.6 :	2.7 :	5000 :	0.319 :	0.00095 :	0.95	:
: 2.7 :	2.8 :	5000 :	0.338 :	0.00101 :	1.01	:
: 2.8 :	2.9 :	5000 :	0.356 :	0.00106 :	1.06	:
: 2.9 :	3.0 :	5000 :	0.374 :	0.00112 :	1.12	:
: 3.0 :	3.1 :	7500 :	0.393 :	0.00078 :	0.78	:
: 3.1 :	3.2 :	7500 :	0.411 :	0.00082 :	0.82	:

Note: The full printout extends to a depth of 16.5 m.

7.4 SETTLEMENT OF FOOTINGS ON STRATIFIED SOILS

The commonly used analysis techniques for predicting settlement are based on soil profiles that are either entirely cohesive or entirely cohesionless. This creates difficulties when engineers must deal with stratified soil profiles.

If the soil profile is predominantly cohesive, then we can compute the settlement using the method for cohesive soils, as described in Section 7.2. The compressibility, C, for the cohesionless portions may be determined using Equations 3.18 and 3.29.

Conversely, if the profile is predominantly cohesionless, then it is better to use Schmertmann's method. The modulus, E, for the cohesive layers may be determined using Equations 3.18 and 3.29.

The more difficult case is when both types of soil are present and both significantly contribute to the settlement of the footing. In such cases, consider using Schmertmann's method to evaluate the cohesionless layers (assigning $E = \infty$ for the cohesive layers) and the method described in Section 7.2 for the cohesive layers (using $C = 0$ for the cohesionless layers), and then adding the results of the two analyses.

7.5 ACCURACY OF SETTLEMENT PREDICTIONS

After studying many pages of formulas and procedures, the reader may develop the mistaken impression that settlement analyses are an exact science. This is by no means true. It is good to recall a quote from Terzaghi (1936):

> Whoever expects from soil mechanics a set of simple, hard-and-fast rules for settlement computations will be deeply disappointed. He might as well expect a simple rule for constructing a geologic profile from a single test boring record. The nature of the problem strictly precludes such rules.

Although much progress has been made since 1936, the settlement problem is still a difficult one. The methods described in this chapter should be taken as guides, not dictators, and should be used with engineering judgment. A vital ingredient in this judgement is an understanding of the sources of error in the analysis. These include:

- Uncertainties in defining the soil profile. This is the largest single cause. There have been many cases of unexpectedly large settlements due to undetected compressible layers, such as peat lenses.
- Disturbance of soil samples.
- Errors in in-situ tests (especially the SPT).
- Errors in laboratory tests.
- Uncertainties in defining the service loads, especially when the live load is a large portion of the total load.
- Construction tolerances (i.e., footing not built to the design dimensions).

- Errors in determining the degree of overconsolidation.
- Inaccuracies in the analysis methodologies.
- Neglecting soil-structure interaction effects.

We can reduce some of these errors by employing more extensive and meticulous exploration and testing techniques, but there are economic and technological limits to such efforts.

Because of these errors, the actual settlement of a spread footing may be quite different from the computed settlement. Figure 7.10 shows 90 percent confidence intervals for spread footing settlement computations.

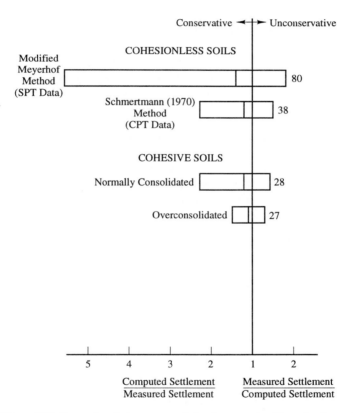

Figure 7.10 Comparison between computed and measured settlements of spread footings. Each bar represents the 90 percent confidence interval (i.e., 90 percent of the settlement predictions will be within this range). The line in the middle of each bar represents the average prediction, and the number to the right indicates the number of data points used to evaluate each method. (Based on data from Burland and Burbidge, 1985; Schmertmann, 1970; Wahls, 1985 and Butler, 1975).

We can draw the following conclusions from this data:

- Settlement predictions are conservative more often than they are unconservative (i.e. they tend to overpredict the settlement more often than they underpredict it). However, the range of error is quite wide.
- Settlement predictions made using the Schmertmann method with CPT data are much more precise than those based on the SPT. (Note that these results are based on the 1970 version of Schmertmann's method. Later refinements, as reflected in this chapter, should produce more precise results).
- Settlement predictions in cohesive soils, especially those that are overconsolidated, are usually more precise than those for cohesionless soils. However, the magnitude of settlement in cohesive soils is often greater.

Many of the soil factors that cause the scatter in Figure 7.10 do not change over short distances, so predictions of differential settlements should be more precise than those for total settlements. Therefore, the allowable differential settlement criteria described in Table 2.2 (which include factors of safety of at least 1.5) reflect an appropriate level of conservatism.

7.6 SETTLEMENT OF SPREAD FOOTINGS ON ROCK

Spread footings on rock that are designed using presumptive bearing values, such as those in Table 6.5 in Chapter 6, will have very little settlement. Thus, it is usually not necessary to perform a settlement analysis. However, it may be appropriate to do so for large or heavily loaded footings supported on poor or moderate quality rock.

Settlement analyses on rock may be based on elastic theory using Equation 7.9. The modulus of elasticity, E, must reflect the properties of the rock mass, including the discontinuities, and thus is difficult to measure. An approximate method is to measure E of a small sample in the laboratory, and then multiply it by α_E to account for the discontinuities. Methods of determining α_E are discussed in Section 6.7 in Chapter 6. Alternatively, we could measure it in-situ using devices such as the *Goodman jack*.

7.7 LOW STRAIN LOAD TESTS

In keeping with the observational approach to geotechnical engineering, it would be valuable to have a means of measuring the load-settlement behavior of footings after they are built but before construction of the superstructure begins. The measured behavior could be compared to the anticipated behavior, thus allowing construction to continue with a much greater degree of confidence.

The Leaning Tower of Pisa

During the Middle Ages, Europeans began to build larger and heavier structures, pushing the limits of design well beyond those of the past. In Italy, the various republics erected towers and campaniles to symbolize their power (Kerisel, 1987). Unfortunately, vanity and ignorance often lead to more emphasis on creative architecture than on structural integrity, and many of these structures collapsed. Although some of these failures were due to poor structural design, many were the result of overloading the soil. Other monuments tilted, but did not collapse. The most famous of these is the campanile in Pisa, more popularly known as the Leaning Tower of Pisa.

Construction of the tower began in the year 1173 under the direction of Bananno Pisano and continued slowly until 1178. This early work included construction of a ring-shaped footing 64.2 ft (19.6 m) in diameter along with three and one-half stories of the tower. By then, the average bearing pressure below the footing was about 6900 lb/ft^2 (330 kPa) and the tower had already begun to tilt. Therefore, the construction activity ceased.

Nearly a century later, in the year 1271, construction resumed under the direction of a new architect, Giovanni Di Simone. Although it probably would have been best to tear down the completed portion and start from scratch with a new and larger foundation, Di Simone chose to continue working on the uncompleted tower, attempting to compensate for the tilt by tapering the successive stories and adding extra weight to the high side. He stopped work in 1278 at the seventh cornice. Finally, the tower was completed with the construction of the belfry during a third construction period, sometime between 1360 and 1370. Altogether, the project had taken nearly 200 years.

The axis of the belfry is inclined at an angle of 3° from the rest of the tower (Mascardi, 1982), which was probably the angle of tilt at that time. However, studies of the tapering of the masonry, reported by Leonards (1979) suggests that the tower has not moved in one direction only, but has at various times tilted in nearly all directions of the compass.

It appears that the rate of tilting gradually slowed until the middle of the nineteenth century when pumping of city water wells lowered the groundwater table and caused the titling to accelerate (Pryke, 1987). The rate slowed again during the early part of the twentieth century, but it is still moving at a rate of about 7 seconds of arc per year. There is evidence to suggest that pumping of groundwater may still be part of the cause of the continued movement (Croce, et. al., 1981). By 1982, the tower was inclined toward the south at an angle of 5.5° (Mascardi, 1982), which means that the top of the 192 ft (58.4 m) tall structure is 18.4 ft (5.6 m) off of being plumb. By then, the average total settlement was over 8.2 ft (2.5 m) and the differential settlement was about 6.6 ft (2 m). The current configuration is shown in Figure 7.11.

The average gross bearing pressure under the completed tower is 10,400 lb/ft^3 (497 kPa), but the tilting has caused its weight to act eccentrically on the foundation, so the bearing pressure is not uniform. It now ranges from 1,300 to 19,600 lb/ft^2 (62 - 930 kPa).

The subsurface conditions below the tower have been investigated, including an exhaustive program sponsored by the Italian government that began in 1965. The profile, shown in Figure 7.12, is fairly uniform across the site and consists of sands underlain by fat clays.

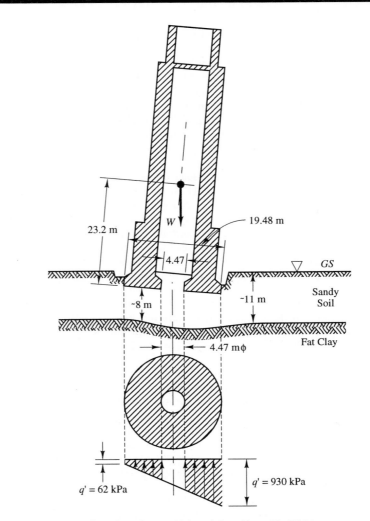

Figure 7.11 Current configuration of tower (Adapted from Terzaghi, 1934a).

Many engineers, including Karl Terzaghi (1934a), have studied the tower and tried to determine the cause of its tilting. Mitchell, Vivatrat, and Lambe (1977) suggest that the differential settlements may have been due to nonuniformities in the soil profile, especially in the upper sand layer. They also believe that a bearing capacity failure would probably have occurred if the entire tower had been completed during the 1100s. Leonards (1979) provided an alternative interpretation that suggests that the tilting is largely due to a local shear type of bearing capacity failure.

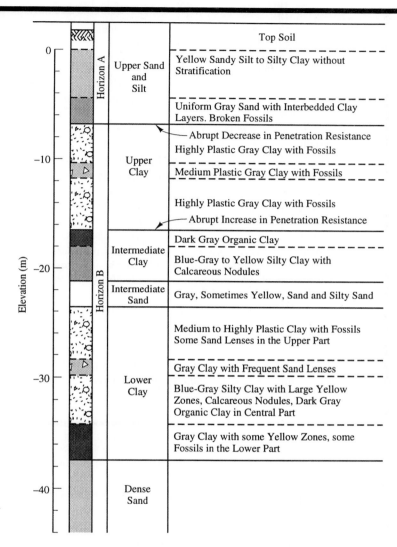

Figure 7.12 Soil profile below tower (Adapted from Mitchell, et al., 1977; Used by permission of ASCE).

Because the tower is continuing to move, there is serious concern that it might topple over, perhaps in the very near future. As a result, the Italians wish to stabilize the structure at its current angle of inclination (to straighten it would destroy a national treasure!). This has attracted the attention of both amateurs and professionals, and the authorities have received countless "solutions", sometimes at the rate of more than 50 per week. Some of them are quite extraordinary, such as the proposal to build a large human statue with an arm 50 m above the ground casually leaning against the south side of the tower or another to attach clutches of balloons held aloft by celestial cherubs.

The more credible proposals include:

- Dismantling the tower and reassembling it on a new and adequate foundation.
- Building a concrete slab on the ground immediately north of the tower, installing anchors, and applying a downward thrust onto the ground. This would consolidate the ground on the north side and slowly right the tower.
- Various schemes of underpinning the foundation.
- Temporarily supporting some or all of the tower's weight using an external bracing system, thus allowing more extensive work on a new foundation and underpinning system. This might even include raising the tower to its original elevation.
- Lifting the tower, moving it approximately 2.5 m to the north, and placing it back on the original foundation. The weight of the tower would then act through the centroid of the foundation, thus evenly distributing the bearing pressure (Pryke, 1987).
- Raising the groundwater table by installing injection wells (partial solution).

The Italians have not yet awarded a contract to remedy the problem and are proceeding cautiously. Their decision is likely to be tempered by the memory of the 1935 effort to seal groundwater leaks by injecting grout into the soil. The tower reacted to this disturbance by tilting at a more rapid rate for several years thereafter (Croce et. al. 1981).

The cost of conducting conventional load tests on production footings would usually be prohibitive, so this type of quality control has not been available. However, a new method, known as the WAK (Wave Activated stiffness (K)) test (Briaud and Lepert, 1990; Lepert, Briaud, and Maxwell, 1991), may permit such measurements to be made on a routine basis.

The WAK test consists of installing two geophones[1] on the footing, and then striking the footing with an instrumented mallet as shown in Figure 7.13. The impact of the mallet causes a very small increment of settlement in the footing. By recording impact force, time, and velocity data and analyzing it using fast Fourier transform software on a portable microcomputer, the first portion of the load vs. settlement curve can be obtained. This test can be performed very rapidly, and could easily be conducted on all the production footings at a site.

---

[1] Geophones are instruments that measure stress waves.

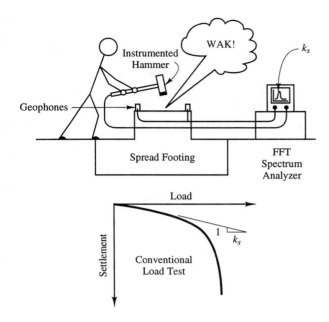

Figure 7.13 Use of the WAK test to measure the load-settlement behavior of spread footings. (Adapted from Briaud and Lepert, 1990; Used by permission of ASCE).

The interpretation of the WAK test data requires a knowledge of the relationship between the slope of the load-settlement curve at small settlements with that at larger settlements. Thus far, a limited amount of research indicates that the slope of the load-settlement curve is constant within the normal working loads of most footings. This, as well as other aspects of the test, needs to be refined before the WAK test can become a routinely used engineering tool.

SUMMARY OF MAJOR POINTS

1. The load on spread footings causes an increase in the vertical effective stress, $\Delta\sigma_v{}'$, in the soil below. This stress increase causes settlement in the soil beneath the footing.

2. The magnitude of $\Delta\sigma_v{}'$ directly beneath the footing is equal to the net bearing pressure, q'. It decreases with depth, and becomes very small at a depth of about $2B$ below square footings or about $6B$ below continuous footings.

3. The distribution of $\Delta\sigma_v{}'$ below a footing may be calculated using Boussinesq's method, Westergaard's method, or the simplified method.

4. The settlement of spread footings can be divided into three components: *distortion settlement*, δ_d, *consolidation settlement*, δ_c, and *secondary compression settlement*, δ_s. The distortion settlement is the result of shearing and lateral movement of the soil, consolidation settlement is the result of a reduction in the volume of the soil voids, and secondary compression settlement is the result of creep in the soil.

5. In cohesive soils, we evaluate distortion, consolidation, and secondary compression settlements separately and sum the results. The consolidation settlement is usually the dominant component. It is determined using a modified one-dimensional consolidation analysis with soil compressibility data from laboratory consolidation tests. The distortion settlement computation is based on elastic theory. Secondary compression is usually negligible, except if the soil is organic.

6. Settlement analyses in cohesionless soils are usually based on the results of in-situ tests, primarily because of the difficulties in obtaining suitable undisturbed samples. Most methods are based on the standard penetration test (SPT) or the cone penetration test (CPT).

7. Settlement estimates based on laboratory consolidation tests of cohesive soils typically range from a 50% overestimate (unconservative) to a 100% underestimate (conservative).

8. Settlement estimates based on CPT data from cohesionless soils typically range from a 50% overestimate (unconservative) to a 100% underestimate (conservative). However, estimates based on the SPT are much less precise.

QUESTIONS AND PRACTICE PROBLEMS

7.1 Explain the physical processes that produce distortion, consolidation, and secondary compression settlements. Which of these three is usually the largest?

7.2 The consolidation settlement computations described in Chapter 3 considered $\Delta\sigma_v'$ to be constant with depth. However, in this chapter, $\Delta\sigma_v'$ decreases with depth. Why?

7.3 Schmertmann's method usually produce more accurate predictions of footing settlements than does the Modified Meyerhof's method. Why?

7.4 Examine the pressure bulbs for square and continuous footings shown in Figure 7.2. Why do those for continuous footings extend deeper than those for square footings?

7.5 A 1500 mm square, 400 mm deep square footing supports a column load of 350 kN. Compute the change in effective vertical stress, $\Delta\sigma_v'$, beneath the center of this footing at a point 500 mm below the bottom of the footing:
   a. Using the simplified method.
   b. Using Boussinesq's method.

7.6 A column that carries a vertical downward load of 120 k is supported on a 5 ft square, 2 ft deep spread footing. The soil below has a unit weight of 124 lb/ft^3 above the groundwater table and 127 lb/ft^3 below. The groundwater table is at a depth of 8 ft below the ground surface.

   a. Develop a plot of the initial vertical stress, σ_{v0}', (i.e., the stress present before the footing was built) vs. depth from the ground surface to a depth of 15 ft below the ground surface.

   b. Using the simplified method, develop a plot of $\Delta\sigma_v'$ vs. depth below the center of the bottom of the footing and plot it on the diagram developed in part a.

   c. Using Boussinesq's method, compute $\Delta\sigma_v'$ at depths of 2 ft and 5 ft below the center of the bottom of the footing and plot them on the diagram.

7.7 A 600 mm wide, 500 mm deep continuous footing carries a vertical downward load of 85 kN/m. Using Boussinesq's method, compute $\Delta\sigma_v'$ at a depth of 200 mm below the bottom of the footing at the following locations:

   • Beneath the center of the footing

   • 150 mm from the center of the footing

   • 300 mm from the center of the footing (i.e., beneath the edge)

   • 450 mm from the center of the footing

Plot the results in the form of a pressure diagram similar to those in Figure 6.1 in Chapter 6.

Hint: Use the principle of superposition.

7.8 A proposed office building will include an 8 ft 6 in square, 3 ft deep spread footing that will support a vertical downward load of 160 k. The soil below this footing is an overconsolidated clay with the following engineering properties: $C = 0.05$, OCR = 2.5, $s_u = 3500$ lb/ft^2, PI = 37, and $\gamma = 113$ lb/ft^3. This soil strata extends to a great depth and the groundwater table is at a depth of 50 ft below the ground surface. The secondary compression settlement is negligible.

   a. Using hand computations, determine the total settlement of this footing.

   b. Using program FTGSETT, determine the total settlement of this footing.

   c. Using program FTGSETT, determine the required footing width to obtain a total settlement of no more than 1.0 in. Select a width that is a multiple of 3 in. Would it be practical to build such a footing?

7.9 A 1.0 m square, 0.5 m deep footing carries a downward load of 200 kN. It is underlain by an overconsolidated clay with the following engineering properties: $C_r = 0.05$, $e = 0.7$, OCR = 2.0, $s_u = 200$ kPa, PI = 55, and $\gamma = 15.0$ kN/m^3 above the groundwater table and 16.0 kN/m^3 below. The groundwater table is at a depth of 1.0 m below the ground surface. The secondary compression settlement is negligible.

   a. Compute the total settlement using hand computations.

   b. Compute the total settlement using program FTGSETT.

   c. Using program FTGSETT, determine the footing width required to produce a total settlement of 25 mm. Select a B that is a multiple of 100 mm.

7.10 A 650 kN column load is supported on a 1.5 m wide by 2.0 m long by 0.5 m deep spread footing. The soil below is a well graded, normally consolidated sand with the following SPT N_{60} values:

Depth (m)	1.0	2.0	3.0	4.0	5.0
N_{60}	12	13	13	18	22

   a. Compute the total settlement using the modified Meyerhof method.
   b. Compute the total settlement 30 years after construction using Burland and Burbidge's method.

7.11 A 190 k column load is to be supported on a 10 ft square, 3 ft deep spread footing underlain by young, normally consolidated sandy soils. The results of a representative CPT sounding at this site are as follows:

Depth (ft)	q_c (kg/cm^2)
0.0 - 6.0	30
6.0 - 10.0	51
10.0 - 18.0	65
18.0 - 21.0	59
21.0 - 40.0	110

The groundwater table is at a depth of 15 ft; the unit weight of the soil is 124 lb/ft^3 above the groundwater table and 130 lb/ft^3 below.

   a. Perform hand computations using Schmertmann's method to compute the total settlement of this footing 30 years after construction.
   b. Use program SCHMERT to compute the total settlement of this footing 30 years after construction.

7.12 A 300 k column load is to be supported on a 10 ft square, 4 ft deep spread footing. Cone penetration tests have been conducted at this site, and the results are shown on page 444. The groundwater table is at a depth of 6 ft, $\gamma = 121$ lb/ft^3, and $\gamma_{sat} = 125$ lb/ft^3.

   a. Compute the settlement of this footing.
   b. The design engineer is considering the use of vibroflotation to densify the soils at this site (see discussion in Chapter 19). This process would increase the q_c values by 70%, and make the soil slightly overconsolidated. The unit weights would increase by 5 lb/ft^3. Compute the settlement of a footing built and loaded after densification by vibroflotation.
   c. Based on the scatter in the CPT results, estimate the differential settlement without vibroflotation. Then, assume the differential settlement with vibroflotation will be 30 percent of the total settlement. Compare these two values and discuss.

Advanced Problems

7.13 A 3 ft square, 2 ft deep footing carries a column load of 28.2 k. An architect is proposing to build a new 4 ft wide, 2 ft deep continuous footing adjacent to this existing footing, as shown in the following figure. The side of the new footing will be only 6 inches away from the side of the existing footing. The new footing will carry a load of 12.3 k/ft.

Develop a plot of $\Delta\sigma_v{}'$ due to the new footing vs. depth along a vertical line beneath the center of the existing footing. This plot should extend from the bottom of the existing footing to a depth of 35 ft below the bottom of this footing.

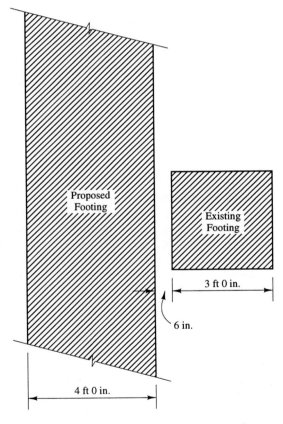

7.14 Using the data from Problem 7.13, $C = 0.08$ and $\gamma = 119$ lb/ft^3, compute the consolidation settlement of the old footing due to the construction and loading of the new footing. The soil is a silty clay, and the groundwater table is at a depth of 8 ft below the ground surface.

8

Spread Footings –
Synthesis of Geotechnical Analyses

> *Your greatest danger is letting the urgent things crowd out the important.*
> From *Tyranny of the Urgent* by Charles E. Hummel[1]

Sometimes geotechnical engineers perform bearing capacity and settlement analyses for specific footings with known loads and thus develop recommended dimensions. However, it is more common to develop general design parameters that apply to a large number of footings to be built at a particular site. The structural engineer then uses these parameters to develop the actual designs. Therefore, we must have a clear and accurate method of synthesizing the geotechnical analyses into usable design parameters.

8.1 FOOTING DEPTH

The minimum acceptable footing embedment depth, D, may be governed by one or more of the following criteria:

- It must extend below the depth of frost penetration.
- If loose or weak surficial soils or fills of unknown quality are present, the footing

---

[1] ©1967 by InterVarsity Christian Fellowship of the USA. Used by permission of InterVarsity Press-USA.

must extend through them and into soils of higher quality.

- In areas of expansive soils, the footing often must penetrate through the depth of seasonal moisture variation.
- If scour is a concern, the footing must extend below the potential scour depth.
- It must be deep enough to accommodate the required concrete thickness, T (see Chapter 10).
- In no event should D be less than 12 in (0.3 m).

Sometimes we also may need to specify a maximum depth. It might be governed by such considerations as:

- Potential undermining of existing foundations, structures, streets, utility lines, etc.
- The presence of soft layers at depth and the desire to support the footings in the stronger near-surface soils.

8.2 FOOTING WIDTH

The geotechnical engineer also must provide design parameters that will dictate the required width, B, of the footings. This must satisfy bearing capacity and settlement considerations as well as constructibility and other practical concerns. There are two primary ways of expressing these criteria: the *allowable bearing pressure method* and the *design chart method*.

Allowable Bearing Pressure Method

As discussed in Chapter 6, engineers have long recognized that the performance of spread footings is a function of the bearing pressure. Because this is a convenient parameter, we often express footing design criteria in this form.

The *net allowable bearing pressure*, q_a'', is the largest net bearing pressure that satisfies both bearing capacity and settlement criteria. In other words, it is equal to the net allowable bearing capacity, q_a', or the q' that produces the greatest acceptable settlement, whichever is less. The geotechnical engineer determines q_a'' and gives it to the structural engineer. Then, it is a simple matter to size the footing so that $q' \le q_a''$.

For small structures, it is common to assign conservative values of q_a'' based on an informal analysis or on presumptive bearing capacity tables. However, much more extensive analyses are appropriate for medium and large structures, especially if they are to be located on poor or marginal sites.

Unfortunately, the advent of allowable bearing pressures in the late 1800s produced many misconceptions, some of which prevail to this day. One of these is the idea that footings that satisfy the allowable bearing pressure criteria will not settle at all (definitely not true) or the idea that there will be no *differential* settlements if all the footings have the same bearing pressure (slightly more reasonable, but still not true). These misconceptions seem to be rooted in the concept of settlement being a black-or-white issue ("either it settles or it doesn't"), when in reality there will always be some settlement, just as a beam will always have some deflection. In both cases, the design objective is to keep the movement within tolerable limits.

The discussions in Chapters 6 and 7 demonstrated how both the bearing capacity and settlement of spread footings are functions of its width, B. There is no single allowable bearing pressure that is applicable to a variety of footing sizes. Thus, we must introduce certain simplifications.

If several footings were built at a site, each with a different B but the same q', the one with the smallest B would have the lowest factor of safety against a bearing capacity failure. As B increases, the ultimate bearing capacity also increases, as shown by Equations 6.13 - 6.15 (except for the special case when $\phi = 0$: then it remains constant). Therefore, the most critical bearing capacity analysis is that for the smallest footing. If we determine its net allowable bearing value, q_a', and use this value for all of the footings, the larger ones will have an even higher factor of safety against a bearing capacity failure.

However, the situation is different when we consider settlement analyses. Each of the footings described in the previous paragraph will have a different settlement and the largest one would settle most. This is because the larger footing stresses the soil to a greater depth, as shown in Figure 8.1. Thus, it strains more of the soil and generates more settlement. For footings on clays loaded to the same q', the settlement is approximately proportional to B, while in sands it is approximately proportional to $B^{0.5}$.

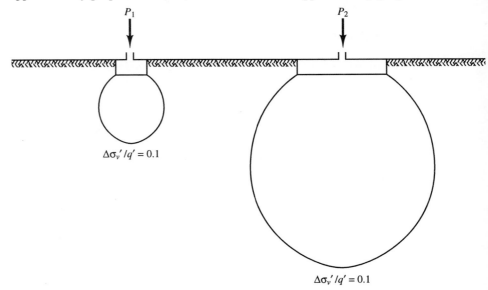

Figure 8.1 Two footings are loaded to the same q', but each has a different width. The larger footing induces stresses to a greater depth in the soil, so it settles more than the smaller footing.

Considering these facts, we can use the following procedure to assign a single $q_a{}''$ to a site:

1. Perform a bearing capacity analysis on the smallest of the proposed footings and determine $q_a{}'$.
2. Perform a settlement analysis on the largest of the proposed footings and determine the largest q' that keeps the settlement within tolerable limits.
3. Use the lower of these two values as "the" $q_a{}''$ for the site. Express it as a multiple of 500 lb/ft^2 or 25 kPa.

Sometimes, engineers assign two values of $q_a{}''$, one for square footings and one for continuous footings.

This method provides design criteria to the structural engineer in the same familiar format as the presumptive bearing values found in building codes. However, because it is based on site-specific soil properties and structural loads, it is more precise. Generally, computed $q_a{}''$ values will be higher than the conservative code values and thus produce more economical designs.

Example 8.1

A certain site is underlain by a sand with $N_{60} = 25$, $\gamma = 18.5$ kN/m^3 and $\phi = 38°$. A proposed building for this site will have 130 to 1000 kN column loads to be supported on 0.5 m deep spread footings. The allowable total settlement is 15 mm. Compute the allowable bearing pressure for square footings.

Solution:

Using Table 6.1, estimate $q_a{}'' \approx 200$ kPa.

$$B = (P/q_a{}'')^{0.5}, \text{ so } B \approx 0.8 \text{ m to } 2.2 \text{ m}$$

Bearing capacity analysis on smallest footing (using Equation 6.13):

$$\sigma_D' = \gamma D = (18.5)(0.5) = 9.2 \text{ kPa}$$

$$
\begin{aligned}
q_u{}' &= 1.3 c N_c + \sigma_D'(N_q - 1) + 0.4 \gamma B N_\gamma \\
&= 0 + (9.2)(61.5 - 1) + 0.4\,(18.5)\,(0.8)\,(82.3) \\
&= 1044 \text{ kPa}
\end{aligned}
$$

Settlement analysis on largest footing (using Equation 7.14):

$$q_a' = \frac{q_u'}{F} = \frac{1044}{3} = 350 \, \text{kPa}$$

$$K_d = 1 + 0.33 \, D/B = 1 + 0.33 \, (0.5/2.2) = 1.07$$

$$\frac{\delta}{B_r} = \frac{0.68 \, q'/\sigma_r}{\bar{N}_{60} \, K_d} \left(\frac{B}{B + B_r} \right)^2$$

$$\frac{15}{300} = \frac{0.68 \, (q'/100)}{(25)(1.07)} \left(\frac{2.2}{2.2 + 0.3} \right)^2$$

$$q' = 250 \, \text{kPa}$$

Settlement controls, so use $q_a'' = \textbf{250 kPa}$.   ⇐ *Answer*

Design Chart Method

The allowable bearing pressure method is sufficient for most small- to medium-size structures. However, larger structures, especially those with a wide range of column loads, warrant a more precise method: the design chart. This added precision helps us reduce both differential settlements and construction costs.

Instead of using a single allowable bearing pressure for all footings, it would be better to use a higher pressure for small ones and a lower pressure for large ones. This would reduce the differential settlements and avoid the material waste generated by the allowable bearing pressure method. This concept is implicit in a design chart such as the one in Figure 8.2.

A design chart is a compilation of many bearing capacity and settlement analyses expressed in a plot of an allowable column load vs. footing width. For example, if a column has a load of 300 k and an allowable settlement of 1.0 in, this chart shows that a footing width of 5 ft 6 in would satisfy the bearing capacity criteria, but 8 ft 0 in is required to keep the settlement below 1.0 in. Thus, settlement controls and B must be 8 ft 0 in. However, a column load of 50 k requires a 2 ft 9 in footing and is governed by bearing capacity. These two footings have net bearing pressures of 4690 and 6610 lb/ft^2, respectively.

These charts clearly demonstrate how the bearing capacity governs the design of narrow footings, whereas settlement governs the design of wide ones.

The advantages of this method over the allowable bearing pressure method include:

- The differential settlements are reduced because the allowable bearing pressure varies with the footing width.

- The selection of design values for total and differential settlement becomes the direct responsibility of the structural engineer, as it should be. (With the allowable bearing pressure method, the structural engineer must give allowable settlement data to the geotechnical engineer who incorporates it into $q_a^{\prime\prime}$).
- The plot shows the load-settlement behavior, which we could use in a soil-structure interaction analysis.

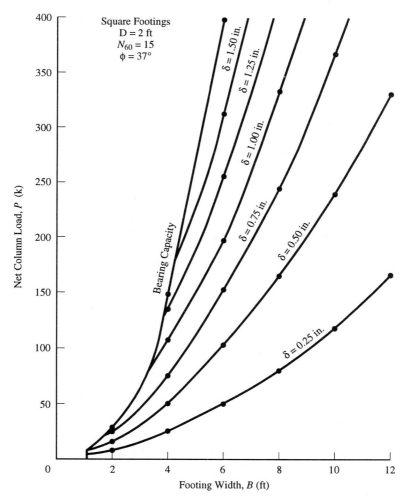

Figure 8.2 A design chart for 2 ft deep square footings on a sand with $N_{60} = 15$ and $\phi = 37°$. The bearing capacity plot includes a factor of safety of 3.

In many structures, the design live loads may never occur, or they may be present for only a short time. This is especially true for office buildings. Therefore, some engineers choose to evaluate settlements using only the dead load plus the long-term live load. However, always evaluate bearing capacity using the dead load plus the full design live load.

Developing Design Charts

Use the following procedure to develop design charts:

1. Determine the footing shape (i.e., square, continuous, etc.) for this design chart. If different shapes are to be used, each must have its own design chart.
2. Conduct a series of bearing capacity analyses using an estimated depth of embedment, D. It is usually easiest to select values of B (footing width), and compute the corresponding allowable column load. Program FTGBC makes the computations much easier and faster. Continue developing data points until the computed allowable column load exceeds the maximum load from the proposed structure.
3. Plot the results of the bearing capacity analyses on a design chart to form the bearing capacity curve.
4. Develop the first settlement curve as follows:
   a. Select a settlement value for the first curve (e.g., 0.25 in).
   b. Select a footing width, B, that is within the range of interest and arbitrarily select a corresponding column load, P_a. Then, compute the settlement of this footing using program FTGSETT (for cohesive soils), SCHMERT (for cohesionless soils) or some other suitable method.
   c. By trial-and-error, adjust the column load until the computed settlement matches the value assigned in step a. Then, plot the point B, P_a on the design chart.
   d. Repeat steps b and c with new values of B until a satisfactory settlement curve has been produced.
5. Repeat step 4 for other settlement values, thus producing a family of settlement curves on the design chart. These curves should encompass a range of column loads and footing widths appropriate for the proposed structure.

8.3 MINIMUM PRACTICAL DIMENSIONS

If the applied loads are small, such as with most one- or two-story wood-frame structures, bearing capacity and settlement analyses may suggest that extremely small footings would be sufficient. However, from a practical perspective, very small footings are not acceptable for the following reasons:

- Construction of the footing and the portions of the structure that connect to it would be difficult.
- Excavation of soil to build a footing is by no means a precise operation. If the footing dimensions were very small, the ratio of the construction tolerances to the footing dimensions would be large, which would create other construction problems.
- A certain amount of flexural strength is necessary to accommodate nonuniformities in the loads and local inconsistencies in the soil, but an undersized footing would have little flexural strength.

At sites where special problems such as expansive soils, collapsing soils, or very erratic soils are not a concern, the minimum dimensions shown in Figure 8.3 are usually sufficient. Local building code minimums also may govern in many cases.

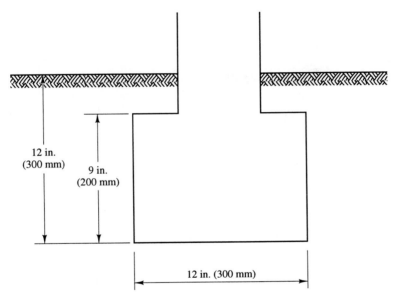

12 in.
(300 mm)

9 in.
(200 mm)

12 in. (300 mm)

Figure 8.3 Minimum dimensions for spread footings at good sites.

8.4 SPREAD FOOTINGS NEAR OR ON SLOPES

Brinch Hansen's bearing capacity formulas are able to consider footings near sloping ground, and we also could compute the settlement of such footings. However, it is best to avoid this condition whenever possible. Special concerns for such situations include:

- The reduction in lateral support makes bearing capacity failures more likely.
- The foundation might be undermined if a shallow (or deep!) landslide were to occur.
- The near-surface soils may be slowly creeping downhill, and this creep may cause the footing to move slowly downslope. This is especially likely in clays.

However, there are circumstances where footings must be built on or near a slope. Examples include abutments of bridges supported on approach embankments, foundations for electrical transmission towers, and some buildings.

Shields, Chandler, and Garnier (1990) produced another solution for the bearing capacity of footings located on sandy slopes. This method, based on centrifuge tests, relates the bearing capacity of footings at various locations with that of a comparable footing with $D = 0$ located on level ground. Figures 8.4 - 8.6 give this ratio for 1.5:1 and 2:1 slopes.

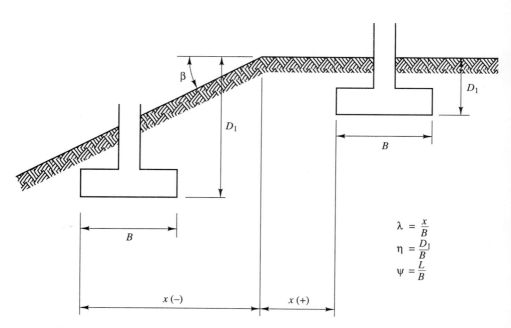

$$\lambda = \frac{x}{B}$$

$$\eta = \frac{D_1}{B}$$

$$\psi = \frac{L}{B}$$

Figure 8.4 Definition of terms for computing bearing capacity of footings near or on sandy slopes (Adapted from Shields, Chandler and Garnier, 1990; Used by permission of ASCE).

The Uniform Building Code (ICBO, 1991a) requires setbacks as shown in Figure 8.7. We can meet these criteria either by moving the footing away from the slope or by making it deeper.

However, this setback criteria does not justify building foundations above unstable slopes. Therefore, we also should perform appropriate slope stability analyses to verify the overall stability.

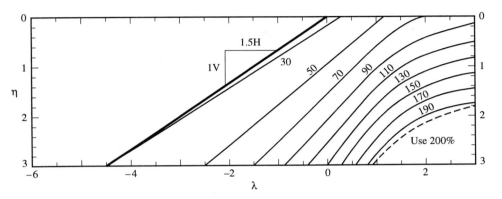

Figure 8.5 Bearing capacity of footings near or on a 1.5H:1V sandy slopes. The contours are the bearing capacity divided by the bearing capacity of a comparable footing located at the surface of level ground, expressed as a percentage. (Adapted from Shields, Chandler and Garnier, 1990; Used by permission of ASCE).

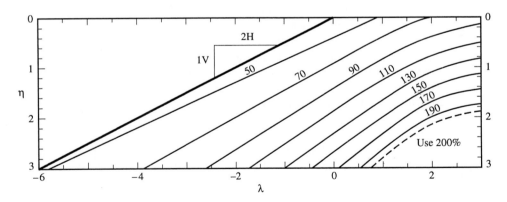

Figure 8.6 Bearing capacity of footings near or on a 2H:1V sandy slopes. The contours are the bearing capacity divided by the bearing capacity of a comparable footing located at the surface of level ground, expressed as a percentage. (Adapted from Shields, Chandler and Garnier, 1990; Used by permission of ASCE).

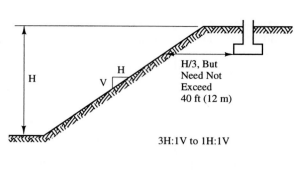

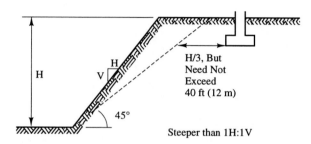

Figure 8.7 Footing setback as required by the Uniform Building Code for slopes steeper than 3 horizontal to 1 vertical. The horizontal distance from the footing to the face of the slope should be at least $H/3$, but need not exceed 40 ft (12 m). For slopes that are steeper than 1 horizontal to 1 vertical, this setback distance should be measured from a line that extends from the toe of the slope at an angle of 45°. (Adapted from the 1991 edition of the *Uniform Building Code*, © 1991, with permission of the publisher, the International Conference of Building Officials).

8.5 SEISMIC AND WIND LOADS

Virtually all structures must be designed to resist significant wind or seismic loads, and these loads eventually reach the foundation. Although seismic (and to a lesser degree, wind) loads are cyclic and dynamic, foundation engineers normally treat them as if they were additional static loads. It also is customary to assume that the design earthquake and the design windstorm will not occur simultaneously.

Shallow Foundations on Cohesive Soils

The shear strength of medium to stiff cohesive soils can decrease during cyclic loading, especially during earthquakes. For example, when the static plus cyclic loads cause strains in the soil of about 50% of the failure strain, the shear strength may be 10 to 20% lower than that under static conditions (Krinitzsky et al., 1993). Greater losses may occur in soft cohesive soils, but they are not likely to support significant foundations.

In spite of this strength loss, engineers often use allowable bearing pressures 1/3 higher than the static values. The justification for this increase is not because of greater soil strength, but an implicitly lower design factor of safety. Therefore, foundations designed with a static factor of safety of 3.0 against a bearing capacity failure may have a seismic factor of safety of about 1.8. This is generally acceptable, as long as the design static factor of safety is not too low.

However, much greater strength losses could occur in sensitive cohesive soils, so the 1/3 increase in allowable bearing pressure may produce unacceptably low factors of safety. Therefore, use wind/seismic bearing pressures equal to or lower than the static values in sensitive soils.

Cyclic loads also may cause additional settlement. However, some nominal settlement-related distress, such as minor cracking, is generally acceptable during a design windstorm or earthquake.

Shallow Foundations on Cohesionless Soils

The bearing capacity of cohesionless soils subjected to wind-induced structural loads should be at least equal to the static bearing capacity. In addition, bearing capacity usually does not control the design of shallow foundations on sand, so the factor of safety is usually high. The additoinal settlement should be tolerable. Therefore, engineers should be able to use allowable bearing pressures 1/3 to 1/2 higher than the static values.

However, cohesionless soils subjected to seismic loads are potentially much more troublesome, especially if they are saturated. The rapid cyclic loading that occur during earthquakes create excess pore water pressures that reduce the shear strength of the soil. This strength loss is greatest when the soil is loose, and the earthquake has a large intensity and a long duration. Sometimes, all of the strength is lost, a condition known as *soil liquefaction* (Seed et al., 1985). Thus, the factor of safety against a bearing capacity failure can be very small, or even zero.

Loose cohesionless soils, whether saturated or not, also can settle during earthquakes. These settlements can be significant. However, settlements in dense cohesionless soils should be small.

Therefore, shallow foundations on dense cohesionless soils generally perform well during earthquakes, while those on loose cohesionless soils sometimes do not. Foundation engineers should carefully consider the possibility of soil liquefaction when loose, saturated soils are present.

SUMMARY OF MAJOR POINTS

1. The required footing depth is governed by both geotechnical and structural concerns.
2. The required footing width must satisfy both bearing capacity and settlement criteria. The *allowable bearing pressure*, q_a'', is the maximum value of q' that satisfies both criteria.
3. For small- to medium-sized structures, we can assign one or two allowable bearing pressures for the entire site. Obtain these design values from bearing capacity and settlement analyses or from presumptive bearing values, such as those described in Chapter 6.
4. For larger structures, we can reduce both differential settlements and construction costs by using higher allowable bearing pressures on small footings and lower pressures on large ones. Design charts implicitly do this, and thus are a useful way to express the results of bearing capacity and settlement analyses.
5. Minimum practical dimensions will often govern the design of lightly loaded footings.
6. Spread footings on or near sloping ground may have a lower allowable bearing pressure.

7. For relatively insensitive cohesive soils and dense cohesionless soils, the allowable bearing pressure may be higher for wind or seismic loading conditions. However, for sensitive cohesive soils or loose cohesionless soils, it might be lower.

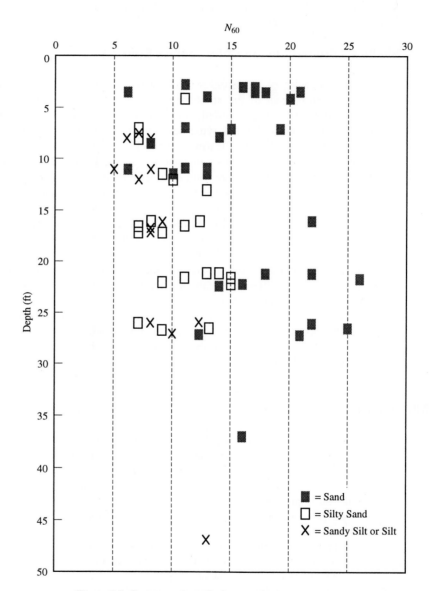

Figure 8.8 Summary of standard penetration test N_{60} values.

CASE STUDY

A large structure is to be built on a normally consolidated alluvial sand. The maximum column loads will be 1000 k. Because the column spacing is wide (50 - 75 ft) and no sensitive finishes will be used, this structure can tolerate up to 1 inch of allowable differential settlement. Consider the possibility of supporting it on spread footings. Assume the bottoms of these footings will be 5 ft below the ground surface.

A field crew has drilled 12 exploratory borings and has conducted standard penetration tests in these borings. The test results are plotted in Figure 8.8.

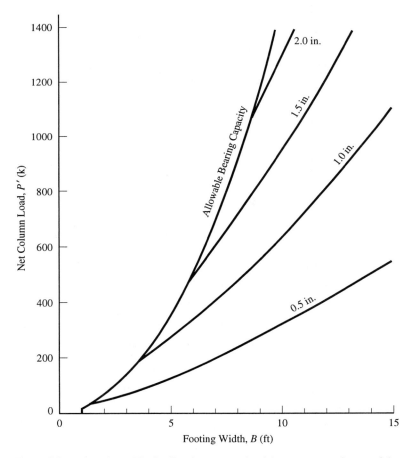

Figure 8.9 Design chart. The family of curves on the right represents estimates of the total settlement; the curve on the left is the allowable bearing capacity. Thus, for a given column load, use this chart to find the footing width that satisfies both bearing capacity and settlement criteria.

Settlement Analysis

The measured N_{60} values seem to vary randomly across the site, both vertically and horizontally. Therefore, we do not need to break the soil into multiple strata and can use a single N_{60} value for design. It seems reasonable to select a value less than the average, but greater than the minimum. Therefore, we will use $N_{60} = 9$.

Using Burland and Burbidge's method, we can develop the settlement curves for the design chart in Figure 8.9. These curves should be slightly conservative because they are based on a conservative N_{60} value. However, because of the scatter in the SPT data, the differential settlement will probably be nearly as large as the total settlement. For design purposes, use $\delta_D / \delta = 0.9$.

Bearing Capacity Analysis

Based on Figure 4.19 in Chapter 4, the friction angle ϕ is about 35°. This soil is primarily sand, so we may use $c = 0$. Using this data and program FTGBC, along with $D = 5$ ft and $F = 3$, we can develop the allowable bearing capacity curve in Figure 8.9.

Conclusions

A project of this magnitude easily justifies the design chart method of presenting bearing capacity and settlement data. Using $\delta_D / \delta = 0.9$ and $\delta_{Da} = 1$ in, we should design for an allowable total settlement, δ_a, of 1.1 in. This means that the largest load (1000 k) will require a 13 ft wide footing. Although this is fairly large, spread footings are economically feasible for this project.

QUESTIONS AND PRACTICE PROBLEMS

8.1 Which method of expressing footing width criteria would be most appropriate for each of the following structures?

    a. A ten-story reinforced concrete building

    b. A one-story wood frame house

    c. A nuclear power plant

    d. A highway bridge

8.2 Explain why an 8 ft wide footing with $q' = 3000$ lb/ft^2 will settle more than a 3 ft wide one with the same q'.

8.3 When would bearing capacity most likely control the design of spread footings? When would settlement usually control?

8.4 A 4 ft square, 2 ft deep spread footing carries a compressive column load of 50 k. The edge of this footing is 1 ft behind the top of a 40 ft tall, 2H:1V descending slope. The soil has the following properties: $c = 200$ lb/ft^2, $\phi = 31°$, $\gamma = 121$ lb/ft^3, and the groundwater table is at a great depth. Compute the factor of safety against a bearing capacity failure and comment on this design.

8.5 The building described in Example 8.1 is to be built on a silty clay with the following engineering properties: $\gamma = 15.1$ kN/m^3 above the groundwater table and 16.5 kN/m^3 below, $s_u = 160$ kPa, $C = 0.020$, plasticity index $= 25$ and overconsolidation ratio $= 3$. The groundwater table is 5 m below the ground surface. The required factor of safety against a bearing capacity failure must be at least 3.

Use programs FTGBC and FTGSETT to compute the allowable bearing pressure for a footing founded 0.5 m below the ground surface. Then, comment on the feasibility of using spread footings at this site.

8.6 A series of columns carrying vertical loads of up to 90 k are to be supported on 3 ft deep square footings. The soil below is a clay with the following engineering properties: $\gamma = 105$ lb/ft^3 above the groundwater table and 110 lb/ft^3 below, $s_u = 3000$ lb/ft^2, $C = 0.03$ in the upper 10 ft and 0.05 below, plasticity index $= 40$, and overconsolidation ratio $= 2$. The groundwater table is 5 ft below the ground surface. The factor of safety against a bearing capacity failure must be at least 3. Use programs FTGBC and FTGSETT to develop a design chart. Consider footing widths of up to 12 ft.

8.7 Several cone penetration tests have been conducted in a young, normally consolidated silica sand. Based on these tests, an engineer has developed the following design soil profile:

Depth (m)	q_c (kg/cm^2)
0 - 2.0	40
2.0 - 3.5	78
3.5 - 4.0	125
4.0 - 6.5	100

This soil has an average unit weight of 18.1 kN/m^3 above the groundwater table and 20.8 kN/m^3 below. The groundwater table is at a depth of 3.1 m.

Using this data with programs FTGBC and SCHMERT, develop a design chart for 1.0 m deep square footings. Consider footing widths of up to 4 m and column loads up to 1500 kN, a factor of safety of 2.5, and a design life of 50 years.

Hint: In a homogeneous soil, the critical shear surface for a bearing capacity failure extends to a depth of approximately B below the bottom of the footing. See Chapter 4 for a correlation between q_c in this zone and ϕ.

8.8 A certain square spread footing for an office building is to support a dead load of 800 kN, a static live load of 500 kN, and a seismic load of 400 kN. However, only 200 kN of the live load is "long-term" live load. Using the design chart from Problem 8.7, determine the required width of this footing such that the total settlement is no more than 20 mm.

9

Mat Foundations

The mere formulation of a problem is far more often essential than its solution, which may be merely a matter of mathematical or experimental skill. To raise new questions, new possibilities, to regard old problems from a new angle requires creative imagination and marks real advances in science.

Albert Einstein

The second type of shallow foundation is the *mat foundation*, as shown in Figure 9.1. A mat is essentially a very large spread footing that usually encompasses the entire footprint of the structure. They are also known as *raft foundations*.

A foundation engineer will often consider a mat when dealing with any of the following conditions:

- The structural loads are so high or the soil conditions so poor that spread footings would be exceptionally large. As a general rule of thumb, if spread footings would cover more than 50% of the building footprint area, a mat or some type of deep foundation will often be more economical.

- The soil is very erratic and prone to excessive differential settlements. The structural continuity and flexural strength of a mat will bridge over these irregularities. The same is true of mats on highly expansive soils prone to differential heaves.

- The structural loads are erratic, and thus increase the likelihood of excessive differential settlements. Again, the structural continuity and flexural strength of the mat will absorb these irregularities.

- Lateral loads are not uniformly distributed through the structure and thus may cause differential horizontal movements in spread footings or pile caps. The continuity of a mat will resist such movements.

- The uplift loads are larger than spread footings can accommodate. The greater weight and continuity of a mat may provide sufficient resistance.

- The bottom of the structure is located below the groundwater table, so waterproofing is an important concern. Because mats are monolithic, they are much easier to waterproof. The weight of the mat also helps resist hydrostatic uplift forces from the groundwater.

Many buildings are supported on mat foundations, as are silos, chimneys, and other types of tower structures. Mats are also used to support storage tanks and large machines. The 75-story Texas Commerce Tower in Houston is one of the largest mat-supported structures in the world. Its mat is 9 ft 9 in (3 m) thick and is bottomed 63 ft (19.2 m) below the street level.

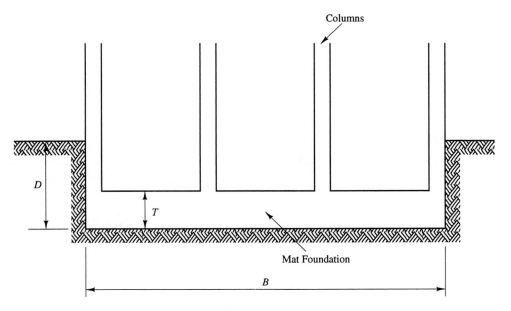

Figure 9.1 A mat foundation.

9.1 DESIGN PHILOSOPHIES

We can divide methods of analysis and design of mat foundations into two broad categories: rigid and nonrigid. Both are primarily intended to produce estimates of differential settlements and flexural stresses in the mat. Neither produces reliable estimates of total settlement. We use the techniques described in Chapter 7 to compute the total settlement.

Rigid Method

The simplest approach is the *rigid method* (also known as the *conventional method*) (Teng, 1962). It assumes that the mat is infinitely rigid (i.e., the flexural deflections are assumed to be negligible) and the soil is a linear elastic material. It also assumes the soil bearing pressure is uniform across the bottom of the footing (if only concentric axial loads are present) or it varies linearly across the footing (if eccentric or moment loads are present), as shown in Figures 6.1e and 6.2 in Chapter 6.

This method defines the distribution of bearing pressure in simple terms, so the structural engineer can easily determine the flexural stresses using conventional structural analyses. For analysis purposes, the mat becomes an inverted and simply loaded two-way slab.

Although this type of analysis is appropriate for spread footings, it does not accurately model mat foundations. Mats are not truly rigid, so the settlement beneath the columns will be greater than that beneath unloaded areas, as shown in Figure 9.2. These differential settlements will cause variations in the soil bearing pressure and corresponding changes in the flexural stresses in the mat.

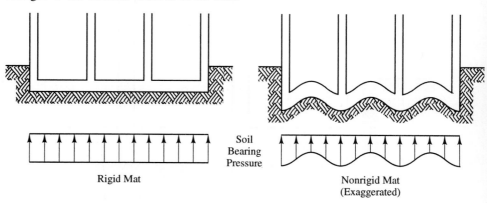

Figure 9.2 The rigid method assumes there are no flexural deflections in the mat, so the distribution of soil bearing pressure is simple to define. However, these deflections are important because they influence the bearing pressure distribution.

Nonrigid Method

Analyses that consider flexural deflections in the mat and the resulting nonuniform soil bearing pressures should produce more precise estimates of the flexural stresses in the mat and thus produce more economical and reliable designs. We will refer to these analyses as *nonrigid methods*.

Figure 9.3 shows that the bearing pressure becomes more concentrated near the columns when the soil is stiffer. Therefore, the use of a nonrigid method is especially important in stiff soils.

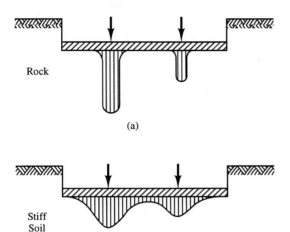

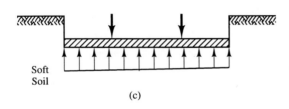

Figure 9.3   Distribution of soil bearing pressure under a mat foundation: (a) Rock or hard soil; (b) Stiff soil; and (c) Soft soil (Adapted from Teng, 1962).

Unfortunately, nonrigid methods introduce two problems:

1. The structural analysis becomes more complex because the soil bearing pressure is no longer a simple geometric shape.
2. We now need to define the relationship between settlement and bearing pressure.

The first problem is easily overcome with modern structural analysis software. However, the second problem is much more difficult. Engineers usually solve it using a *beam on elastic foundation analysis*.

9.2 BEAM ON ELASTIC FOUNDATION ANALYSES

Beam on elastic foundation analyses define the relationship between bearing pressure and settlement using the *modulus of subgrade reaction*, k_s (also known as the *coefficient of subgrade reaction*, or the *subgrade modulus*):

$$k_s = \frac{q}{\delta}$$ (9.1)

Where:

q = gross bearing pressure

δ = settlement

The modulus k_s has units of force per length cubed. Although we use the same units to express unit weight, k_s is not the same as the unit weight and they are not numerically equal.

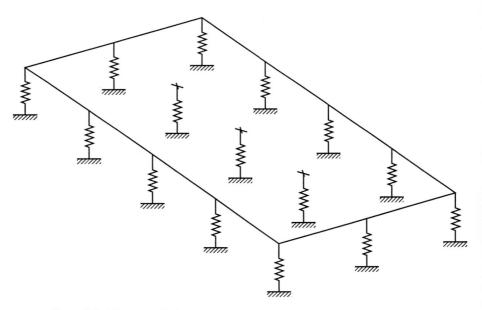

Figure 9.4 A beam on elastic foundation analysis uses a "bed-of-springs" analogy to model the relationship between bearing pressure and settlement.

We then model the soil as a bed of springs, each with a stiffness k_s, as shown in Figure 9.4. The sum of the spring forces must equal the applied structural loads plus the weight of the mat:

$$P + W_f = \int q\,dA = \int \delta k_s\,dA$$ (9.2)

Where:

P = weight of the structure plus any other applied vertical loads

W_f = weight of the mat

A = mat-soil contact area

dA = tributary area for each spring

Areas of the mat that settle most will have correspondingly larger spring forces. When placed in a structural analysis, these forces will create shears and moments in the mat and thus form the basis for a structural design. See Scott (1981) and Hetényi (1974) for more information on these analyses. This concept has been attributed to Winkler (1867), so the analytical model is sometimes called a *Winkler foundation*.

Determining the Modulus of Subgrade Reaction

Although the beam on elastic foundation analysis is superior to a rigid mat analysis, its accuracy depends on our ability to define the modulus of subgrade reaction, k_s, of the soil. Unfortunately, this task is not as simple as it might first appear because k_s is not a fundamental soil property. Its magnitude also depends on many other factors, including the following:

- **The width of the loaded area** — A wide mat will settle more than a narrow one with the same q because it mobilizes the soil to a greater depth as shown in Figure 8.1 in Chapter 8. Therefore, each has a different k_s.
- **The shape of the loaded area** — The stresses below long narrow loaded areas are different from those below square loaded areas as shown in Figure 7.2 in Chapter 7. Therefore, k_s will differ.
- **The depth of the loaded area below the ground surface** — At greater depths, the change in stress in the soil due to q is a smaller percentage of the initial stress, so the settlement is also smaller and k_s is greater.
- **The position on the mat** — To model the soil accurately, k_s needs to be larger near the edges of the mat and smaller near the center.
- **Time** — Much of the settlement of mats on deep compressible soils will be due to consolidation and thus may occur over a period of several years. Therefore, it may be necessary to consider both short-term and long-term cases.

Actually, there is no single k_s value, even if we could define these factors because the q-δ relationship is nonlinear. Therefore, the most we could hope for is some equivalent k_s as shown in Figure 9.5, or possibly some nonlinear q-δ function.

Engineers have often used plate load tests[1] to measure k_s in situ. However, the

---

[1] See the discussion of plate load tests in Section 6.3 of Chapter 6.

test results must be adjusted to compensate for the differences in width, shape, and depth of the plate and the mat. Terzaghi (1955) proposed a series of correction factors, but the extrapolation from a small plate to a mat is so great that these factors are not very reliable. Plate load tests also include the dubious assumption that the soils within the shallow zone of influence below the plate are comparable to those in the much deeper zone below the mat. Therefore, plate load tests generally do not provide good estimates of k_s.

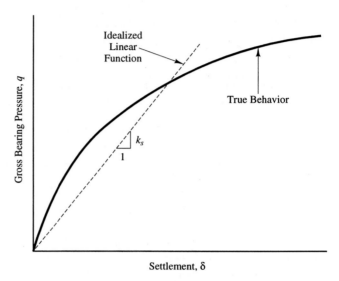

Figure 9.5 The q-δ relationship is nonlinear, so k_s must represent some "equivalent" linear function.

Vesić (1961) proposed the following relationship between k_s and the modulus of elasticity of the soil:

$$k_s = \frac{0.65\,E}{B\,(1 - v_p^2)} \sqrt[12]{\frac{E\,B^4}{E_f I_f}} \qquad (9.3)$$

Where:

B = width of mat

E = modulus of elasticity of soil

v_p = Poisson's ratio of soil

E_f = modulus of elasticity of foundation $\approx 4.5 \times 10^8$ lb/ft^2 $\approx$ 23,000 MPa

I_f = moment of inertia of foundation for bending in a vertical plane = $BT^3/12$

T = thickness of mat

Another alternative is to compute the average mat settlement using the techniques described in Chapter 7 and express the results in the form of k_s using Equation 9.1.

Because it is difficult to develop accurate k_s values, it may be appropriate to conduct a parametric studies to evaluate its effect on the mat design. ACI (1988) suggests varying k_s from one-half the computed value to 5 or 10 times the computed value. Fortunately, we often find that the design is not very sensitive to global changes in k_s. However, as discussed in the next section, it can be very sensitive to variations in k_s across the mat.

9.3 ADVANCED SOIL-STRUCTURE INTERACTION ANALYSES

Beam on elastic foundation analyses are simplifications of the real interaction between mat foundations and soil (Hain and Lee, 1974; Horvath, 1983). As a result, they do not always produce accurate estimates of the flexural stresses or differential settlements in the mat. For example, according to the analysis, a mat with a uniformly distributed structural load would settle uniformly. In reality, the settlement would be greater in the center and less along the edges. Clays are especially prone to this kind of "dishing." Thus, the analysis can be unconservative.

The primary problem with the beam on elastic foundation model is that it assumes each spring acts independent of the others. In contrast, a load acting on a real soil influences both the soil directly below it and the nearby soil, as shown in Figure 7.2 in Chapter 7. This lack of interaction between the springs generates more error than does our uncertainty in selecting k_s.

We can improve the precision by doubling the k_s values along the perimeter of the mat. ACI (1988) found that this method increased the computed moments in a test mat by 18 to 25%. We could expand this concept by using multiple values of k_s, with the lowest (softest) in the center of the mat and the highest (stiffest) along the edge (ACI, 1988). These techniques would be especially useful when the bottom of the mat is near the ground surface because this is when end effects are most pronounced.

A more promising method involves the use of a soil-structure finite element model. This method divides both the mat and the soil into small elements, each with defined elastic properties. A computer with finite element analysis software then computes the stresses and displacements. An even more complete model also would include the superstructure, and thus consider soil-mat-superstructure interaction.

Although engineers routinely use finite element models for the mat, these generally represent the soil using beam on elastic foundation "springs." Finite element models that encompass both the mat and the soil are not yet common. However, this method may enjoy wider use in the future as more powerful computers and appropriate software become more readily available.

9.4 BEARING CAPACITY

Because of their large width, mat foundations on cohesionless soils will not have bearing capacity problems. However, bearing capacity might be important with cohesive soils, especially if undrained conditions prevail. The Fargo Grain Silo failure described in Chapter 6 is a notable example of a bearing capacity failure in a saturated clay.

We can evaluate bearing capacity using the analysis techniques described in Chapter 6. It is good practice to design the mat so the bearing pressure at all points is less than the allowable bearing capacity.

SUMMARY OF MAJOR POINTS

1. Mat foundations are essentially large spread footings that usually encompass the entire footprint of a structure. They are often an appropriate choice for structures that are too heavy for spread footings.

2. The oldest and simplest method of analyzing mats is the rigid method. It assumes that the mat is infinitely rigid and the soil is a linear elastic material. Although both assumptions simplify the analysis, neither is true.

3. Nonrigid analyses are superior because they consider the flexural deflections in the mat and the corresponding redistribution of the soil bearing pressure.

4. Most nonrigid analyses model the soil reaction using the beam on elastic foundation method (also known as a Winkler foundation). It is most easily visualized as a bed of springs. Each spring acts independent of the others and has a stiffness k_s (known as the modulus of subgrade reaction).

5. The modulus of subgrade reaction is difficult to determine. Fortunately, the mat design is often not overly sensitive to global changes in k_s. Parametric studies are often appropriate.

6. Beam on elastic foundation analyses can give a moderately reliable estimate of the flexural stresses and differential settlements in a mat, but they do not give reliable estimates of the total settlement. Evaluate total settlements using the techniques described in Chapter 7.

7. The analysis can be improved by using stiffer "springs" (i.e., higher k_s values) near the edges of the mat and softer ones near the center. This more accurately models the true soil response and has a greater impact than small global changes in k_s.

8. More advanced analyses might use soil-structure finite element methods. These are not yet in common use, but will probably become more popular in the future as computer hardware and software become more readily available.

9. Bearing capacity is not a problem with cohesionless soils, but can be important in cohesive soils.

QUESTIONS AND PRACTICE PROBLEMS

9.1 Explain the reasoning behind the statement "mat foundations on cohesionless soils will not have bearing capacity problems" in Section 9.4.

9.2 How has the development of powerful and inexpensive digital computers affected the analysis and design of mat foundations? What changes do you expect in the future as this trend continues?

9.3 A mat foundation supports 42 columns for a building. These columns are spaced on a uniform grid pattern. How would the moments and differential settlements change if we used a nonrigid analysis with a constant k_s in lieu of a rigid analysis?

9.4 A 90 ft wide, 4 ft thick reinforced concrete mat is supported on a soil with $E = 800,000$ lb/ft^2 and $v_p = 0.45$. Compute k_s using Vesić's formula.

9.5 According to a settlement analysis conducted using the techniques described in Chapter 7, a certain mat will have a total settlement of 40 mm if the average gross bearing pressure is 100 kPa. Compute the average k_s.

9.6 A 30.0 m diameter, 10.0 m tall reinforced concrete water tank is to be built on the following soil profile:

Depth (m)	Soil Classification	Unit Weight, γ (kN/m^3)	Compressibility, C
0.0- 4.0	Compacted silty sand fill, SM	20.0 above groundwater 21.5 below groundwater	Assume incompressible
4.0 - 10.5	Normally consolidated Medium silty clay, CL	14.2	0.15
> 10.5	Dense glacial till	22.0	Assume incompressible

The groundwater table is at a depth of 3.6 m.

The tank will be supported on a circular mat foundation. The combined weight of the empty tank and the mat will be 6700 kN. When full, the depth of water in the tank will be 9.0 m.

Compute the total settlement beneath the center and along the perimeter of the full tank, assuming the mat is perfectly flexible. Then, compute the total settlement of the tank assuming the mat is perfectly rigid.

Hint: Compute $\Delta\sigma'$ using Boussinesq's method on a square having an area equal to that of the circle, then use the settlement computation methods described in Chapter 7.

10

Shallow Foundations –
Structural Design

Foundations ought to be twice as thick as the wall to be built on them; and regard in this should be had to the quality of the ground, and the largeness of the edifice; making them greater in soft soils, and very solid where they are to sustain a considerable weight.

The bottom of the trench must be level, that the weight may press equally, and not sink more on one side than on the other, by which the walls would open. It was for this reason the ancients paved the said bottom with Tivertino, *and we usually put beams or planks, and build on them.*

The foundations must be made sloping, that is, diminished in proportion as they rise; but in such a manner, that there may be just as much set off one side as on the other, that the middle of the wall above may fall plumb upon the middle of that below: Which also must be observed in the setting off of the wall above ground; because the building is by this method made much stronger than if the diminutions were done any other way.

Sometimes (especially in fenny places, and where the columns intervene) to lessen the expence, the foundations are not made continued, but with arches, over which the building is to be.

It is very commendable in great fabricks, to make some cavities in the thickness of the wall from the foundation to the roof, because they give vent to the winds and vapours, and cause them to do less damage to the building. They save expence, and are of no little use if there are to be circular stairs from the foundation to the top of the edifice.

The First Book of Andrea Palladio's Architecture (1570)
as translated by Isaac Ware (1738)

The plan dimensions and minimum embedment depth of a shallow foundation are primarily geotechnical concerns. Chapters 6 - 9 discussed methods of determining these dimensions. Once they have been set, the next step is to develop an appropriate structural design to give the foundation enough integrity to safely transmit the design loads from the structure to the ground. The structural design process for reinforced concrete foundations includes:

- Selecting a concrete with an appropriate strength.
- Selecting an appropriate grade of reinforcing steel.
- Determining the required foundation thickness, T.
- Determining the size, number, and spacing of the reinforcing bars.
- Designing the connection between the superstructure and the foundation.

The structural design aspects of foundation engineering are far more codified than are the geotechnical aspects. These codes are based on the results of research, the performance of existing structures, and the professional judgment of experts. Unfortunately, there is a tendency to view codes as recipes for successful designs and substitutes for engineering judgment, neither of which is true. They should be treated as guides that specify minimum requirements, not something to be followed blindly.

Engineers in North America use the *Building Code Requirements for Reinforced Concrete [ACI 318-89]* for most projects. This code is published by the American Concrete Institute (ACI, 1989). The most notable alternative to ACI is the *Standard Specifications for Highway Bridges* published by the American Association of State Highway and Transportation Officials (AASHTO, 1981). It is used primarily for highway structures. Engineers in other parts of the world use different codes, many of which are very similar to ACI.

Rather than attempting to describe all of these codes, this book will concentrate on the 1989 edition of the ACI code (also known as ACI-318-89). References to sections of this code are shown in brackets [ ]. This chapter is limited to aspects of the code that relate to spread footings, and is not an exhaustive coverage. For additional information, see MacGregor (1992).

This chapter primarily addresses the structural design of conventional spread footings. The design of combined footings and mats is only a secondary topic.

10.1 ULTIMATE STRENGTH DESIGN OF REINFORCED CONCRETE

There are two ways to design reinforced concrete: the *working stress method* and the *ultimate strength method*. The working stress method consists of determining the strength of the concrete and reinforcing steel, dividing those strengths by a factor of safety to obtain the allowable stresses, and designing the structural members so the working stresses do not exceed the allowable stresses. The bearing capacity analyses described in Chapter 6 use this kind of analysis. In contrast, the ultimate strength method consists of magnifying the working loads using *load factors* to obtain the ultimate loads and designing the structural member such that this ultimate load does not exceed the ultimate capacity of the member. Because the ultimate strength method reflects the most recent research and is the preferred method, it is the only one we will use here.

Compute the *ultimate load* (or *factored load*), P_u, for ultimate strength analyses using Equations 10.1 - 10.6. Design the foundation using the formula that produces the largest P_u [9.2]. Note that ACI does not require designing structures to resist the maximum wind and earthquake loads simultaneously.

$$P_u = 1.4P_d + 1.7P_l \tag{10.1}$$

$$P_u = 0.75 (1.4P_d + 1.7P_l + 1.7P_w) \tag{10.2}$$

$$P_u = 0.9P_d + 1.3P_w \tag{10.3}$$

$$P_u = 0.75 (1.4P_d + 1.7P_l + 1.87P_e) \tag{10.4}$$

$$P_u = 0.9P_d + 1.43P_e \tag{10.5}$$

$$P_u = 1.4P_d + 1.7P_l + 1.7P_h \tag{10.6}$$

Where:

P_u = ultimate normal load (or factored normal load)
P_d = normal dead load
P_l = normal live load
P_w = normal wind load
P_e = normal earthquake load
P_h = normal earth pressure load

Compute ultimate moment loads in the same fashion by simply substituting M for P in Equations 10.1 - 10.6. Compute ultimate shear loads using V.

The next step is to compute the nominal normal load capacity of the member, P_n. In theory, this load would cause the member to fail. The structure must be designed such that the factored normal load, P_u, does not exceed the nominal load capacity, P_n, multiplied by a strength reduction factor, ϕ:

$$P_u \leq \phi P_n \qquad (10.7)$$

For moment and shear loads, the parallel relationships are:

$$M_u \leq \phi M_n \qquad (10.8)$$

$$V_u \leq \phi V_n \qquad (10.9)$$

The ϕ factor reflects uncertainties that result from construction tolerances and variations in material strength. The appropriate value of ϕ, as shown in Table 10.1, depends on the type of loading and other concerns.

TABLE 10.1 ϕ FACTORS FOR ULTIMATE STRENGTH DESIGN OF REINFORCED CONCRETE [9.3]

Design Situation	Strength Reduction Factor, ϕ
Flexure without axial load	0.90
Axial tension with or without flexure	0.90
Axial compression with or without flexure	With spiral reinforcement: 0.75 Without spiral reinforcement: 0.70 (in some circumstances, ϕ can be as high as 0.90. See ACI 9.3.2.2.)
Shear and torsion	0.85
Bearing on concrete	0.70

The most common practice is to use working stress methods to evaluate bearing capacity and settlement, and ultimate strength methods to develop structural designs. Thus, bearing capacity and settlement analyses use unfactored loads, P, V, and M, whereas the structural designs use the factored loads, P_u, V_u and M_u. This can be a source of confusion. However, a simple way to know which load to use is to remember that problems relating to the interaction between the footing and the soil, such as bearing capacity or settlement analyses, use unfactored loads, whereas those that relate to internal structural design, such as sizing reinforcing bars, use factored loads.

Design of Flexural Members

Concrete is strong in compression, but weak in tension. Therefore, engineers add reinforcing steel, which is strong in tension, to form *reinforced concrete*. This reinforcement is necessary in members subjected to pure tension, and those that must resist *flexure* (bending). Reinforcing steel may consist of either *deformed bars* (more commonly known as *reinforcing bars*, or *rebars*) or *welded wire fabric*. However, wire fabric is rarely used in foundations.

  Manufacturers produce reinforcing bars in various standard diameters, typically ranging between 3/8 in (9.5 mm) and 2 1/4 in (57.2 mm). In the United States, standard bars are identified using the *bar size* designations in Table 10.2.

  Rebars are available in various strengths, depending on the steel alloys used to manufacture them. The most common bars are designated as grade 40 or grade 60, which correspond to yield strengths, f_y, of 40 and 60 k/in^2 (275 and 400 MPa), respectively.

TABLE 10.2 DESIGN DATA FOR STEEL REINFORCING BARS

Bar Size Designation	Grades	Nominal Dimensions			
		Diameter, d_b		Cross-Sectional Area, A_s	
		(in)	(mm)	(in^2)	(mm^2)
#3	40, 60	0.37	9.5	0.11	71
#4	40, 60	0.50	12.7	0.20	127
#5	40, 60	0.62	15.9	0.31	198
#6	40, 60	0.75	19.1	0.44	285
#7	60	0.87	22.2	0.60	388
#8	60	1.00	25.4	0.79	507
#9	60	1.13	28.6	1.00	641
#10	60	1.27	31.8	1.27	792
#11	60	1.41	34.9	1.56	958
#14	60	1.69	44.5	2.25	1550
#18	60	2.26	57.2	4.00	2560

  The primary design problem for flexural members is as follows: Given a factored moment, M_u, and a factored shear force, V_u, determine the necessary dimensions of the member and the necessary size and location of reinforcing bars. Fortunately, flexural design in foundations is simpler than that for some other structural members because geotechnical concerns dictate some of the dimensions.

  The amount of steel required to resist flexure depends on the *effective depth, d,* which is the distance from the extreme compression fiber to the centroid of the tension reinforcement, as shown in Figure 10.1.

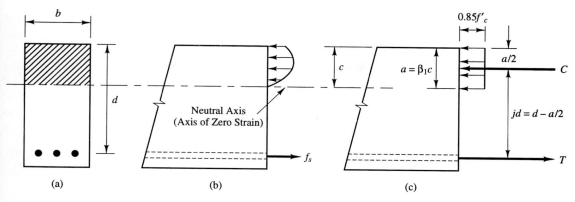

Figure 10.1 The reinforcing bars are placed in the portion of the member that is subjected to tension. (a) Cross-section, (b) actual stress distribution, and (c) equivalent rectangular stress distribution. The effective depth, d, is the distance from the extreme compression fiber to the centroid of the tension reinforcement (Adapted from MacGregor, 1992).

The nominal moment capacity of a flexural member as shown in Figure 10.1 is:

$$M_n = A_s f_y \left(d - \frac{a}{2} \right) \tag{10.10}$$

$$a = \frac{\rho \, d \, f_y}{0.85 f_c'} \tag{10.11}$$

$$\rho = \frac{A_s}{b \, d} \tag{10.12}$$

Setting $M_{uc} = \phi \, M_n$ and solving for A_s gives:

$$A_s = \left(\frac{f_c' \, b}{1.17 f_y} \right) \left(d - \sqrt{d^2 - \frac{2.35 \, M_{uc}}{\phi f_c' \, b}} \right) \tag{10.13}$$

Where:

A_s = cross sectional area of reinforcing steel

f_c' = 28-day compressive strength of concrete

f_y = yield strength of reinforcing steel

ρ = steel ratio

b = width of flexural member

d = effective depth

ϕ = 0.9 for flexure in reinforced concrete

M_{uc} = ultimate moment at critical section

The flexural stresses in most spread footings are low, so it is conservative to set $a = 0.15d$. Thus, Equation 10.10 becomes:

$$M_n = 0.93\, A_s\, f_y\, d \qquad (10.14)$$

Again setting $M_{uc} = \phi\, M_n$ and solving for A_s gives:

$$A_s = \frac{1.08\, M_{uc}}{\phi\, f_y\, d} \qquad (10.15)$$

For the low stresses in conventional spread footings, Equation 10.15 gives a steel area no more than 5% greater than that from Equation 10.13. However, some combined footings and mats may have greater flexural stresses, and therefore should be evaluated using the more precise Equation 10.13. Equation 10.15 may be unconservative when the flexural stresses are large.

Two additional considerations also enter the design process: minimum steel and maximum steel. Unfortunately, the ACI code is ambiguous regarding the proper minimum steel requirement for footings. For "flexural members", $\rho_{min} = 200/f_y$ [10.5.1], whereas "structural slabs of uniform thickness" need a ρ_{min} of only 0.0018 to 0.0020 [10.5.3 and 7.12.2]. The later criterion requires about half the steel, so the interpretation of the code on this point can have a significant effect on the final design. Because footings more closely resemble slabs, MacGregor (1992) recommends using that criterion, which is as follows:

For grade 40 or grade 50 steel $\rho_{min} = 0.0020$
For grade 60 steel $\rho_{min} = 0.0018$

The maximum steel requirement [10.3] is intended to maintain sufficient ductility. It never governs the design of simple footings, but may be of concern in combined footings or mats.

We can supply the required area of steel, computed using Equation 10.13 or 10.15, using any of several combinations of bar size and number of bars. This selection must satisfy the following minimum and maximum spacing requirements:

• The clear space between bars must be at least equal to d_b, 1 in (25 mm), or 4/3

times the nominal maximum aggregate size [3.3.2 and 7.6.1], whichever is greater.
- The spacing of the reinforcement must not exceed $3d$ or 18 in (450 mm), whichever is less.

The data in Tables 10.3 and 10.4 may be used to develop a design.

TABLE 10.3 TOTAL AREA OF REINFORCING STEEL, A_s (in^2)

Bar Size	Number of Bars								
	1	2	3	4	5	6	7	8	9
#3	0.11	0.22	0.33	0.44	0.55	0.66	0.77	0.88	0.99
#4	0.20	0.39	0.59	0.79	0.98	1.18	1.37	1.57	1.77
#5	0.31	0.61	0.92	1.23	1.53	1.84	2.15	2.45	2.76
#6	0.44	0.88	1.33	1.77	2.21	2.65	3.09	3.53	3.98
#7	0.60	1.20	1.80	2.41	3.01	3.61	4.21	4.81	5.41
#8	0.79	1.57	2.36	3.14	3.93	4.71	5.50	6.28	7.07
#9	1.00	2.00	3.00	4.00	5.00	6.00	7.00	8.00	9.00
#10	1.27	2.53	3.80	5.07	6.33	7.60	8.87	10.13	11.40
#11	1.56	3.12	4.68	6.25	7.81	9.37	10.93	12.49	14.05

TABLE 10.4 TOTAL AREA OF REINFORCING STEEL, A_s (mm^2)

Bar Size	Number of Bars								
	1	2	3	4	5	6	7	8	9
#3	70	140	210	280	360	430	500	570	640
#4	130	250	380	510	630	760	890	1010	1140
#5	200	400	590	790	990	1190	1390	1580	1780
#6	280	570	850	1140	1420	1710	1990	2280	2560
#7	390	780	1160	1550	1940	2330	2720	3100	3490
#8	510	1010	1520	2030	2530	3040	3550	4050	4560
#9	640	1280	1920	2560	3210	3850	4490	5130	5770
#10	790	1580	2370	3170	3960	4750	5540	6330	7130
#11	960	1920	2870	3830	4790	5750	6710	7660	8620

10.2 SELECTION OF MATERIALS

Unlike the geotechnical engineer, who usually has little or no control over the engineering properties of the soil, the structural engineer can, within limits, select the engineering properties of the structural materials. In the context of spread footing design, we must select an appropriate concrete strength, f_c', and reinforcing steel strength, f_y.

When designing a concrete superstructure, engineers typically use concrete that has $f_c' = 3000$ to 5000 lb/in^2 (20 to 35 MPa). In very tall structures, f_c' might be as large as 10,000 lb/in^2 (70 MPa). The primary motive for using high strength concrete in the superstructure is that it reduces the section sizes that allows more space for occupancy and reduces the weight of the structure, thereby reducing the dead loads on the lower members.

However, with spread footings, saving size and weight is much less important. The plan dimensions of the footing are governed by bearing capacity and settlement concerns and will not be affected by changes in the strength of the concrete; only the thickness, T, will change. Even then, the required excavation depth, D, may or may not change because it might be governed by other factors. In addition, saving weight in a footing is of little concern because it is the lowest structural member and does not affect the dead load on any other members. In fact, additional weight may actually be a benefit in that it increases the uplift capacity.

Because of these considerations, and because of the additional materials and inspection costs of high strength concrete, spread footings are usually designed using an f_c' of 2000 to 3000 lb/in^2 (15 - 20 MPa). For footings that carry relatively large loads, perhaps greater than about 500 k (2000 kN), higher strength concrete might be justified to keep the footing thickness within reasonable limits, perhaps using an f_c' as high as 5000 lb/in^2 (35 MPa).

Since the flexural stresses in footings are small, grade 40 steel ($f_y = 40$ k/in^2 (275 MPa)) is usually adequate. However, grade 40 is readily available only in sizes up through #6, and grade 60 steel ($f_y = 60$ k/in^2 (400 MPa)) may be required on the remainder of the project. Therefore, some engineers use grade 60 steel in the footings.

10.3 BASIS FOR DESIGN METHODS

Before the twentieth century, the design of spread footings was based primarily on precedent. Engineers knew very little about how footings behaved, so they followed designs that had worked in the past.

The first major advance in our understanding of the structural behavior of reinforced concrete footings came as a result of full-scale load tests conducted at the University of Illinois by Talbot (1913). He tested 197 footings in the laboratory and studied the mechanisms of failure. These tests highlighted the importance of shear in footings.

During the next five decades, other individuals in the United States, Germany, and elsewhere conducted additional tests. These tests produced important experimental information on the flexural and shear resistance of spread footings and slabs as well as the response of new and improved materials. Richart's (1948) tests were among the most significant of these. He tested 156 footings of various shapes and construction details by placing them on a bed of automotive coil springs that simulated the support from the soil and loaded them using a large testing machine until they failed. Whitney (1957) and Moe (1961) also made important contributions.

A committee of engineers (ACI-ASCE, 1962) synthesized this data and developed the analysis and design methodology that engineers now use. Because of the experimental nature of its development, this method uses simplified, and sometimes arbitrary, models of the behavior of footings. It is also conservative.

As often happens, theoretical studies have come after the experimental studies and after the establishment of design procedures. For example, Jiang (1983) and Rao and Singh (1987) have conducted such studies. Although work of this type has had some impact on engineering practice, it is not likely that the basic approach will change soon. Engineers are satisfied with the current procedures for the following reasons:

- Spread footings are inexpensive, and the additional costs of a conservative design are small.
- The additional weight that results from a conservative design does not increase the dead load on any other member.
- The construction tolerances for spread footings are wider than those for the superstructure, so additional precision in the design probably would be lost during construction.
- Although perhaps crude when compared to some methods available to analyze superstructures, the current methods are probably more precise than the geotechnical analyses of spread footings and therefore are not the weak link in the design process.
- Spread footings have performed well from a structural point-of-view. Failures and other difficulties have usually been due to geotechnical or construction problems, not bad structural design.
- The additional weight of conservatively designed spread footings provides more resistance to uplift loads.

The current design methods emphasize two modes of failure: shear and flexure. A shear failure, shown in Figure 10.2, causes a section of concrete to come out of the bottom of the footing. Thus, this type of failure is actually a combination of tension and shear on inclined failure surfaces. We resist this mode of failure by providing an adequate footing thickness, T.

A flexural failure is shown in Figure 10.3. We can analyze this mode of failure by treating the footing as an inverted cantilever beam and resisting the flexural stresses by placing tensile steel reinforcement near the bottom of the footing.

Figure 10.2 "Shear" failure in a spread footing loaded in a laboratory (Talbot, 1913). Observe how this failure actually is a combination of tension and shear.

Figure 10.3 Flexural failure in a spread footing loaded in a laboratory (Talbot, 1913).

10.4 DESIGN OF SQUARE FOOTINGS

This section considers the design of square footings supporting single columns that carry primarily downward vertical loads.

In most reinforced concrete design problems, it is customary to do the flexural analysis first. However, with spread footings, it is most expedient to do the shear analysis first. This is because it is not cost-effective to use shear reinforcement (stirrups) in most spread footings and it is customary to neglect the shear strength of the flexural steel. The only source of shear resistance is the concrete, so the effective depth, d, must be large enough to provide sufficient shear capacity. We then perform the flexural analysis using this value of d.

Designing for Shear

ACI defines two modes of shear failure, *one-way shear* (also known as *beam shear* or *wide-beam shear*) and *two-way shear* (also known as *diagonal tension shear*). In the context of spread footings, these two modes correspond to the failures shown in Figure 10.4. Although the failure surfaces are actually inclined, as shown in Figure 10.2, engineers use these idealized vertical surfaces to simplify the computations.

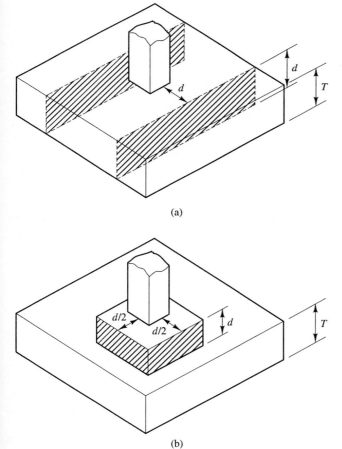

(a)

(b)

Figure 10.4   The two modes of shear failure: (a) one-way shear; and (b) two-way shear.

Various investigators have suggested different locations for the idealized critical shear surfaces, as shown in Figure 10.4. The ACI code [11.12.1.2] specifies that they are to be located a distance d from the face of the column for one-way shear and a distance $d/2$ for two-way shear. If the column sits on a capital, then these distances are measured from the edge of the capital.

The nominal shear load capacity, V_n, is [11.1]:

$$V_n = V_c + V_s \qquad (10.16)$$

Where:

V_c = nominal shear load capacity of concrete

V_s = nominal shear load capacity of reinforcing steel

For spread footings, it is customary to neglect V_s and rely only on the concrete for shear resistance.

The nominal one-way shear load capacity [11.3.1.1] is[1]:

$$V_n = V_c = 0.54 \, b_w \, d \, \sigma_r \sqrt{f_c' / \sigma_r} \qquad (10.17)$$

Where:

b_w = length of critical shear surface

d = effective depth

σ_r = reference stress = 2000 lb/ft^2 = 100 kPa = 14 lb/in^2 = 0.1 MPa

B_r = reference width = 1.0 ft = 12 in = 0.30 m = 300 mm

For square or rectangular footings, use $b_w = 2B$.

To determine the nominal two-way shear load capacity [11.12], compute ψ using Equations 10.18 - 10.20, and then use the lowest value in Equation 10.21. Usually, Equation 10.20 governs[2].

$$\psi = 2 + \frac{4}{\beta_c} \qquad (10.18)$$

---

[1] The ACI formula has been rewritten with the reference stress, σ_r, to form Equation 10.17. Thus, this formula may be used with any consistent set of units (see Section 1.6 in Chapter 1).

[2] The ACI formula has been rewritten with the reference stress, σ_r, to form Equation 10.21. Thus, this formula may be used with any consistent set of units (see Section 1.6 in Chapter 1).

$$\psi = \frac{\alpha_s d}{b_0} + 2 \qquad\qquad (10.19)$$

$$\psi = 4 \qquad\qquad (10.20)$$

$$V_n = V_c = 0.27\ \psi\ b_0\ d\ \sigma_r\ \sqrt{f_c'/\sigma_r} \qquad\qquad (10.21)$$

$$b_0 = 2c_1 + 2c_2 + 4d \qquad\qquad (10.22)$$

Where:

β_c = ratio of long side to short side of column

b_0 = perimeter of critical shear surface (see Figure 10.5)

α_s = 40 for footings

c_1, c_2 = plan dimensions of column (for circular columns, $c_1 = c_2 = $ diameter)

σ_r = reference stress = 2000 lb/ft^2 = 100 kPa = 14 lb/in^2 = 0.1 MPa

For concentrically loaded square or rectangular footings with no applied moment loads, two-way shear will control and it is the only mode we need to check. However, both modes should be checked for other conditions.

To visualize the two-way shear analysis, we can consider the footing to be divided into two blocks, one inside the shear surface and the other outside, as shown in Figure 10.5. The factored load is applied to the top of the inner block and is transferred to a uniform pressure acting on the base of both blocks. For design purposes, we need to know the factored shear force, V_u, acting on the shear surfaces, and this force would equal the force acting on the bottom of the outer block:

$$V_u = P_u \left(\frac{\text{base area of outer block}}{\text{total base area}} \right)$$

$$= P_u \left(\frac{BL - (c_1 + d)(c_2 + d)}{BL} \right) \qquad\qquad (10.23)$$

Because both V_u and V_n are functions of d, we must first estimate d, then determine whether $V_u \leq \phi V_n$ and revise d as necessary. Setting d equal to the width of the column is usually a good first estimate.

b_0 Is the Perimeter of the Inner
Block as Seen from the Top

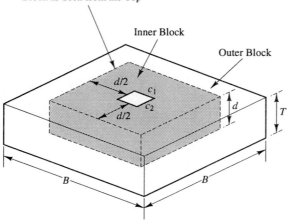

Figure 10.5 The two-way shear analysis.

The minimum acceptable footing thickness, T, is then:

$$T = d + d_b + 3 \text{ in} \qquad\qquad \text{(10.24 English)}$$

$$T = d + d_b + 75 \text{ mm} \qquad\qquad \text{(10.24 SI)}$$

The 3 in (75 mm) of cover between the rebars and the soil provides corrosion protection, allows for irregularities in the excavation, and accommodates possible contamination of the lower part of the concrete. We may wish to specify additional cover between the rebar and the soil at sites where it may be difficult to keep the bottom of the excavation smooth (e.g., loose sands). In some cases, it may be expedient to place a thin layer of lean concrete in the bottom of the excavation before setting the steel to provide a smooth working surface.

Footings are typically excavated using backhoes, so the contractor is not able to make a precise excavation. Therefore, there is no need to be overly precise when specifying T. Round it out to a multiple of 3 in or 100 mm. As a result, d will always be equal to a multiple of 3 in (or 100 mm) less one bar diameter less 3 in (or 75 mm).

If the column load is very light, the computed value of T may be very small. Ease of construction, good practice, and ACI [15.7 and 7.7.3.1] dictate that T should never be less than 9 in or 0.2 m.

Example 10.1 - Part A

A 21 inch square reinforced concrete column carries a vertical dead load of 380 k and a vertical live load of 270 k. It is to be supported on a square spread footing that will be founded on a soil with a net allowable bearing pressure of 6000 lb/ft^2. Determine the required width, B, thickness, T, and effective depth, d.

Solution:

Compute the required footing width (note that it is based on the unfactored load):

$$B = \sqrt{\frac{P_d + P_l}{q_a'}} = \sqrt{\frac{380 + 270}{6}} = 10.41 \quad \rightarrow \quad \text{Use 10 ft 6 in} = 126 \text{ in}$$

Compute the ultimate load (using Equation 10.1):

$$P_u = 1.4 P_d + 1.7 P_l = (1.4)(380) + (1.7)(270) = 991 \text{ k}$$

Because of the large applied load and because this is a large spread footing, we will use $f_c' = 4000$ lb/in^2 and $f_y = 60$ k/in^2.

Determine required thickness based on a shear analysis (using Equations 10.18 - 10.24):

Try $T = 24$ in:

$$d = T - 1 \text{ bar diameter} - 3 \text{ in} = 24 - 1 - 3 = 20 \text{ in}$$

$$b_0 = 2c_1 + 2c_2 + 4d = 2(21) + 2(21) + 4(20) = 164 \text{ in}$$

Find the most critical (lowest) ψ :

$$\psi = 2 + \frac{4}{\beta_c} = 2 + \frac{4}{1} = 6$$

$$\psi = \frac{\alpha_s d}{b_0} + 2 = \frac{(40)(20)}{164} + 2 = 6.88$$

$$\psi = 4 \quad \text{(governs)}$$

Using $\psi = 4$ and Equation 10.21, compute the nominal shear capacity:

$$V_n = V_c = 0.27 \, \psi \, b_0 \, d \, \sigma_r \sqrt{f_c'/\sigma_r}$$

$$= \frac{(0.27)(4)(164)(20)(14)\sqrt{4000/14}}{1000 \text{ lb/k}}$$

$$= 838 \text{ k}$$

$$\phi V_n = (0.85)(838) = 712 \text{ k}$$

$$V_u = P_u \left(\frac{B^2 - (c_1 + d)(c_2 + d)}{B^2} \right)$$

$$= 991 \left[\frac{126^2 - (21 + 20)(21 + 20)}{126^2} \right]$$

$$= 886 \text{ k}$$

$$V_u > \phi V_n \quad \textbf{NG}$$

Try $T = 27$ in:

$$d = T - 1 \text{ bar diameter} - 3 \text{ in} = 27 - 1 - 3 = 23 \text{ in}$$

$$b_0 = 4(21 + 23/2 + 23/2) = 176 \text{ in}$$

Computations parallel to those conducted above will demonstrate that $\psi = 4$.

$$V_n = V_c = 0.27 \; \psi \; b_0 \; d \; \sigma_r \sqrt{f_c' / \sigma_r}$$

$$= \frac{(0.27)(4)(164)(23)(14)\sqrt{4000/14}}{1000 \text{ lb/k}}$$

$$= 1040 \text{ k}$$

$$\phi V_n = (0.85)(1040) = 884 \text{ k}$$

$$V_u = P_u \left(\frac{B^2 - (c_1 + d)(c_2 + d)}{B^2} \right)$$

$$= 991 \left[\frac{126^2 - (21 + 23)(21 + 23)}{126^2} \right]$$

$$= 870 \text{ k}$$

$$V_u \le \phi V_n \quad \textbf{OK}$$

$\therefore$ Use **B = 10 ft 6 in; T = 27 in; d = 23 in** $\Leftarrow$ *Answer*

Designing for Flexure

A square footing bends in two perpendicular directions as shown in Figure 10.6a, and therefore might be designed as a *two-way slab* using methods similar to those that might be applied to a floor slab that is supported on all four sides. However, for practical purposes, it is customary to design footings as if they were a *one-way slab* as shown in Figure 10.6b. This conservative simplification is justified because of the following:

- The full-scale load tests on which this analysis method is based were interpreted this way.
- It is appropriate to design foundations more conservatively than the superstructure.
- The flexural stresses are low, so the amount of steel required is nominal and often governed by ρ_{min}.
- The additional construction cost due to this simplified approach is nominal.

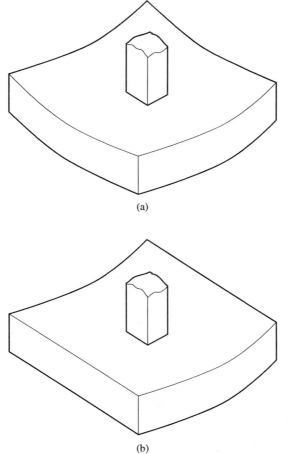

(a)

(b)

Figure 10.6 (a) A spread footing is actually a two-way slab, bending in both the "north-south" and "east-west" directions; (b) For purposes of analysis, engineers assume that the footing is a one-way slab that bends in one axis only.

Once we know the amount of steel needed to carry the applied load in one-way bending, we place the same steel area in the perpendicular direction. In essence the footing is reinforced twice, which provides about 60% more reinforcement than required by a more rigorous two-way analysis.

The usual procedure for designing flexural members is to prepare a moment diagram and select an appropriate amount of steel for each portion of the member. However, for simple spread footings, we again simplify the problem and design all the steel for the moment that occurs at the *critical section*. The location of this section for various types of columns is shown in Figure 10.7.

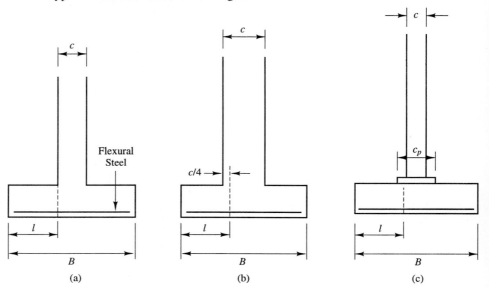

Figure 10.7 Location of critical section for bending: (a) with a concrete column; (b) with a masonry column; and (c) with a steel column.

We can simplify the computations by defining a distance l, measured from the critical section to the outside edge of the footing. In other words, l is the cantilever distance. It is computed using the formulas in Table 10.5.

The factored bending moment at the critical section, M_{uc}, is:

$$M_{uc} = \frac{P_u \, l^2}{2\,B} + \frac{2\,M_u\,l}{B} \tag{10.25}$$

Where:

P_u = factored compressive load from column

M_u = factored moment load from column

TABLE 10.5 DESIGN CANTILEVER DISTANCE FOR USE IN
DESIGNING REINFORCEMENT IN SPREAD FOOTINGS [15.4.2].

Type of Column	l
Concrete	$(B - c)/2$
Masonry	$(B - c/2)/2$
Steel	$(2B - (c + c_p))/4$

1. ACI does not specify the location of the critical section for timber
   columns, but in this context, it seems reasonable to treat them in
   the same way as concrete columns.
2. If the column has a circular, octagonal, or other similar shape, use
   a square with an equivalent cross-sectional area.
3. B = footing width; c = column width; c_p = base plate width

Equation 10.25 is based on the assumption that P_u acts through the centroid of the
footing. If it is eccentric, compute an equivalent moment using Equation 6.7 from
Chapter 6 and use this moment in the second term of the equation.

After computing M_u, find the steel area, A_s, and reinforcement ratio, ρ, using
Equations 10.15 and 10.12. Check if the computed ρ is less than ρ_{min}. If so, then use
ρ_{min}. Rarely will ρ be larger than 0.0040. This light reinforcement requirement develops
because we made the effective depth d relatively large to avoid the need for stirrups.

The required area of steel for each direction is:

$$A_s = \rho\, B\, d \tag{10.26}$$

Carry the flexural steel out to a point 3 in (75 mm) from the edge of the footing
as shown in Figure 10.7. It is not necessary to check the development length.

Example 10.1 - Part B:

Using the results from Part A, design the required flexural steel.

Solution:

Find the required steel area (using Table 10.5 and Equations 10.25, 10.15, and 10.12):

$$l = \frac{B - c}{2} = \frac{126 - 21}{2} = 52.5 \text{ in}$$

$$M_{uc} = \frac{P_u\, l^2}{2\, B} + 0 = \frac{(991{,}000)(52.5)^2}{(2)(126)} = 10{,}800{,}000 \text{ in-lb}$$

$$A_s = \frac{1.08\,M_{uc}}{\phi f_y d} = \frac{(1.08)(10,800,000)}{(0.9)(60,000)(23)} = 9.39 \text{ in}^2$$

$$\rho = \frac{A_s}{dB} = \frac{9.39}{(23)(126)} = 0.0032$$

$$\rho_{min} = 0.0018$$

$$\rho > \rho_{min}$$

Use 12 #8 bars each way ($A_s = 9.42$ in^2)   ⇐ *Answer*

Lightly Loaded Footings

Although the principles described above apply to footings of all sizes, when the applied load is small, practical minimums begin to govern the design. For example, if P_u is less than about 90 k (400 kN), the ACI minimum T of 9 in (0.2 m) controls. Thus, there is no need to conduct a shear analysis, only to compute a T smaller than the minimum. In the same vein, if P_u is less than about 30 k (130 kN), the minimum steel requirement ($\rho = 0.0018$) governs, so there is no need to conduct a flexural analysis. Often, these minimums also apply to footings that support larger loads.

If the entire base of the footing is within a 45° frustum, as shown in Figure 10.8, we can safely presume that very little or no tensile stresses will develop. No reinforcement is required in such cases, but it may be good to include nominal reinforcement (perhaps $\rho \approx 0.0018$) to accommodate unanticipated stresses.

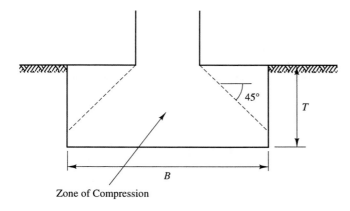

Figure 10.8 Zone of compression in small footings.

Structures built before the era of reinforced concrete often had large unreinforced footings that met the B/T criterion, but today only small footings, say, those with B less than about 3 ft (1 m), could be economically designed with these proportions.

10.5 DESIGN OF CONTINUOUS FOOTINGS

The structural design of continuous footings is very similar to that for square footings. The differences, described below, are primarily the result of the differences in geometry.

Designing for Shear

As with square footings, the depth of continuous footings is governed by shear criteria. However, whereas two-way shear governed before, one-way shear now governs. The critical surfaces for evaluating one-way shear are located a distance d from the face of the wall as shown in Figure 10.9.

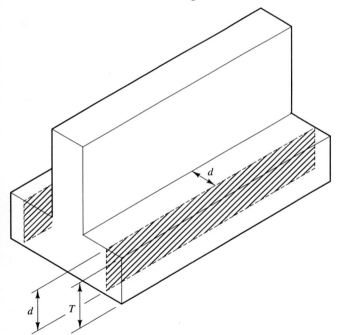

Figure 10.9 Location of idealized critical shear surfaces for one-way shear in a continuous footing.

The ultimate shear force acting on a unit length of the footing is:

$$V_u/b = \left(P_u/b\right)\left(\frac{B - c - 2d}{B}\right) \qquad (10.27)$$

Where:

V_u/b = factored shear force on critical shear surface per unit length of footing

P_u/b = ultimate applied compressive load per unit length of footing

   c = width of wall

   b = unit length of footing (usually 1 ft or 1 m)

Based on Equation 10.17, the shear load capacity on the surfaces shown in Figure 10.9 is:

$$V_n/b = 1.1\, d\, \sigma_r \sqrt{f_c'/\sigma_r}$$ (10.28)

Where:

V_n = nominal shear load capacity per unit length of footing

   d = effective depth

   σ_r = reference stress = 14 lb/in^2 = 100 kPa

Setting $V_u = \phi\, V_n$, equating Equations 10.27 and 10.28, and solving for d gives:

$$d = \frac{(P_u/b)\,(B-c)}{1.1\,\phi\,B\,\sigma_r\sqrt{f_c'/\sigma_r} + 2\,P_u/b}$$ (10.29)

Then, compute the footing thickness, T, using the criterion described earlier. Be careful to use consistent units in Equation 10.29 (i.e., all units should cancel). When using English units, this normally requires expressing P_u/b in units of lb/in.

Designing for Flexure

Nearly all continuous footings should have longitudinal reinforcing steel (i.e., running parallel to the wall). This steel helps the footing resist flexural stresses from non-uniform loading, soft spots in the soil, or other causes. Temperature and shrinkage stresses also are a concern. Therefore, place a nominal amount of longitudinal steel in the footing to produce a steel ratio, ρ, of 0.0015 - 0.0020. If large differential heaves or settlements are likely, we may need to use additional longitudinal reinforcement. Chapter 20 includes a discussion of this issue.

Transverse steel (that which runs perpendicular to the wall) is another issue. Most continuous footings are narrow enough so the entire base is within a 45° frustum, as shown in Figure 10.8. Thus, they do not need transverse steel. However, wider footings should include transverse steel designed to resist the flexural stresses at the critical section as defined in Table 10.5. The factored moment at this section is:

$$M_{uc}/b = \frac{(P_u/b)\, l^2}{2\,B} + \frac{2\,(M_u/b)\, l}{B} \qquad (10.30)$$

Where:

M_{uc}/b = factored moment at critical section per unit length of footing

P_u/b = ultimate applied compressive load per unit length of footing

M_u/b = ultimate applied moment load perpendicular to wall per unit length of footing

Compute the required transverse steel area per unit length, A_s/b, using a modified version of Equation 10.15:

$$A_s/b = \frac{1.08\,(M_{uc}/b)}{\phi\, f_y\, d} \qquad (10.31)$$

Example 10.2

A 200 mm wide concrete block wall carries a vertical dead load of 120 kN/m and a vertical live load of 88 kN/m. It is to be supported on a continuous spread footing that is to be founded at a depth of at least 500 mm below the ground surface. The allowable bearing pressure of the soil beneath the footing is 200 kPa. Develop a structural design for this footing.

Solution:

Compute the required footing width using the unfactored loads:

$$B = \frac{P_d/b + P_l/b}{q_a'} = \frac{120 + 88}{200} = 1.04 \quad \rightarrow \quad \text{Use } 1.1 \text{ m}$$

Use $f_c' = 15$ MPa; $f_y = 275$ MPa (grade 40)

Compute the ultimate load (using a modified form of Equation 10.1):

$$P_u/b = 1.4\,P_d/b + 1.7\,P_l/b = (1.4)(120) + (1.7)(88) = 318 \text{ kN/m}$$

Compute the required thickness using a shear analysis (using Equation 10.29):

$$d = \frac{(P_u/b)(B - c)}{1.1\, \phi\, B\, \sigma_r \sqrt{f_c'/\sigma_r} + 2\, P_u/b}$$

$$= \frac{(318 \text{ kN/m})(1.1 - 0.2 \text{ m})}{(1.1)(0.85)(1.1 \text{ m})(100 \text{ m})\sqrt{15 \text{ MPa}/0.1} + (2)(318 \text{ kN/m})}$$

$$= 0.151 \text{ m}$$
$$= 151 \text{ mm}$$

For ease of construction, place the longitudinal steel below the lateral steel. Assuming #4 bars (diameter = 12.7 mm), the footing thickness, T, is:

$$T = d + (1/2)\,(\text{diam. of lat. steel}) + \text{diam. of long steel} + 75 \text{ mm}$$
$$= 151 + 12.7/2 + 12.7 + 75$$
$$= 245 \text{ mm} \quad \rightarrow \quad \text{Use 300 mm}$$

$$d = 300 - 12.7/2 - 12.7 - 75 = 206 \text{ mm}$$

Design the lateral steel (using Table 10.5 and Equations 10.30 and 10.31):

$$l = \frac{B - c/2}{2} = \frac{1.1 - 0.2/2}{2} = 0.50 \text{ m} = 500 \text{ mm}$$

$$M_{uc}/b = \frac{(P_u/b)\, l^2}{2B} + 0 = \frac{(318)(0.50)^2}{2(1.1)} = 36.1 \text{ kN-m/m}$$

$$A_s/b = \frac{1.08\,(M_{uc}/b)}{\phi f_y d} = \frac{1.08\,(36.1)}{(0.9)(275{,}000)(0.206)} \times \frac{10^6 \text{ mm}^2}{\text{m}^2} = 765 \text{ mm}^2/\text{m}$$

$$\rho = \frac{A_s/b}{d} = \frac{765 \text{ mm}^2/\text{m}}{(206 \text{ mm})(1000 \text{ mm/m})} = 0.0037$$

$$\rho_{min} = 0.0020$$

$$\rho > \rho_{min}$$

Use #5 bars @ 250 mm OC ($A_s = 792 \text{ mm}^2/\text{m} > 765 \text{ mm}^2/\text{m}$)

Design the longitudinal steel:

$$A_s = \rho b d = (0.0015)(1100)(206) = 340 \text{ mm}^2$$

Use 3 #4 bars ($A_s = 380 \text{ mm}^2$)

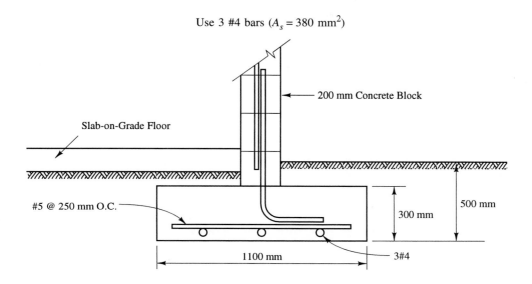

Lightly Loaded Footings

Many continuous footings are lightly loaded and their design is governed by practical minimums rather than the analyses just described. If the ultimate load is less than about 10 k/ft (150 kN/m), then the minimum T of 9 in (0.2 m) will govern and a shear analysis is not necessary. Likewise, if the ultimate load is less than about 4 k/ft (60 kN/m), ρ_{min} will govern and a flexure analysis is not necessary. Often, these minimums also apply to footings that support larger loads. Continuous footings with $B/T < 2$ do not require lateral steel, but all should have longitudinal steel.

10.6 DESIGN OF RECTANGULAR FOOTINGS

Rectangular footings with width B and length L that support only one column are similar to square footings. Design them as follows:

1. Check both one-way shear (Equation 10.17) and two-way shear (Equation 10.21) using the critical shear surfaces shown in Figure 10.10a. Determine the minimum required d and T to satisfy both.
2. Design the long steel (see Figure 10.10b) by substituting L for B in Table 10.5 and Equation 10.25, and using Equation 10.26 with no modifications. Distribute this

steel evenly across the footing as shown in Figure 10.10c.

3. Design the short steel (see Figure 10.10b) using Table 10.5 and Equation 10.25 with no modifications, and substituting L for B in Equation 10.26.

4. Since the central portion of the footing takes a larger portion of the short-direction flexural stresses, place more of the short steel in this zone [15.4.4]. To do so, divide the footing into inner and outer zones, as shown in Figure 10.10c. The portion of the total short steel area, A_s, to be placed in the inner zone is E:

$$E = \frac{2}{L/B + 1} \qquad (10.32)$$

Distribute the balance of the steel evenly across the outer zones.

10.7 DESIGN OF COMBINED FOOTINGS

Combined footings are those that carry more than one column. Their loading and geometry is more complex, so it is appropriate to conduct a more rigorous structural analysis. The rigid method, described in Chapter 9, is appropriate for most combined footings. It uses a soil bearing pressure that varies linearly across the footing, thus simplifying the computations. Once the soil pressure has been established, MacGregor (1992) suggests designing the longitudinal steel using idealized beam strips ABC, as shown in Figure 10.11. Then, design the transverse steel using idealized beam strips AD. See MacGregor (1992) for a complete design example.

Large or heavily loaded combined footings may justify a beam on elastic foundation analysis, as described in Chapter 9.

10.8 DESIGN OF MAT FOUNDATIONS

Choose the required thickness, T, of a mat foundation using a two-way shear analysis similar to that for square footings. Set $V_u = P_u$ for the heaviest column and solve for d using Equation 10.21.

Chapter 9 discussed methods of computing the distribution of soil bearing pressure and computing flexural stresses in the mat. We can use this information to determine the necessary reinforcement. Shukla (1984) gives a method of conducting these computations by hand, but most engineers now use finite element computer programs.

Prestressed or Post-Tensioned Slabs

Thin prestressed or post-tensioned slab-on-grade floors are a special type of mat foundation. They are typically 4 to 6 in (100 - 150 mm) thick and have been used for residential and light commercial structures, especially those on expansive soils, and for

One-Way Shear Surface

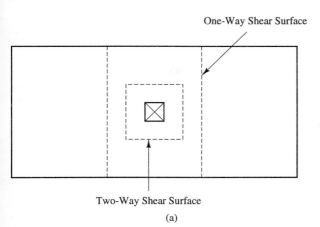

Two-Way Shear Surface

(a)

Long Steel

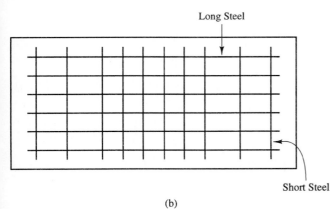

Short Steel

(b)

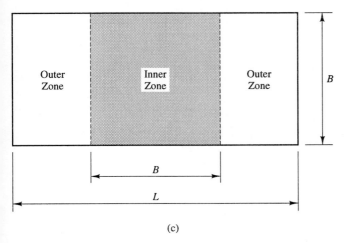

Outer
Zone

Inner
Zone

Outer
Zone

B

B

L

(c)

Figure 10.10   Structural design of rectangular footings: (a) critical shear surfaces; (b) long steel and short steel; and (c) distribution of short steel.

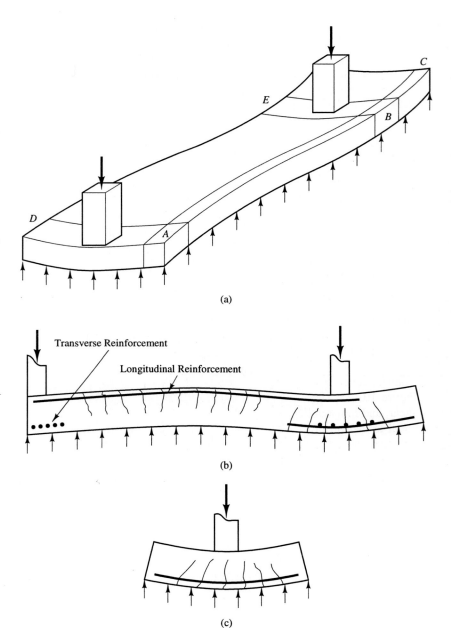

Figure 10.11 Structural design of combined footings: (a) idealized beam strips; (b) longitudinal beam strip; and (c) transverse beam strip (Adapted from MacGregor, 1992).

industrial floors that must support heavy equipment. PTI (1980) presents a procedure for designing post-tensioned slabs.

These slabs are much more rigid than isolated footings or standard slab-on-grade floors. Therefore, they help spread isolated structural loads or localized soil settlements or heaves over a greater area, thus reducing differential settlements. They are also less likely to crack. However, they are not nearly as rigid as conventional mats and do not necessarily eliminate distortions in the structure. Engineers have observed differential movements large enough to cause doors and windows to jam.

10.9 CONNECTIONS WITH THE SUPERSTRUCTURE

One last design feature that needs to be addressed is the connection between the footing and the superstructure. Connections are often the weak link in structures, so this portion of the design should not be taken lightly. A variety of connection types are available, each intended for particular construction materials and loading conditions. The design of proper connections is especially important in seismic areas (Dowrick, 1987).

Connections with Columns

Columns may be made of concrete, masonry, steel, or wood, and each has its own concerns when designing connections.

Concrete and Masonry

Connect concrete or masonry columns to their footing [15.8] using *dowels*, as shown in Figure 10.12. These dowels are simply pieces of reinforcing bars that transmit axial, shear, and moment loads. Use at least four dowels with a total area of steel, A_s, at least equal to that of the column steel or 0.005 times the cross-sectional area of the column, whichever is greater. They may not be larger than #11 bars [15.8.2.3] and must extend into the footing for a distance at least equal to the development length.

Check the bearing strength of the footing [10.15.1] to verify that it is able to support the axial column load. This is especially likely to be a concern if the column carries large compressive stresses that might cause something comparable to a bearing capacity failure inside the footing. To check this possibility, compare the factored column load, P_u, and compare it to the nominal bearing strength, P_{nb}:

$$P_{nb} = 0.85 f_c' A_1 s \qquad (10.33)$$

Then, determine whether the following statement is true:

$$P_u \leq \phi P_{nb} \qquad (10.34)$$

Where:

P_u = factored column load

P_{nb} = nominal bearing strength

f_c' = 28-day compressive strength of concrete

$s = (A_1 + A_2)^{0.5} \leq 2$ if the frustum in Figure 10.13 fits entirely within the footing

$s = 1$ if the frustum in Figure 10.13 does not fit entirely within the footing

A_1 = cross-sectional area of the column

$A_2 = (c_1 + 4d)(c_2 + 4d)$ as shown in Figure 10.13

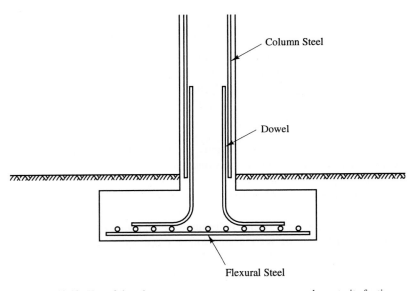

Figure 10.12 Use of dowels to connect a concrete or masonry column to its footing.

If P_u is greater than the ultimate bearing strength, use a higher strength concrete (greater f_c') in the footing or design the dowels as compression steel.

Since the footing and column are poured separately, there is a weak shear plane along the cold joint. Therefore, the dowels must transmit all of the shear load. The minimum required dowel steel area is:

$$A_s = \frac{V_u}{\phi f_y \mu} \tag{10.35}$$

Where:

A_s = minimum required dowel steel area

V_u = ultimate shear load

$\phi = 0.85$ for shear

f_y = yield strength of reinforcing steel

$\mu = 0.6$ if the cold joint not intentionally roughened or 1.0 if the cold joint is roughened by heavy raking or grooving [11.7.4.3]

However, the ultimate shear load, V_u, cannot exceed $0.2\,\phi\,f_c'\,A_c$, where f_c' is the compressive strength of the column concrete, and A_c is the cross-sectional area of the column.

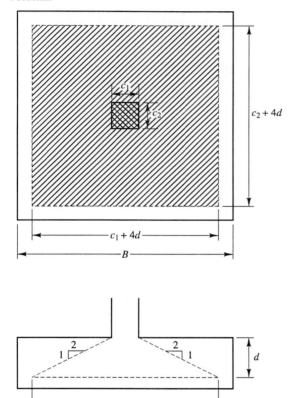

Figure 10.13 Application of frustum to find s and A_2 to compute the nominal bearing strength.

Most designs require a lap splice to connect the dowels and the vertical column steel. However, these splices have not always performed well during earthquakes. Cyclic horizontal movements can create large moments in the bottom of the column, especially in bridges and other structures that do not have the benefit of shear walls. This has led to spalling of the concrete, loss of bond in the lap splice, and collapse of the column. Therefore, use welded or threaded splices in columns subject to large seismic loads.

Steel

Steel columns often carry large compressive stresses, so the connection with the footing must transfer these stresses smoothly and efficiently without overstressing the concrete. Lightweight pipe columns, such as those sometimes used in residential and light commercial construction, might simply be embedded into the concrete, as shown in Figure 10.14a. Heavier columns are supported by *anchor bolts* and *base plates* as shown in Figure 10.14b. The contractor places the anchor bolts in the wet concrete and installs the base plate later. These bolts also provide a means of precisely aligning and leveling the column.

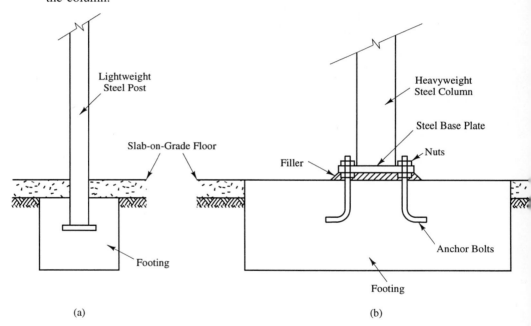

Figure 10.14 Methods of connecting a steel column and its footing; (a) embedding the column, and (b) a base plate and anchor bolts.

Wood

Wood columns, often called *posts*, usually carry light loads and require relatively simple connections. The most common type is a metal bracket, as shown in Figure 10.15. These are set in the wet concrete immediately after it is poured. The manufacturers determine the allowable loads and tabulate them in their catalogs.

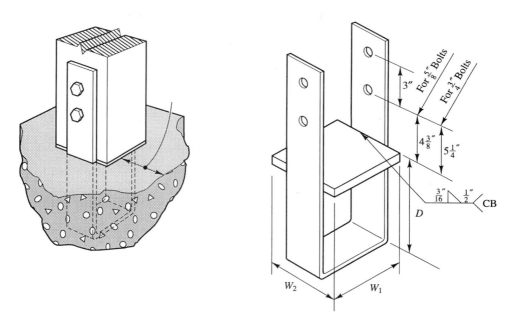

Figure 10.15 Steel post base for securing a wood post to a spread footing (Courtesy Simpson Strong-Tie Co., Inc.).

It is poor practice to simply imbed a wooden post into the footing. Although at first this would be a very strong connection, in time the wood will rot and become weakened.

Connections with Walls

The connection between a concrete or masonry wall and its footing is a simple one. Simply extend the vertical wall steel into the footing [15.8.2.2], as shown in Figure 10.16. For construction convenience, design this steel with a lap joint immediately above the footing.

The design of vertical steel in concrete retaining walls is discussed in Chapter 24. Design procedures for other walls are beyond the scope of this book.

Connect wood-frame walls to the footing using anchor bolts, as shown in Figure 10.17. Normally, standard building code criteria govern the size and spacing of these bolts. For example, the *Uniform Building Code* (ICBO, 1991a) specifies 1/2 in (12 mm) nominal diameter bolts embedded at least 7 in (175 mm) into the concrete. It also specifies bolt spacings of no more than 6 ft (1.8 m) on center.

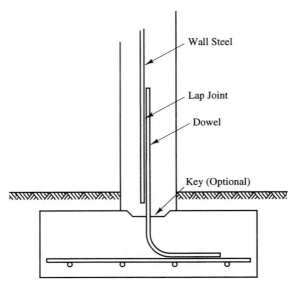

Figure 10.16   Connection between a concrete or masonry wall and its footing.

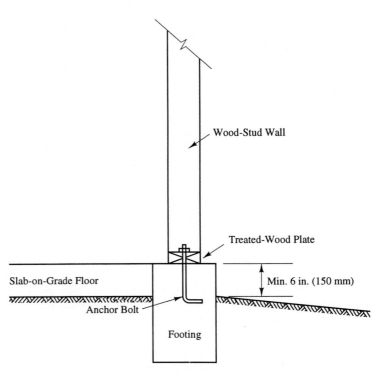

Figure 10.17 Use of anchor bolts to connect a wood-frame wall to a footing.

Sometimes we must supply a higher capacity connection between wood frame walls and footings, especially when large uplift loads are anticipated. Steel holdown brackets, such as that shown in Figure 10.18, are useful for these situations.

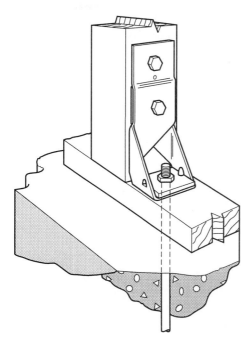

Figure 10.18 Use of steel holdown bracket to connect a wood-frame wall with large uplift force to a footing (Courtesy Simpson Strong-Tie Co., Inc.).

Figure 10.19 House that separated from its foundation during the 1989 Loma Prieta Earthquake (Photo by C. Stover, U.S. Geological Survey).

Many older wood-frame buildings have inadequate connections between the structure and its foundation. Figure 10.19 shows one such structure that literally fell off its foundation during the 1989 Loma Prieta Earthquake in Northern California.

Some wood-frame buildings have failed during earthquakes because the *cripple walls* collapsed. These are short wood-frame walls that connect the foundation to the floor. Shepherd and Delos-Santos (1991) provide a method of retrofitting cripple walls.

SUMMARY OF MAJOR POINTS

1. For a given applied load, the plan dimensions and minimum depth of a spread footing are governed by geotechnical concerns, whereas the thickness and reinforcement requirements are governed by structural concerns.

2. The ultimate strength design method consists of magnifying the service load (to obtain the ultimate load) and comparing it to the ultimate capacity of the member. In a properly designed structural member, the ultimate load is no greater than the capacity multiplied by a strength reduction factor, ϕ.

3. A spread footing design must consider both shear and flexural failure modes. A shear failure consists of the column or wall punching through the footing, while a flexural failure occurs when the footing has insufficient cantilever strength.

4. Since we do not wish to use stirrups (shear reinforcement), conduct the shear analysis first. Select an effective depth, d, in the footing that provides enough shear resistance in the concrete to resist the shear force induced by the applied load. This analysis ignores the shear strength of the flexural steel.

5. Once the shear analysis is completed, conduct a flexural analysis to determine the amount of steel required to provide the needed flexural strength. Since d is large, the required steel area will be small, and it is often governed by ρ_{min}.

6. For square footings, use the same flexural steel in both directions. Thus, the footing is reinforced twice.

7. For continuous footings, the lateral steel, if needed, is based on a flexural analysis. Use nominal longitudinal steel to resist nonuniformities in the load and to accommodate inconsistencies in the soil bearing pressure.

8. Design rectangular footings similar to square footings, but group a greater portion of the short steel near the center.

9. Practical minimum dimensions will often govern the design of lightly loaded footings.

10. The connection between the footing and the superstructure is very important. Use dowels to connect concrete or masonry structures. For steel columns and wood-frame walls, use anchor bolts. For wood posts, use specially manufactured brackets.

QUESTIONS AND PRACTICE PROBLEMS

10.1 A column carries the following vertical downward loads: dead load = 400 kN, live load = 300 kN, wind load = 150 kN, and earthquake load = 250 kN. Compute the ultimate load, P_u, to be used in an ultimate strength design.

10.2 A flexural member has a moment due to the dead load of 200 ft-k and a moment due to the live load of 150 ft-k. The computed nominal moment capacity, M_n, is 600 ft-k. Is the design of this member satisfactory?

10.3 Why are spread footings usually made of low-strength concrete?

10.4 Explain the difference between the shape of the actual shear failure surfaces in footings with those used for analysis and design.

10.5 A 400 mm square column that carries an ultimate vertical downward load of 450 kN is supported on a 1.5 m square footing. The effective depth of the concrete in this footing is 500 mm. Compute the ultimate shear force that acts on the critical section for two-way shear failure in the footing.

10.6 A 16 in square concrete column reinforced with 8 #6 bars carries vertical dead and live loads of 150 and 100 k, respectively. It is to be supported on a square footing with $f_c' = 3000$ lb/in^2 and $f_y = 60$ k/in^2. The soil has a net allowable bearing pressure of 4500 lb/ft^2. Because of frost heave considerations, the bottom of this footing must be at least 30 inches below the ground surface. Develop a complete structural design for this footing and show it in a sketch.

10.7 A 100 mm diameter steel post carries vertical dead and live loads of 30 and 20 kN, respectively. It is to be supported on a square footing and embedded as shown in Figure 10.14a. The soil has a net allowable bearing pressure of 200 kPa. Select appropriate concrete and steel properties and develop a structural design for this footing.

10.8 A 12 in wide concrete block wall carries vertical dead and live loads of 13.0 and 12.1 k/ft, respectively. The vertical reinforcement in this wall consists of one #6 bar at 24 in on center. It is to be supported on a continuous footing made of 2000 lb/in^2 concrete and 40 k/in^2 steel. The soil has a net allowable bearing pressure of 4000 lb/ft^2. The local building code requires that the bottom of this footing be at least 24 inches below the ground surface. Develop a complete structural design and show it in a sketch.

10.9 A 400 mm diameter concrete column carries vertical dead and live loads of 980 and 825 kN, respectively. It is to be supported on a 2.0 m × 3.5 m rectangular footing. The concrete in the footing will have $f_c' = 20$ MPa and $f_y = 400$ MPa. The building will have a slab-on-grade floor, so the top of the footing must be at least 150 mm below the finish floor elevation. Develop a complete structural design and show it in a sketch.

Part C

Deep Foundation
Analysis and Design

11

Deep Foundations

For these reasons, Caesar determined to cross the Rhine, but a crossing by means of boats seemed to him both too risky and beneath his dignity as a Roman commander. Therefore, although construction of a bridge presented very great difficulties on account of the breadth, depth and swiftness of the stream, he decided that he must either attempt it or give up the idea of a crossing. The method he adopted in building the bridge was as follows. He took a pair of piles a foot and a half thick, slightly pointed at the lower ends and of a length adapted to the varying depth of the river, and fastened them together two feet apart. These he lowered into the river with appropriate tackle, placed them in position at right angles to the bank, and drove them home with pile drivers, not vertically as piles are generally fixed, but obliquely, inclined in the direction of the current. Opposite these, forty feet lower down the river, another pair of piles was planted, similarly fixed together, and inclined in the opposite direction to the current. The two pairs were then joined by a beam two feet wide, whose ends fitted exactly into the spaces between the two piles forming each pair . . . A series of these piles and transverse beams was carried right across the stream and connected by lengths of timber running in the direction of the bridge . . . Ten days after the collection of the timber had begun, the work was completed and the army crossed over.

From *The Conquest of Gaul*, translated by S.A. Handford, revised by Jane F. Gardner (Penguin Classics 1951, 1982) copyright © The Estate of S.A. Handford, 1951, revisions copyright © Jane F. Gardner, 1982). Used with permission.

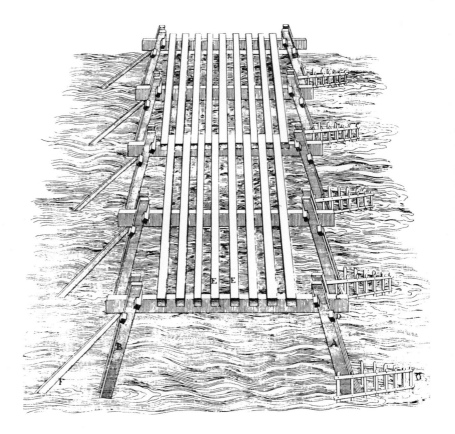

Figure 11.1 Caesar's bridge over the Rhine as envisioned by Palladio in 1570.

Although spread footings are the most common type of foundation, engineers often encounter circumstances where other types are more appropriate. Examples include the following:

- The upper soils are so weak and/or the structural loads so high that spread footings would be too large. A good rule-of-thumb for buildings is that spread footings cease to be economical when the total plan area of the footings exceeds one-half of the building footprint area.
- The upper soils are subject to scour or undermining. This would be especially of concern with midstream foundations for bridges.
- The foundation must penetrate through water (e.g. foundations for a pier).
- A large uplift capacity is required (the uplift capacity of a spread footing is limited to its dead weight).
- A large lateral load capacity is required.
- There will be a future excavation adjacent to the foundation, and this excavation would undermine shallow foundations.

In some of these circumstances, a mat foundation may be appropriate, but the most common alternative to spread footings is some type of *deep foundation.*

   A deep foundation is one that transmits some or all of the applied load to soils well below the ground surface, as shown in Figure 11.2. These foundations typically extend to depths on the order of 50 ft (15 m) below the ground surface, but they can be much longer, perhaps extending as deep as 150 ft (45 m). Even greater lengths have been used in some offshore structures, such as oil drilling platforms. Since soils usually improve with depth, and this method mobilizes a larger volume of soil, deep foundations are often able to carry very large loads.

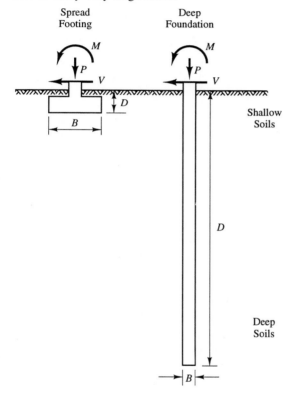

Figure 11.2 Deep foundations transfer the structural loads to deeper soil strata.

11.1 CLASSIFICATION OF DEEP FOUNDATIONS

Engineers and contractors have developed many kinds of deep foundations, each of which is best suited to certain loading and soil conditions. Unfortunately, people use many different names to identify these designs. Different individuals often use the same terms to mean different things and different terms to mean the same thing. This confusion reigns in both verbal and written communications, and is often the source of misunderstanding, especially to the newcomer. In this book, we will use terms that appear to be most common in North America.   Other terms and alternative definitions are in Appendix D.

We will divide deep foundations in to three broad groups, as shown in Figure 11.3. These are *piles*, *drilled shafts*, and *other types*.   These three groups are discussed in Chapters 12-14, 15, and 16, respectively.

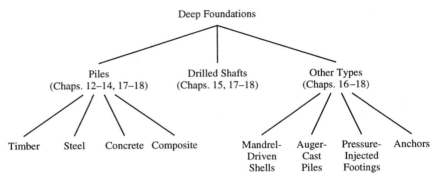

Figure 11.3 Classification of deep foundations.

11.2 LOAD TRANSFER

From 1929 to 1949, engineers and builders often used the rule of thumb: "If in doubt about the foundation, drive piles" (Davisson, 1989).   They recognized that deep foundations were more reliable and had higher load capacities. However, they had only a vague idea of the load bearing capacity of these deep foundations. Designs were based primarily on precedent, common sense, and crude pile driving formulas. Thus, some deep foundations were overbuilt, while others were underbuilt.

In-depth research and development on pile behavior, which began during the 1950s, has helped us predict load capacities with greater certainty, and thus produce more economical and reliable designs. The following chapters discuss the principal results of this research and development.   Nevertheless, there are still many aspects of deep foundation behavior we do not fully understand, so there continue to be many uncertainties in our analyses and designs.

Classification of Applied Loads

Structures can impose normal, shear, moment, and torsion loads onto the foundations, as shown in Figure 2.1 of Chapter 2.   When designing deep foundations, we divide these loads into two categories: *axial loads* and *lateral loads*. Axial loads are those that act parallel to the axis of the foundation; lateral loads are those that act perpendicular to the axis.   Thus, if the axis is vertical, the applied uplift and downward loads from the structure induce axial loads in the foundation, whereas applied shear and moment loads induce lateral loads. This chapter and Chapters 12 - 16 focus on axial loads. Chapter 17 discusses lateral loads. Torsional loads are rarely a concern in deep foundation design.

Transferring Axial Loads Into the Ground

Deep foundations transfer applied axial loads to the ground via two mechanisms: *skin friction* and *end bearing*. Both are shown in Figure 11.4. The skin friction resistance is the result of sliding friction along the side of the foundation and adhesion between the soil and the foundation. In contrast, the end bearing capacity (also known as *point bearing capacity*) is the result of compressive loading between the bottom of the foundation and the·soil, and thus is similar to load transfer in spread footings.

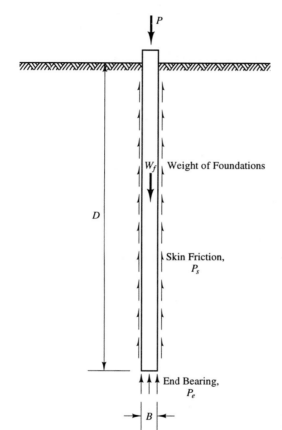

Figure 11.4 Transfer of axial loads from a deep foundation into the ground by skin friction and end bearing.

Formulation of Downward Load Capacity

Skin friction and end bearing are fundamentally different modes of resistance, so it is customary to evaluate each of them separately. Thus, the net allowable downward capacity, P_a, is computed as follows:

$$P_a = \frac{P_e + P_s - W_f}{F} \tag{11.1}$$

We can simplify this formula by using the net end bearing resistance, P_e':

$$P_e' = P_e - W_f \tag{11.2}$$

$$P_a = \frac{P_e' + P_s}{F} \tag{11.3}$$

Rewriting in terms of the unit end bearing and skin friction resistance gives:

$$P_a = \frac{q_e' A_e + \Sigma f_s A_s}{F} \tag{11.4}$$

Where:

   P_a = allowable downward axial load
   P_e = end bearing capacity
   P_s = skin friction capacity
   W_f = weight of the foundation (considering buoyancy, if necessary)
   q_e' = net unit end bearing resistance
   A_e = end bearing contact area
   f_s = unit skin friction resistance
   A_s = skin friction contact area
   F = factor of safety

The unit skin friction resistance typically varies with depth, thus the need for dividing the foundation into sections, finding the skin friction in each section, and summing. However, some analysis methods use the *average* unit skin friction resistance along the length of the foundation, $\bar{f}_s$. When using these methods, Equation 11.4 becomes:

$$P_a = \frac{q_e' A_e + \bar{f}_s A_s}{F} \tag{11.5}$$

In theory, a foundation failure requires full mobilization of both the skin friction and end bearing capacities. However, the load-displacement relationships for each are different. The skin friction resistance is fully developed after only 0.2 - 0.4 in (5 - 10 mm) of downward displacement (i.e., foundation settlement). Much more displacement is required to fully mobilize the end bearing capacity (typically, about 10% of the foundation diameter). Therefore, most deep foundations carry most of the working load by skin friction. Most of the end bearing capacity is in reserve, thus forming the factor of safety.

However, foundations that penetrate through weak soils to bedrock or some other hard strata have very little skin friction resistance, and thus carry most of the service loads by end bearing.

Formulation of Uplift Load Capacity

Compute the net uplift capacity in a similar fashion, but drop the end bearing term and add the foundation weight:

$$P_{au} = 0.9\,W_f + \frac{R\sum f_s A_s}{F} \qquad (11.6)$$

Where:

P_{au} = net allowable upward axial load

R = reduction factor

F = factor of safety (use higher value for uplift - see text below Table 11.1)

If some or all of the foundation is beneath the groundwater table, subtract the buoyancy force from the unsubmerged weight to obtain W_f. The buoyancy force equals the submerged volume of the foundation multiplied by the unit weight of water.

For foundations with a D/B ratio greater than 6, set the reduction factor R equal to 1. This means that the ultimate skin friction resistance of long foundations is equal in both upward and downward loading. However, for shorter foundations, a cone of soil may form, as shown in Figure 11.5, thus reducing the skin friction resistance (Kulhawy, 1991). This appears to happen only when both of the following conditions are met:

- $D/B < 6$
- $\chi > 1$

Where:

$\chi = \overline{\beta} = \overline{\alpha}\,\overline{s}_u / \sigma_D'$

$\overline{\beta}$ = average β factor along length of foundation (See Chapters 13, 15 and 16)

$\overline{\alpha}$ = average α factor along length of foundation (See Chapters 13, 15 and 16)

$\overline{s}_u$ = average undrained shear strength along length of foundation

σ_D' = vertical effective stress at tip of foundation

If so, then:

$$R = \frac{2 + \chi}{3\chi} \qquad\qquad (11.7)$$

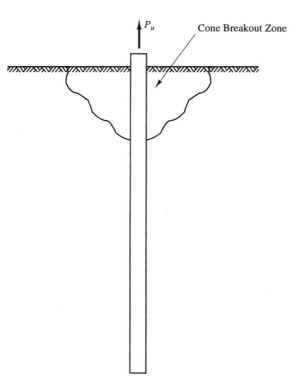

Figure 11.5   Cone breakout condition adjacent to a deep foundation loaded in uplift (Kulhawy, 1991).

Cyclic up-and-down loads can be more troublesome than static loads of equal magnitude. Turner and Kulhawy (1990) studied deep foundations in sands and found that cyclic loads smaller than a certain critical level do not cause failure, but enough cycles larger than that level cause the foundation to fail in uplift. The greatest differences between static and cyclic capacities seem to occur in dense sands and in foundations with large depth-to-diameter ratios.

Foundations with Enlarged Bases

Deep foundations with enlarged bases, such as belled drilled shafts or pressure injected footings, gain additional uplift capacity, as shown in Figure 11.6. This additional capacity can be large, but it is difficult to quantify. O'Neill (1987) suggested the following relationship for cohesive soils:

$$P_{ub} = (s_u N_u + \sigma_D)(\pi/4)(B_b^2 - B_s^2) \qquad (11.8)$$

$$N_u = 2(D/B_b - 0.5) \le 9 \qquad (11.9)$$

Where:

P_{ub} = uplift capacity contributed by the enlarged base

s_u = undrained shear strength in soil above the base

σ_D = total stress at the bottom of the base

B_b = diameter of the enlarged base

B_s = diameter of the shaft

Equation 11.6 then becomes:

$$P_{au} = 0.9\, W_f + \frac{P_{ub} + R\sum f_s A_s}{F} \qquad (11.10)$$

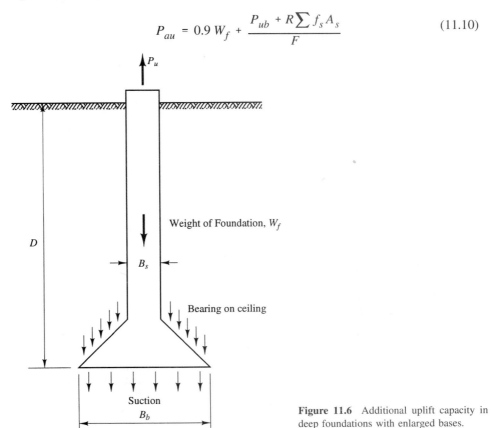

Figure 11.6 Additional uplift capacity in deep foundations with enlarged bases.

Since the uplift pressure from the enlarged base interacts with the skin friction resistance of the lower portion of the shaft, O'Neill recommends neglecting the skin friction between the bottom of the foundation and a distance $2B_b$ above the bottom.

If the bottom is below the groundwater table, suction forces might produce additional uplift resistance. Although they might be large, especially for short-term loading (i.e., less than 1 minute), it is best to ignore them until additional research better defines their character and magnitude.

Dickin and Leung (1990) suggested the following relationship for cohesionless soils:

$$P_{au} = 0.9\,W_f + \frac{N_u \sigma_D' A_b}{F} \tag{11.11}$$

Where:

P_{au} = net allowable uplift capacity

W_f = weight of foundation (considering buoyancy, if necessary)

N_u = breakout factor (from Figure 11.7)

σ_D' = vertical effective stress at bottom of foundation

A_b = area of bottom of enlarged base

F = factor of safety

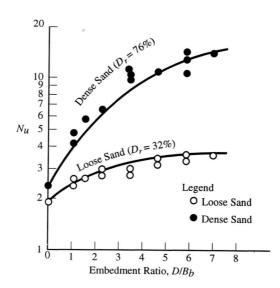

Figure 11.7 Breakout factor, N_u, for foundations with enlarged bases in cohesionless soils (Adapted from Dickin and Leung, 1990).

Factor of Safety

The design factor of safety, F, depends on many factors, including the following:

- The type and importance of the structure.
- The spatial variability of the soil.
- The thoroughness of the subsurface exploration program.
- The type and number of soil tests performed.
- The availability of on-site or nearby full-scale load test results.
- The anticipated level of construction inspection and quality control.
- The probability of the design loads actually occurring during the life of the structure.

To select F, classify the structure using Pugsley's (1966) system:

- *Monumental structures* have a design life exceeding 100 years, such as large bridges or extraordinary buildings.
- *Permanent structures* have design lives between about 25 and 100 years, such as ordinary rail and highway bridges and most large buildings.
- *Temporary structures* will be in place for a short time (perhaps less than 25 years), such as temporary industrial facilities.

Then, lump the remaining factors into an assessment of *control*:

- *Good control* implies that the subsurface conditions are uniform at the construction site and have been defined by a thorough program of subsurface exploration and testing, that nearby or on-site load test results are available. Good control also includes construction monitoring and, in the case of piles, Case Method analyses[1] on selected piles. Most projects do not have this degree of control.
- *Normal control* is a more common situation. This is similar to good control, except that no load test data are available or the subsurface conditions are not as uniform. For pile projects, the field engineer should have a bearing graph obtained from a wave equation analysis. On more difficult sites, one or more Case Method analyses may be appropriate.
- *Poor control* also is common. This implies that no load test data are available and that the soil conditions are more erratic. However, good subsurface data have been obtained from an exploration and testing program. In addition, construction inspection and quality control might not be as thorough as in the good control case.

---

[1] Case Method and wave equation analyses are discussed in Chapter 13.

- *Very poor control* implies that the soil conditions are much more erratic and difficult to define and the exploration and testing program was more limited.

Table 11.1 gives typical design factors of safety based on these assessments. Consider them to be guides, not absolute dictates, so do not hesitate to modify them as necessary.

TABLE 11.1 TYPICAL FACTORS OF SAFETY FOR DESIGN OF DEEP FOUNDATIONS FOR DOWNWARD LOADS

Classification of Structure	Acceptable Probability of Failure	Design Factor of Safety, F			
		Good Control	Normal Control	Poor Control	Very Poor Control
Monumental	10^{-5}	2.3	3.0	3.5	4.0
Permanent	10^{-4}	2.0	2.5	2.8	3.4
Temporary	10^{-3}	1.4	2.0	2.3	2.8

Expanded from Reese and O'Neill, 1989.

It is good practice to use higher factors of safety for analyses of uplift loads because uplift failures are much more sudden and catastrophic. For example, some pile foundations in Mexico City failed in uplift during the 1985 earthquake and heaved 10 ft (3 m) out of the ground. Therefore, most engineers use design values 1.5 to 2.0 times higher than those in Table 11.1.

The actual factor of safety for both downward and uplift loading (i.e., the real capacity divided by the real load) is usually much higher than the design F used in the formulas. This is because of the following:

- We usually interpret the soil strength data conservatively.
- The actual service loads are probably less than the design loads, especially in buildings.
- The as-built dimensions of the foundation may be larger than planned.
- Some (but not all!) of the analysis methods are conservative.

12

Pile Foundations – Methods and Applications

Amsterdam, die oude Stadt, is gebouwed op palen
Als die stad eens ommevelt, wie zal dat betalen?

An old Dutch nursery rhyme that translates to:

The old town of Amsterdam is built on piles
If it should fall down, who would pay for it?

Piles are long, slender, prefabricated structural members driven into the ground to form a foundation. Engineers use them both on land and in the sea to support many kinds of structures. Piles are made from a variety of materials and in different diameters and lengths according to the needs of each project. They can resist both axial (downward or uplift) and lateral (shear and moment) loads.

This chapter discusses the types of pile foundations, construction methods, and applications. The following chapters also include discussions of pile foundations:

- Chapter 13—Axial load capacities
- Chapter 14—Other geotechnical considerations
- Chapter 17—Lateral load capacities
- Chapter 18—Structural design

Although some engineers also use the word *pile* to describe certain types of cast-in-place deep foundations, we will use it only to describe prefabricated deep foundations. Chapters 15 and 16 discuss cast-in-place methods.

12.1 HISTORY

Mankind has used pile foundations for more than 2000 years. Alexander the Great utilized them in the city of Tyre in 332 BC, and the Romans used them extensively. Bridge builders in China during the Han Dynasty (200 BC - AD 200) also used piles. These early builders drove their piles into the ground using weights hoisted and dropped by hand (Chellis, 1961). By the Middle Ages, builders used pulleys and leverage to raise heavier weights.

Construction methods improved more quickly during the Industrial Revolution, especially when steam power became available. Larger and more powerful equipment was built, thus improving pile driving capabilities. These improvements continued through the twentieth century.

Pile materials also have become better. The early piles were always made of wood, and thus were limited in length and capacity. Fortunately, the advent of steel and reinforced concrete in the 1890s enabled the construction of larger and stronger piles, and better driving equipment made it possible to install them. Without these improved foundations, many of today's major structures would not have been possible.

Today, pile foundations can support very high loads, even in hostile environments. Perhaps the most impressive are those for offshore oil drilling platforms. These are as large as 10 ft (3 m) in diameter and must resist large lateral loads due to wind, wave, and earthquake forces.

12.2 TYPES OF PILES

Most piles are now made from wood, concrete, or steel. Each material has its advantages and disadvantages and is best suited for certain applications. We must consider many factors when selecting a pile type, including the following:

- **The applied loads**—Some piles, such as timber, are best suited for low to medium loads, whereas others, such as steel, may be most cost-effective for heavy loads.
- **The required diameter**—Most pile types are available only in certain diameters.
- **The required length**—Highway shipping regulations and practical pile driver heights generally limit the length of pile segments to about 60 ft (18 m). Therefore, longer piles must consist of multiple segments spliced together during driving. Some types of piles are easily spliced, whereas others are not.
- **The local availability of each pile type**—Some pile types may be abundant in certain geographic areas, whereas others may be scarce. This can significantly

affect the cost of each type.

- **The durability of the pile material in a specific environment**—Certain environments may cause piles to deteriorate, as discussed in Chapter 2.
- **The anticipated driving conditions**—Some piles are tolerant of hard driving, while others are more likely to be damaged.

Timber Piles

Timber piles have been used for thousands of years and continue to be a good choice for many applications. They are made from the trunks of straight trees and resemble telephone poles, as shown in Figure 12.1. Because trees are naturally tapered, these piles are driven upside down, so the largest diameter is at the top, as shown in Figure 12.2.

Figure 12.1 Groups of timber piles. Those in the foreground have been cut to the final top elevation (Photo courtesy of National Timber Piling Council).

Many different species of trees have been used to make timber piles. Today, most of those driven in North America are either Southern Pine or Douglas Fir. These trees are tall and straight and are abundant enough that the materials cost is low. They typically have butt diameters in the range of 6 to 16 in (150 - 400 mm) and lengths between 20 and 60 ft (6 - 20 m), but greater lengths are sometimes available (up to 80 ft

(24 m) in Southern Pine and 125 ft (38 m) in Douglas Fir). The branches and bark must be removed and it is sometimes necessary to trim the pile slightly to give it a uniform taper. Table 18.1 in Chapter 18 gives standard dimension requirements and ASTM D25 gives detailed specifications.

Although it is possible to splice lengths of wood piling together to form longer piles, this is a slow and time-consuming process that makes the piles much more expensive. Therefore, if longer piles are necessary, use some other material.

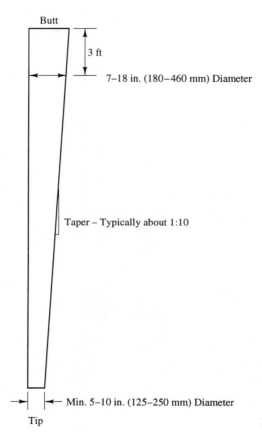

Butt

3 ft

7–18 in. (180–460 mm) Diameter

Taper – Typically about 1:10

Min. 5–10 in. (125–250 mm) Diameter

Tip

Figure 12.2 Typical timber pile.

Most timber piles are designed to carry downward axial loads of 20 to 100 k (100 400 kN). Their primary advantage is low construction cost, especially when suitable trees are available nearby. They are often used on waterfront structures because of their resistance to impact loads, such as those from ships.

When continually submerged, timber piles can have a very long life. For example, when the Campanile in Venice fell in 1902, its timber piles, which were driven in AD 900, were found in good condition and were reused (Chellis, 1962). However, when placed above the groundwater table, or in cyclic wetting conditions, timber piles are susceptible to decay, as discussed in Chapter 2. Therefore, they are nearly always treated with a preservative before installation.

When used in marine environments, timber piles are subject to attack from various marine organisms as well as abrasion from vessels and floating debris. In cooler waters (in North America, generally north of 40 degrees latitude), piles with heavy creosote treatment will usually remain serviceable for decades (ASCE, 1984). However, in warmer waters, biological attack is more of a problem and other chemical treatments are usually necessary. Even when treated, their usable life in such environments is often limited to about 10 years.

These piles are also susceptible to damage during driving. Repeated hard hammer blows can cause *splitting* and *brooming* at the butt and damage to the tip. It is often possible to control these problems by:

- Using lightweight hammers with appropriate cushions between the hammer and the pile
- Using steel bands near the butt (usually necessary only with Douglas Fir)
- Using a steel shoe on the tip, as shown in Figure 12.3
- Predrilling (see discussion in Section 12.3)

However, sometimes even these measures are not sufficient to prevent damage. Therefore, timber piles are best suited for light driving conditions, such as friction piles in loose sands and soft to medium clays. They are usually not good for dense or hard soils or as end bearing piles.

Figure 12.3 Use of steel tip points to reduce damage to timber piles during driving
(Photo courtesy of Associated Pile and Fitting Corp.)

Steel Piles

By the 1890s, steel had become readily available and many structures were being built of this new material. The use of steel piling was a natural development. Today, steel piles are very common, especially in situations that require high capacity foundations.

Because of their high strength and ductility, steel piles can be driven through hard soils and carry large loads. They also have the highest tensile strength of any major pile type, so they are especially attractive for applications with large applied tensile loads. There are standard steel pile sections and they are readily available in most areas.

Steel piles are easy to splice, so they are often a good choice when the required length is greater than about 60 ft (18 m). The contractor simply drives the first section, then welds on the next one, and continues driving. Special steel splicers can make this operation faster and more efficient. Hunt (1987) reported the case of a spliced steel pile driven to the extraordinary depth of 700 ft (210 m). They are also easy to cut, which can be important with end-bearing piles driven to irregular rock surfaces.

Steel piles have the disadvantages of being expensive to purchase and noisy to drive. In certain environments, they may be subject to excessive corrosion, as discussed in Chapter 2.

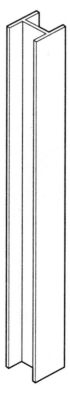

Figure 12.4 A steel H-pile.

H-Piles

Special rolled steel sections, known as HP sections, or simply *H-piles*, are made specifically to be used as piles. These sections are similar to WF (wide flange) shapes as shown in Figures 12.4 and 12.5. The primary difference is that the web is thinner than the flanges in wide flange members, while they have equal thicknesses in H-piles. Dimensions and other relevant information for standard steel H-piles are listed in Table 18.2 in Chapter 18. These piles are typically 50 to 150 ft (15 - 50 m) long and carry working axial loads of 80 to 400 k (350 - 1800 kN).

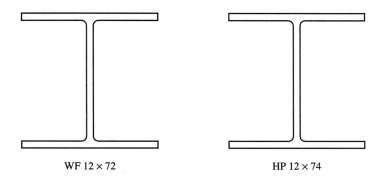

WF 12 × 72   HP 12 × 74

Figure 12.5 Comparison between typical wide flange (WF) and H-pile (HP) sections.

H-piles are *small displacement piles* because they displace a relatively small volume of soil as they are driven. This, combined with their high strength and moment of inertia, make them an excellent choice for hard driving conditions. They are often driven to bedrock and used as end bearing piles. If the pile will encounter hard driving, it may be necessary to use a hardened steel point to protect its tip, as shown in Figure 12.6.

Pipe Piles

Steel pipe sections are also commonly used as piles, as shown in Figure 12.7. They are typically 8 to 36 in (200 - 1000 mm) in diameter, 100 to 150 ft (30 - 50 m) long, and carry axial loads of 100 to 1500 k (450 - 7000 kN). A wide variety of diameters and wall thicknesses are available, and some engineers have even reclaimed used steel pipelines and used them as piles. Special sizes also can be manufactured as needed and pipe piles as large as 10 ft (3 m) in diameter with 3 in (75 mm) wall thickness have been used in offshore projects. Table 18.3 in Chapter 18 lists some of the more common sizes.

Pipe piles have a larger moment of inertia than H-piles, so they may be a better choice if large lateral loads are present.

Figure 12.6 Hardened steel tip welded to the bottom of a steel H-pile to protect it during hard driving (Photo courtesy of Associated Pile and Fitting Corp.)

Figure 12.7 A 16 inch (406 mm) diameter steel pipe pile.

Pipe piles may be driven with a *closed-end* or with an *open-end*. A closed-end pipe has a flat steel plate or a conical steel point welded to the bottom. These are *large displacement piles* because they displace a large volume of soil. This increases the load capacity, but makes them more difficult to drive. Conversely, an open-end pipe has nothing blocking the bottom and soil enters the pipe as it is being driven. The lower portion of small diameter open-end pipe piles will often become jammed with soil, thus forming a *soil plug*. Thus, an open-end pipe pile displaces less soil than a closed-end one, but more than an H-pile.

Closed-end pipe piles can be inspected after driving because the inside is visible from the ground surface. Thus, it is possible to check for integrity and alignment.

Special steel pipe piles are also available, such as the *monotube pile*, which is tapered and has longitudinal flutes.

Concrete Piles

Concrete piles are precast reinforced concrete members driven into the ground. This category does not include techniques that involve casting the concrete in the ground: They are covered in Chapters 15 and 16.

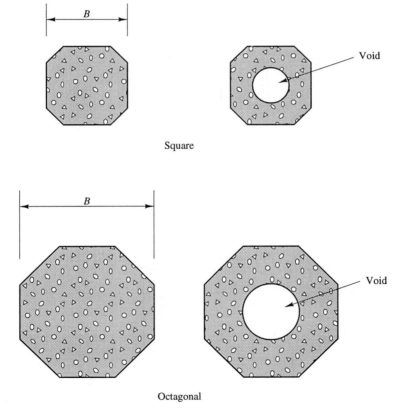

Figure 12.8 Cross-sections of typical concrete piles.

Concrete piles usually have a square or octagonal cross section, as shown in Figure 12.8, although other shapes have been used (ACI, 1980). They are typically 10 to 24 in (250 - 600 mm) in diameter, 40 to 100 ft (12 - 30 m) long, and carry working axial loads of 100 to 800 k (450 - 3500 kN). A few offshore projects have been built using much larger concrete piles.

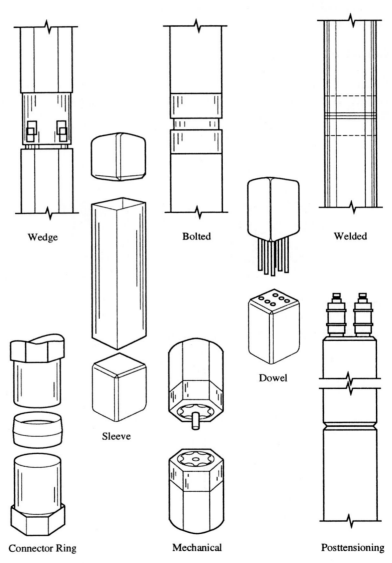

Figure 12.9 Typical splices for concrete piles (Courtesy of Precast/Prestressed Concrete Institute).

Although conventionally-reinforced concrete piles were once common, prestressed piles have almost completely replaced them. These improved designs have much more flexural strength and are therefore less susceptible to damage during handling and driving. Prestressing is usually a better choice than post-tensioning because it allows piles to be cut, if necessary, without losing the prestress force.

Several methods are available to splice concrete piles, as shown in Figure 12.9. Although these techniques are generally more expensive than those for splicing steel piles, they can be cost-effective in some situations. However, unlike steel, concrete piles are difficult and expensive to cut. Therefore, they are best suited for use as friction piles that do not meet *refusal* during driving (refusal means that the pile cannot be driven any further, so it becomes necessary to cut off the upper portion).

Concrete piles do not tolerate hard driving conditions as well as steel, and are more likely to be damaged during driving. Nevertheless, concrete piles are very popular because they are often less expensive than steel piles, yet still have a large load capacity.

Composite Piles

Concrete-Filled Steel Pipe Piles

Sometimes steel pipe piles are filled with concrete after driving. These will have more uplift capacity due to their greater weight (see Equation 11.6 in Chapter 11) and enhanced shear and moment capacity because of the strength of the concrete. However, there will be little, if any, usable increase in the downward load capacity because a pipe with sufficient wall thickness to withstand the driving stresses will probably have enough capacity to resist the applied downward loads. The net downward capacity may even be less because of the additional weight of the concrete in the pile.

Plastic-Steel Composite Piles

A plastic-steel composite pile consists of a steel pipe core surrounded by a plastic cover as shown in Figure 12.10. The plastic cover is typically made of recycled material, thus making this design attractive from a resource conservation perspective (Heinz, 1993).

Plastic-steel composite piles have been used successfully in waterfront applications (see Figure 12.11), where their resistance to marine borers, decay, and abrasion along with their higher strength make them superior to timber piles. Although the materials cost for plastic-steel composites is higher, their longer life and resource conservation benefits make them an attractive alternative to timber piles.

12.3 CONSTRUCTION METHODS AND EQUIPMENT

The construction of deep foundations is much more complex than that of shallow foundations, and the construction methods have a much greater impact on their performance. Therefore, design engineers must understand how contractors build pile foundations.

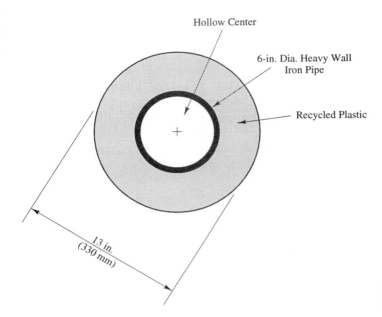

Figure 12.10 Cross-section of a typical plastic-steel composite pile.

Figure 12.11 Installation of a plastic-steel composite pile in a waterfront application. The existing piles are timber (Photo courtesy of Plastic Pilings, Inc.).

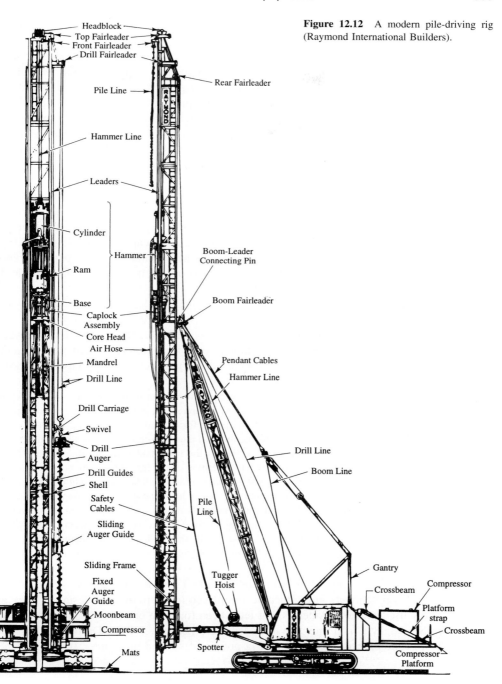

Figure 12.12 A modern pile-driving rig (Raymond International Builders).

Pile-Driving Rigs

The major piece of equipment used to install piles is the *pile-driving rig* (or simply the *pile driver*). Its function is to raise and temporarily support the pile while it is being driven and to support the pile hammer. Early rigs were relatively crude, but modern pile drivers, such as the one in Figure 12.12, are much more powerful and flexible.

Usually, vertical tracks, called *leads*, guide the hammer as the pile descends into the ground. Hydraulic or cable-operated actuators allow the operator to move the leads into the desired alignment. Sometimes, contractors omit the leads and simply place the hammer on top of the pile, which is aligned by other means.

Hammers

The *pile hammer* is the device that provides the impacts necessary to drive the pile. Repeated blows are necessary, so the hammer must be capable of cycling quickly. It also must deliver sufficient energy to advance the pile, while not being powerful enough to break it. The selection of a proper hammer is one of the keys to efficient pile driving.

Drop Hammers

The original type of pile hammer was the *drop hammer*. They consisted of a weight that was hoisted up, and then dropped directly onto the pile. These hammers became much larger and heavier during late nineteenth century as described by Powell (1884):

> The usual method of driving piles is by a succession of blows given by a heavy block of wood or iron, called a ram, monkey or hammer, which is raised by a rope or chain, passed over a pulley fixed at the top of an upright frame, and allowed to fall freely on the head of the pile to be driven. The construction of a pile-driving machine is very simple. The guide frame is about the same in all of them: the important parts are the two upright timbers, which guide the ram in its descent. The base of the framing is generally planked over and loaded with stone, iron, or ballast of some kind, to balance the weight of the ram. The ram is usually of cast-iron, with projecting tongue to fit the grooves of frame. Contractors have all sizes of frames, and of different construction, to use with hand or steam power, from ten feet to sixty feet in height. The height most in use is one of twenty feet, with about a twelve hundred pound ram. In some places the old hand-power method has to be used to avoid the danger of producing settling in adjoining buildings from jarring.

These hammers could deliver only about 3 to 12 blows per minute.

Drop hammers have since been replaced by more modern designs. They are now rarely used in North America, but have been found to be effective in the very soft clays in Scandinavia, and thus are still used there (Broms, 1981). Drop hammers also form part of the *pressure injected footing* process described in Chapter 16.

Steam, Pneumatic and Hydraulic Hammers

New types of hammers began to appear in the late 1800s. These consisted of a self-contained unit with a ram, anvil, and raising mechanism, as shown in Figure 12.13. These hammers had slightly larger weights, but much shorter strokes than the drop hammers. For example, the "Nasmyth steam pile-driver" of the 1880s had 3000 - 5000 lb (1400 - 2300 kg) rams with a stroke of about 3 ft (900 mm). Although these hammers delivered less energy per blow, they were more efficient because they cycled much more rapidly (about 60 blows/min for the Nasmyth hammer).

The early self-contained hammers used steam to raise the ram. This steam was produced by an on-site boiler. *Steam hammers* are still in use. Later, *pneumatic hammers* (powered by compressed air) and *hydraulic hammers* (powered by high-pressure hydraulic fluid) were introduced.

All three types can be built as a *single acting hammer* or as a *double acting hammer*. Single acting hammers raise the ram by applying pressure to a piston, as shown in Figure 12.13a. When the ram reaches the desired height, typically about 3 ft (0.9 m), an exhaust valve opens and the hammer drops by gravity and impacts the anvil. When compared to other types, this design is characterized by a low impact velocity and heavy ram weights.

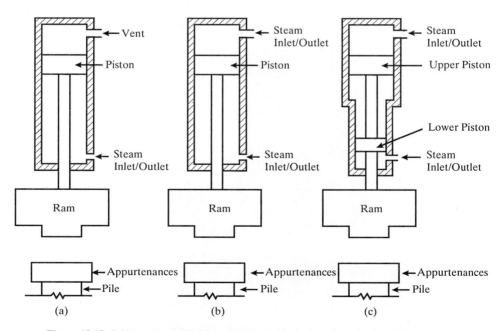

Figure 12.13 Self-contained pile driving hammers: (a) single acting; (b) double acting; and (c) differential.

A double acting hammer, shown in Figure 12.13b, uses pressure for both the upward and downward strokes, thus delivering a greater impact than would be possible by gravity alone. The impact energy depends to some degree upon the applied pressure and therefore can be controlled by the operator. These hammers usually have shorter strokes and cycle more rapidly than single acting hammers. Practical design limitations prevent these hammers from delivering as much energy as comparable single acting hammers, so they are principally used for driving sheet piles.

A *differential hammer*, shown in Figure 12.13c, is similar to a double acting hammer in that it uses air, steam, or hydraulic pressure to raise and lower the ram, but it differs in that it has two pistons with differing cross-sectional areas. This allows differential hammers to use the heavy rams of single acting hammers and operate at the high speed and with the controllability of double acting hammers.

Steam and pneumatic differential hammers cycle slowly under soft driving conditions and faster as the penetration resistance increases. The reverse is true of hydraulic hammers.

Diesel Hammers

A diesel hammer, shown in Figure 12.14, is similar to a diesel internal combustion engine. The ram falls from a high position and compresses the air in the cylinder below. At a certain point in the stroke, diesel fuel is injected and the air-fuel mixture is further compressed until the ram impacts the anvil. Combustion occurs about this time, forcing the ram up and allowing another cycle to begin.

Figure 12.14 A diesel pile hammer. The auger to the left is for predrilling.

Diesel hammers are either of the open-top (single acting) or closed-top (double acting) type. The closed-top hammer includes a bounce chamber above the ram that causes the hammer to operate with shorter strokes and at higher speeds than an open-top hammer with an equivalent energy output.

The operator can monitor the energy output of a diesel hammer by noting the rise of the ram (in an open-top hammer) or the bounce chamber pressure (in a closed-top hammer). Diesel hammers develop their maximum energy under hard driving conditions and may be difficult to operate under soft conditions because of insufficient hammer rebound. They typically deliver 40 to 55 blows per minute.

Although diesel hammers have been popular for many years, the exhaust is a source of air pollution, so air quality regulations now restrict their use in some areas.

Vibratory Hammers

A *vibratory hammer* (Warrington, 1992) is not a hammer in the same sense as those discussed earlier. It uses rotating eccentric weights to create vertical vibrations, as shown in Figure 12.15. When combined with a static weight, these vibrations force the pile into the ground. The operating frequency of vibratory hammers may be as high as 150 Hz and can be adjusted to resonate with the natural frequency of the pile.

Figure 12.15 A vibratory pile hammer (Photo courtesy of The Association of Drilled Shaft Contractors).

Vibratory hammers are most effective when used with piles being driven into sandy soils. They operate more quickly and with less vibration and noise than conventional impact hammers. However, they are ineffective in clays or soils containing obstructions such as boulders.

Appurtenances

A pile driving system also includes other components, as shown in Figure 12.16. Except for drop hammers, the ram does not directly strike the top of the pile. Such an arrangement would damage both the pile and the hammer. Instead, the ram hits a steel *striker plate*. It then transmits the impact energy through a *hammer cushion* (also known as a *capblock*) to a *drive head* (also known as a *drive cap*, *bonnet*, *hood*, or *helmet*). The drive head is attached directly to the pile except for a concrete pile where a *pile cushion* is placed between them.

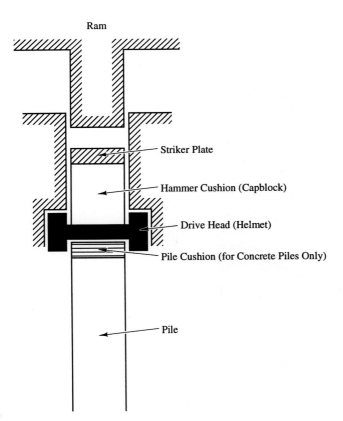

Figure 12.16 Pile driving appurtenances.

The cushions soften the sharp blow from the hammer by spreading it out over a longer time. Ideally, they should do this without absorbing too much energy. The traditional hardwood hammer cushions are not nearly as efficient as the more modern aluminum-micarta ones. Pile cushions are usually made of plywood.

The optimal selection of the pile hammer and appurtenances is part of the key to efficient pile driving. Wave equation analyses, discussed in Chapter 13, can be very useful in this regard.

Predrilling, Jetting, and Spudding

When driving piles in soils with hard strata, it is sometimes necessary to use predrilling, jetting, or spudding.

Predrilling means drilling a vertical hole, and then driving the pile into this hole. The diameter of the predrill hole must be less than that of the pile to assure firm contact with the soil. Predrilling also reduces the heave and lateral soil movement sometimes associated with pile driving. The predrill hole does not necessarily need to extend for the entire length of the pile. This technique is most commonly used in clayey soils.

To use *jetting*, the contractor pumps high pressure water through a pipe to a nozzle located at the pile tip. This loosens the soil in front of the pile, thus allowing it to advance with very few or no hammer blows. Jetting is useful in sandy and gravelly soils, but is ineffective in clays. It is most often used to quickly penetrate through sandy layers to reach deeper bearing strata.

Spudding consists of driving hard metal points into the ground, and then removing them and driving the pile into the resulting hole. This method is much less common than predrilling or jetting.

12.4 PILE ARRANGEMENTS AND GEOMETRIES

Occasionally, engineers use single isolated piles to support a structure. For example, an large outdoor sign on a single steel column might be supported on a single steel pipe pile. However, isolated piles can create problems in buildings and other structures.

The primary problem relates to the construction tolerances associated with pile driving. The designer should anticipate a tolerance of plus-or-minus several inches (≈ 20 cm), which is much greater than the tolerance in building the superstructure. Therefore, the centerlines of the columns will not be aligned with the centerlines of single isolated piles, as shown in Figure 12.17. This eccentricity will cause unanticipated moments and deflections in both the column and the pile.

There are three ways to avoid this problem:

- Provide grade beams to connect each of the piles, as shown in Figure 12.18a. These beams provide a moment to resist the eccentric loads, and thus relieve most of the added flexural stresses on the piles and columns.

- Support each column with two piles and provide grade beams along the weak axis only, as shown in Figure 12.18b.
- Support each column with three or more piles, as shown in Figure 12.18c. The potential eccentricities are small compared to the pile spacing, so the induced moments will be small. Therefore, no grade beams are necessary.

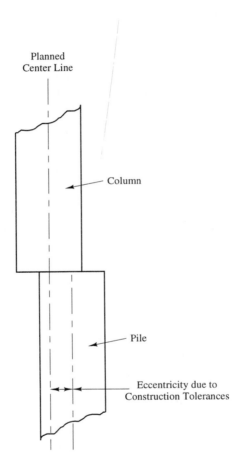

Figure 12.17 Unanticipated eccentricities between columns and single piles caused by construction tolerances.

12.5 CHANGES IN THE SOIL DURING CONSTRUCTION

The process of building pile foundations generates large stresses and deformations in the soil. These changes are much greater than those beneath spread footings or mat foundations and are important because they alter the engineering properties of the soil.

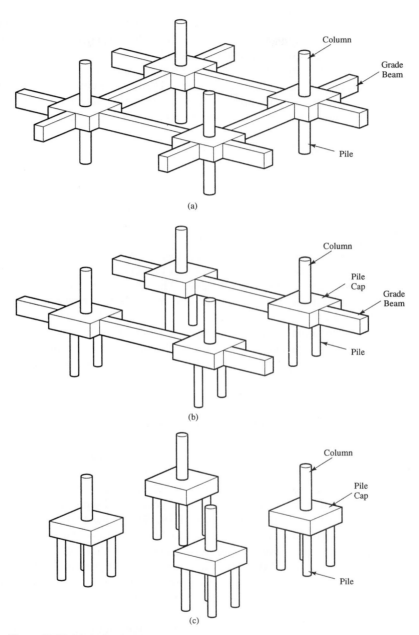

Figure 12.18 Methods of avoiding eccentricity-induced problems: (a) two-way grade beams; (b) two-pile groups with one-way grade beams; and (c) groups of three or more piles.

Chapters 5 - 9 discussed methods of using soil properties to predict the behavior of spread footings and mats. These methods are reliable because soil samples obtained before construction also represent the post-construction behavior. However, analyses of pile foundations are fundamentally more difficult because these same soil samples no longer reflect the post-construction behavior.

We do not fully understand these changes, and current deep foundation analysis techniques cannot rationally consider them. Therefore, we must use empirical and semi-empirical corrections plus appropriately conservative factors of safety. This is an important source of uncertainty in deep foundation design.

Changes in Cohesive Soils

Distortion

As a pile is driven into the ground, the soil below the tip must move out of the way. This motion causes both shear and compressive distortions. Additional distortions occur as a result of sliding friction along the side of the advancing pile. They are greatest around large displacement piles, such as closed-end steel pipe piles. For example, Cooke and Price (1973) observed the distortion in London Clay as a result of driving a 6.6 in (168 mm) diameter closed-end pipe pile. The soil within a radius of 1.2 pile diameters from the edge of the pile was dragged down, while that between 1.2 and 9 diameters moved upward.

This remolding of the clay changes its structure and reduces its strength to a value near the residual strength. Nevertheless, current analysis techniques are based on the peak strength and implicitly consider the difference between peak and residual strength. An analysis based on the residual strength might be more reasonable, but no such method has yet been perfected.

Compression and Excess Pore Water Pressures

Pile driving also compresses the adjoining soils. If saturated clays are present, this compression generates large excess pore water pressures, as discussed in Chapter 3. The ratio of the excess pore water pressure, u_e, to the original vertical effective stress, σ_v', may be as high as 1.5 to 2.0 near the pile, gradually diminishing to zero at a radius of 30 to 40 diameters, as shown in Figure 12.19. The greatest compression occurs near the pile tip, so u_e/σ_v' in that region may be as high as 3 to 4 (Airhart et al., 1969). These high pore water pressures dramatically decrease the shear strength of the soil, which makes it easier to install the pile, but temporarily decreases its load bearing capacity.

The presence of excess pore water pressures is always a transient condition because they also cause the water to flow away; in this case, radially away from the pile. Thus, the pore water pressures eventually return to the hydrostatic condition. If the pile is timber, some of the water also might be absorbed into the wood. This, along with thixotropic effects[1] and consolidation, eventually restores or even increases the strength of the clay. We can observe this increase in strength by resetting the pile hammer a couple of days after driving the pile and noting the increase in the blow count, a phenomenon known as *set up* or *freeze*.

In most clays, the excess pore water pressures that develop around a single isolated pile completely dissipate in about 1 month. Therefore, the full load capacity develops very quickly, as shown in Figure 12.20. However, in pile groups, the excess pore water pressures develop throughout a much larger zone of soil and may require a year or more to dissipate.

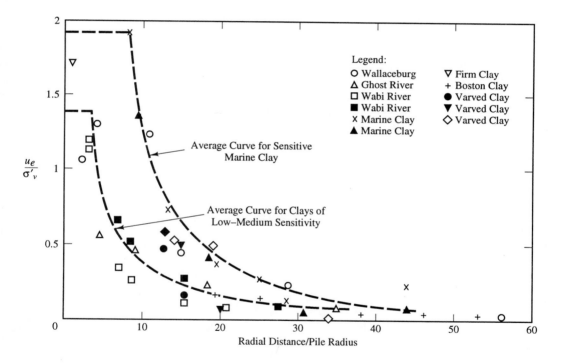

Figure 12.19 Summary of measured excess pore water pressures, u_e, in the soil surrounding isolated piles driven into clay (Adapted from *Pile Foundation Analysis and Design* by Poulos and Davis, Copyright ©1980 by John Wiley and Sons, Inc. Reprinted by permission of John Wiley & Sons, Inc.).

Loss of Contact Between Pile and Soil

Piles wobble during driving, thus creating gaps between them and the soil. Soft clays will probably flow back into this gap, but stiff clays will not. Tomlinson (1987) observed

---

[1] Thixotropy is the hardening of a disturbed material (in this case, soil) that sometimes occurs as it is allowed to rest.

such gaps extending to a depth of 8 to 16 diameters below the ground surface. Therefore, the skin friction in this zone may be unreliable, especially in stiff clays.

Changes in Cohesionless Soils

Soil compression from the advancing pile also generates excess pore water pressures in loose saturated sands. However, sands have a much higher hydraulic conductivity (permeability) than do clays, so these excess pore water pressures dissipate very rapidly. Thus, the full pile capacity develops almost immediately.

Some local dilation (soil expansion) can occur when driving piles through very dense sands. This temporarily generates negative excess pore water pressures that increase the shear strength and make the pile more difficult to drive. This effect is especially evident when using hammers that cycle rapidly.

Usually, we are most interested in the permanent changes that occur in the sand during pile driving. The impact and vibrations from the pile will cause particle rearrangement, crushing, and densification. In loose sands, these effects are especially pronounced, and engineers sometimes use piles purely for densification. However, dense sands will likely require predrilling or jetting to install the pile, both of which loosen the sand and partially or fully negate this effect.

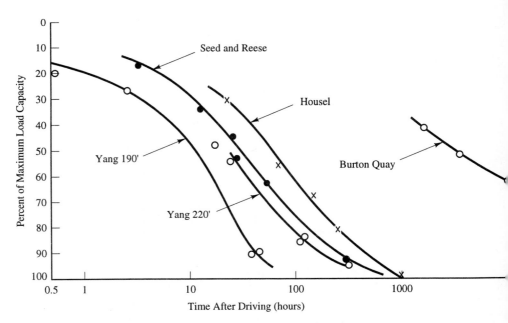

Figure 12.20 Increase in load capacity of an isolated pile with time (Adapted from Soderberg, 1962).

The sand in the center of pile groups is influenced by more than one pile, and therefore becomes denser than the sand near the edge of the group. This, in turn, probably causes the center piles to carry a larger share of the total load.

SUMMARY OF MAJOR POINTS

1. Piles are long, slender, prefabricated structural members driven into the ground to form a foundation. Most are made of steel, concrete, or wood, and can be used to support many kinds of structures.

2. Contractors use pile hammers mounted on pile drivers to install piles. Various hammer types and capacities are available to match the needs for each project.

3. It may be necessary to predrill, jet, or spud hard soils before driving piles. These operations can adversely affect the load bearing capacity of the completed pile, but may be necessary to permit installation without breaking the pile.

4. Because of the wide construction tolerances, columns placed on single piles may have problems with unanticipated eccentricities and the resulting moments and deflections. Engineers avoid these problems by using three or more piles for each column or by connecting the columns with grade beams.

5. Pile driving produces large compressive and shear strains in the soil, thus changing its engineering properties. Therefore, tests conducted on samples obtained before construction may not represent the post-construction conditions.

6. Pile driving in saturated clays produces large excess pore water pressures. These temporarily reduce the strength of the soil, thus making pile driving easier. Once these excess pore water pressures dissipate, the clay regains its strength and the pile load capacity increases, a phenomenon known as *freeze* or *set-up*.

QUESTIONS AND PRACTICE PROBLEMS

12.1   Discuss the advantages and disadvantages of timber, steel, and concrete piles.

12.2   What type or types of piles would be appropriate to support a heavy structure on an undulating bedrock surface located 25 to 40 m below the ground surface. Assume the skin friction in the overlying soils provides less than 20% of the total axial load capacity. Explain.

12.3   Why are most concrete piles prestressed instead of being conventionally reinforced?

12.4   In the context of pile construction, what are cushions, when are they used, and what is their purpose?

12.5   What is predrilling and when might it be used? What would happen if the predrill diameter or length was excessive?

12.6 Pile foundations that support buildings usually have at least three piles for each column. Why?

12.7 Engineers have observed that the axial load capacity of piles driven into saturated clay improves with time. What physical mechanism causes this behavior? Why don't we observe it in piles driven into clean sands?

12.8 Engineers often ignore the skin friction resistance in the upper several feet of piles driven into clays, especially stiff clays. Why?

12.9 Pile driving in loose sands without predrilling tends to densify these soils. What effect does this densification have on the load bearing capacity of such piles?

12.10 What problems might occur if a contractor selects a pile hammer that is too small (i.e., one that does not deliver enough energy)? What if the hammer is too large?

12.11 Sometimes, pile driving contractors and design engineers have conflicting interests during the construction of pile foundations. These conflicts often are associated with the use of certain construction methods or with unforeseen subsurface conditions. Give some examples of possible conflicts, and explain each party's perspective.

13

Pile Foundations –
Axial Load Capacity

Everything should be made as simple as possible, but no simpler.
Albert Einstein

This chapter discusses methods of assessing the axial load capacity of pile foundations. *Axial loads* are those that act parallel to the pile's longitudinal axis. Thus, if the pile is vertical, the axial loads consist of the applied downward and upward normal loads. *Lateral loads*, which are those that act perpendicular to the pile axis (i.e., shear and moments loads), are discussed in Chapter 17.

Pile foundations typically support heavier and more important structures, so it is appropriate to devote more time and effort to determine their axial load capacities. The methods of predicting these capacities may be divided into three broad categories:

- Full-scale load tests.
- Analyses based on soil properties obtained from laboratory or in-situ tests or from direct correlations with in-situ test results. These are known as *static methods*.
- Analyses based on pile driving dynamics. These are known as *dynamic methods*.

These methods are shown in the tree diagram in Figure 13.1 and are discussed in more detail later in this chapter. Each has advantages and disadvantages, and none is perfect. Therefore, engineers often use more than one method, and base final pile designs on a synthesis of the results.

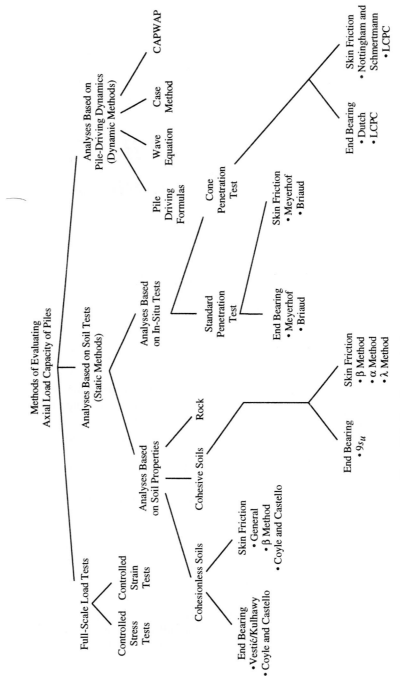

Figure 13.1 Tree diagram of axial load capacity analysis methods.

13.1 FULL-SCALE LOAD TESTS

The most precise way to determine the axial load capacity is to drive a full-size prototype pile at the site of the proposed production piles and load it to failure. All other methods determine the capacity indirectly, and therefore are less precise. However, load tests also are much more expensive, and thus must be used more judiciously.

A variety of equipment and procedures have been used to conduct load tests (Crowther, 1988), and these differences can influence the results. The proper interpretation of the test results is also a point of debate, thus introducing more variability. As a result, there is no single "correct" capacity for most piles. Nevertheless, engineers judge the accuracy of all other methods by comparing them to full scale load tests.

Equipment

To conduct a pile load test, we must have a means of applying the desired loads to the pile. Early tests often accomplished this by progressively stacking dead weights on top of the pile. Precast concrete blocks, pig iron, and water tanks have been used with varying degrees of success. Unfortunately, this method can be dangerous because it is difficult to place large weights without creating excessive eccentricities that can cause them to collapse. Therefore, this method is rarely used today.

A safer alternative is to support the weights, as shown in Figure 13.2, and use a hydraulic jack to provide the test load. This system is much more stable and less prone to collapse.

Another method of developing the test load is to drive *reaction piles* on each side of the test pile and connect them with a beam, as shown in Figure 13.3. A hydraulic jack located between the beam and the test pile provides the downward load. Most load tests use this method. For uplift tests, place the jack between the beam and one of the reaction piles.

Traditionally, engineers have measured the applied load by calibrating the hydraulic jack and monitoring the pressure of the hydraulic fluid during the test. However, even when done carefully, this method is subject to errors of 20% or more (Fellenius, 1980). Therefore, it is best to place a load cell (an instrument that measures force) between the jack and the pile and use it to measure the applied load.

A load test also must include a means of measuring the displacement of the pile. This most frequently consists of dial gages mounted on *reference beams*, as shown in Figures 13.2 and 13.3. It also is wise to include a backup system, such as a surveyor's level.

Conducting the Test

There are two categories of load tests: *controlled stress tests* and *controlled strain tests*. The former uses predetermined loads (the independent variable) and measured movements (the dependent variable), while the latter uses the opposite approach. ASTM D1143 describes both procedures.

It is best to conduct load tests after the excess pore water pressures have dissipated and any set-up or relaxation has occurred. This typically requires a delay of at least 2 days in sands and at least 30 to 90 days in clays.

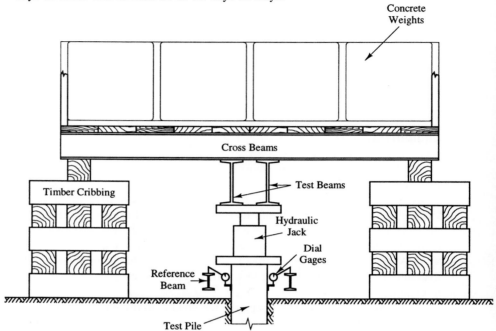

Figure 13.2 The use of a hydraulic jack reacting against dead weight to develop the test load.

Controlled Stress Tests

Most load tests use controlled stress methods. These are also known as a *maintained load (ML) test*. The field crew applies the test load in increments and allows the pile to move under each increment; they then record the corresponding movements and generate a load-displacement curve.

For many years, *slow ML tests* have been the most common type of pile load tests. These tests typically use load increments of 25, 50, 75, 100, 125, 150, 175, and 200% of the proposed design load and hold each increment until the pile stops moving or until the rate of movement is acceptably small. This process generally requires holding each load for at least 1 or 2 hours, sometimes more. Thus, these tests may require 24 hours or more to complete.

More recently, the *quick ML test* is beginning to dominate. This method is similar to the slow test except that each load increment is held for a predetermined time interval regardless of the rate of pile movement at the end of that interval. Typically, each load increment is about 10% of the anticipated design load and is held for 2 1/2 to 15 minutes.

Crowther (1988) suggests holding each increment for 5 minutes. This process continues until reaching about 200% of the anticipated design load and generally requires 2 to 5 hours to complete. This may be the best method for most piles.

Figure 13.3 This hydraulic jack is loading the steel pipe pile in compression. The jack reacts against the beam above, which loads the reaction piles (background) in tension. Notice the reference beams and dial gages in the foreground. These are used to measure the pile settlement.

Controlled Strain Tests

Controlled strain tests are the opposite of controlled stress tests because they use movement as the independent variable and load as the dependent variable. These require more sophisticated loading systems and therefore are less common than controlled stress tests.

One method of controlling strain is the *constant rate of penetration test*, which presses the pile into the ground at a constant rate. As the test progresses, the loads and settlements are measured to develop a load-settlement curve. Typically, the rate of penetration is 0.01 to 0.05 in/min (0.25 to 1.25 mm/min) for clays and 0.03 to 0.10 in/min (0.75 to 2.5 mm/min) for sands. The test duration is on the order of 1 hour.

In contrast, the *constant settlement increment test* uses increments of load chosen so that each increment produces a specified settlement. The required load is different for each increment and must be determined in the field as the test progresses.

Quality Control

Unfortunately, crews that conduct load tests sometimes use less than satisfactory workmanship, so the test results are not always reliable. This is especially true with slow ML tests. Peck's (1958) review of load tests demonstrates some of these problems, including the following:

- Cases where an increase in the downward load caused an upward deflection.
- Lack of smoothness in the load-settlement curve.
- Misleading load-settlement curves due to selection of unusual plotting scales.

Therefore, it is important to have a formal quality control program.

Interpretation of the Test Results

Once we have obtained the "correct" load-settlement curve, it is necessary to determine the magnitude of the failure load. For piles in soft clays, this is relatively straightforward. In these soils, the load-settlement curve has a distinct *plunge*, as shown by curve A in Figure 13.4, and the ultimate capacity is the load that corresponds to this plunge. However, piles in sands, intermediate soils, and stiff clays have a sloped curve with no clear point of failure, as shown by curve B in Figure 13.4.

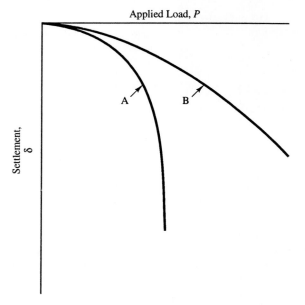

Figure 13.4 Typical load-settlement curves: Curve A is typical in soft clayey soils (note plunge); and curve B is typical of intermediate, stiff clay and sandy soils (ever-increasing load).

Many different methods have been proposed to interpret the latter type of load-settlement curve (Fellenius, 1990), and these methods often produce significantly different results. For example, Fellenius (1980) used nine different methods to analyze the results of a load test and found that the computed ultimate capacity varied from 362 k to 470 k. Another method, found in early editions of the AASHTO[1] specifications, produces an ultimate capacity of only 100 k!

Davisson's Method (1973) is one of the most popular. It defines the ultimate capacity as that which occurs at a settlement of $0.012B_r + 0.1\ B/B_r + PD/(AE)$ as shown in Figure 13.5. The last term in this formula is the elastic compression of a pile that has no skin friction.

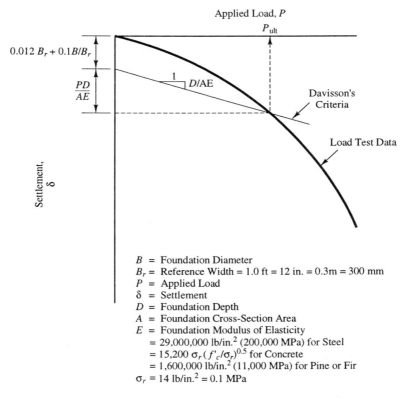

B = Foundation Diameter
B_r = Reference Width = 1.0 ft = 12 in. = 0.3m = 300 mm
P = Applied Load
δ = Settlement
D = Foundation Depth
A = Foundation Cross-Section Area
E = Foundation Modulus of Elasticity
    = 29,000,000 lb/in.^{2} (200,000 MPa) for Steel
    = 15,200 $\sigma_r (f'_c/\sigma_r)^{0.5}$ for Concrete
    = 1,600,000 lb/in.^{2} (11,000 MPa) for Pine or Fir
σ_r = 14 lb/in.^{2} = 0.1 MPa

Figure 13.5 Davisson's method of interpreting pile load test data.

Davisson's method seems to work best with data from quick ML tests. It may produce overly conservative results when applied to data from slow ML tests.

Some engineers in the United States prefer to express pile capacities using tons (the 2000 lb variety), whereas others use kips. This book uses only kips to maintain consistency with the remainder of structural and geotechnical engineering practice.

---

[1] The American Association of State Highway and Transportation Officials.

Example 13.1

The load-settlement data shown in the figure were obtained from a full-scale load test on a 400 mm square, 17 m long concrete pile ($f_c' = 40$ MPa). Use Davisson's method to compute the ultimate downward load capacity.

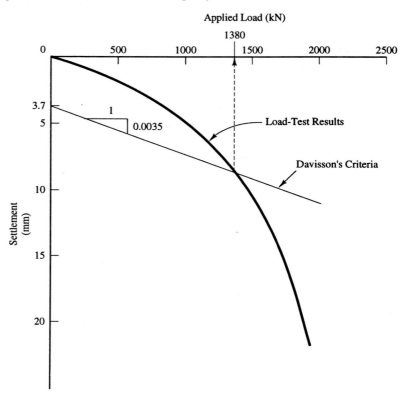

Solution

$$E = 15{,}200 \, \sigma_r \left(\frac{f_c'}{\sigma_r}\right)^{0.5} = (15{,}200)(100 \text{ kPa})\left(\frac{40 \text{ MPa}}{0.10 \text{ MPa}}\right)^{0.5} = 30.4 \, x \, 10^6 \text{ kPa}$$

$$0.012 B_r + \frac{0.1B}{B_r} + \frac{PD}{AE} = (0.012)(300) + \frac{(0.1)(400)}{300} + \frac{P(17{,}000)}{(400^2)(30.4 \times 10^6)(1 \times 10^{-6} \text{ m}^2/\text{mm}^2)}$$

$$= 3.7 \text{ mm} + 0.0035 \, P$$

Plotting this line on the load-displacement curve produces $P_{\text{ult}} = \textbf{1380 kN}$. $\Leftarrow$ ***Answer***

When and Where to Use Pile Load Tests

Full-scale load tests are expensive, but the cost is often justifiable because the data are more precise than other methods of analysis. Therefore, we can use a lower design factor of safety, as shown in Table 11.1, and thus reduce construction costs. Therefore, these tests are most cost-effective when one or more of the following conditions are present:

- Many piles are to be driven, so even a small savings on each pile will significantly reduce the overall construction cost.
- The soil conditions are erratic or unusual (static analyses are less reliable in these soils).
- The pile is supported in soft or medium cohesive soils (failure is much more dramatic in these soils).
- The structure is especially important or especially sensitive to settlements.
- The engineer has little or no experience in the area.
- The piles must resist uplift loads (the consequences of a failure are much greater).

Nearly all large pile foundation projects should include at least one full-scale load test. However, it is not practical to test every pile, even for the largest and most important projects. Therefore, we only can test representative piles and extrapolate the results to other piles at that site. Engel (1988) suggested the guidelines in Table 13.1 for determining the required number of load tests for typical projects. He also suggested that load tests are usually not necessary if the allowable design load, P_a, is less than 90 k (400 kN) per pile.

TABLE 13.1 GUIDELINES FOR DETERMINING THE REQUIRED NUMBER OF FULL-SCALE LOAD TESTS (Engel, 1988)

Summation of Length of All Piles at the Site		Required Number of Load Tests
(ft)	(m)	
0 - 6,000	0 - 1,800	0
6,000 - 10,000	1,800 - 3,000	1
10,000 - 20,000	3,000 - 6,000	2
20,000 - 30,000	6,000 - 9,000	3
30,000 - 40,000	9,000 - 12,000	4

Adapted from Engel (1988).

Because only a limited number of load tests can be performed on any given site, it may be wise to supplement them with *indicator piles*. These are driven at various locations across the site to verify that the driving conditions (and therefore the load capacities) are similar to those at the test pile.

Ideally, the load tests are conducted well before construction of the production piles. This allows time to finalize the design, develop a contract to build the foundation, and fabricate the production piles. Unfortunately, this also requires an extra mobilization and demobilization of the pile driver and crew, which is very expensive. Alternatively, the engineer could design and fabricate the piles without the benefit of a load test, and then *proof test* some of the first production piles. This method is faster and saves the extra mobilization costs, but may generate last-minute changes in the design.

The state-of-practice for pile load tests varies significantly from one geographic area to another. In some areas, they rarely are performed, even for very large structures, whereas elsewhere they are required for nearly all pile-supported structures. Sometimes these differences in practice are the result of local soil and geologic conditions or building code requirements. However, in other cases, the underutilization or overutilization of load tests seems to be primarily the result of custom and habit.

13.2 ANALYSES BASED ON SOIL TESTS (STATIC METHODS)

The second category of methods assess axial load capacities based on soil properties, such as the friction angle, ϕ, or the undrained shear strength, s_u, or on the results of in-situ tests, such as the standard penetration test or cone penetration test. These are known as *static methods*. Section 13.3 discusses methods based on soil properties; Section 13.4 discusses those based directly on in-situ test results.

These analyses are similar to those described in Chapters 6 - 8 for spread footings. Unfortunately, this task is fundamentally more difficult for piles because of the following:

- The pile driving process changes the soil properties, as described in Section 12.5. Thus, soil properties obtained from undisturbed samples do not necessarily reflect the post-construction conditions.
- The interaction between piles and soil is more complex than that for spread footings.

Although future research may provide more insight, both problems currently introduce large uncertainties in the analyses. Nevertheless, these analyses are much less expensive than load tests, and thus are useful for the following purposes:

- To serve as a preliminary analysis for planning a pile load test program.
- To extend the results of load tests to other pile lengths, diameters, and types.
- To design piles for relatively small or medium-size projects where a conservative design is less expensive than a load test.

These analysis methods evaluate skin friction and end bearing separately and combine them using Equations 11.4, 11.5 or 11.6 from Chapter 11. The remainder of this section and Sections 13.3 and 13.4 discuss methods of computing the parameters for these formulas, especially unit end bearing resistance, $q_e{}'$, and the unit skin friction resistance, f_s. These methods often give different results, so it may be appropriate to compute the capacity using more than one and compare the results using engineering judgment.

Computing Contact Areas A_e and A_s

The end bearing and skin friction contact areas, A_e and A_s, are usually easy to compute. However, be careful to properly consider *soil plugging* when working with open-ended steel pipe piles or steel H-piles. When driving open-ended steel pipe piles, the inside of the pile may become plugged with soil, as shown in Figure 13.6. This is especially likely to occur with small diameter piles. If such a plug does form, then A_e is the same as if it were closed-ended. However, if only a partial plug forms, then A_e is correspondingly less, perhaps about one-half of the total end area.

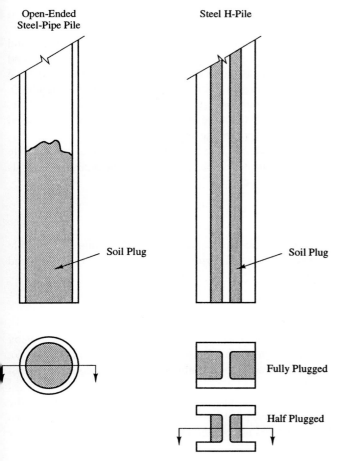

Figure 13.6 Soil plugging in open-ended steel pipe piles and steel H-piles.

For H-piles, soil plugging affects both end bearing and skin friction. For analysis purposes, we usually can assume the end of the pile will be half plugged as shown in Figure 13.6. However, for skin friction computations, it is both conservative and customary to assume the pile is fully plugged. Therefore, for H-piles, use A_e equal to half the cross-sectional area of the steel and soil plug, and compute A_s based on the perimeter of the pile with a full plug.

13.3 ANALYSES BASED ON SOIL PROPERTIES

We will define "analyses based on soil properties" as those that use basic soil parameters such as the friction angle, ϕ, or the undrained shear strength, s_u. An engineer might obtain these parameters from laboratory and/or in-situ tests. Analysis methods that use in-situ test results directly, such as SPT N value, are discussed in Section 13.4.

Soil Classification

Geotechnical engineers often divide soils into two broad categories: cohesionless and cohesive, as discussed in Chapter 3. Clean sands are the ideal cohesionless soil, and saturated clays are the ideal cohesive soil. Because these two types of soil often behave differently (especially their differing drainage conditions under normal rates of loading), foundation engineers usually develop separate analysis methods for each type. Therefore, the design engineer must first determine the type of soil at the site of a proposed foundation, then select analysis methods appropriate for that type.

Unfortunately, not all soils clearly fall into one category. For example, intermediate soils, such as clayey sands, can be difficult to classify. Therefore, the analysis methods are probably less reliable in these "hard-to-classify" soils.

Cohesionless Soils

Very little, if any, excess pore water pressure is present in cohesionless soils once the pile driving is finished. Therefore, evaluate the end bearing and skin friction capacities based on the drained strength of the soil.

End Bearing

Many engineers have attempted to express the net unit end bearing capacity, $q_e{}'$, using formulas similar to Terzaghi's formula for spread footings (Equation 6.15 in Chapter 6). Because D is large and B is small, the second term dominates and N_q becomes the most important bearing capacity factor.

The magnitude of N_q depends on the assumed shape of the shear surfaces near the pile tip. Several different shear patterns have been proposed, and each has its own function for N_q. Several of these functions are plotted in Figure 13.7. Unfortunately, for any given friction angle, the corresponding values of N_q vary by more than one order of magnitude!

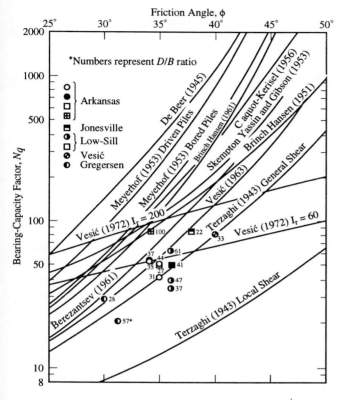

Figure 13.7 Several functions of N_q vs. ϕ have been proposed (Adapted from Coyle and Castello, 1981; Used with permission of ASCE).

Equation 6.15 also indicates that $q_e{'}$ increases almost linearly with depth, which does not correspond with the behavior of real piles as observed in pile load tests. It seems that $q_e{'}$ does increase with depth, but at an ever decreasing rate, as shown in Figure 13.8 (Kulhawy, 1984; Kraft, 1991).

To account for this difference between theory and observed behavior, many analysis methods include a maximum limit on $q_e{'}$, thus producing functions similar to that shown in Figure 13.8. Although these functions are useful for design purposes, they suggest a lack of understanding of the underlying processes that control end bearing. A more complete analysis would need to consider the following effects:

- The conventional bearing capacity analysis assumes that a general shear failure will occur, such as that which controls the bearing capacity of most spread footings. However, the mode of failure below a pile tip is much more complex, and more closely resembles a combination of local and punching shear. To evaluate this mode of failure, it is necessary to consider both the strength and the compressibility of the soil (the bearing capacity formulas in Chapter 6 consider only strength).

- The friction angle, ϕ, varies with the effective stress, σ'. This is especially noticeable at high stresses such as those that are present beneath a pile. Therefore, the actual friction angle in the soil may be less than that measured at lower stresses in the laboratory. This effect becomes more pronounced as the pile penetrates deeper into the soil.

- Even soil deposits that may appear to be normally consolidated usually have an overconsolidated crust near the ground surface. This crust, which may be the result of desiccation and other mechanisms, increases the end bearing capacity near the ground surface.

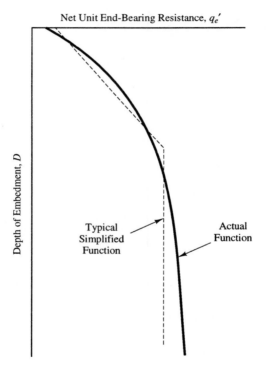

Figure 13.8 Distribution of net unit end bearing resistance, q_e', with depth in an apparently "homogeneous" soil deposit and proposed functions to describe this behavior.

Vesić/Kulhawy Method

Vesić (1977) provided a more complete framework for evaluating end bearing when he defined the *rigidity index*, I_r, of the soil as follows:

$$I_r = \frac{E}{2(1 + \nu_p)\,\sigma_D'\,\tan\phi} \tag{13.1}$$

Where:

E = modulus of elasticity of soil

ν_p = Poisson's ratio of soil

$\sigma_D{'}$ = vertical effective stress at the pile tip elevation

ϕ = friction angle

The magnitude of I_r is typically between about 10 and 400. Soils with relatively high values of I_r tend to fail in general shear, as modeled by conventional bearing capacity theory. However, low values of I_r suggest that soil compressibility is important and the local or punching shear failure modes govern.

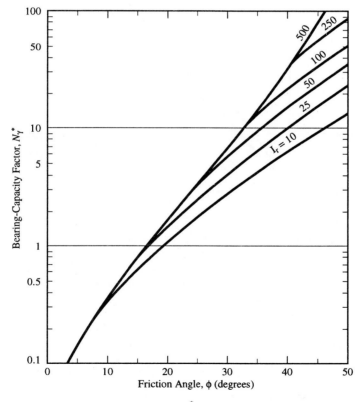

Figure 13.9 Bearing capacity factor $N_\gamma{}^{*}$ (Adapted from Kulhawy et al., 1983; Copyright © 1983 Electric Power Research Institute. Reprinted with permission).

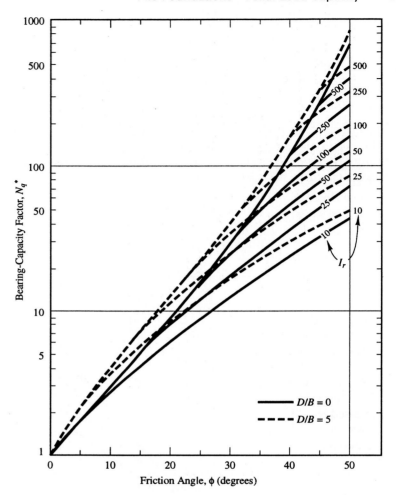

Figure 13.10 Bearing capacity factor N_q^* (Adapted from Kulhawy et al., 1983; Copyright © 1983 Electric Power Research Institute. Reprinted with permission).

    Building on Vesić's work, Kulhawy et al. (1983) defined the net unit end bearing capacity, q_e', as follows[2]:

$$q_e' = B\gamma N_\gamma^* + \sigma_D'(N_q^* - 1) \qquad (13.2)$$

---

    [2] Kulhawy wrote this formula in terms of the gross bearing pressure; the author modified it to produce the net pressure.

Where:

B = pile diameter

N_γ^*, N_q^* = bearing capacity factors as defined in Figures 13.9 and 13.10

γ = unit weight of soil immediately beneath pile tip (use γ_b if tip is located below the groundwater table)

When D/B exceeds 5, which is true for nearly all deep foundations, the first factor in the formula becomes negligible and may be ignored.

The Vesić/Kulhawy approach computes the net end bearing pressure that would create a bearing capacity failure beneath the pile. However, the pile displacement (i.e., settlement) required to produce such a failure may be greater than most structures can tolerate. In contrast, most other methods implicitly include some limitation on settlement. Therefore, Equation 13.2 usually produces relatively high q_e' values.

Coyle and Castello's Method

Coyle and Castello (1981) developed an empirical method for evaluating end bearing capacity in sands that considers pile settlement. This method defines failure as the load that produces a settlement of 10% of the pile diameter. Although this settlement also would be excessive in most cases, the working loads in the pile will be no more than about one-third of the ultimate loads, so the actual settlement of an end-bearing pile will be correspondingly less, probably on the order of 0.5 in (12 mm). If the pile has a significant skin friction resistance, as most of them do, then the settlement under service loads may be even smaller.

This method is based on the results of 16 instrumented pile load tests and is expressed as a correlation between q_e', pile length, and the soil friction angle, ϕ, as shown in Figure 13.11 It does not consider soil compressibility or groundwater conditions. For design, do not use a q_e' greater than 300,000 lb/ft^2 (15,000 kPa).

Skin Friction

The full unit skin friction capacity, f_s, develops when the shear stress along the soil-pile interface exceeds the shear strength. This requires a pile displacement of only 0.2 - 0.4 in (5 - 10 mm).

A simple sliding friction model describes this mechanism as follows:

$$f_s = \sigma_h' \tan\phi_s \tag{13.3}$$

Where:

σ_h' = horizontal effective stress (i.e., perpendicular to the pile axis)

$\tan\phi_s$ = coefficient of friction between the soil and the pile

ϕ_s = soil-pile interface friction angle

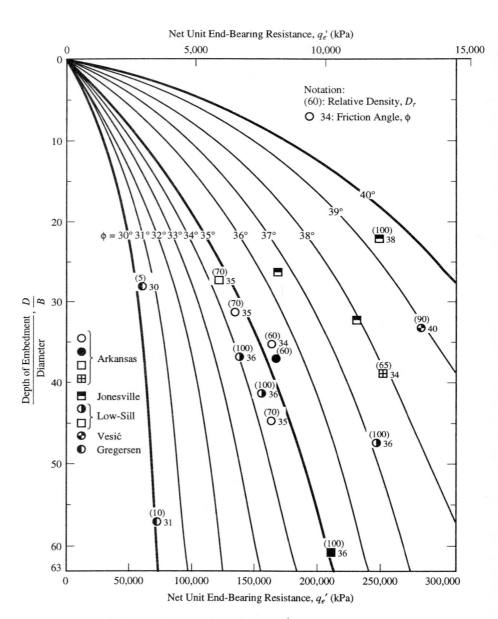

Figure 13.11 Net unit end bearing resistance, q_e', in sands (Adapted from Coyle and Castello, 1981; Used with permission of ASCE).

Researchers have used laboratory tests to measure the coefficient of friction, $\tan \phi_s$. Table 13.2 gives typical values in terms of the ratio ϕ_s/ϕ, where ϕ is the friction angle of the soil. Saturated dense soils have ϕ_s/ϕ ratios on the high end of these ranges, whereas dry loose soils tend to the low end. Uncoated steel piles, such as conventional pipe or H-piles, have ϕ_s/ϕ values between the smooth and rough steel values.

TABLE 13.2 ϕ_s/ϕ VALUES FOR THE INTERFACE BETWEEN PILES AND SOIL

Pile and Soil Types	ϕ_s/ϕ
Sand/rough concrete (i.e., cast-in-place concrete)	1.0
Sand/smooth concrete (i.e., precast concrete)	0.8 - 1.0
Sand/rough steel (i.e., corrugated steel)	0.7 - 0.9
Sand/smooth steel (i.e., steel coated with bitumen)	0.5 - 0.7
Sand/timber	0.8 - 0.9

Adapted from Kulhawy et al., 1983.

As discussed in Chapter 3, the ratio between the horizontal and vertical effective stresses is defined as the coefficient of lateral earth pressure, K:

$$K = \frac{\sigma_h'}{\sigma_v'} \qquad (13.4)$$

Pile driving induces significant changes in the surrounding soils, as discussed in Chapter 12. Therefore, the coefficient of lateral earth pressure, K, is generally not equal to K_0, the coefficient of lateral earth pressure in the ground before construction. The ratio K/K_0 depends on many factors, including the following:

- **High displacement vs. low displacement piles**—High displacement piles, such as closed-end steel pipes, displace much more soil than do low displacement piles, such as a steel H-piles, and therefore have a much higher K/K_0 ratio. However, some of this gain may be lost over time as creep effects tend to relax the locally high horizontal stresses.
- **Soil consistency**—Dense soils provide more resistance to distortion, which results in a greater K/K_0 ratio.

- **Special construction techniques**—Predrilling or jetting loosens the soil, thus reducing the K/K_0 ratio.

The largest possible value of K is K_p, the coefficient of passive earth pressure,[3] which is equal to $\tan^2 (45 + \phi/2)$. Kulhawy et al. (1983) suggested the K/K_0 ratios in Table 13.3.

TABLE 13.3 RATIO OF COEFFICIENT OF LATERAL EARTH PRESSURE AFTER CONSTRUCTION TO THAT BEFORE CONSTRUCTION

Foundation Type and Method of Construction	K/K_0
Jetted pile	1/2 - 2/3
Drilled shaft	2/3 - 1
Pile - small displacement	3/4 - 1 1/4
Pile - large displacement	1 - 2

Adapted from Kulhawy et al., 1983.

Thus, Equation 13.3 becomes:

$$f_s = \sigma_v' \, K_0 \left(\frac{K}{K_0} \right) \tan \left[\phi \left(\frac{\phi_s}{\phi} \right) \right] \tag{13.5}$$

The most difficult factor to assess in Equation 13.5 is K_0. In a truly normally consolidated soil, we could compute K_0 using Equation 3.12, but such soils are rare. Most soils that appear to be normally consolidated actually have an overconsolidated crust as described earlier. This crust often extends from the ground surface to a depth of 10 to 20 ft (3 - 6 m) or more. It is relatively easy to detect in cohesive soils by examining consolidation test results or s_u/σ_v' ratios, but it is very difficult to detect in cohesionless soils. As a result, K_0 in "normally consolidated" soil deposits is typically quite high near the ground surface and gradually become smaller with depth, as shown in Figure 13.12 (Kulhawy, 1984, 1991).

In more classical overconsolidated soil deposits (i.e., those precompressed by removal of overburden), the overconsolidation ratio, OCR, decreases with depth, so K_0 also decreases with depth. As discussed in Chapter 3, K_0 is higher in overconsolidated soils.

The most satisfactory way to assess K_0 is to measure it in-situ using tests such as the pressuremeter (PMT), dilatometer (DMT), or K_0 stepped blade. Unfortunately, none of these tests is widely available. Another way is to use a correlation with the friction

---

[3] Passive earth pressures are discussed in more detail in Chapter 23.

angle and overconsolidation ratio (OCR), such as Equation 3.13 in Chapter 3. Unfortunately, the OCR also is very difficult to evaluate in cohesionless soils. This is true even when interbedded cohesive soils are present because the OCR of these soils may be different, especially if desiccation has occurred.

Hopefully, future research and experience with in-situ tests will help engineers define K with more certainty, and thus improve our ability to predict skin friction resistance. Until then, we must rely on more empirical methods.

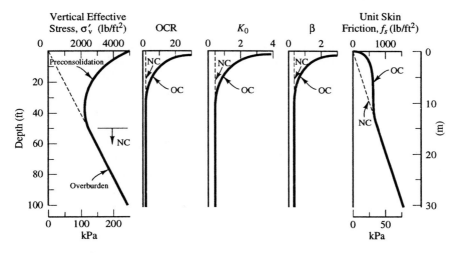

Figure 13.12 Typical distributions of parameters that influence skin friction resistance (Adapted from Kulhawy, 1991; Used with permission of Van Nostrand Reinhold Publishers). NC = normally consolidated; OC = overconsolidated.

The β Method

Because of the difficulties in measuring K_0, engineers often combine the K and $\tan \phi_s$ terms as follows:

$$\beta = K \tan \phi_s \tag{13.6}$$

$$f_s = \beta \, \sigma_v' \tag{13.7}$$

This format is known as *the β method* (Burland, 1973). Engineers can use site-specific β values back-calculated from full-scale load tests, or generic β values from the technical literature. The latter were compiled from pile load tests and semiempirically correlated them with other parameters. For example, Bhushan (1982) suggested the following formula for large displacement piles in sands:

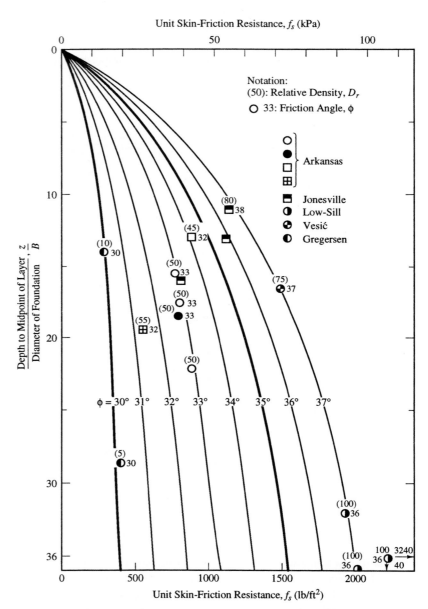

Figure 13.13 Unit skin friction resistance, f_s, in cohesionless soils (Adapted from Coyle and Castello, 1981; Used with permission of ASCE).

$$\beta = 0.18 + 0.65 D_r \qquad\qquad (13.8)$$

Where:

D_r = relative density of the sand expressed in decimal form

Coyle and Castello's Method

Coyle and Castello (1981) also developed a function for skin friction. They empirically correlated the unit skin friction, f_s, with depth and friction angle, as shown in Figure 13.13. The z/B ratio in this plot is the depth to the midpoint of the strata divided by the diameter of the pile.

Example 13.2

A 400 mm diameter, 18.0 m long open-end pipe pile with 10 mm thick walls is to be driven into a sandy soil, as shown in the figure. It will be filled with concrete and support a "permanent" structure and be designed and built using "poor" control. Use Coyle and Castello's method to compute the allowable downward and uplift load capacities.

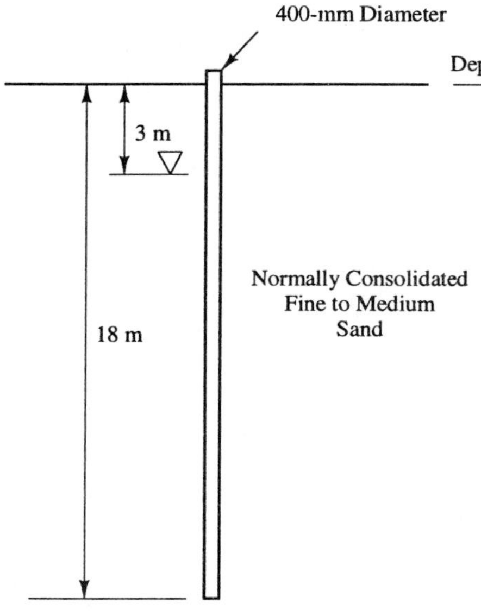

400-mm Diameter

Depth (m)	N_{60}
1	6
2	8
4	9
6	14
10	11
13	17
17	23
20	32
22	31

3 m

18 m

Normally Consolidated
Fine to Medium
Sand

Solution

End Bearing:

To use Coyle and Castello's method, we need the friction angle, ϕ. Since no direct measurements of ϕ are available, we must obtain it using the correlation between N_{60} and ϕ in Figure 4.20 in Chapter 4.

Estimate $\gamma = 18.0$ kN/m^3 above the groundwater table, 21.0 kN/m^3 below the groundwater table (from Table 3.3) in Chapter 3.

$$N_{60} \text{ at tip} \approx 23$$

$$\sigma_D{}' \text{ at tip} = (18)(3) + (21)(15) - (9.8)(15) = 222 \text{ kPa}$$

$$\text{Therefore, use } \phi = 36° \text{ (from Figure 4.19)}$$

$$D/B = 18/0.4 = 45$$

$$q_e{}' = 10,300 \text{ kPa} \text{ (from Figure 13.11)}$$

Assume that this pile will become completely plugged. Therefore, use the full end area:

$$A_e = \frac{\pi (0.4)^2}{4} = 0.126 \text{ m}^2$$

$$P_e{}' = q_e{}' A_e = (10,300)(0.126) = 1300 \text{ kN}$$

Skin Friction:

Depth (m)	N_{60}	σ' (kPa)	ϕ (Deg)
1.0	6	18	-
2.0	8	36	-
4.0	9	64	36
6.0	14	88	36
10.0	11	132	33
13.0	17	166	34
17.0	23	211	36

The friction angle is fairly constant with depth, so evaluate the skin friction using only one soil layer with $\phi = 34°$. Compute f_s at the midpoint of this layer.

$$z/B = 9/0.4 = 22.5$$

$$f_s = 55 \text{ kPa} \quad \text{(from Figure 13.13)}$$

$$A_s = \pi (0.4)(18) = 22.6 \text{ m}^2$$

$$P_s = f_s A_s = (55)(22.6) = 1240 \text{ kN}$$

Downward Capacity:

Table 11.1 in Chapter 11 suggests a factor of safety of 2.8. However, because of the uncertainties in the standard penetration test (see discussion in Chapter 4), use $F = 3.5$. Using Equation 11.3:

$$P_a = \frac{P_e' + P_s}{F} = \frac{1300 + 1240}{3.5} = \textbf{726 kN} \quad \Leftarrow\textbf{\textit{Answer}}$$

Uplift Capacity:

$$\gamma_{\text{steel}} = 77 \text{ kN/m}^3; \quad \gamma_{\text{concrete}} = 23.6 \text{ kN/m}^3; \quad \gamma_{\text{water}} = 9.8 \text{ kN/m}^3$$

$$W_f = \frac{\pi (0.400^2 - 0.380^2)}{4}(18)(77) + \frac{\pi (0.380)^2}{4}(18)(23.6) - \frac{\pi (0.400)^2}{4}(15)(9.8)$$
$$= 47 \text{ kN}$$

$$D/B > 6, \text{ so } R = 1 \quad \text{(per Chapter 11)}$$

$$F = (3.5)(2) = 7 \quad \text{(per Chapter 11)}$$

$$P_{au} = 0.9 W_f + \frac{R \sum f_s A_s}{F} = 0.9(47) + \frac{(1)(1240)}{7} = \textbf{219 kN} \quad \Leftarrow\textbf{\textit{Answer}}$$

Special or Unusual Soils

Most of the design methods for piles in cohesionless soils are based on tests performed in silicious sands located in temperate climates, such as the eastern United States and western Europe. Therefore, be cautious when attempting to use these methods with other types of soil, especially those that form in other climates. For example, *calcareous sands*[4] may be strong in the laboratory, yet produce much less skin friction and end bearing resistance than a silicious sand with the same friction angle. *Cemented sands* also demand special attention.

Cohesive Soils

As discussed in Chapter 12, the excess pore water pressures induced by pile driving dissipate much more slowly in cohesive soils, so some time must pass before these piles develop their full load capacity. Usually, the foundation is loaded slowly as construction of the structure progresses, so most of the excess pore water pressures generated during driving probably dissipate before the full dead load is applied. Therefore, if we are interested only in the dead load capacity, it would be appropriate to base the load capacity analyses on the drained strength of the soil.

However, most foundations are subjected to both dead loads and live loads. The application of a live load sometime during the life of the structure causes the pile to move down slightly, thus compressing the soil beneath the pile tip and generating new excess pore water pressures. Therefore, when significant live loads are present (which includes nearly all practical problems), the end bearing capacity computations should be based on the undrained strength of the soil.

The circumstances controlling the skin friction resistance are different from those that control end bearing. Once the pore water pressures induced by pile driving have dissipated, the soil will have attained its drained strength. Any additional downward movement of the pile does not further compress the soil adjacent to the shaft, so little or no new excess pore water pressures develop. Therefore, the drained strength controls the skin friction capacity.

End Bearing

Compute the unit end bearing resistance, q_e', from the undrained shear strength, s_u, as follows:

$$q_e' = N_c^* s_u \qquad (13.9)$$

Where:

q_e' = net unit end bearing resistance

N_c^* = bearing capacity factor

s_u = undrained shear strength of soil within about ± 2 pile diameters of pile tip

---

[4] Calcareous sands are those that contain calcium carbonate. Their engineering properties are often quite different than those of the more common silicious sands, which are predominantly silica (quartz).

It is customary to use $N_c^* = 9$, even though greater and lesser values have been measured in load tests. The end bearing capacity in clays is generally not a large percentage of the total capacity, so a more complex analysis is not warranted.

Be especially cautious when working with fissured clays. The spacing of these fissures is often large compared to the size of the soil samples, so test results may represent the soil between the fissures rather than the entire soil mass. In such cases, reduce s_u accordingly (perhaps to a value of about 0.75 times the measured value).

Skin Friction

Because of the loss of contact between the upper part of the pile and the soil, as discussed in Section 12.5, neglect the skin friction resistance in the upper 7 ft (2 m), especially in stiff clays.

Analyses Based on Drained Strength

Although we often think of the skin friction resistance in cohesive soils as an *adhesion* phenomenon (i.e., a bonding action), it is more accurate to view it as sliding friction (Azzouz et al., 1990). Therefore, drained strength analyses in cohesive soils are fundamentally the same as those for cohesionless soils (Equations 13.5 - 13.8).

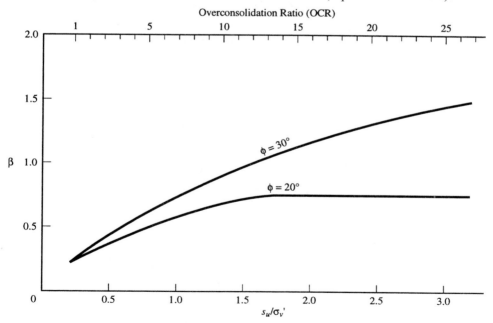

Figure 13.14 Design values of β for clays (Adapted from Randolph and Wroth, 1982 and Equation 3.38). When working with open-ended steel pipe piles that do not form a soil plug, reduce these values by 20%.

We could determine the friction angle between the pile and soil, ϕ_s, using the ϕ_s/ϕ values in Table 13.2 and the coefficient of lateral earth pressure, K, using the K/K_0 ratios in Table 13.3. However, the preconstruction coefficient of lateral earth pressure, K_0, is once again the most difficult parameter to assess.

Because methods of assessing K_0 and its impact on pile capacity are not yet well developed, engineers usually perform drained analyses using the β method (Equation 13.7). Randolph and Wroth (1982) developed β values for clays, as shown in Figure 13.14. This figure shows how overconsolidated soils have a higher β (because of their higher K_0) than normally consolidated soils. However, for design purposes, β should be no greater than $\tan^2(45 + \phi/2) \tan \phi$, where ϕ is the drained friction angle.

Example 13.3

A certain overconsolidated clay has a s_u/σ_v' ratio of 0.61 and a drained friction angle of 22°. Its unit weight is 110 lb/ft³ and the groundwater table is at a depth of 5 ft.

Using the β method, compute the allowable downward axial capacity of a 50 ft long, 12 in square prestressed concrete pile that will support a "permanent" structure. The foundation is to be built with "poor" control.

Solution:

Skin friction:

$$\text{From Figure 13.14, } \beta = 0.45$$

$$\sigma_v' \text{ at } 25\,\text{ft} = \sum \gamma h - u = (110)(25) - (62.4)(20) = 1500 \text{ lb/ft}^2$$

$$f_s = \beta \sigma_v' = (0.45)(1500) = 676 \text{ lb/ft}^2$$

$$A_s = (4)(1)(50) = 200 \text{ ft}^2$$

End Bearing:

$$\sigma_v' \text{ at } 50\,\text{ft} = \sum \gamma h - u = (110)(50) - (62.4)(45) = 2690 \text{ lb/ft}^2$$

$$s_u = (0.61)(2690) = 1640 \text{ lb/ft}^2$$

$$q_e^{/} = 9\,s_u = (9)(1640) = 14{,}800 \text{ lb/ft}^2$$

$$A_e = 1.00 \text{ ft}^2$$

Capacity:

From Table 11.1, $F = 2.8$

Using Equation 11.4:

$$P_a = \frac{q_e^{/} + \Sigma f_s A_s}{F} = \frac{(14{,}800)(1.00) + (676)(200)}{(2.8)(1000\,\text{lb/k})} = \textbf{53.6 k} \quad \Leftarrow\!\!\textit{Answer}$$

Analyses Based on Undrained Strength

Although a drained strength analysis is theoretically more correct, it also is possible to analyze skin friction resistance based on empirical correlations with the undrained strength, s_u. These correlations implicitly "convert" the undrained strength to the drained strength.

This may seem to be a roundabout way to compute skin friction, and indeed it is. Nevertheless, it continues to be widely used because of the large base of experience and because the tests required to obtain s_u (typically, unconfined compression tests, vane shear tests, or UU triaxial tests) are inexpensive. Although this is currently the most popular way to evaluate skin friction in cohesive soils, it may eventually be replaced with analyses based on drained strength.

This type of analysis is known as *the α method* because it defines the unit skin friction resistance using the *adhesion factor*, α:

$$f_s = \alpha\,s_u \tag{13.10}$$

Many factors affect α, including:

• Remolding of the clay during construction
• Consolidation of the clay during construction
• Dragdown of one soil strata into another during pile driving
• The method of determining s_u
• Type, diameter, length, and taper of pile

- Predrilling
- Jetting

The best way to determine α for a given soil is to conduct on-site pile load tests and compare measured pile capacity with the average s_u value from laboratory or in-situ tests. Then, use this information to extrapolate the pile load test results to piles of other lengths or diameters.

When no site-specific load test data are available, determine α from generic α functions. They usually are expressed solely as a function of s_u (Sladen, 1992), as shown in Figure 13.15. Note the wide scatter between these functions and the experimental data.

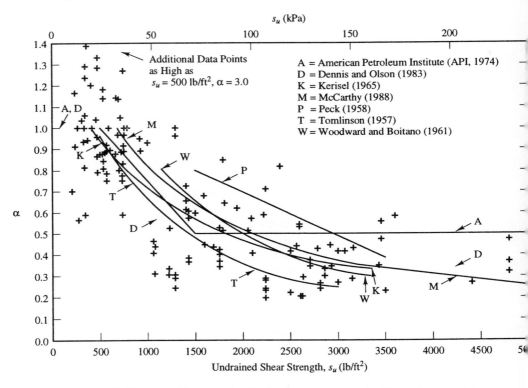

Figure 13.15 Proposed functions for adhesion factor, α, compared with actual α vs s_u data obtained from load tests (Load test data adapted from Vesić, 1977).

The API (1981) α function, shown in Figure 13.15, is one of the most common. According to Dennis and Olson (1983), it computes capacities that, on average, are correct, but the scatter is wide, sometimes being as low as 1/3 or as high as 3 times the actual capacity.

Tomlinson (1987) studied the load-transfer behavior of piles in clays and suggested that engineers also consider the dragdown of soils during driving.[5]   In stratified soil profiles, the upper strata may be drawn well into lower strata, as shown in Figure 13.16. If the upper strata is weaker than the lower, this dragdown effect reduces the capacity of the pile, so the α factor in the lower strata also must be reduced.   Conversely, the α factor in the lower strata must increase if the upper soils are stronger.   Tomlinson developed the α functions in Figures 13.17 - 13.19 to account for this effect.

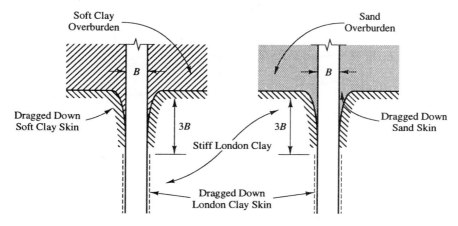

Figure 13.16 Dragdown of soil near soil interfaces as observed in London (Adapted from Tomlinson, 1987, Used with permission of Chapman and Hall).

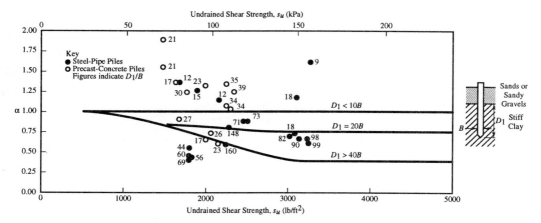

Figure 13.17 Adhesion factor for use in the α method with stiff clays overlain by sands or gravels (Adapted from Tomlinson, 1971 and 1987, Used with permission of Thomas Telford, and Chapman and Hall). Does not include the skin friction resistance in the sand or gravel strata.

---

[5] Do not confuse this dragdown of soils with the downdrag forces discussed in Chapter 19.

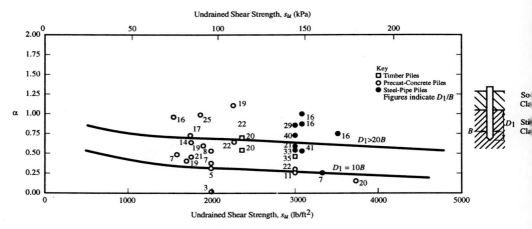

Figure 13.18 Adhesion factor for use in the α method with stiff clays overlain by soft clays (Adapted from Tomlinson, 1971 and 1987, Used with permission of Thomas Telford, and Chapman and Hall). Conduct a separate calculation to compute the skin friction resistance in the soft clay strata.

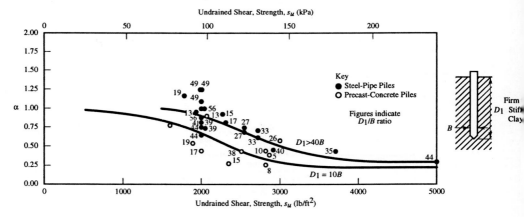

Figure 13.19 Adhesion factor for use in the α method with stiff clays without any overlying strata (Adapted from Tomlinson, 1971 and 1987, Used with permission of Thomas Telford, and Chapman and Hall). This data implicitly considers soil-pile separation, so it should be used with the entire contact area (i.e., do not ignore the upper 2 m).

Example 13.4

A 400 mm diameter, 16.0 m long steel pipe pile is to be driven into the following soil profile:

Depth (m)	Soil Classification	s_u (kPa)
0 - 4.0	Soft to medium clay	25
4.0 - 10.0	Stiff silty clay	75
10.0 - 14.0	Stiff sandy clay	90
14.0 - 25.0	Very stiff sandy clay	180

The groundwater table is at a depth of 5.0 m.

The pile will have a 10 mm wall thickness and will be fitted with a 90 kg conical steel tip welded to the bottom. It will support an office building and will be built using "normal" control. Compute the allowable downward and uplift load capacities.

Solution:

End Bearing:

Using Equation 13.9

$$q_e' = N_c^* s_u$$
$$= (9)(180)$$
$$= 1620 \text{ kPa}$$

$$A_e = \frac{\pi B^2}{4}$$
$$= \frac{\pi (0.400 \text{ m})^2}{4}$$
$$= 0.126 \text{ m}^2$$

$$q_e' A_e = (1620)(0.126)$$
$$= 204 \text{ kN}$$

Skin Friction:

Using the API method (Figure 13.15)

Depth (m)	α (Fig. 13.15)	f_s (kPa)	A_s (m^2)	$f_s A_s$ (kN)
0 - 2.0	-	-	-	0
2.0 - 4.0	0.99	24.7	2.51	62.0
4.0 - 10.0	0.50	37.5	7.54	282.7
10.0 - 14.0	0.50	45.0	5.03	226.3
14.0 - 16.0	0.50	90.0	2.51	225.9
			$\Sigma f_s A_s$	797

Using Tomlinson's method (Figure 13.18)

Depth (m)	α (Fig. 13.18)	f_s (kPa)	A_s (m^2)	$f_s A_s$ (kN)
0 - 2.0	-	-	-	0
2.0 - 4.0	1.00	25.0	2.51	62.7
4.0 - 10.0	0.70	52.5	7.54	395.8
10.0 - 14.0	0.68	61.2	5.03	307.8
14.0 - 16.0	0.60	108.0	2.51	271.1
			$\Sigma f_s A_s$	1037

Using the average of these two analyses:

$$f_s A_s = \frac{797 + 1037}{2} = 917 \text{ kN}$$

Downward Capacity (using Equation 11.4):

$$F = 2.5 \quad \text{(per Table 11.1)}$$

$$P_a = \frac{q_e' A_e + \Sigma f_s A_s}{F}$$

$$= \frac{204 + 917}{2.5}$$

$$= \textbf{448 kN} \quad \Leftarrow \textit{Answer}$$

Uplift Capacity:

Compute weight of foundation using $\gamma_{steel} = 77.1$ kN/m^3

$$\text{Volume} = \pi \left(\frac{B_0^2 - B_i^2}{4} \right) D$$

$$= \pi \left(\frac{0.400^2 - 0.380^2}{4} \right) 16.0$$

$$= 0.196 \text{ m}^3$$

$$W_{point} = \frac{(90)(9.8)}{1000}$$
$$= 0.9 \text{ kN}$$

$$\text{Buoyancy} = \pi \left(\frac{0.400^2}{4} \right) (11)(9.8 \text{ kN/m}^3)$$

$$= 13.5 \text{ kN}$$

$$W_f = (77.1 \text{ kN/m}^3)(0.196 \text{ m}^3) + 0.9 \text{ kN} - 13.5 \text{ kN}$$
$$= 2.5 \text{ kN}$$

Compute the uplift capacity using Equation 11.6:

$$F = (2.5)(1.75) = 4.4 \quad \text{(per Chapter 11)}$$

$$D/B > 6, \text{ so } R = 1 \quad \text{(per Chapter 11)}$$

$$P_{au} = 0.9 \, W_f + \frac{R \sum f_s A_s}{F}$$

$$= (0.9)(2.5) + \frac{(1)(917)}{4.4}$$

$$= \textbf{211 kN} \quad \Leftarrow \textit{Answer}$$

Note: Filling the pile with sand or concrete would slightly increase the uplift capacity.

The λ *Method*

The λ method (Vijayvergiya and Focht, 1972; Kraft, Focht and Amerasinghe, 1981) is a combined drained and undrained analysis. It computes the average unit skin friction as follows:

$$\bar{f}_s = \lambda(\bar{\sigma}'_v + 2\bar{s}_u) \tag{13.11}$$

Where:

$\bar{f}_s$ = average unit skin friction resistance
λ = frictional capacity coefficient (dimensionless)
$\bar{\sigma}'_v$ = average vertical effective stress between ground surface and pile tip
$\bar{s}_u$ = average undrained shear strength between ground surface and pile tip

Figure 13.20 gives experimental values of λ as a function of pile length.

This analysis uses a single λ factor based on the total embedded length of the pile and average values of s_u and σ'_v. Do not divide the soil into layers or compute f_s for each layer. Therefore, this method may be used only with profiles that are almost entirely clay or silt.

The larger values of λ for shorter piles probably reflect the higher degree of overconsolidation that is typically present near the top of soil deposits, as discussed earlier. Longer piles contact deeper soils that are typically less overconsolidated. Thus, λ is smaller for longer piles.

The λ method has been used primarily on large offshore projects with very long piles. It may not be appropriate for piles shorter than about 50 ft (15 m).

Rock

Piles driven to bedrock generally have large downward load capacities. However, because rock is a more difficult material to evaluate, the ultimate capacity is harder to predict. Fortunately, even conservative estimates may be acceptable for design.

Methods of computing the end bearing capacity are discussed in Section 6.11 in Chapter 6. These include presumptive bearing pressures and empirical correlations with rock properties.

The skin friction resistance is even more difficult to predict, and very little instrumented load test data are available. Poulos and Davis (1980) suggested the following formula:

$$f_s = 0.05\, q_{uc} \leq 0.05 f_c' \tag{13.12}$$

Where:

f_s = unit skin friction resistance

q_{uc} = unconfined compressive strength of rock

f_c' = 28-day compressive strength of concrete (for concrete piles)

Equation 13.12 is not valid for highly fractured rock or for expansive shales.

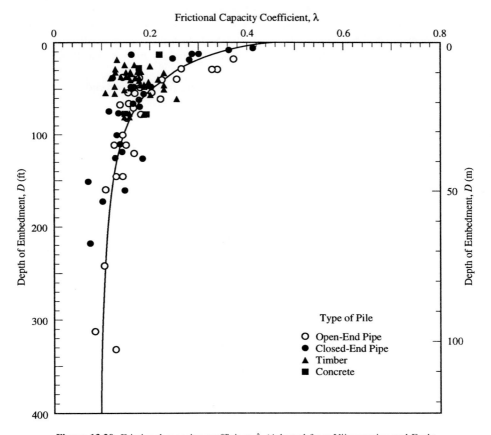

Figure 13.20 Frictional capacity coefficient, λ (Adapted from Vijayvergiya and Focht, 1972 and Kraft, Focht and Amerasinghe, 1981: Used by permission of ASCE and the Offshore Technology Conference).

13.4 ANALYSES BASED ON IN-SITU SOIL TESTS

There are two ways to compute axial load capacities from in-situ tests: The first is to convert the test results to standard engineering properties, such as ϕ or s_u, and use the analysis methods described in Section 13.3. The other way is to use direct correlations between the test results and pile capacity, as described in this section.

Analysis methods based on in-situ tests are especially attractive in cohesionless soils because they are difficult to sample and test in the laboratory.

Analyses Based on Standard Penetration Test Results

We can use data from the standard penetration test (SPT) to predict pile load capacities in cohesionless soils. However, these predictions are only moderately reliable, largely because of the inconsistencies in the SPT test procedure as discussed in Chapter 4. Therefore, consider using a factor of safety slightly higher than those in Table 11.1 in Chapter 11.

Meyerhof (1976) proposed the following formulas for end bearing:

For sands and gravels:

$$q_e' = 0.40 \, N_{60}' \frac{D}{B} \sigma_r \leq 4.0 \, N_{60}' \sigma_r \qquad (13.13)$$

For nonplastic silts:

$$q_e' = 0.40 \, N_{60}' \frac{D}{B} \sigma_r \leq 3.0 \, N_{60}' \sigma_r \qquad (13.14)$$

The upper limits of Equations 13.13 and 13.14 apply whenever $D/B \geq 10$ (sands and gravels) or $D/B \geq 7.5$ (silts). Nearly all piles have greater D/B ratios, so the upper limits nearly always control.

He also proposed formulas for skin friction:

For large displacement piles in cohesionless soils:

$$f_s = \frac{\sigma_r}{50} N_{60} \qquad (13.15)$$

For small displacement piles in cohesionless soils:

$$f_s = \frac{\sigma_r}{100} N_{60} \qquad (13.16)$$

Where:

q_e' = net unit end bearing resistance

f_s = unit skin friction resistance

σ_r = reference stress = 2000 lb/ft^2 = 100 kPa

B = pile diameter

D = pile embedment depth

N_{60} = SPT N value corrected for field procedures only

$N_{60}{}'$ = SPT N value corrected for field procedures and overburden stress

For end bearing computations, the SPT N value should represent the conditions near the pile tip, especially within the range of one diameter above the tip to two diameters below the tip. For skin friction computations, it should represent the conditions adjacent to the pile segment being considered.

For piles with more than 1% taper, Meyerhof recommended multiplying the f_s values from Equations 13.15 and 13.16 by a factor of 1.5.

Briaud et al. (1985) proposed another set of formulas:

$$q_e{}' = 19.7\,\sigma_r\,(N_{60})^{0.36} \tag{13.17}$$

$$f_s = 0.224\,\sigma_r\,(N_{60})^{0.29} \tag{13.18}$$

The standard penetration test does not seem to be a reliable indicator of pile capacities in cohesive soils.

Example 13.5

Reanalyze the pile in Example 13.2 using Meyerhof's method and the Briaud et al. method.

Solution using Meyerhof:

End Bearing:

Use Equations 4.3 and 4.4 to compute $N_{60}{}'$

• At 17 m, $\sigma' = 211$ kPa,

$$C_N = \frac{2}{1 + \sigma_v{}'/\sigma_r} = \frac{2}{1 + 211/100} = 0.64$$

$$N_{60}{}' = C_N\,N_{60} = (0.64)(23) = 15$$

• At 20 m, $\sigma' = 244$ kPa,

$$C_N = \frac{2}{1 + \sigma_v'/\sigma_r} = \frac{2}{1 + 244/100} = 0.90$$

$$N_{60}' = C_N N_{60} = (0.90)(32) = 29$$

The pile tip is at a depth of 18 m. Therefore, based on the computed values above, use $N_{60}' = 18$ in Equation 13.13:

$$D/B \geq 10, \text{ so } q_e' = 4.0 \, N_{60}' \, \sigma_r = (4.0)(18)(100) = 7200 \text{ kPa}$$

$$A_e = \frac{\pi (0.4)^2}{4} = 0.126 \text{ m}^2$$

$$P_e' = q_e A_e = (7200)(0.126) = 907 \text{ kN}$$

Skin Friction:

The soil profile appears to be fairly homogeneous, so we can average all of the N_{60} values along the shaft and compute a single average f_s. If the profile was more complex, it would be necessary to divide the pile into segments and compute the skin friction capacity of each segment separately.

Using Equation 13.15:

$$N_{60} = \frac{6+8+9+14+11+17+23}{7} = 12.5$$

$$f_s = \frac{\sigma_r}{50} N_{60} = \frac{100}{50} (12.5) = 25.0 \text{ kPa}$$

$$A_s = \pi (0.4)(18) = 22.6 \text{ m}^2$$

$$P_s = f_s A_s = (25.0)(22.6) = 565 \text{ kN}$$

Using a factor of safety of 3.5 in Equation 11.3:

$$P_a = \frac{P_e' + P_s}{F} = \frac{907 + 565}{3.5} = \mathbf{421\ kN} \quad \Leftarrow Answer$$

Solution using Briaud et al. (Equations 13.17 and 13.18):

End bearing:

$$q_e' = 19.7\ \sigma_r\ (N_{60})^{0.36} = 19.7\ (100)\ (20)^{0.36} = 5800\ kPa$$

$$P_e' = q_e A_e = (5800)\ (0.126) = 731\ kN$$

Skin friction:

$$f_s = 0.224\ \sigma_r\ (N_{60})^{0.29} = 0.224\ (100)\ (12.5)^{0.29} = 47\ kPa$$

$$P_s = f_s A_s = (46.6)\ (22.6) = 1050\ kN$$

Overall:

$$P_a = \frac{P_e' + P_s}{F} = \frac{731 + 1050}{3.5} = \mathbf{509\ kN} \quad \Leftarrow Answer$$

Summary

Compare the results from this example and that from Example 13.2:

Coyle and Castello's method	$P_a = 726$ kN
Meyerhof's method	$P_a = 421$ kN
Briaud et al.'s method	$P_a = 509$ kN

In this case, Coyle and Castello's method is probably less reliable than the others because we had to use another correlation to convert the SPT N values to ϕ values. Therefore, it may be appropriate to use $P_a = 500$ kN for design.

Analyses Based on Cone Penetration Test Results

The modern cone penetration test (CPT) was originally developed in the Netherlands during the 1930s as a means of predicting pile load capacities. The cone is similar to a model pile just as a plate load test is similar to a model footing. Because the process of conducting a cone penetration test is so similar to driving piles, analysis methods based on this data should be good predictors of pile capacity.

End Bearing

The end bearing capacity correlates very well with the cone resistance, q_c. As a result, engineers have used the CPT to compute end bearing for many years, especially in sands.

Dutch Method

The Dutch method of computing end bearing in cohesionless soils (Heijnen, 1974; DeRuiter and Beringen, 1979) is as follows:

1. Consider the cone resistance, q_c, between a depth of $8B$ above the pile tip and $4B$ below the pile tip, as shown in Figure 13.21. The strength of this soil controls the end bearing capacity.

2. Determine q_{c1} by computing the average q_c value along the line **abcd**, as shown in Figure 13.21. Note that point **b** is at a depth x below the proposed pile tip and $0.7B \leq x \leq 4B$. Select the x that produces the minimum q_{c1}.

   An isolated value of q_c that is much smaller than the typical q_c values will have a significant impact on q_{c1}. If this isolated low value truly represents a weak zone in the soil, then the end bearing capacity will be correspondingly less. However, if it is a spurious reading, then ignore it, thus producing a higher q_{c1}.

3. Trace the path **defgh** as shown in Figure 13.21, and determine q_{c2} by computing the average q_c value along this line.

4. Using Table 13.4, determine the correction factor, ω, to account for gravel content or overconsolidation.

TABLE 13.4 ω VALUES FOR USE IN EQUATION 13.19

Soil Condition	ω
Sand with OCR = 1	1.00
Very gravelly coarse sand; sand with OCR = 2 to 4	0.67
Fine gravel; sand with OCR = 6 to 10	0.50

Adapted from DeRuiter and Beringen, 1979.
OCR = Overconsolidation ratio

5. Compute the net unit end bearing capacity, q_e', as follows:

$$q_e' = \frac{\omega\,(q_{c1} + q_{c2})}{2} \leq 300{,}000 \text{ lb/ft}^2 \ (15{,}000 \text{ kPa}) \qquad (13.19)$$

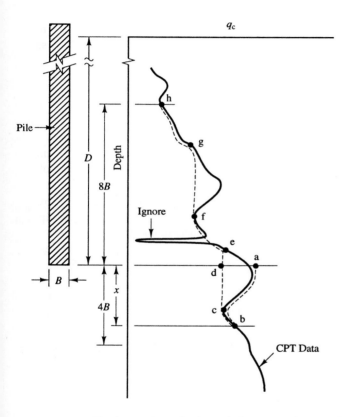

Figure 13.21   Dutch procedure for computing end bearing based on CPT (Adapted from Heijnen, 1974).

Engineers have had much less experience using the CPT to compute end bearing capacity in cohesive soils. However, Nottingham and Schmertmann (1975) conducted tests on model piles and found the Dutch method also works well for normally consolidated or slightly overconsolidated cohesive soils so long as $s_u < 1000$ lb/ft^2 (50 kPa). For stiffer soils, Schmertmann (1978) recommended multiplying the result from Equation 13.19 by the reduction factor shown in Figure 13.22. He also recommend multiplying the computed end bearing capacity by 0.60 when using a mechanical cone in a cohesive soil.

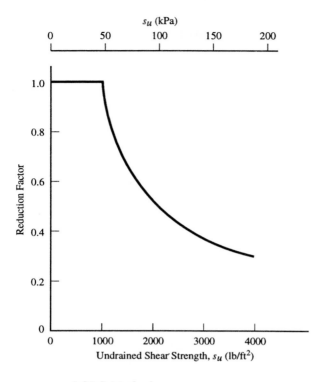

Figure 13.22 Reduction factor for use of the Dutch method with cohesive soils (Adapted from Schmertmann, 1978).

LCPC Method

Engineers from the Laboratoire Central des Ponts et Chaussées (LCPC) in France also have developed a CPT-based method (Bustamante and Gianeselli, 1982; Briaud and Miran, 1991). This method applies to a wide variety of soil conditions and considers both piles and cast-in-place foundations (the latter are described in Chapters 15 and 16).

The net unit end bearing capacity is:

$$q_e' = q_{ca} k_c \tag{13.20}$$

Where:

q_{ca} = equivalent cone bearing resistance at pile tip

k_c = cone end bearing factor (from Table 13.5)

The equivalent cone bearing resistance, q_{ca}, is the average q_c value from 1.5 pile diameters above to 1.5 diameters below the pile tip. If the q_c values in this region are erratic, it may be necessary to apply some engineering judgment. Bustamante and

Gianeselli suggest cropping exceptionally high and exceptionally low q_c values, as shown in Figure 13.23.

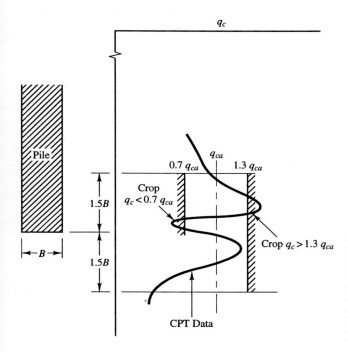

Figure 13.23 When computing q_{ca}, crop q_c values that are less than 0.7 q_{ca} or greater than 1.3 q_{ca} as shown (Adapted from Bustamante and Gianeselli, 1982; Proceedings, Second European Symposium on Penetration Testing, ©A.A. Balkema Publishers, Brookfield, VT; Used with permission).

TABLE 13.5 k_c VALUES FOR LCPC METHOD

Soil Type	Cone End Bearing Factor, k_c	
	Drilled Shafts	Piles
Clays and silts	0.375	0.600
Sands and gravels	0.150	0.375
Chalk	0.200	0.400

Adapted from Briaud and Miran, 1991.

Skin Friction

Nottingham and Schmertmann's Method

It also seems reasonable to expect that the skin friction resistance would correlate well with the local side friction, f_{sc}, obtained from the cone friction sleeve. Nottingham and

Schmertmann (1975) have developed such a method based on the results of tests on model piles.

To use this method, divide the pile into segments and assign an appropriate f_{sc} to each segment. Then, compute the skin friction capacity using the following formulas:

For cohesionless soils with $z < 8B$:

$$f_s = \alpha_s' \left(\frac{z}{8B} \right) f_{sc}$$ (13.21)

For cohesionless soils with $z \geq 8B$:

$$f_s = \alpha_s' \, f_{sc}$$ (13.22)

For cohesive soils:

$$f_s = \alpha_c' \, f_{sc}$$ (13.23)

Where:

f_s = unit skin friction resistance

z = depth from ground surface to midpoint of segment

B = pile diameter

α_s', α_c' = Nottingham adhesion factor (from Figure 13.24)

f_{sc} = local side friction[6]

D = penetration of pile below ground surface

Note that α_s' is based on the overall D/B ratio of the pile. Do not assign a different value to each pile segment. Conversely, α_c' is based on the average f_{sc} in each pile segment.

Nottingham and Schmertmann used their method to compute skin friction and the Dutch method to compute end bearing on 17 full-scale piles in Florida and Georgia and found that the predicted capacity ranged from 40% less to 23% more than that measured by a load test. On average, the predicted capacity was 11% less than the measured capacity. Therefore, they recommended using factors of safety of about 2.25 when using an electric cone and 3.0 when using a mechanical cone.

---

[6] The q_c and f_{sc} measurements from the CPT are often expressed in units of kg/cm^2. Thus, it is necessary to convert them to lb/ft^2 or kPa.

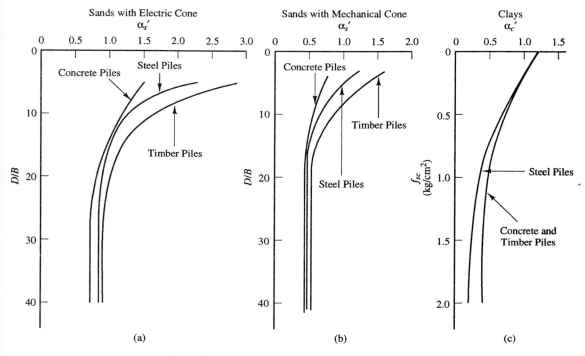

Figure 13.24 α_s' and α_c' factors for Nottingham and Schmertmann's skin friction formula:
a) sands with electric cone; b) sands with mechanical cone; and c) clays.

LCPC Method

The LCPC method also includes provisions for computing skin friction resistance based on the cone bearing resistance, q_c. Although cone bearing is a less direct analogy to skin friction, the field measurements of q_c are generally more precise than those for f_{sc}, especially in sands.

To use this method, first classify the foundation using Table 13.6. Then, select the friction curve number using Table 13.7. Finally, determine f_s using Figure 13.25 or 13.26. The original references also include methods of computing skin friction in chalk, but these are not included here.

TABLE 13.6 CLASSIFICATION OF FOUNDATION TYPE (Adapted from Briaud and Miran, 1991)

Type	Description	Comments
1	Drilled shaft built using the open hole method	Applicable only for cohesive soils above the groundwater table.
2	Drilled shaft built using drilling mud	
3	Drilled shaft built using casing	
4	Auger cast pile	See Chapter 16.
5	Hand excavated pier	Excavated without using drilling machines.
6	Micropile—Type I (dia. < 10 in (250 mm))	Excavated using casing, and then filled with concrete under pressure.
7	Screwed-in pile	A corrugated pipe is pushed and screwed into the ground, and then filled with concrete while the pipe is screwed back out. Not applicable for cohesionless soils below the groundwater table.
8	Concrete coated steel pile	A steel pile is driven into the ground using an oversized protecting shoe. As the driving proceeds, concrete is injected through a hose near the shoe, thus producing a concrete coating around the pile.
9	Concrete pile	Installed using a pile hammer or a vibratory pile hammer.
10	Steel pile	Installed using a pile hammer.
11	Post-tensioned concrete tube pile	Cylindrical concrete tube sections, 2 - 3 ft (600 - 900 mm) diameter.
12	Pressure injected footing with compacted shaft	See Chapter 16.
13	Pressure injected footing with cased shaft	See Chapter 16.
14	Jacked concrete pile	1 - 2 ft (300 - 600 mm) diameter.
15	Jacked steel pile	Installed using a hydraulic jack.
16	Micropile—Type II (dia. < 10 in (250 mm))	Reinforcing cage placed before concrete.
17	High pressure injected pile (dia. > 10 in (250 mm))	The injection system must produce a high pressure in the concrete.

TABLE 13.7 SELECTION OF DESIGN SKIN FRICTION CURVE FOR LCPC METHOD (Adapted from Briaud and Miran, 1991)

Foundation Type (Table 13.6)	Design Curve for Clays and Silts (Figure 13.25)					Design Curve for Sands and Gravels (Figure 13.26)				
	1	2	3	4	5	1	2	3	4	5
1	a,b	b,i	i							
2	a,b	b,i	i			m,q	q	u	v	
3	j	j				m,q	q	u	v	
4	a	c				m	r			
5	a	c								
6	a	h	h			m	n	s		
7	g	g	g			m	n	s		
8	a	c				m			w	
9	a	c	f			m	p	t		
10	a,d	e				m,o	o	s		
11	a	c				m	n			
12	a		k			m	n		w	
13	a	c				m	n	s		
14	a	c				m	p	t		
15	a,d	e				m,o	o	s		
16	a				l					x
17	a				l			s		x

Notes:

a.   For $q_c < 0.7$ MPa (7 kg/cm^2).

b.   For $q_c > 0.7$ MPa (7 kg/cm^2). Use design values between Curves 1 and 2 when using drilling tools without teeth or those with oversized blades (which can produce remolded soils along the sides of the hole), when drilling deep holes below the groundwater table that require multiple cleanings, or when there is a long delay between drilling and placing concrete. Use Curve 1 if no load test data are available.

c.   For $q_c > 1.2$ MPa (12 kg/cm^2).

d.   For highly plastic soils when no load test data are available.

e.   For $q_c > 1.2$ MPa (12 kg/cm^2) in low plasticity soils; or in highly plastic soils with load test data.

f.   For low plasticity soils with $q_c > 2.5$ MPa (25 kg/cm^2) or those with sand and/or gravel layers or boulders.

g.   If the rate of penetration is fast, use Curve 1; if it is slow and $q_c < 2.5$ MPa (25 kg/cm^2), use Curve 2; if it is slow and $q_c > 4.5$ MPa (46 kg/cm^2), use Curve 3.

h.   For $q_c > 1.2$ MPa (12 kg/cm^2). Use Curve 2 if no load test data are available, and Curve 3 if data are available.

i   For $q_c > 1.2$ MPa (12 kg/cm^2). Applicable when the shaft is carefully drilled using an auger equipped with teeth, and then immediately filled with concrete. Use Curve 3 if construction is carefully supervised and the sides of the shaft are grooved. Otherwise, use Curve 2.

j.   For $q_c > 1.2$ MPa (12 kg/cm^2). For shafts above the groundwater table, use Curve 2. For shafts that extend below the groundwater table and pumped dry before placing concrete, but have no load test data, use Curve 1.

k.   For 1.2 MPa (12 kg/cm^2) $< q_c < 2.0$ MPa (20 kg/cm^2).

l.   For $q_c > 2.0$ MPa (20 kg/cm^2) and construction using selective and repetitive injections at low flow rates with supporting load test data.

m.   For $q_c < 3.5$ MPa (36 kg/cm^2).

n.   For fine sands with $q_c > 3.5$ MPa (36 kg/cm^2).

o.   For fine sands with $q_c > 3.5$ MPa (36 kg/cm^2). Use Curve 2 if load test data are available; otherwise use Curve 1.

p.   Use Curve 1 for $q_c < 7.5$ MPa (76 kg/cm^2); Curve 2 for $q_c > 7.5$ MPa (76 kg/cm^2).

q.   If $D < 100$ ft (30 m), the soil is a fine sand, $q_c > 5.0$ MPa (51 kg/cm^2), and if load test data are available, use Curve 2. With all of the above and $D > 100$ ft (30 m), use a design value between Curves 1 and 2. If no load test data are available, use Curve 1.

r.   For sands with some cohesion and $q_c > 5.0$ MPa (51 kg/cm^2).

s.   For coarse gravelly sand or gravel with $q_c > 7.5$ MPa (76 kg/cm^2).

t.   For coarse gravelly sand or gravel with $q_c > 7.5$ MPa (76 kg/cm^2). Use Curve 4 if justified by load test data; otherwise use Curve 3.

u.   For coarse gravelly sand or gravel with $q_c > 7.5$ MPa (76 kg/cm^2) and $D < 100$ ft (30 m).

v.   For gravel with $q_c > 4.0$ MPa (41 kg/cm^2).

w.   For gravelly sand and gravel with $q_c > 7.5$ MPa (76 kg/cm^2).

x.   For $q_c > 5.0$ MPa (51 kg/cm^2) and construction using selective and repetitive injections at low flow rates with supporting load test data.

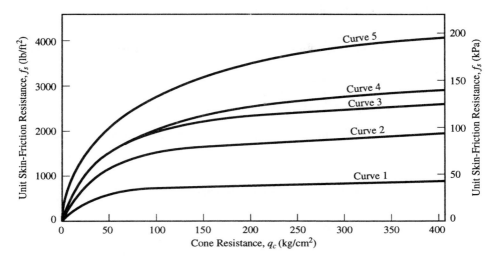

Figure 13.25 Skin friction curves for clays and silts (Adapted from Briaud and Miran, 1991). Note that there is no direct reference to Curve 4.

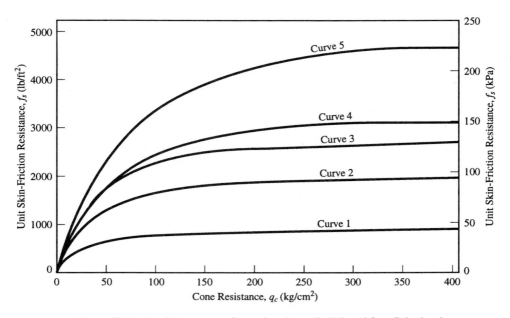

Figure 13.26 Skin friction curves for sands and gravels (Adapted from Briaud and Miran, 1991).

Example 13.6 (Adapted from Nottingham and Schmertmann, 1975)

The CPT data shown on the next page represent the soil conditions at a proposed construction site. This data was obtained using a mechanical cone. Based on Figure 4.23 in Chapter 4, these test results indicate that the upper 4.5 m is sand, and it is underlain by 3.4 m of clay, then additional sand.

Compute the allowable load capacity of an 18 in (457 mm) square, 34 ft (10.36 m) long prestressed concrete pile that is to be driven into this soil.

Solution using the Dutch and Nottingham & Schmertmann's methods:

End Bearing (Dutch method):

Try several x values to determine the minimum value of q_{c1}.

x/B	x (m)	$D+x$ (m)	q_{c1} Calculation	q_{c1} (kg/cm^2)
0.7	0.32	10.68	$[115 + 3(105)]/4$	107.5
1.0	0.46	10.82	$[115 + 105 + 2(108) + 2(105)]/6$	107.7
1.5	0.69	11.05	$[115 + 105 + 108 + 5(99)]/8$	* 102.9
2.0	0.91	11.27	$[115 + 105 + 108 + 99 + 2(108) + 4(99)]/10$	103.9
2.5	1.14	11.50	$[115 + 105 + 108 + 99 + 108 + 2(114) + 108 + 4(99)]/12$	105.6
3.0	1.37	11.73	$[115 + 105 + 108 + 99 + 108 + 114 + 2(117) + 114 + 108 + 4(99)]/14$	107.2
3.5	1.60	11.96	$[115 + 105 + 108 + 99 + 108 + 114 + 117 + 4(111) + 108 + 4(99)]/16$	107.1
4.0	1.83	12.19	$[115 + 105 + 108 + 99 + 108 + 114 + 117 + 111 + 6(105) + 4(99)]/18$	105.7

Therefore, $q_{c1} = 102.9$ kg/cm^2

$D - 8B = 10.36$ m $- 8(18$ in$)(1$ m$/39.4$ in$) = 6.70$ m = upper limit of q_{c2} averaging.

$q_{c2} = \{5(99) + 2(93) + 85 + 3(70) + 50 + 0.6[8 + 2(7) + 3(6)]\}/18 = 58.3$ kg/cm^2
(The 0.6 factor in this equation reflects the use of the mechanical cone in a cohesive soil.)

No data are available to define the degree of overconsolidation. However, in a natural soil deposit at a depth of 35 ft, the OCR might be in the range of 2 to 3. Therefore, $\omega = 0.67$ seems to be a reasonable value for design.

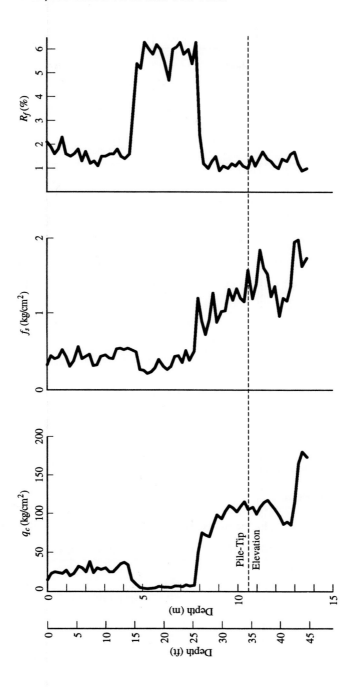

$$q_e' = \frac{\omega\,(q_{c1} + q_{c2})}{2} = \frac{0.67\,(102.9 + 58.3)}{2} = 54.0 \text{ kg/cm}^2 = 110 \text{ k/ft}^2$$

$$A_e = 1.5^2 = 2.25 \text{ ft}^2$$

Skin Friction (Nottingham and Schmertmann's method):

Layer Depth (m)	z (m)	f_{sc} (kg/cm^2)	α_s'	α_c'	f_s (lb/ft^2)	A_s (ft^2)	P_s (k)
0 - 3.66	1.83	0.42	0.44		189	72.1	13.6
3.66 - 4.5	4.08	0.53	0.44		378	16.5	7.9
4.5 - 7.9	6.20	0.36		0.85	627	66.9	41.9
7.9 - 10.36	9.13	1.08	0.44		973	48.4	47.1
						Total	110.5

Note how the depth of the first layer is equal to $8B$.

Using Equation 11.4:

$$P_a = \frac{q_e'\,A_e + \Sigma f_s A_s}{F} = \frac{(110)(2.25) + 110.5}{3} = \textbf{119 k} \quad \Leftarrow\!\textit{Answer}$$

Solution using LCPC method:

End Bearing:

Compute q_{ca} for the soil between depths of 31.7 and 36.3 ft (9.7 - 11.1 m) per Figure 13.23.

$$q_{ca} = \frac{107.0 + 102.0 + 109.0 + 115.0 + 105.0 + 108.0 + 99.0}{7} = 106.4 \text{ kg/cm}^2$$

$$k_c = 0.375 \quad \text{(per Table 13.5)}$$

$$q_e' = q_{ca}k_c$$
$$= (106.4)(0.375)$$
$$= 39.9 \text{ kg/cm}^2$$
$$= 81.7 \text{ k/ft}^2$$

$$A_e = 1.5^2 = 2.25 \text{ ft}^2$$

Skin Friction:

Foundation type 9 (per Table 13.6).

Depth (m)	Soil Type	q_c (kg/cm^2)	Figure No.	Curve No.	f_s (lb/ft^2)	A_s (ft^2)	P_s (k)
0 - 4.5	Sand	27.4	13.26	2	1000	88.6	88.6
4.5 - 7.9	Clay	6.5	13.25	1	100	66.9	6.7
7.9 - 10.36	Sand	89.5	13.26	2	1500	48.4	72.6
						Total	167.9

Using Equation 11.4:

$$P_a = \frac{q_e' A_e + \Sigma f_s A_s}{F} = \frac{(81.7)(2.25) + 167.9}{3} = \textbf{117 k} \quad \Leftarrow\textbf{\textit{Answer}}$$

13.5 ANALYSES BASED ON PILE DRIVING DYNAMICS (DYNAMIC METHODS)

The third major category of methods includes those based on the dynamics of pile driving. These are known as *dynamic analyses*. They determine the static load capacity based on the effort required to drive the pile. In general, piles that are more difficult to drive should have a greater static load capacity.

Pile Driving Formulas

The earliest attempts at developing dynamic methods were based on empirical correlations between hammer weight, blow count (number of blows per foot of penetration), and other factors with the static capacity. These are collectively known as *pile driving formulas*. Hundreds of pile driving formulas have been advanced over the years, some of them dating as far back as the 1850s. Although these formulas have different formats, all share a common methodology of computing the pile capacity based on the driving energy delivered by the hammer. They use the principle of conservation of energy to compute the work performed during driving after attempting to consider the various losses and inefficiencies in the driving system. These effects are determined from pile load tests and incorporated into the formulas through empirical correction factors.

Pile driving formulas are convenient because the engineer can compute the capacity of each pile as it is driven by simply determining the final blow count.

The basic relationship common to all pile driving formulas is:

$$P_a = \frac{W_r h}{s F}$$
(13.24)

Where:

P_a = net allowable vertical load capacity

W_r = hammer ram weight

h = hammer stroke (the distance the hammer falls)

s = pile set (penetration) per blow at the end of driving

F = factor of safety

The *Sanders formula* of 1851 uses Equation 13.24 with a factor of safety as high as 8. This relatively high value is partially a true factor of safety and partially a method of accounting for the energy losses.

One of the most popular pile driving formulas is the one first published over 100 years ago in the journal *Engineering News* (Wellington, 1888). It has since become known as the *Engineering News Formula*:

$$P_a = \frac{W_r h}{F (s + c)}$$
(13.25)

Based on load test data, Wellington recommended using a c coefficient of 1 in (25 mm) to account for the difference between the theoretical set and the actual set. However, his data base included only timber piles driven with drop hammers. Some engineers use $c = 0.1$ in (2 mm) for single-acting hammers, although this was not part of the original formula. He also recommended using a factor of safety of 6.

Wellington apparently had much confidence in his work when he stated that his formula was:

> ... first deduced as the correct form for a theoretically perfect equation of the bearing power of piles, barring some trifling and negligible elements to be noted; and I claim in regard to that general form that it includes in proper relation to each other every constant which ought to enter into such a theoretically perfect practical formula, and that it cannot be modified by making it more complex ... (Wellington, 1892)

The Engineering News Formula has been used quite extensively since then and has routinely been extrapolated to other types of piles and hammers.

In 1961, the Michigan Highway Department conducted a series of pile driving tests (Housel, 1966). One of the objectives of these tests was to evaluate the accuracy of the Engineering News Formula. Following the driving and load testing of 88 piles, the engineers found that the formula overpredicted the pile capacities by a factor of 2 to 6. Thus, piles designed using $F = 6$ actually had a factor of safety between 1 and 3.

Based on their test results, the Michigan engineers developed the *Modified Engineering News Formula:*

$$P_a = \frac{0.0025\, E\, (W_r + e^2\, W_p)}{(s + 0.1)\, (W_r + W_p)} \tag{13.26}$$

Where:

P_a = allowable pile load (k)

E = rated hammer energy per blow (ft-lb)

W_p = weight of pile plus driving appurtenances (lb)

W_r = weight of hammer ram (lb)

s = pile set (in/blow)

e = coefficient of restitution (based on pile and hammer types)

Accuracy of Pile Driving Formulas

Pile driving formulas are attractive, and they continue to be widely used in practice. Unfortunately, the accuracy of these methods is less than impressive. Cummings (1940) was one of the first to describe their weaknesses. Since then, many engineers have objected to the use of these formulas and many lively discussions have ensued. Terzaghi's (1942) comments are typical (Used by permission of ASCE):

> In spite of their obvious deficiencies and their unreliability, the pile formulas still enjoy a great popularity among practicing engineers, because the use of these formulas reduces the design of pile foundations to a very simple procedure. The price one pays for this artificial simplification is very high. In some cases the factor of safety of a foundation designed on the basis of the results obtained by means of the pile formula is excessive and in other cases significant settlements have been experienced.
>
> ... On account of their inherent defects, all the existing formulas are utterly misleading as to the influence of vital conditions, such as the ratio between the weight of the pile and the hammer, on the result of the pile driving operations. In order to obtain reliable information concerning the effect of the impact of the hammer on the penetration of the piles it is necessary to take into consideration the vibrations that are produced by the impact.
>
> ... Newton himself warned against the application of his theory to problems involving the impact produced "by the stroke of a hammer."

In another article, Peck (1942) suggested marking various pile capacities on a set of poker chips, selecting a chip at random, and using that capacity for design. His data suggest that even this method would be more accurate than pile driving formulas.

Not everyone agreed with Cummings, Terzaghi, and Peck, so this topic was the subject of heated discussions, especially during the 1940s. However, comparisons between pile load tests and capacities predicted by pile driving formulas have clearly demonstrated the inaccuracies in these formulas. Some of this data are presented in Figure 13.27 in the form of 90% confidence intervals. All of these piles were driven into soils that were primarily or exclusively sand. Predictions of piles driven into clay would be much worse because of freeze effects.

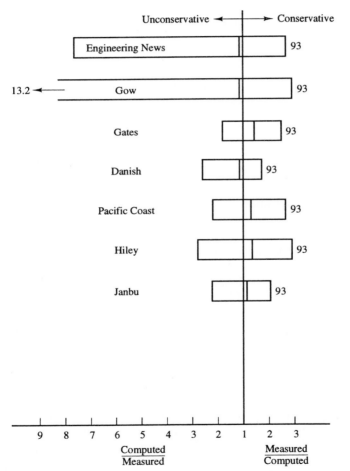

Figure 13.27 Ratio of measured pile capacity (from pile load tests) to capacities computed using various pile driving formulas. The bars represent the 90% confidence intervals, and the line near the middle of each bar represents the mean. The number to the right of each bar is the number of data points (based on data from Olson and Flaate, 1967).

Although the principle of conservation of energy is certainly valid, pile driving formulas suffer because it is very difficult to accurately account for all of the energy losses in a real pile driving situation. The sources of these uncertainties include the following:

- The pile, hammer, and soil types used to generate the formula may not be the same as those at the site where it is being used. This is probably one of the major reasons for the inaccuracies in the original Engineering News Formula.
- The formulas do not account for freeze effects.
- The hammers do not always operate at their rated efficiencies.
- The energy absorption properties of cushions can vary significantly.
- The formulas do not account for flexibility in the pile.
- There is no simple relationship between the static and dynamic strength of soils.

As a result of these many difficulties, the time has come to completely abandon the use of all pile driving formulas. The author agrees with Davisson (1979) when he stated "... it is hoped that such formulas have been purged from practice," yet is aware that they continue to be used. The wave equation approach, discussed in the next section, is far superior.

Wave Equation Analyses

It is unfortunate that the pile driving formulas are so unreliable, because it would be very useful to define pile capacity in terms of hammer blow counts. Such a function would provide a convenient means of construction control that could easily be applied to every pile installed at a given site. To satisfy this need, engineers studied the dynamics of pile driving in more detail and eventually developed another type of dynamic analysis: the *wave equation method*. This method provides a more accurate function of capacity vs. blow count, helps optimize the driving equipment, and computes driving stresses.

The Wave Equation Model

Pile driving formulas consider the pile to be a rigid body subject to classical Newtonian physics. This is a poor model because the impact energy from the hammer actually travels down the pile as a stress wave. Isaacs (1931) appears to have been the first to suggest the advantages of evaluating piles based on wave propagation. An analysis that more closely resembles the actual dynamics of pile driving should produce more accurate results.

The one-dimensional propagation of stress waves in a long slender rod, such as a pile, is described by the *one-dimensional wave equation*:

$$\frac{d^2u}{dt^2} = \frac{E}{\rho}\frac{d^2u}{dx^2} \tag{13.27}$$

Where:

    x = a position on the rod

    t = time

    u = displacement of the rod at point x

    E = modulus of elasticity of the rod

    ρ = mass density of the rod

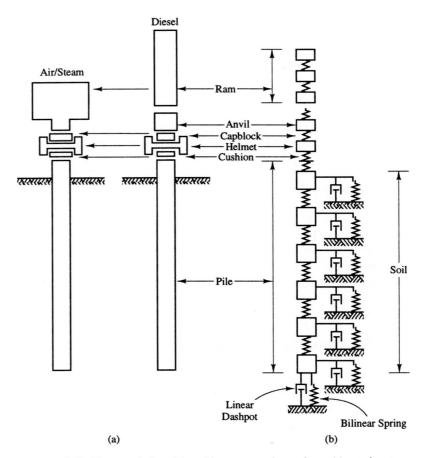

Figure 13.28 The numerical model used in wave equation analyses: (a) actual system;
(b) numerical model (Adapted from Goble, 1987).

    This formula is manageable when the boundary conditions are simple, but it becomes much more difficult with the complex boundary conditions associated with real piles. Therefore, for practical problems, we must use numerical methods to solve it, and this was not possible until digital computers became available. Smith experimented with numerical solutions soon after the Second World War (Smith, 1951) in what appears to have been one of the first civilian applications of computers. He later refined this work (Smith, 1960, 1962), thus forming the basis for modern wave equation analyses of piles. Although Smith's model is actually a simulation of the pile system, not a true solution of Equation 13.27, it is still commonly called a wave equation analysis.

Smith's numerical model divides the pile, hammer and driving accessories into discrete elements, as shown in Figure 13.28. Each element has a mass that equals the true mass of the corresponding portion of the real system. These elements are connected with springs that have the same stiffness as the corresponding element. For example, if the pile is divided into 1 foot long elements, the corresponding masses and springs in the numerical model pile would correspond to the mass and stiffness of 1 foot of piling.

Some portions of the pile system are more difficult to model than others. For example, the hammer is more difficult to model than is the pile. This is especially true of diesel hammers because their energy output varies with the hardness of the driving. Fortunately, some of the newer programs include improved hammer models.

The method also models the interface between the pile and the soil using a series of springs and dashpots along the sides of the elements and on the bottom of the lowest element to model the skin friction and end bearing resistance. The springs model resistance to driving as a function of displacement; the dashpots model resistance as a function of velocity.

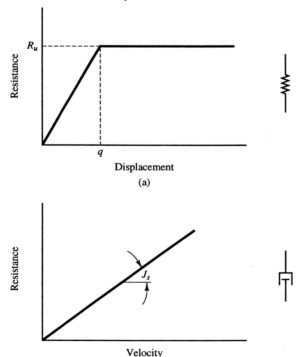

Figure 13.29 Smith's model of the soil-pile interface: (a) bilinear springs and (b) linear dashpots.

Smith proposed using a bilinear elastic-plastic spring and a linear dashpot as shown in Figure 13.29. The spring resistance increases until it reaches a displacement q, known as the *quake*. At that point, it reaches the *ultimate resistance*, R_u, and becomes completely plastic. The dashpot resistance is a linear function of the velocity and is defined by the *Smith damping factor*, J_s. Although this is a simplified view of the forces that act between the pile and the soil, it seems to work well for practical problems. More complex spring and dashpot functions do not significantly improve the analysis.

The design values of q, R_u, and J_s for each element are based on the soil type and other factors. Researchers have developed recommended design values by comparing wave equation analyses with pile load test results. Therefore, it is best to think of them as experimental calibration coefficients, not physical soil properties.

Smith suggested using a quake value of 0.1 in (2.5 mm) in all cases. Others have modified his suggestion slightly and used a quake value of 0.1 in per foot of pile diameter (0.008 mm/mm). Although these guidelines often produce acceptable results, there are situations where the quake is much larger, perhaps as much as 0.5 in (150 mm). Failure to recognize these special situations, and the use of a quake value that is too small can result in pile breakage.

Once assembled, the numerical model simulates the pile driving by imparting a downward velocity to the hammer ram elements equal to the impact velocity. This generates a stress impulse that travels to the bottom of the pile, generating displacements and resistances along the way. Upon reaching the bottom, the impulse reflects back, thus continuing until all of the energy is dissipated. In the real pile, this process occurs very quickly, and the stress waves dissipate before the next hammer blow occurs.

Once the stresses have dissipated, the pile will have advanced some distance, known as the *set*. The model predicts the magnitude of the set, but often expresses it as the inverse of the set, which is the *blow count* (i.e., hammer blows/ft). A *bearing graph*, as shown in Figure 13.30, is a plot of ultimate load bearing capacity vs. blow count for a certain pile type and size driven with a certain hammer at a certain site. It may be necessary to develop several bearing graphs to represent piles of different diameters, lengths, or types, or for different hammers.

Once the bearing graph (or graphs) have been obtained, they may be used as a means of construction control. For example, if a certain pile has a design ultimate capacity of 100 k, the field engineer simply refers to the appropriate bearing graph and determines the necessary blow count.

Freeze (Setup) Effects

As discussed earlier, piles driven into saturated clays produce excess pore water pressures that temporarily decrease their load capacity. The capacity returns when these pressures dissipate, a process known as *freeze* or *setup*. Thixotropic effects also contribute to pile freeze.

We are primarily interested in the load capacity after freeze has occurred, yet dynamic analyses based on data obtained during driving only give the prefreeze capacity.

Therefore, engineers use *retap* blow counts in these soils. These are obtained by bringing the pile driver back to the pile sometime after it has been driven (perhaps a few days) and driving it an additional couple of inches.

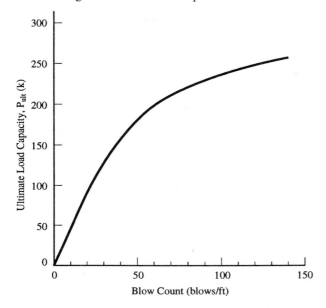

Figure 13.30   A typical bearing graph obtained from a wave equation analysis.

Analysis of Driving Stresses and Selection of Optimal Driving Equipment

If the hammer is too small, it will not efficiently drive the pile; if it is too large, the pile may be overstressed and break. Fortunately, wave equation analyses also compute the driving stresses in the pile, and they can be compared to the allowable stresses from Chapter 18 to select the optimal hammer and driving accessories. This is known as a *driveability analysis*.

Wave Equation Analysis Software

The first publicly available wave equation software was the TTI program developed at Texas A&M University (Edwards, 1967). In 1976, researchers at the Case Institute of Technology[7] developed the WEAP (Wave Equation Analysis of Piles) program. It has been revised on several occasions and is in the public domain. The WEAP program has since formed the basis for other more advanced proprietary programs.

----

[7] Now known as Case Western Reserve University.

The Case Method

During the 1960s and 1970s, researchers at the Case Institute of Technology developed another dynamic method of determining the axial load capacity of piles: the *Case method*. It is based on an analysis of dynamic forces and accelerations measured in the field while the pile is being driven (Rausche, Goble, and Likins, 1985; Hannigan, 1990).

The Pile Driving Analyzer

Field equipment for measuring the forces and accelerations in a pile during driving was developed during the 1960s and became commercially available in 1972. The methodology is now standardized and is described in ASTM standard D4945.

This equipment includes three components:

- A pair of *strain transducers* mounted near the top of the pile
- A pair of *accelerometers* mounted near the top of the pile
- A *pile driving analyzer* (PDA)

The strain transducers and accelerometers are shown in Figure 13.31, and a pile driving analyzer is shown in Figure 13.32.

The pile driving analyzer monitors the output from the strain transducers and accelerometers as the pile is being driven, and evaluates this data as follows:

- The strain data, combined with the modulus of elasticity and cross-sectional area of the pile, gives the axial force in the pile.
- The acceleration data integrated with time produces the particle velocity of the waves travelling through the pile.
- The acceleration data, double integrated with time produces the pile set per blow.

Using this data, the PDA computes the Case method capacity, using the method described later, and displays the results immediately. It also can store the field data on a floppy disk to provide input for a CAPWAP analysis, as described later in this chapter.

Wave Propagation

The hammer impact creates a compressive wave pulse that travels down the pile. As it travels, the pulse induces a downward (positive) particle velocity. If the pile has only end bearing resistance (no skin friction), the pulse reflects off the bottom and travels back up as another compression wave. This reflected wave produces an upward moving (negative) particle velocity.

The time required for the wave to travel to the bottom of the pile and return is $2D_2/c$, where:

D_2 = distance from strain transducers and accelerometers to the pile tip

c = wave velocity in the pile

The time $2D_2/c$ is very short compared to the interval between hammer blows. Therefore, the PDA can observe the effects of a single blow.

The plots of force and particle velocity near the top of this end bearing pile (as measured by the PDA) are similar to those in Figure 13.33. Note the arrival of the return pulse at time $2D_2/c$. These plots are called *wave traces*.

Figure 13.31 Strain gage and accelerometer mounted near the top of a pile to provide input to the pile driving analyzer (Photo courtesy of Pile Dynamics, Inc.).

Figure 13.32 A pile driving analyzer (photo courtesy of Pile Dynamics, Inc.).

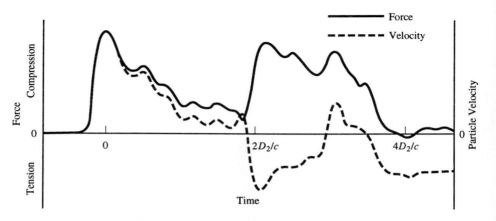

Figure 13.33 Typical plots of force and particle velocity near the top of the pile vs. time for an end bearing pile (Hannigan, 1990; Used with permission of Deep Foundations Institute).

Now, consider another pile that also has no skin friction resistance, but much less end bearing resistance than the previous one. Now, more of the energy in the downward moving wave is expended in advancing the pile, so the reflected wave has a smaller force and a smaller velocity, as shown in the wave trace in Figure 13.34. Therefore, the shape of the wave trace at time $2D_2/c$ reflects the bearing resistance.

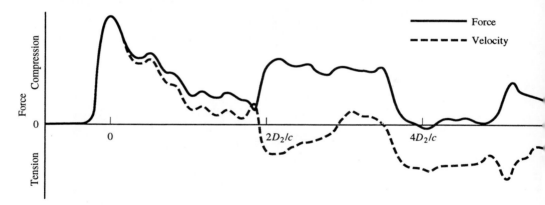

Figure 13.34 Typical plots of force and particle velocity near the top of the pile vs. time for an end bearing pile with less end bearing capacity than that shown in Figure 13.33 (Hannigan, 1990; Used with permission of Deep Foundations Institute).

Finally, consider a friction pile with very little end bearing resistance. As the compression wave pulse travels down the pile, it encounters the skin friction resistance. Each increment of resistance generates a reflected wave that travels back up the pile, so the wave trace measured near the top will be similar to that in Figure 13.35. The time scale corresponds to the depth below the instruments, and the vertical distance between the force and velocity plots reflects the soil resistance at various depths.

The wave trace also provides pile integrity data. For example, if the pile breaks during driving, the fracture produces a reflected wave that changes the wave trace recorded by the pile driving analyzer (Rausche and Goble, 1978).

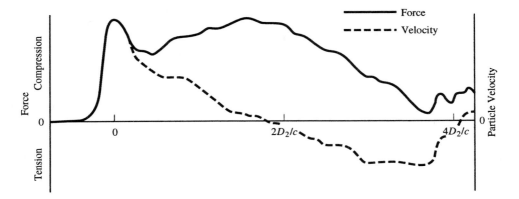

Figure 13.35 Typical plots of force and particle velocity near the top of the pile vs. time for a friction pile (Hannigan, 1990; Used with permission of Deep Foundations Institute).

Case Method Analyses

The Case method is an analytical technique for determining the static pile capacity from wave trace data (Hannigan, 1990). The PDA is programmed to solve for pile capacity using this method and gives the results of this computation immediately.

The Case method computations include an empirical correlation factor, j_c, that can be determined from an on-site pile load test. Thus, engineers can use this method to extend pile load test results to indicator piles or selected production piles. However, for most projects, it would not be cost-effective to obtain PDA measurements on all of the production piles.

It is also possible to use the Case method without an on-site pile load test by using j_c values from other similar soils. This approach is less accurate, but still very valuable.

CAPWAP Analyses

The Case method, while useful, is a simplification of the true dynamics of pile driving. The empirically obtained damping factor, j_c, calibrates the analysis, so the final results are no better than the engineer's ability to select the proper value. In contrast, a wave equation analysis utilizes a much more precise numerical model, but suffers from weak estimates of the actual energy delivered by the hammer. Fortunately, the strengths and weaknesses of these two methods are complimentary, so we can combine them to form an improved analysis (Rausche, Moses, and Goble, 1972). This combined analysis is known as CAPWAP (CAse Pile Wave Analysis Program).

The numerical model used in CAPWAP is essentially the same as that in Figure 13.29 except that the hammer and accessories are removed and replaced with force-time and velocity-time data obtained from the pile driving analyzer. By conducting enough iterations, the analysis produces values of R_u (the ultimate resistance in the soil "springs"), q (the quake), and j_c (the Case method damping factor). Unfortunately, the CAPWAP analysis is difficult to perform. Perhaps only 20 people in the world are qualified to conduct these analyses. Therefore, most people contract the analysis to a specialist.

A CAPWAP analysis performed on PDA data from an indicator pile could be used as follows:

- To provide more accurate input parameters for a wave equation analysis that could then be used to select the optimal driving equipment as well as to produce a bearing graph.
- To provide a site-specific Case method damping factor, j_c, for use in PDA analyses of selected production piles.
- To obtain quantitative measurements of pile setup (Fellenius et al., 1989).

CAPWAP analyses are not a substitute for pile load tests. However, they may reduce the required number of tests.

13.6 SYNTHESIS AND PRACTICAL APPLICATION

Because there are so many methods of predicting pile load capacities, it often is difficult to decide which ones to use for a particular project. These choices depend on the size of the project, the subsurface conditions, local experience, and other factors. Often, engineers use different methods at different stages of the project.

Initial Design

Initial pile designs are almost always based on one of the static methods using either laboratory or in-situ tests. Unfortunately, static methods, especially those based on laboratory tests, are not very precise. Briaud, et al. (1986) compared the computed capacities of 98 piles in Mississippi with static load tests. The results of their study, reinterpreted and shown in Figure 13.36, show that all of these methods have a wide margin of uncertainty. Although this study suggested that analyses based on pressuremeter tests (PMT) were the most precise, those based on the cone penetration test (CPT) seem to have more promise for most design situations.

We can reduce the risk of grossly overestimating or underestimating the pile capacity by using more than one static analysis method (e.g., the α method, and a CPT based method) and comparing the results.

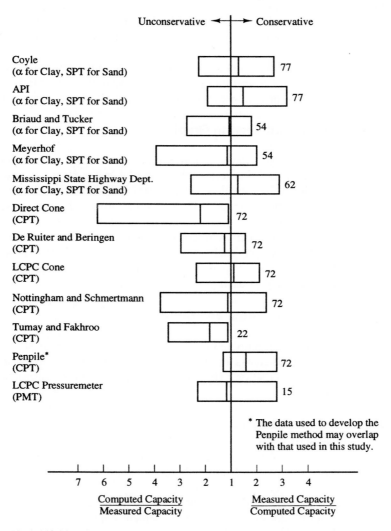

Figure 13.36 Comparison of capacities computed from static analysis methods with those measured in full-scale load tests. The bar for each method represents the 90% confidence interval, the line in the middle represents the average prediction, and the number to the right represents the number of data points (based on data from Briaud et al., 1986). The best methods are those that have narrow bands centered on a ratio of 1.

Full-Scale Load Tests

If the project is sufficiently large (or the soil conditions are sufficiently unusual), then the next step is to build one or more prototype piles based on the initial design criteria and conduct full-scale load tests. The engineer then uses the pile load test results to calibrate the initial design criteria (i.e., develop site-specific parameters, such as β and q_e'), and uses this data to produce the final pile design. This procedure should produce load capacity predictions that are much more precise than the uncalibrated predictions shown in Figure 13.36.

A load test is the most accurate way to determine pile capacity, but it also is the most expensive way. Sometimes engineers choose to use Case Method and/or CAPWAP analyses to reduce or eliminate the number of pile load tests, thus reducing costs.

Design Charts and Bearing Graphs

Usually, engineers must develop design criteria that apply to an entire site. It is often convenient to express this criteria in the form of a design chart, as shown in Figure 13.37. These charts should be a synthesis of analyses and load tests. Usually, the geotechnical engineer develops the design charts, and then the structural engineer uses them to size the individual foundations.

Just prior to construction, it also is helpful to conduct a wave equation analysis and develop a bearing graph, such as the one in Figure 13.30. This graph should be calibrated to any on-site full-scale load test data. Then, field engineers can use this graph to judge the adequacy of each pile as it is driven.

Construction Monitoring and Control

Pile foundations are usually not built exactly as shown on the design drawings. Natural variations in the soils across the site and unanticipated construction problems often dictate changes during construction. Therefore, a formal program of construction monitoring and control is essential.

Construction monitoring also provides a confirmation of the design capacities. At a minimum, the field engineer should use a bearing graph from a wave equation analysis as described earlier and use it to compare the design capacities with the final blow count for each pile. It is also very helpful to use a pile driving analyzer on selected piles and analyze the results using a Case method analysis. For more critical or difficult projects, a CAPWAP analysis may be warranted.

Figure 13.38 demonstrates the value of using the Case Method to confirm pile capacities during construction. The 90 percent confidence interval for this method is much narrower than those for static analyses (see Figure 13.36) or for pile driving formulas (see Figure 13.27). This means the Case Method produces excellent results, and thus significantly reduces the uncertainties associated with predicting load capacities. The Engineering News Formula, shown here and in Figure 13.27, is much less precise.

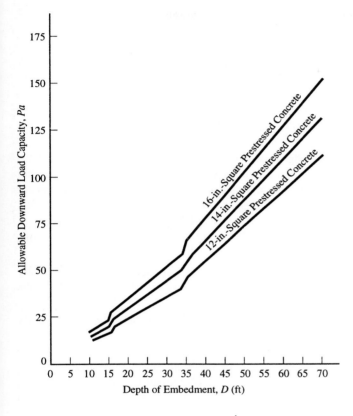

Figure 13.37 A typical pile design chart. The jogs at 15 and 35 feet are due to changes in end bearing capacity as the pile tip enters stronger soil strata.

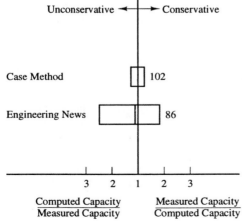

Figure 13.38 Comparison of capacities computed during construction using the Case Method and the Engineering News Formula with those measured by full-scale load tests (based on data from Goble, Likins, and Rausche, 1975). The ends of the bars represent 90% confidence intervals, the line in the middle represents the mean prediction, and the number on the right side is the number of data points.

SUMMARY OF MAJOR POINTS

1. The most reliable way to determine the axial load capacity is to conduct a full-scale load test on a prototype pile. However, these tests are expensive and may not be cost-effective for small to medium-size projects.

2. There are two types of load tests. Constant stress tests apply the load in known increments and measure the corresponding settlements. Controlled strain tests apply the load at a constant rate of strain and measure the loads that correspond to certain settlements. Most load tests use the constant stress method. The load-settlement curve will vary depending on the test method.

3. There are several ways to interpret load-settlement plots from full-scale load tests. These methods produce different values of the ultimate capacity.

4. The second way to determine the axial load capacity is to perform a static analysis. These methods are based on laboratory or in-situ soil tests. Static methods are not as precise as load tests because we do not fully understand the changes that occur in the soil during pile driving or the way piles transfer load into the ground. However, these analyses are less expensive to perform.

5. Most static methods apply only to cohesionless soils or only to cohesive soils. Thus, it is important to identify which type of soil is present.

6. Analyses based on the cone penetration test (CPT) appear to be especially promising, especially in cohesionless soils. This may be because the CPT probe is similar to a model pile.

7. The third way is to use dynamic methods. These predict the static load capacity based on the dynamics of pile driving. These methods include pile driving formulas (not recommended), wave equation analyses (a computer model), the Case method (based on measurements made in the field during construction), and CAPWAP analyses (a combination of the wave equation and Case methods. Dynamic methods can be used to supplement pile load tests and as a construction control.

8. The uncertainties in pile capacity predictions can be reduced by using more than one method of analysis and integrating the results.

QUESTIONS AND PRACTICE PROBLEMS

13.1 The piles for a certain monumental construction project will cost 165 monetary units per zup (a unit of length). This price includes both materials and installation. If no load test is performed, then the design of these piles should be based on the criteria for "normal" control as described in Table 11.1 in Chapter 11. However, if a load test is performed, then we can use the "good" control criteria. A pile load test would cost 120,000 monetary units. Determine the minimum project size, expressed in total zups of piling, required to justify the cost of a load test.

13.2 The results of pile load tests are usually considered to be the "correct" load capacity, and all other analysis methods are compared to this standard. However, there are many ways to conduct load tests, and many ways to interpret them. Therefore, can we truly establish a single "correct" capacity for a pile? Explain.

13.3 A 250 mm square, 15 m long prestressed concrete pile ($f_c' = 40$ MPa) was driven at a site in Amsterdam as described by Heijnen and Janse (1985). A full-scale load test conducted 31 days later produced the following load-settlement curve:

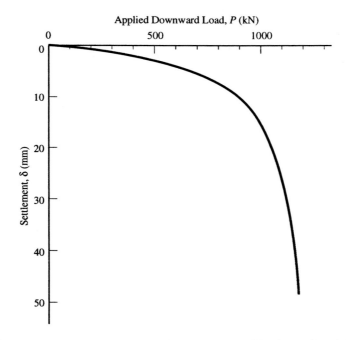

Using Davisson's method, compute the ultimate downward load capacity of this pile.

13.4 A proposed highway bridge will be supported by 180 piles, each about 18 m long. The design load per pile is greater than 1000 kN. The soil conditions are neither unusually good, nor unusually bad, and you have limited experience with other pile foundations in the area. Based on this limited information, determine the number of pile load tests required for this project.

13.5 How did Coyle and Castello develop their design curves for q_e' and f_s (Figures 13.11 and 13.13)?

13.6 Skin friction analyses that use a full effective stress analysis (Equation 13.5) are theoretically more correct than the others described in Section 13.3, yet this formula is rarely used. Why? Could additional research make it more viable for practical design problems? Explain.

13.7   Why is the cone penetration test a good source of data for pile designs?

13.8   Why is the Engineering News Formula (Equation 13.25) a poor predictor of pile capacity? Which method has made this formula obsolete?

13.9   What useful information can be obtained from a wave equation analysis?

13.10   At what point or points in the design and construction process would it be helpful to use a pile driving analyzer?

13.11   The skin friction capacity of piles depends on many factors.   Which of these do not explicitly appear in Coyle and Castello's method?

13.12   A 14 in square, 45 ft long prestressed concrete pile is to be driven into the following soil profile:

Depth (ft)	Soil Classification	Friction Angle, ϕ (deg)
0 - 10.0	Silty sand (SM)	33
10.0 - 16.5	Sandy silt (ML)	31
16.5 - 25.0	Fine to medium sand (SW)	35
25.0 - 65.0	Well-graded sand (SW)	37

This pile will support part of an industrial building with a 40 year design life, and will be built using "poor" control. The groundwater table is at a depth of 15 ft. Compute the allowable downward and uplift load capacities.

13.13   Using the soil profile, pile type, and structure described in Problem 13.12, determine the required pile length to support a 95 k compressive load.

13.14   The soil profile beneath a proposed construction site is as follows:

Depth (m)	Soil Classification	Undrained Shear Strength, s_u (kPa)
0 - 2.5	Stiff silty clay (CL)	80
2.5 - 6.7	Soft clay (CL)	15
6.7 - 15.1	Medium clay (CL)	30
15.1 - 23.0	Stiff clay (CL)	100

Develop a plot of allowable downward load capacity vs. depth for a 350 mm square concrete pile. Consider pile embedment depths between 5 and 20 m and use a factor of safety of 3.0.

13.15 An HP 13×87 pile (see Table 18.2 in Chapter 18 for dimensions) is embedded 45 ft into a cohesive soil. The unit weight of this soil is 100 lb/ft^3 above the groundwater table (which is 12 ft below the ground surface) and 112 lb/ft^3 below. The soil in the vicinity of the pile tip has an undrained shear strength of 2800 lb/ft^2. According to a pile load test, the ultimate downward load capacity is 143 k.

You wish to compute a site-specific β factor for HP 13x87 piles to be used in the design of other piles at this site. Based on these test results, what is that β factor?

Hint: Compute β based on the average $\sigma_v{}'$ and the average measured f_s.

13.16 A 250 mm diameter, open-end steel pipe pile penetrates 10.0 m into a sandy soil. This soil has an average N_{60} of 21 and an average $N_{60}{}'$ of 19. Compute the allowable downward load capacity using a factor of safety of 4.0. Assume this pile becomes fully plugged.

13.17 A 12 inch square prestressed concrete pile is to be driven 45 ft into the soil described by the electric CPT results in Figure 4.22 in Chapter 4. Compute the allowable downward load capacity using a factor of safety of 2.8 and each of the following methods:

a. Dutch method for end bearing, and Nottingham and Schmertmann's method for skin friction. Assume the silty sand at the pile tip has OCR = 2.
b. LCPC method.

Advanced Questions and Problems

13.18 The pile described in Problem 13.3 (Heijnen and Janse, 1985) was driven into the following soil profile (described from the ground surface downward):

Elevation (m)	Soil Description
+1.12 to -2.1	Old sandy fill
-2.1 to -11.1	Soft clay, silt, and peat
Below -11.1	Sand

The ground surface is at elevation +1.12 m and the groundwater table is approximately at elevation +0.4 m.

Both SPT and electric CPT tests have been conducted at this site, the results of which are shown on pages 424 and 425. Afterwards, the pile described in Problem 13.3 was driven into the ground using a Delmag D12 single-acting diesel hammer (W_r = 12.2 kN; h = 2.61 m). The hammer blow counts recorded during driving are also shown.

Using this data, compute the ultimate downward axial downward load capacity using each of the following methods:

a. Based on CPT data
b. Based on SPT data
c. Using the Engineering News formula

Compare your computed capacities with the load test results in Problem 13.3 and discuss.

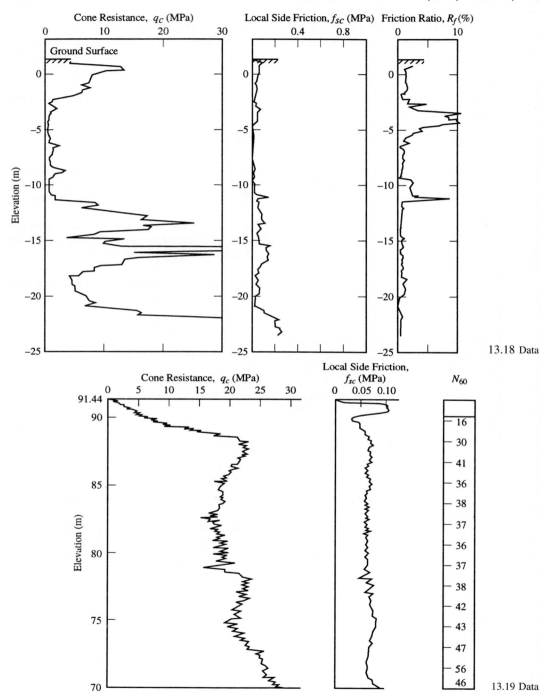

13.18 Data

13.19 Data

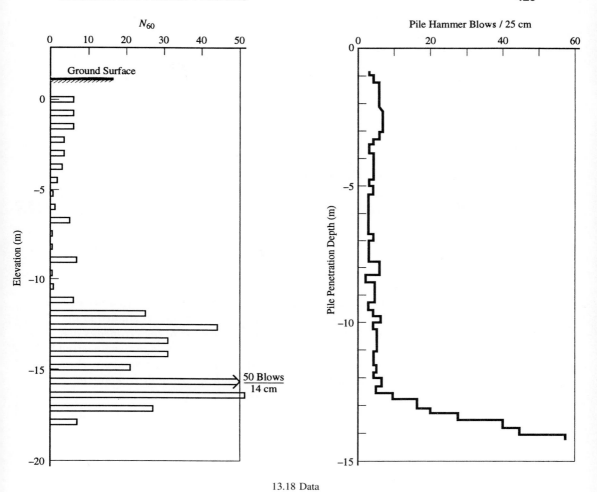

13.18 Data

13.19 A series of twelve exploratory borings and twelve CPT soundings have been conducted at the site of an electrical power plant near Sikeston, Missouri. The soil at this site consists of poorly graded alluvial fine sands overlain by about 3 m of silty clay. The groundwater table is at a depth of 1.5 to 3.0 m below the ground surface. Representative electric CPT and SPT data for this site are presented in the figure on page 424.

Develop a design chart for downward loading on 400 mm square concrete, 400 mm diameter open-end steel pipe, and 305 mm H section piles. Consider depths of embedment between 8 and 17 m. Assume that some predrilling will be required for the concrete and pipe piles, that the pipe pile becomes fully plugged, and "normal" control.

(Data adapted from Buhr et al., 1982; Proceedings, Second European Symposium on Penetration Testing, ©A.A. Balkema Publishers, Brookfield, VT; Used with permission).

13.20 Using the data from Problem 13.19, develop a design chart for uplift loading.

14

Pile Foundations –
Other Geotechnical Considerations

*In reality, soil mechanics is only one of the bodies of knowledge
upon which the foundation engineer may draw. If studied to the
exclusion of other aspects of the art, it leads to the erroneous and
dangerous impression that all problems in foundation engineering
are susceptible of direct scientific solution. Unfortunately, the
vagaries of nature and the demands of economy combine to
eliminate this possibility.*

From *Foundation Engineering* by Peck, Hanson, and Thornburn,
Copyright ©1974 by John Wiley & Sons, Inc. Reprinted by
permission of John Wiley & Sons, Inc.

14.1 GROUP EFFECTS

The analyses in Chapter 13 addressed only single isolated piles. However, piles almost
always are installed in groups of three or more, as shown in Figure 14.1. Pile groups are
better than single isolated piles because of the following:

- A single pile usually does not have enough capacity.
- Piles are *spotted* or located with a low degree of precision, and can easily be 6 in
  (150 mm) or more from the desired location. If a column for a building, which is
  located with a much greater degree of precision, were to be supported on a single
  pile, the centerlines would rarely coincide and the resulting eccentricity would
  generate unwanted moments in both the pile and the column. However, if the

column is supported on three or more piles, any such eccentricities are much less significant.

- Multiple piles provide redundancy, and thus can continue to support the structure even if one pile is defective.
- The lateral soil compression during pile driving is greater, so the coefficient of lateral earth pressure, K, and the skin friction capacity increase.

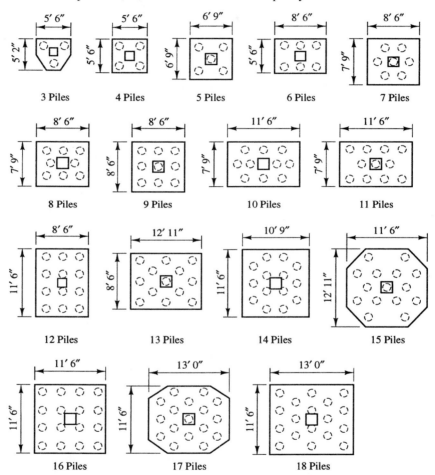

Figure 14.1 Typical configurations of pile groups (Adapted from CRSI, 1992).

Piles in groups for onshore structures are typically spaced between 2.5 and 3.0 diameters on-center. Closer spacings provide much less allowance for errors in positioning and alignment. Typically, the closest acceptable spacing is 2.0 to 2.5 diameters or 2 ft (0.6 m) on-center, whichever is greater. It is also good to avoid excessively wide spacings because they tend to require uneconomically large pile caps.

Group Efficiency

The piles in a group and the soil between them interact in a very complex fashion, and the capacity of the group is not necessarily equal to the capacity of a single isolated pile multiplied by the number of piles. The *group efficiency* depends on several factors, including the following:

- The number, length, diameter, arrangement, and spacing of the piles.
- The load transfer mode (skin friction vs. end bearing).
- The construction procedures used to install the piles.
- The sequence of installation of the piles.
- The soil type.
- The elapsed time since the piles were driven.
- The interaction, if any, between the pile cap and the soil.
- The direction of the applied load.

Engineers compute the axial load capacity of pile groups using a *group efficiency factor*, η, as follows[1]:

$$P_{ag} = \frac{\eta N \left(P_e' + P_s \right)}{F} \tag{14.1}$$

Where:

P_{ag} = net allowable capacity of pile group

η = group efficiency factor

N = number of piles in group

P_e' = net end bearing capacity of a single isolated pile

P_s = skin friction capacity of a single isolated pile

F = factor of safety

Converse and Labarre were among the first to address this issue, but they had little or no test data available to develop a formula for computing η. Therefore, they were forced to rely on assumed relationships between the pile group geometry and group efficiency, thus forming the basis for the *Converse-Labarre Formula* (Bolin, 1941):

---

[1] Some authorities suggest applying the efficiency factor to the entire pile capacity, as shown here, while others apply it only to the skin friction component.

$$\eta = 1 - \theta \frac{(n-1)m + (m-1)n}{90mn} \tag{14.2}$$

Where:

m = number of rows of piles

n = number of piles per row

$\theta = \tan^{-1}(B/s)$ (expressed in degrees)

B = diameter of a single pile

s = center-to-center spacing of piles

Although pile group efficiency formulas have been widely used, they were based on very little or no hard data and are not very accurate. This was well illustrated in 1977 when several deep foundation authorities were asked to predict the load capacity of a group of piles. The range of their predictions, which included a variety of techniques, and the actual capacity are shown in Figure 14.2. Although the mean of the predictions is close to the actual capacity, the scatter is on the order of ±50%. It seems that these techniques are poor models of the real behavior of pile groups.

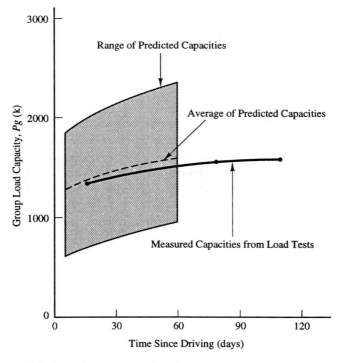

Figure 14.2 Comparison between 11 predictions of the load capacity of a pile group and the measured capacity (Adapted from Vesić, 1980).

Results of Model and Full-Scale Load Tests

Some researchers have conducted model load tests to study the performance of pile groups. Although these tests provide some insight, scale effects, especially the lower effective stresses in the model compared to the real soil, make the results difficult to interpret. We can alleviate some of these problems by performing model tests inside of a centrifuge, which has the net effect of increasing the apparent unit weight of the soil and thereby increases the effective stresses, but few such tests have yet been performed.

Full-scale load tests overcome the problems with model tests. Unfortunately, these are very expensive because of the large load frames, jacks, and other equipment required to conduct the test. Therefore, very few have been done (DiMillio et al., 1987a, 1987b). They are certainly much too expensive to use as a routine design tool. Full-scale tests also lack the versatility of model tests.

Individual vs. Block Failure Modes

When the piles are widely spaced, the soil between stays relatively stationary as the piles move down. However, more closely spaced piles will engage the soil between them and they move together as a large single pile. Both modes of failure are shown in Figure 14.3. Generally, block failure controls only when the spacing/diameter (s/B) ratio is less than about 2. Most pile groups have wider spacings, so block failure is usually not a concern.

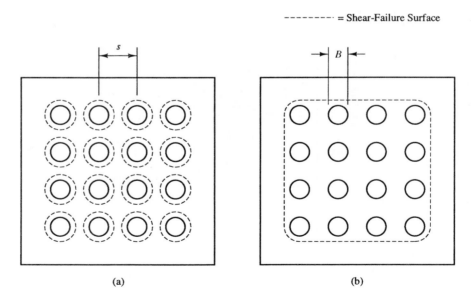

Figure 14.3 Types of failures in pile groups: (a) individual; (b) block.

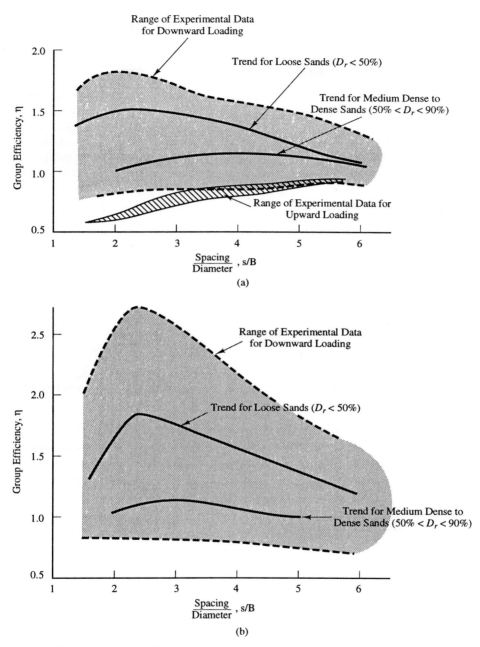

Figure 14.4 Group efficiencies from tests of model pile groups in cohesionless soils subjected to vertical loads: (a) groups of 4 piles; (b) groups of 9 - 16 piles (Adapted from O'Neill, 1983; Used with permission of ASCE).

Tests in Cohesionless Soils

Group efficiency factors from several model tests of piles in cohesionless soils are shown in Figure 14.4. O'Neill (1983) drew the following conclusions from these diagrams:

- In loose cohesionless soils, η is always greater than 1 and reaches a peak at $s/B \approx 2$. It also seems to increase with the number of piles in the group.
- In dense cohesionless soils with $2 < s/B < 4$ (the normal range), η is usually slightly greater than 1 so long as the pile was installed without predrilling or jetting. However, either of these construction techniques can significantly reduce the group efficiency.

The results of full-scale load tests in cohesionless soils, shown in Figure 14.5, also suggest η values greater than 1, except when predrilling or jetting was used.

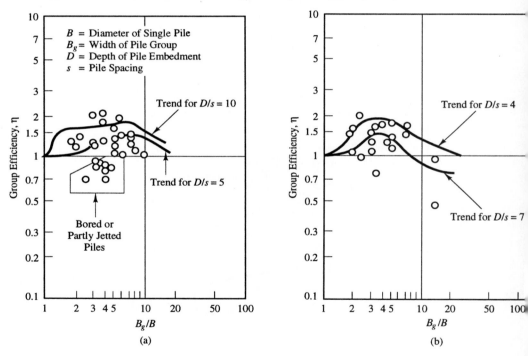

Figure 14.5 Group efficiencies from tests of full-scale pile groups in cohesionless soils subjected to vertical loads: (a) with pile cap suspended above the ground; (b) with pile cap in contact with the ground (Adapted from O'Neill, 1983; Used with permission of ASCE).

The high values of η in cohesionless soils seem to be primarily due to the radial consolidation that occurs during driving and the resulting increase in lateral stress. Less consolidation occurs if predrilling or jetting is used, so η is lower for those groups.

Tests in Cohesive soils

Group efficiency factors from several model tests of piles in cohesive soils are shown in Figure 14.6. These results are much different from those for cohesionless soils because η is now always less than 1.

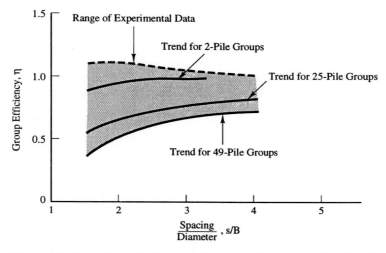

Figure 14.6 Group efficiencies from model pile groups in cohesive soils subjected to vertical loads (Adapted from O'Neill, 1983; Used with permission of ASCE).

The results of full-scale tests in cohesive soils, shown in Figure 14.7, indicate group efficiencies in the same range as the model tests as long as the pile cap is suspended above the ground. However, the group efficiency increases when the cap is in contact with the ground. This geometry seems to promote a large block failure, thus increasing the capacity. However, the settlement required to reach this capacity is quite large.

The efficiency of pile groups in cohesive soils is largely governed by excess pore water pressures. Although the magnitude of these pressures is not significantly higher than those near single piles, they encompass a larger volume of soil and therefore dissipate much more slowly, as shown in Figure 14.8. For single piles, nearly all of the excess pore water pressures dissipate within days or weeks of pile driving, whereas in groups they may persist for a year or more.

Guidelines for Practice

No satisfactory methods of assessing group action have yet been developed, so we must use engineering judgment when analyzing pile groups. Consider the following guidelines:

- Try to determine whether a block failure will control the design. If the perimeter of the pile group is greater than the sum of the perimeters of the individual piles, then a block failure is not likely. Even if this criterion is not met, block failure still might not control. Model pile tests suggest that block failure is critical only when the piles are at relatively close spacings (s/B less than about 2), so it is usually not a concern. However, if it appears that it does control, use $\eta = $ (perimeter of pile group)/(Σ perimeters of individual piles).

- The capacity of pile groups in cohesive soils will be temporarily reduced as a result of the formation of excess pore water pressures. The short-term group efficiency factor, η, will probably be in the range of 0.4 to 0.8, but it will increase with time. If $s/d > 2$ (virtually all practical cases), η will eventually reach a value of about 1. The rate at which it rises depends primarily on the dissipation of excess pore water pressures. Small groups will probably reach $\eta \approx 1$ within 1 - 2 months, which may be faster than the rate of loading, whereas larger groups may require a year or more. In some cases, it may be appropriate to install piezometers to monitor the dissipation of excess pore water pressures.

    Highly sensitive clays will probably have much lower group efficiency factors. Unfortunately, very little data seem to be available.

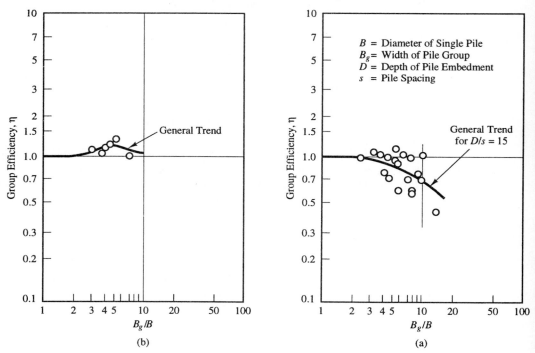

Figure 14.7 Group efficiencies from full-scale pile groups in cohesive soils subjected to vertical loads: (a) with the pile cap suspended above the ground; (b) with the pile cap in contact with the ground (Adapted from O'Neill, 1983; Used with permission of ASCE).

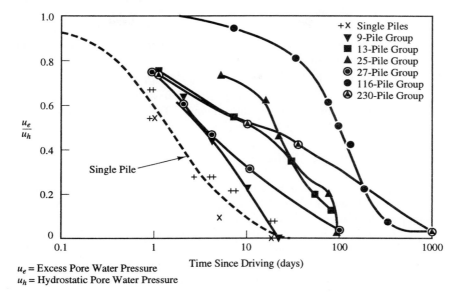

u_e = Excess Pore Water Pressure
u_h = Hydrostatic Pore Water Pressure

Figure 14.8 Measured dissipation of excess pore water pressures in soil surrounding full-scale pile groups (Adapted from O'Neill, 1983; Used with permission of ASCE).

• Pile groups in cohesionless soils will achieve their maximum capacity almost immediately because the excess pore water pressures dissipate very rapidly. The group efficiency factor will be at least equal to 1 (as long as $s/B > 2$) and will often be greater than 1, especially with closely spaced high-displacement piles. For design purposes, use group efficiency factors no greater than 1.25.

If predrilling was required before the pile was driven, as is often the case in dense cohesionless soils, then the sand will have been loosened and the group efficiency factor may be less than 1. Therefore, it is wise to use as little predrilling or jetting as possible.

Hopefully, additional research on this topic will eventually produce more definitive design guidelines.

14.2 SETTLEMENT

Isolated piles designed using the methods described in Chapter 12 usually settle no more than 0.5 in (12 mm) when subjected to the working loads. This settlement is acceptable for nearly all structures. Pile groups may have slightly greater, but still acceptable, settlements. Therefore, engineers usually do not conduct settlement analyses for pile foundations.

Nevertheless, a settlement analysis may be warranted if the engineer has one or more of the following concerns:

- The structure is especially sensitive to settlement, so the engineer must develop a quantitative estimate.
- One or more highly compressible strata are present.
- The engineer must express the pile response in terms of an equivalent "spring" located at the bottom of the column. This analytical model is used in some sophisticated structural analyses.

Imaginary Footing Method

A simple way to compute the settlement of a pile group is to replace it with an imaginary footing, as shown in Figure 14.9, then use the methods similar to those in Chapter 7 to compute the settlement of this imaginary footing.

For friction piles, place the imaginary footing at a depth of two thirds the pile embedment. For end bearing piles, place it at the pile tip elevation. When both skin friction and end bearing are significant, place it somewhere between these two positions.

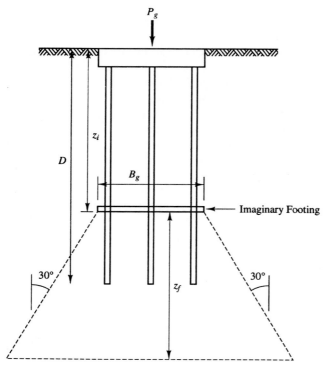

Figure 14.9 Use of an imaginary footing to compute the settlement of pile groups.

Assume the load beneath the imaginary footing spreads out over a 30° frustum as shown in Figure 14.9. Therefore, the increase in the vertical stress in the soil at a depth z_f is:

$$\Delta\sigma_v' = \frac{P_g}{(B_g + 2z_f \tan 30°)(L_g + 2z_f \tan 30°)} \qquad (14.3)$$

Where:

$\Delta\sigma_v'$ = increase in vertical effective stress

P_g = downward load acting on pile group

B_g = width of pile group = width of imaginary footing

L_g = length of pile group = length of imaginary footing

z_f = depth below bottom of imaginary footing

Use the procedures described in Chapter 7 to compute the settlement of the imaginary footing. Then, add the elastic compression of the piles:

$$\delta_e = \frac{P z_i}{A E} \qquad (14.4)$$

Where:

δ_e = settlement due to elastic compression of piles

P = downward load on each pile

z_i = depth to bottom of imaginary footing

A = cross-sectional area of a single pile

E = modulus of elasticity of pile

    = 29,000,000 lb/in^2 (200,000 MPa) for steel

    = 15,200 $\sigma_r (f_c'/\sigma_r)^{0.5}$ for concrete

    = 1,600,000 lb/in^2 (11,000 MPa) for Southern Pine or Douglas Fir

σ_r = 14 lb/in^2 = 0.10 MPa

Be sure to consider settlements due to causes other than the structural loads on the piles. For example, the construction process may include lowering the groundwater table, which increases the effective stress in the soil and thus create settlement.

Empirical Formulas

Meyerhof (1976) developed the following empirical method of computing the settlement of pile groups in cohesionless soils:

$$\delta = \frac{0.17 \, B_r \, q_e' \, I \, \sqrt{B_g/B_r}}{\sigma_r \, N_{60}'} \qquad (14.5)$$

$$\delta = \frac{q_e' \, B_g \, I}{2 \, q_c} \qquad (14.6)$$

$$I = 1 - \frac{z_i}{8 \, B_g} \geq 0.5 \qquad (14.7)$$

Where:

$\quad \delta$ = settlement of pile group

$\quad B_r$ = reference width = 1.0 ft = 0.3 m = 12 in = 300 mm

$\quad q_e'$ = equivalent net bearing pressure = $P_g / (B_g \, L_g)$

$\quad B_g$ = width of pile group

$\quad L_g$ = length of pile group

$\quad \sigma_r$ = reference stress = 2000 lb/ft^2 = 100 kPa

$\quad N_{60}'$ = SPT N value within a depth of z_i to $z_i + B_g$ with overburden correction

$\quad q_c$ = CPT cone bearing within a depth of z_i to $z_i + B_g$

Meyerhof's formulas assume the compressibility of the soil beneath the imaginary footing is constant or increases with depth.

He compared 16 SPT-based predictions (Equation 14.5) and 18 CPT-based predictions (Equation 14.6) with observed settlements of real pile groups, and found that the observed settlements were never more than 0.3 in (8 mm) greater than the predicted settlements.

t-z Method

A more precise analytical model would need to consider the following characteristics:

- The differences between the load-displacement relationships for skin friction and end bearing (see discussion in Section 11.2 of Chapter 11 and Figures 15.22 - 15.25 in Chapter 15).
- The form of these load-displacement relationships.
- The elastic compression of the pile.

The *t-z method* attempts to consider these effects by using a numerical model similar to that shown in Figure 13.28b in Chapter 13. This is a static model, so it has "springs" but no "dashpots."   Each of the skin friction and end bearing "springs" is defined by a *t-z curve* (Kraft, Ray and Kagawa, 1981), where *t* is the load and *z* is the settlement of that pile segment.   The "springs" within the pile reflect its elastic deformation properties.   Thus, as axial loads are applied to the top of the pile, the numerical model predicts the corresponding settlements.

Commercial software is available to conduct *t-z* analyses. These programs include model *t-z* curves developed from pile load tests.

This method is not widely used, primarily because settlement is not a problem with most deep foundations.   However, it is a good way to analyze settlement-sensitive problems.

14.3 SEISMIC DESIGN

Although failures of deep foundations subjected to static loads are rare, there have been some failures during earthquakes.   The most common failure modes appear to be as follows:

- Lateral spreads due to soil liquefaction, and the corresponding lateral movement of deep foundations (Youd, 1989; Bartlett and Youd, 1992).   For example, lateral pile movements of up to 3.5 m caused damage to 266 highway and railroad bridges during the 1964 Alaska earthquake.

- Structural failure in and around the junction of pile caps and batter piles, as shown in Figure 17.3 in Chapter 17.   Batter piles are those driven at an angle from the vertical so they can resist horizontal loads.   Unfortunately, this configuration produces a much more rigid foundation system that is more prone to damage during earthquakes.   For example, wharves on San Francisco Bay that had batter piles generally did not perform well during the 1989 Loma Prieta Earthquake, whereas those without batter piles generally survived with little or no damage.

- Settlement due to liquefaction-induced losses in skin friction and end bearing capacities. This caused significant structural damage during the 1964 Niigata, Japan earthquake.

- Settlement and/or lateral movement due to seismic loads combined with soil softening. This was observed in Mexico City during the 1985 earthquake.

- Overloading produced by resonance between the structure and the earthquake ground motions (also observed in Mexico City).

The liquefaction-related problems are beyond the scope of this book. See Bartlett and Youd (1992), and Lam and Martin (1986) for more information. The batter pile problems can be avoided by using only vertical piles and using them to resist lateral loads, as described in Chapter 17.

Geotechnical Design in Non-Liquefiable Soils

There are two primary ways to analyze and design deep foundations to resist seismic loads: the *pseudostatic method* and the *dynamic method.*

The vast majority of foundation designs are based on a pseudostatic analysis. This method ignores the cyclic nature of the earthquake loads, and treats them as if they were additional static loads. This approach simplifies the analysis, and makes it possible to use the static design methods described in Chapters 11 - 17 to evaluate seismic loads. We call it a pseudostatic method because it uses static analysis methods to assess dynamic loads.

It is common to base these analyses on soil strengths one-third higher than the static strengths, even though the real seismic strengths are lower. Thus, seismic analyses have an implicitly lower factor of safety, as discussed in Section 8.5 in Chapter 8. This practice is generally acceptable in most soils, especially those with load-displacement behavior as shown in Figure 13.4b in Chapter 13. However, it may cause problems when any of the following circumstances are present:

- The soil is a sensitive clay. These soils lose a greater portion of their strength during cyclic loading.
- The soil is a saturated, loose to medium dense sand. Even if these soils do not liquefy, they can develop large excess pore water pressures during earthquakes and thus loose much of their shear strength.
- The foundation has a plunging load-displacement curve, as shown in Figure 13.4a in Chapter 13. This behavior, which occurs in soft clays, is more likely to produce excessive settlements if the foundation is overloaded.
- The foundation must sustain uplift loads. Uplift failures are potentially more catastrophic, especially when the foundation is loaded with cyclic up-and-down forces. This was illustrated by uplift failures of deep foundations in Mexico City during the 1985 earthquake.

In these situations, it probably is best to use design soil strengths equal to or less than the static strengths. Be especially cautious when combinations of these problems exist together (i.e., uplift loading in sensitive clays).

Another way to evaluate seismic loads is to conduct a dynamic analysis[2]. This method uses a numerical model of the soil-foundation system similar to that used for wave equation analyses (Figure 13.28 in Chapter 13). In theory, dynamic analyses provide better representations of the true response to seismic loads. However, they rarely are used in practice (Hadjian et al., 1992) and generally are reserved for research projects or very critical structures, such as nuclear power plants.

Most engineers believe dynamic foundation analyses are not necessary for the vast

---

[2] These analyses are intended to model the dynamic response of the foundation. Dynamic analyses intended to develop response spectra for use in the design of the superstructure are an entirely different matter.

majority of structures for the following reasons:

- Although specialized software is available, dynamic analyses are more difficult and expensive to perform.
- The input parameters, especially those that describe the soil, are difficult and expensive to define.
- Except for the soil liquefaction and batter pile problems, deep foundations have performed well during earthquakes. Both of these problems can be solved without complex dynamic analyses. Thus, there is no demonstrated need to invest in more elaborate analysis methods.
- Engineers are not likely to use lighter foundations, even if justified by a dynamic analysis.

Although dynamic analyses may become more popular in the future, they currently appear to be well beyond the level of sophistication needed for routine foundation designs.

Seismic Structural Loading Due to Resonance

Earthquake ground motions have certain characteristic frequencies, and these characteristics change as the waves travel through different rock and soil masses. Thus, the motions that reach a structure depend on the subsurface conditions. This phenomenon was especially evident during the 1985 Mexico and 1989 Loma Prieta earthquakes, when structures founded on soft soils were much more heavily damaged than those on hard soils or bedrock.

Sometimes, the soil modifies the ground motions such that the predominant periods are the same as the natural periods of certain structures, thus producing resonance. For example, ground motions in certain parts of Mexico City caused resonance in 9 to 16-story buildings during the 1985 earthquake.

When resonance occurs, each cycle of earthquake shaking imparts additional energy into the structure, causing greater motions and larger loads on their foundations. These loads may be significantly larger than those from heavier, non-resonating structures. Special structural analysis methods are available to estimate the magnitude of these loads.

Girault (1987) suggested that controlled movements of foundations subjected to seismic loads are beneficial, especially beneath resonating structures, because they help absorb much of the energy imparted by the earthquake. Therefore, a ductile foundation system helps protect the superstructure.

14.4 PILE-SUPPORTED MATS

Mat foundations, discussed in Chapter 9, are essentially large spread footings that encompass the entire footprint of the structure. Because of their large width, mat foundations often have a sufficient factor of safety against a bearing capacity failure. However, they may be prone to excessive settlements. Often, the solution to this problem is to use a floating foundation, as described in Chapter 17. However, when this is not possible, a *pile-supported mat*, as shown in Figure 14.10, may be appropriate (Fleming et al., 1985).

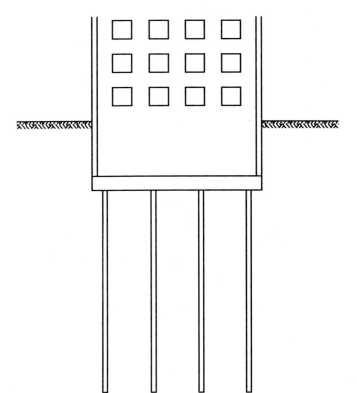

Figure 14.10   A pile-supported mat foundation.

We could design this type of foundation using only the piles to resist the structural loads and treating the mat simply as a very large pile cap. However, it may be possible to obtain a more economical design by also considering the soil resistance acting on the mat, as described by Poulos and Davis (1980).

SUMMARY OF MAJOR POINTS

1. Piles are almost always installed in groups. However, the capacity of a group is not necessarily equal to that of a single pile multiplied by the number of piles in the group. Account for this using group efficiency factors.

2. Unfortunately, our knowledge of group efficiency mechanisms is limited, so it is difficult to develop reliable design values of the group efficiency factor.

3. The settlement of pile foundations is usually tolerable, so engineers generally do not attempt to compute it. However, it may be appropriate to check the settlement if the structure is especially sensitive, if highly compressible soils are present, or if the structural engineer needs an equivalent "spring" to represent the foundation.

4. Settlement analyses may be performed using imaginary footings, empirical methods, or t-z analyses.

5. The seismic load capacity of deep foundations may be evaluated using either a pseudostatic analysis, which treats the seismic load as if it were an additional static load, or a dynamic analysis. The pseudostatic analyses are preferred for most practical design problems.

6. Pseudostatic analyses often use soil strength values that are larger than the static values. This practice produces an implicitly lower factor of safety.

7. Some structures combine mat and pile foundations to form a pile-supported mat.

QUESTIONS AND PRACTICE PROBLEMS

14.1 What is "block failure" in a group of piles?

14.2 What is a typical group efficiency factor for piles driven into loose cohesionless soils without predrilling or jetting? How does predrilling and jetting affect this factor? Why?

14.3 The group efficiency factor for piles in saturated cohesive soils is low soon after driving, and increases with time. Why does this occur?

14.4 Based on a pile load test, a certain pile has a net end bearing capacity of 200 k and a skin friction capacity of 500 k. A series of these piles are to be driven in a 3×4 group, which will have an estimated group efficiency factor of 1.15. Compute the net allowable downward load capacity of this pile group using a factor of safety of 3.0.

14.5 A 3.5 m square, 5×5 pile group supports a downward load of 12,000 kN. Each pile in the group is 300 mm square, 20 m long concrete ($f_c' = 40$ MPa). The load transfer consists of 75% skin friction and 25% end bearing. The soil profile at this site is as follows:

0 - 25 m: Stiff clay, $\gamma = 18.6$ kN/m^3, $\gamma_{sat} = 19.7$ kN/m^3, $C = 0.06$, $C=0.04$, $E_u =200$MPa
> 25 m: Dense sand and gravel

The groundwater table is at a depth of 10 m. Compute the settlement of this pile group.

Advanced Problem

14.6 A group of five closed-end steel pipe piles were driven into a sandy hydraulic fill at Hunter's Point in San Francisco, California (DiMillio, et al., 1987a). A single isolated pile also was driven nearby. Each pile had an outside diameter of 10.75 in and a length of 30 ft. The group piles were placed 3 to 4 ft on-center, and their pile cap was elevated above the ground surface.

The upper 4.5 ft of the soil was predrilled to a diameter larger than the piles, and the top of the completed piles extended 5 ft above the ground surface. Therefore, only 20.5 ft of each pile was in contact with the soil. No other predrilling or jetting was done.

An extensive subsurface investigation was conducted before these piles were installed. This included SPT, CPT, DMT and other tests. The CPT results are shown in the figure below.

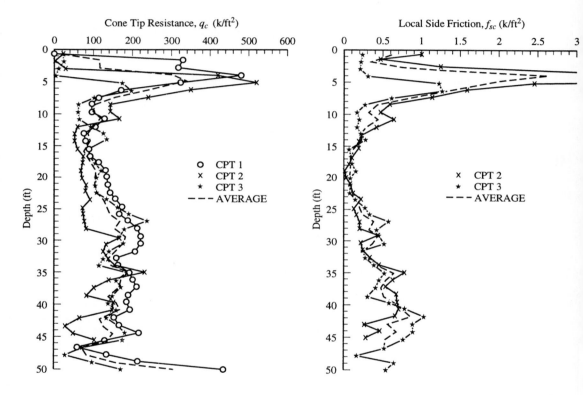

a. Using this CPT data, compute the ultimate downward load capacity of the single pile.

b. Based on a pile load test, the ultimate downward load capacity of the single pile was 80 k (based on Davisson's method). Other methods of reducing the load test data gave ultimate load capacities of 80 to 117 k. How accurate was your prediction?

c. Using this CPT data, compute the ultimate downward load capacity of the pile group.

d. Based on a group pile load test, the ultimate downward load capacity of the pile group was 432 to 573 k, depending on the method of reducing the load test data. How accurate was your prediction?

15

Drilled Shaft Foundations

There is no glory in the foundations.
Karl Terzaghi, 1951

Drilled shafts are another type of deep foundation. The fundamental difference between piles and drilled shafts is that piles are prefabricated members driven into the ground, whereas drilled shafts are cast-in-place.

The construction procedure in competent soils, known as the *dry method*, is generally as follows:

1. Using a *drill rig*, excavate a cylindrical hole (the *shaft*) into the ground to the required depth, as shown in Figure 15.1a.
2. Fill the lower portion of the shaft with concrete as shown in Figure 15.1b.
3. Place a prefabricated reinforcing steel cage inside the shaft as shown in Figure 15.1c.
4. Fill the shaft with concrete as shown in Figure 15.1d.

Alternative construction procedures for use in difficult soils are discussed later in this chapter.

Engineers and contractors also use other terms to describe to this type of deep foundation, including the following:

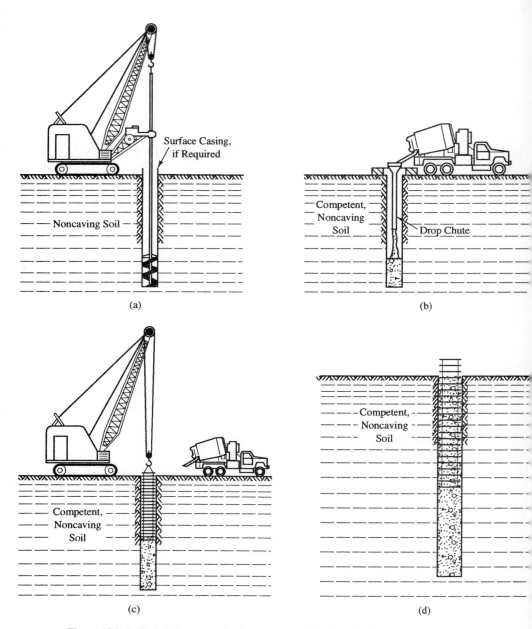

Figure 15.1 Drilled shaft construction in competent soils using the dry method: (a) Drilling the shaft; (b) Starting to place the concrete; (c) Placing the reinforcing steel cage; and (d) Finishing the concrete placement (Reese and O'Neill, 1988).

- *Pier*
- *Drilled pier*
- *Bored pile*
- *Cast-in-place pile*
- *Caisson*
- *Drilled caisson*

However, drilled shafts are not the same as certain other methods that also involve cast-in-place concrete, such as *auger-cast piles*, *pressure injected footings*, *step-taper piles*, and *grouted anchors*. They are covered in Chapter 16.

15.1 HISTORY

The quality of soils usually improves with depth, so it often is helpful to excavate through weak surface soils to support structures on deeper bearing materials. Even the ancient Greeks understood the value of removing poor quality soils (Kerisel, 1987).

During the late nineteenth and early twentieth centuries, builders began to modify the traditional techniques of reaching good bearing soils. Many of the greatest advances occurred in the large cities of the Great Lakes region. As taller and heavier buildings began to appear, engineers realized that traditional spread footing foundations would no longer suffice. Following many years of problems with excessive settlements, they began to use foundation systems consisting of a single hand-dug shaft below each column.

General William Sooy-Smith was one of the pioneers in this new technology. His *Chicago Well Method*, developed in 1892, called for the contractor to hand-excavate a cylindrical hole about 3.5 ft (1 m) in diameter and 2 to 6 ft (0.5 to 2 m) deep, as shown in Figure 15.2. To prevent caving, they lined its wall with vertical wooden boards held in place with steel rings, and then repeated the process until reaching the bearing stratum. Finally, they enlarged the base of the excavation to form a bell and filled it with concrete up to the ground surface.

Hand excavation methods were slow and tedious, so the advent of machine-dug shafts was a natural improvement. The early equipment was similar to that used to drill oil wells, so much of the early development occurred in oil-producing areas, especially Texas and California. A few examples of horse and engine-driven drills appeared between 1900 and 1930, but they had very limited capabilities. By the late 1920s, manufacturers were building practical truck-mounted engine-driven drill rigs, such as the one in Figure 15.3, thus bringing drilled shaft construction into its maturity.

During the next 35 years, manufacturers and contractors developed larger and more powerful equipment along with a variety of cutting tools. They also borrowed construction techniques from the oil industry, such as casing and drilling mud, to deal with difficult soils. By the 1960s, drilled shafts had become a strong competitor to driven piles.

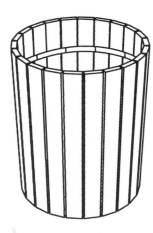

Figure 15.2 Early shaft construction using the Chicago Well Method. This is one set of wooden shores; the excavation would continue with additional sets until reaching the desired depth.

Figure 15.3 Early truck mounted drill rig, circa 1925 (photo courtesy of the Association of Drilled Shaft Contractors).

Today, drilled shafts support structures ranging from one story wood frame buildings to the largest skyscrapers. For example, the Sears Tower in Chicago is supported on 203 drilled shafts, some of them 100 ft (30 m) deep.

The advantages of drilled shaft foundations include the following:

- The costs of mobilizing and demobilizing a drill rig are much less than those for a pile driver. This is especially important on small projects, where they represent a larger portion of the total costs.

- The construction process generates less noise and vibration, both of which are especially important when working near existing buildings.
- Engineers can observe and classify the soils excavated during drilling and compare them with the anticipated soil conditions.
- Contractors can easily change the diameter or length of the shaft during construction to compensate for unanticipated soil conditions.
- The foundation can penetrate through soils with cobbles or boulders, especially when the shaft diameter is large. It is also possible to penetrate many types of bedrock.
- It is usually possible to support each column with one large shaft instead of several piles, thus eliminating the need for a pile cap.

The disadvantages include the following:

- Successful construction is very dependent on the contractor's skills, much more so than with spread footings or even driven piles. Poor workmanship can produce weak foundations that may not be able to support the design load. Unfortunately, most of these defects are not visible. This is especially important because a single drilled shaft does not have the benefit of redundancy that is present in a group of driven piles.
- Driving piles pushes the soil aside, thus increasing the lateral stresses in the soil and generating more skin friction capacity. However, shaft construction removes soil from the ground, so the lateral stresses remain constant or decrease. Thus, a shaft may have less skin friction capacity than a pile of comparable dimensions. However, this effect is at least partially offset by rougher contact surface between the concrete and the soil and the correspondingly higher coefficient of friction.
- Pile driving densifies the soil beneath the tip, whereas shaft construction does not. Therefore, the unit end bearing capacity in shafts may be lower.
- Full-scale load tests are very expensive, so the only practical way to predict the axial load capacity is to use semiempirical methods based on soil properties. We typically have no independent check. However, the Osterberg load test device and high-strain dynamic impact tests, discussed later in this chapter, may overcome this problem.

15.2 MODERN CONSTRUCTION TECHNIQUES

Contractors use different equipment and techniques depending on the requirements of each project (Greer and Gardner, 1986). The design engineer must be familiar with these methods to know when and where drilled shafts are appropriate. The construction method also influences the shaft's load capacity, so the engineer and contractor must cooperate to assure compatibility between the design and construction methods.

Drilling Rigs

A typical modern truck-mounted drilling rig, such as the one shown in Figure 15.4, is able to drill shafts up to 72 inches (1800 mm) in diameter and 80 ft (24 m) deep. These rigs could complete most projects.[1] Specialized rigs, such as those in Figures 15.5 and 15.6, are available for difficult or unusual projects.

Figure 15.4   Typical drilling rig for constructing drilled shafts (Photo courtesy of the   Association   of   Drilled   Shaft Contractors).

Drilling Tools

Contractors have different drilling tools, each suited to a particular subsurface condition or drilling technique. The *flight auger*, shown in Figure 15.7, is most common. These helix-shaped augers are typically 18 to 36 in (500 to 1000 mm) in diameter, but some are as large as 12 ft (3600 mm) in diameter.

---

[1] Typical drilled shaft foundations are 18 to 36 in (500 - 1000 mm) in diameter and 20 to 60 ft (6 - 18 m) deep, but dimensions far outside this range are not uncommon.

Figure 15.6 Extremely large rig capable of drilling 26 ft (8 m) diameter holes 200 ft (60 m) deep. (Photo courtesy of Anderson Drilling Co., Lakeside, CA).

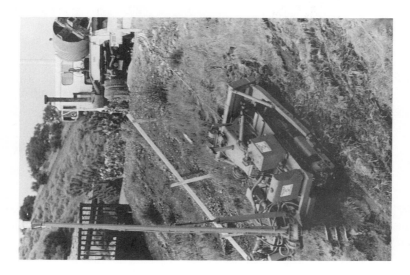

Figure 15.5 Small track-mounted drilling rig capable of working on a hillside. (Photo courtesy of the Association of Drilled Shaft Contractors).

Figure 15.7 Typical flight auger (Photo courtesy of the Association of Drilled Shaft Contractors).

The drill rig rotates the auger into the ground until it fills with soil. Then, it draws the auger out and spins it around to remove the cuttings, as shown in Figure 15.8. This process repeats until the shaft reaches the desired depth.

Figure 15.8 Spinning the auger to remove the cuttings (Photo courtesy of the Association of Drilled Shaft Contractors).

Conventional flight augers are effective in most soils and soft rocks. However, when encountering difficult conditions, the contractor has the option of switching to special augers or other tools. For example, augers with hardened teeth and pilot *stingers* are effective in hardpan or moderately hard rock. Spiral-shaped rooting tools help loosen cobbles and boulders, thus allowing the hole to advance under conditions that might cause refusal in a driven pile. Some of these special tools are shown in Figure 15.9.

Figure 15.9 Special flight augers for difficult subsurface conditions (Photo courtesy of the Association of Drilled Shaft Contractors).

Other drilling tools include the following:

- *Bucket augers* that collect the cuttings in a cylindrical bucket that is raised out of the hole and emptied. They are especially useful in running sands.
- *Belling buckets* that have extendable arms to enlarge the bottom of the shaft. These enlargements are called *bells* or *underreams*.
- *Core barrels* that cut a circular slot, creating a removable cylindrical core. They are especially useful in hard rock.
- *Multiroller bits* to cut through hard rock.
- *Cleanout buckets* to remove the final cuttings from a hole and produce a clean bottom suitable for end bearing.

Some of these tools are shown in Figures 15.10 and 15.11.

Figure 15.10 Various drilling tools (Photo courtesy of the Association of Drilled Shaft Contractors).

Drilling Techniques in Firm Soils

In firm soils, contractors use the *dry method* (also known as the *open-hole method*) to build the shaft, as shown in Figure 15.1. These holes usually advance quickly using conventional flight augers and remain open without any special support. After checking the open hole for cleanliness and alignment, it is a simple matter to insert the steel reinforcing cage and dump concrete in from the top. Some contractors use a tremie or a concrete pump to deliver the concrete. Open-hole shafts in firm soils are very common because of their simplicity and economy of construction and their good reliability.

It also is possible to excavate stiff soils below the groundwater table using the open-hole method. Usually, the contractor simply pumps the water out as the hole advances and places the concrete in the dewatered shaft.

Figure 15.11 A belling bucket used to produce a bell at the bottom of a shaft (Photo courtesy of the Association of Drilled Shaft Contractors).

Drilling Techniques in Caving or Squeezing Soils

A hole is said to be *caving* when the sides collapse before or during concrete placement. This is especially likely in clean sands below the groundwater table. *Squeezing* refers to the sides of the hole bulging inward, either during or after drilling, and is most likely in soft clays and silts or highly organic soils. Either of these conditions could produce *necking* in the shaft (a local reduction in its diameter) or soil inclusions in the concrete, as shown in Figure 15.12, both of which could have disastrous consequences.

The two most common construction techniques for preventing these problems are the use of *casing* or the use of *drilling fluid.*

The casing method, shown in Figure 15.13, uses the following procedure:

1. Drill the hole using conventional methods until encountering the caving strata.
2. Insert a steel pipe (the casing) into the hole and advance it past the caving strata as shown in Figure 15.13a. Contractors do this using vibratory hammers such as the one in Figure 12.15. The diameter of this casing is usually 2 to 6 in (50 - 150 mm) less than the diameter of the upper part of the shaft.
3. Drill through the casing and into the non-caving soils below using a smaller diameter auger as shown in Figure 15.13b.
4. Place the reinforcing steel cage and the concrete through the casing and extract the casing as shown in Figure 15.13c. This is a very critical step, because premature extraction of the casing can produce soil inclusions in the shaft.

There are many variations to this method, including the option of leaving the casing in place and combining the casing and slurry methods.

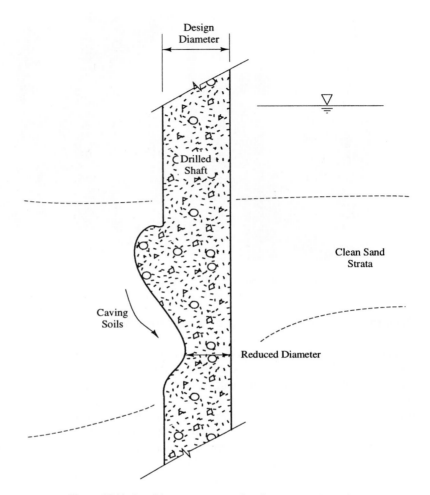

Figure 15.12 Possible consequences of caving or squeezing soils.

The drilling fluid method (also known as the *slurry method*) is shown in Figure 15.14. It uses the following procedure:

1. Drill a *starter hole*, perhaps 10 ft (3 m) deep.
2. Fill the starter hole with a mixture of water and bentonite clay to form a *drilling mud* or *slurry*. This material will have a consistency comparable to pea soup and keeps the hole open because of the hydrostatic pressure it applies to the soil.
3. Advance the hole by passing the drilling tools through the slurry as shown in Figure 15.14a. Continue to add water and bentonite as necessary.

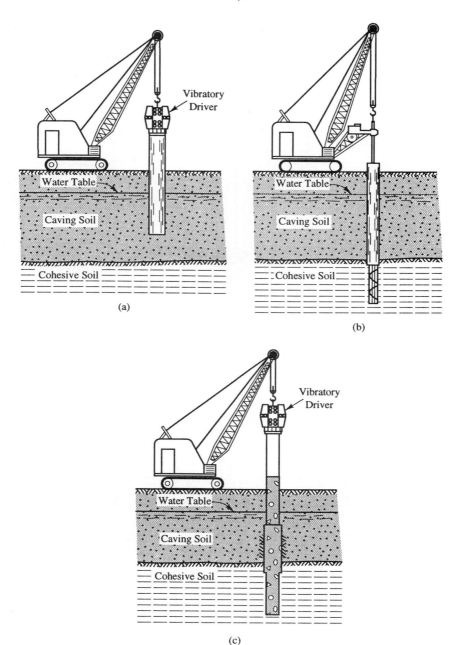

Figure 15.13 Using casing to deal with caving or squeezing soils: (a) Installing the casing; (b) Drilling through and ahead of the casing; and (c) Placing the reinforcing steel and concrete, and removing the casing (Reese and O'Neill, 1988).

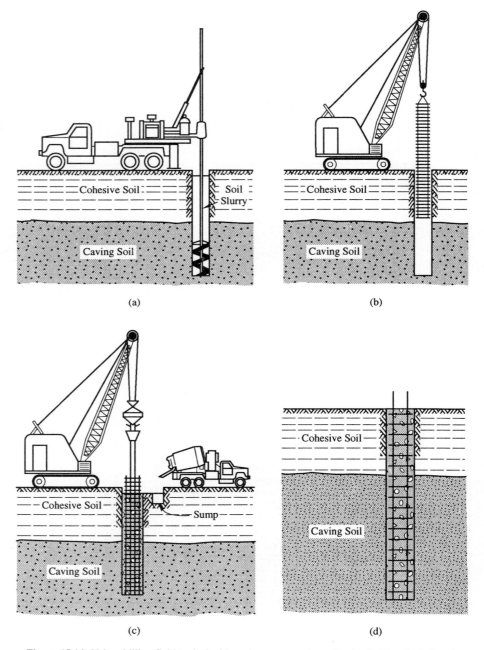

Figure 15.14 Using drilling fluid to deal with caving or squeezing soils: (a) Drilling the hole using slurry; (b) Installing the reinforcing steel cage through the slurry; (c) Placing the concrete using a tremie pipe and recovering slurry at the top; and (d) The completed foundation (Reese and O'Neill, 1988).

4. Insert the reinforcing steel cage directly into the slurry as shown in Figure 15.14b.

5. Fill the hole with concrete using a *tremie pipe* that extends to the bottom as shown in Figure 15.14c. The concrete pushes the slurry to the ground surface, where it is captured.

This method is quite effective, but slow and messy.

Do not be concerned about the quality of the bond between the rebar and the concrete. Although the rebar is first immersed in slurry, research has shown that the bond is satisfactory. However, the slurry can form a cake on and in the surrounding soil, thus reducing the skin friction resistance.

Underreamed Shafts

An *underreamed shaft* (also known as a *belled shaft*) is one with an enlarged base, as shown in Figure 15.15. Usually, the ratio of the underream diameter to the shaft diameter (B_b/B_s) should be no greater than 3. Contractors build underreams using special belling buckets, such as the one shown in Figure 15.11.

The larger base area of underreamed shafts increases their end bearing capacity, and thus they are especially useful for shafts bottomed on strong soils or rock. However, the displacement required to mobilize the full end bearing is typically on the order of 10% of the base diameter, which may be more than the structure can tolerate.

Underreamed shafts also have greater uplift capacities due to bearing between the ceiling of the underream and the soil above. Chapter 11 discusses methods of computing the uplift capacity.

Unfortunately, the construction of underreamed shafts can be hazardous to the workmen. The bottom of the underream must be cleaned of loose soil before placing concrete, and this task is typically done by hand. The extra equipment, such as casing and protective cages, required to protect workers that descend into the shaft is very expensive. Thus, underreamed shafts are not as common as they once were.

Concrete

Concrete for drilled shafts must have sufficient slump to flow properly and completely fill the hole. Using concrete that is too stiff creates voids that weaken the shaft. Typically, the slump should be between 4 and 9 in (100 - 220 mm), with the lower end of that range being most appropriate for large-diameter dry holes with widely spaced reinforcement and the high end for concrete placed under drilling fluid. Sometimes it is appropriate to include concrete admixtures to obtain a high slump while retaining sufficient strength.

Some people have experimented with expansive cements in drilled shaft concrete. These cements cause the concrete to expand slightly when it cures, thus increasing the lateral earth pressure and skin friction resistance. So far, this has been only a research topic, but it may become an important part of future construction practice.

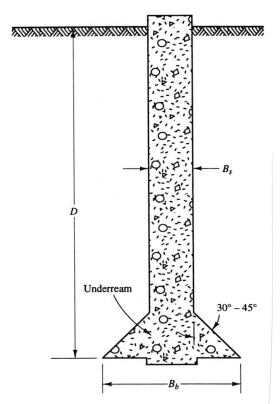

Figure 15.15 An underreamed drilled shaft.

15.3 DESIGNING TO RESIST AXIAL LOADS

The primary load on a drilled shaft is usually axial, either downward or uplift. Therefore, it is necessary to select a diameter and length that produce sufficient skin friction and end bearing capacities.

Full Scale Load Tests

The most reliable way to determine the ultimate load capacity is to conduct a full-scale load test, as discussed in Chapter 13. However, because of their larger dimensions, drilled shafts have much greater capacities than most piles, so the test frame, jack, and other test equipment must be correspondingly larger, and the costs can become prohibitive. Therefore, engineers rarely perform conventional full-scale load tests on drilled shafts. They are economically justified only on very large projects or for research purposes when the benefits extend beyond a single project (see the side bar "Full-Scale Instrumented Load Tests").

Fortunately, Osterberg (1984) in the United States and DaSilva in Brazil have developed a method that may overcome the high cost of conducting load tests. It eliminates the expensive test frame and substitutes a hydraulic pancake jack at the bottom of the shaft, as shown in Figure 15.16. Once the concrete is in place, the operator pumps hydraulic fluid into the jack and keeps track of both pressure and volume. The jack expands and pushes up on the shaft (Osterberg, 1984). A dial gage measures this movement. Thus, we obtain a plot of skin friction capacity vs. axial movement.

This device also includes a tell-tale rod that extends from the bottom of the pancake jack to the ground surface. It measures the downward movement at the bottom, and thus produces a plot of end bearing capacity vs. axial movement.

These two plots continue until the shaft fails in skin friction or end bearing. If an end bearing failure occurs first, we must extrapolate the data to obtain the ultimate skin friction. However, if a skin friction failure occurs first, it may be possible to add a nominal static load to the top, and then continue the test to find the end bearing capacity.

This method measures the skin friction in uplift, so it may be necessary to adjust the measured values using Equation 11.6 in Chapter 11 to estimate the downward capacity.

After completing the test, it is possible to fill the jack with grout and use the shaft to carry structural loads.

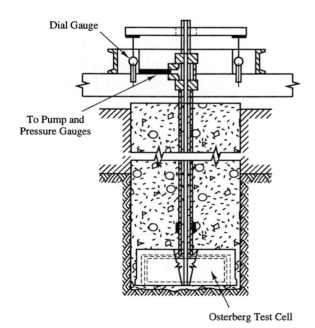

Osterberg Test Cell

Figure 15.16 Osterberg load test device (Courtesy of Loadtest, Inc.).

Full-Scale Instrumented Load Tests

The behavior of drilled shafts under load is very complex, and there is much we do not understand about the underlying mechanisms. Therefore, we can not develop reliable design methods based exclusively on theoretical soil mechanics. It is necessary to conduct load tests to study the behavior of shafts and calibrate the analysis methods.

When conducting these tests, we are interested in more than simply the load vs. settlement curve at the top of the shaft (although this is certainly the most fundamental data). We would also like to know how this load is transferred into the soil. How much skin friction develops and how much end bearing? How is the skin friction distributed along the shaft?

We can find the answers to these questions by *instrumenting* the shaft. This consists of installing load cells or strain gages at various depths before placing the concrete. These are connected by wires to a readout unit at the ground surface and allow the engineer to monitor the loads at various points in the shaft as the test progresses. A typical instrumented shaft is shown in Figure 15.17.

For example, if the stress measured by load cell 1 is the same as that measured by load cell 5, then we conclude that all of the applied load is resisted by end bearing. Conversely, if the measured load gradually decreases with depth and is zero at load cell 5, then all of the resistance is skin friction. In reality, it would dissipate with depth in some non-uniform fashion that would reflect the actual transfer of load from the shaft to the soil. By using this information, we can compute the unit skin friction and end bearing resistances and use this information to develop an analysis method.

These tests also have confirmed that service loads are carried primarily by skin friction, as discussed in Chapter 11.

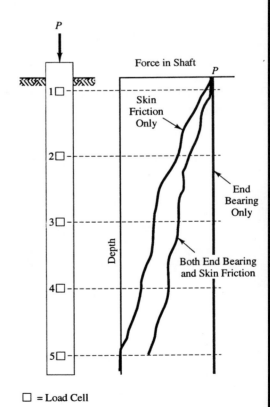

□ = Load Cell

Figure 15.17 Typical instrumented drilled shaft.

High Strain Dynamic Impact Tests

Some engineers have attempted to determine the static load capacity of drilled shaft foundations by dropping a large weight onto the top of the completed shafts, measuring the induced waves using a pile driving analyzer, and conducting a CAPWAP analysis similar to that described in Chapter 13 (Seitz and Rausche, 1987). This method seems to be usable only if it induces large strains in the concrete. Therefore, it requires large weights, perhaps on the order of 40 k (20 Mg), and they must be dropped from heights of 12 ft (4 m) or more.

Another similar method, the *statnamic test*, is being developed in Canada. It consists of placing a static weight of approximately 10 k (5 Mg) on top of the shaft with a slow-burning explosive material between the weight and the shaft (Briaud, 1991). Detonation of the explosives generates a compression wave in the shaft. Instruments measure these waves, and then the engineer uses a CAPWAP analysis to determine the static capacity.

Both of these methods appear promising, but we need additional research before they become routine design tools.

Presumptive Bearing Pressures

Small drilled shafts that support lightweight structures may not justify extensive analysis and design effort. Therefore, conservative presumptive bearing pressures may be appropriate. The Uniform Building Code (ICBO, 1991a) allows a presumptive end bearing pressure, q_e', equal to the presumptive values for spread footings (see Table 6.1 in Chapter 6), or a unit skin friction resistance, f_s, of $q_e'/6$ (to a maximum of 500 lb/ft^2 (25 kPa)). However, ICBO does not allow shafts to be designed for both skin friction and end bearing using only presumptive bearing pressures.

Designs based on presumptive bearing pressures may be very conservative for larger structures, and they may be unreliable on soft soils. Either situation justifies a site-specific study and an analysis based on soil properties at that site.

Analyses Based on Soil Properties

The most common methods of designing drilled shafts are based on soil properties obtained from field or laboratory tests. Usually, this is the only method available to the engineer.

These analyses are similar to those for piles, as described in Chapter 13, and are subject to the same kinds of uncertainties. We compute the unit end bearing and skin friction resistances, q_e' and f_s, and insert them into Equations 11.4 - 11.6 in Chapter 11. However, the methods of computing these factors are different than those for piles because the construction methods are different. Section 15.4 describes methods developed specifically for drilled shafts.

15.4 USING SOIL PROPERTIES TO COMPUTE AXIAL LOAD CAPACITIES

Engineers have developed theoretical, semi-empirical, and empirical methods to compute q_e' and f_s from laboratory or in-situ test data. Most of these methods have been calibrated using full-scale load tests.

When computing the contact areas, A_e and A_s, remember the as-built shaft diameter may be larger than the design diameter. This difference depends on the construction methods and on the soil and groundwater conditions. Although most engineers choose to conservatively ignore this effect, some use an estimate of the as-built diameter, as indicated in Table 15.1, to compute the load capacity.

TABLE 15.1 TYPICAL RATIOS OF AS-BUILT TO DESIGN DIAMETERS FOR DRILLED SHAFTS

Soil or Rock Type	Construction Method	Typical Ratio of As-Built Diameter/Design Diameter
Soft clay	Casing	1.10
Stiff clay	Casing	1.10
Soft clay	Open hole	1.00 - 1.05
Stiff clay	Open hole	1.00 - 1.05
Sand	Slurry	1.00 - 1.15
Sand	Vibrated casing	1.00
Stiff soil with cobbles	Open hole	1.10 - 1.15
Loess	Open hole	1.00 - 1.05
Flowing mud	Slurry	Overpour as much as 100%
Residual soil	Open hole	Overpour as much as 50%
Competent rock	Core barrel	1.00
Cavernous rock	Core barrel	Overpour as much as 100%

Adapted from Stewart and Kulhawy, 1981.

Overpour is the volume of concrete placed beyond the as-designed volume. It is expressed as a percentage of the design volume.

Cohesionless Soils

End Bearing

The unit end bearing capacity for drilled shafts in cohesionless soils will be less than that for piles because of the following:

- The soil is disturbed by the augering process.
- The soil compression that occurs below piles is not present.
- There is a temporary stress relief while the hole is open.
- The diameter and depth of influence are greater.

These and other factors have not been well defined, so the simple and conservative empirical formula suggested by Reese and O'Neill (1989) seems to represent an appropriate level of sophistication:

$$q_e' = 0.60 \; \sigma_r \; N_{60} \leq 90{,}000 \; lb/ft^2 \; (4500 \; kPa) \qquad (15.1)$$

Where:

σ_r = reference stress = 2000 lb/ft^2 = 100 kPa

N_{60} = mean SPT N_{60} value for the soil between the base of the shaft and a depth equal to two times the base diameter below the base. Do not apply an overburden correction.

q_e' = net unit end bearing resistance

If the base of the shaft is more than 50 in (1200 mm) in diameter, the value of q_e' from Equation 15.1 could result in settlements greater than 1 in (25 mm), which would be unacceptable for most buildings. To keep settlements within tolerable limits, reduce the value of q_e' to q_{er}' as follows and use q_{er}' in Equation 11.4 or 11.5 of Chapter 11:

$$q_{er}' = 4.17 \; \frac{B_r}{B_b} \; q_e' \qquad B_b \geq 50 \; in \; (1200 \; mm) \qquad (15.2)$$

Where:

B_r = reference width = 1.0 ft = 0.3 m = 12 in = 300 mm

B_b = base diameter of drilled shaft

Skin Friction

The principles governing skin friction resistance are similar to those for piles, as described in Chapter 13, with the following important differences:

- The process of augering the shaft will leave a rough soil surface, so the ϕ_s/ϕ ratio will be higher than for piles.
- The construction process relieves lateral stresses in the ground, whereas pile driving increases these stresses. Therefore, the K/K_0 ratio for shafts will be lower than for piles.

• If drilling mud is used, some of it may infiltrate into the soil and form a *slurry cake*, thus reducing its coefficient of friction. This possibility depends on the quality of workmanship, soil type, and other factors.

Use the β method to compute f_s in cohesionless soils:

$$f_s = \beta \, \sigma_v' \tag{15.3}$$

$$\beta = K \, \tan\phi_s \tag{15.4}$$

Where:

    f_s = unit skin friction resistance

    σ_v' = vertical effective stress at midpoint of soil layer

    K = coefficient of lateral earth pressure

    ϕ_s = soil-shaft interface friction angle

We can use Tables 15.2 and 15.3 to compute ϕ_s and K from the soil friction angle, ϕ, and the preconstruction coefficient of lateral earth pressure, K_0. However, K_0 is difficult to determine. Another alternative is to compute β directly using Reese and O'Neill's (1989) empirical function:

$$\beta = 1.5 - 0.135 \sqrt{\frac{z}{B_r}} \qquad 0.25 \leq \beta \leq 1.20 \tag{15.5}$$

Where:

    σ_v' = vertical effective stress at the midpoint of strata

    B_r = reference width = 1.0 ft = 0.3 m = 12 in = 300 mm

    z = depth from the ground surface to the midpoint of the strata

TABLE 15.2 TYPICAL ϕ_s/ϕ VALUES FOR DESIGN OF DRILLED SHAFTS

Construction Method	ϕ_s/ϕ
Open hole or temporary casing	1.0
Slurry method — minimal slurry cake	1.0
Slurry method — heavy slurry cake	0.8
Permanent casing	0.7

Adapted from Kulhawy, 1991.

TABLE 15.3 TYPICAL K/K_0 VALUES FOR DESIGN OF DRILLED SHAFTS

Construction Method	K/K_0
Dry construction with minimal sidewall disturbance and prompt concreting	1
Slurry construction - good workmanship	1
Slurry construction - poor workmanship	2/3
Casing under water	5/6

Adapted from Kulhawy, 1991, and Stas and Kulhawy, 1984.

Equation 15.5 is based on the results of 41 full-scale load tests. The decrease in β with depth corresponds to the observed behavior of deep foundations in the field and seems to reflect the greater overconsolidation (and therefore higher K) in the near-surface soils.

Example 15.1

The drilled shaft shown in the figure below will be built using a temporary casing. This shaft will support an office building and be designed and built with "normal" controls. Compute its allowable downward load capacity.

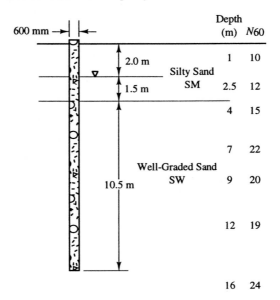

Solution

The unit weights of these soils, γ, have not been given (probably because it was not possible to obtain a suitably undisturbed sample of these sandy soils). We can't compute the load capacity without this information, so we must estimate γ for each strata using typical values from Table 3.3 in Chapter 3:

Silty sand above groundwater table: $\gamma \approx 17 \text{ kN/m}^3$
Silty sand below groundwater table: $\gamma \approx 20 \text{ kN/m}^3$
Sand below groundwater table: $\gamma \approx 20 \text{ kN/m}^3$

Based on Table 15.1, the as-built diameter will be about equal to the design diameter, so use $B = 0.6$ m for design.

Using Equation 15.5 to compute skin friction:

Strata	z (m)	β	$\sigma_v{'}$ (kPa)	f_s (kPa)	A_s (m²)	P_s (kN)
Silty sand above GWT	1.00	1.20	17.0	20.4	3.77	77
Silty sand below GWT	2.75	1.09	41.5	45.2	2.83	128
Sand (3.5 - 9 m)	6.25	0.88	77.3	68.0	10.37	705
Sand (9 - 13.3 m)	11.15	0.68	126.8	86.2	8.10	699
						1609

Note how we compute z and $\sigma_v{'}$ at the midpoint of each strata.

Use Equation 15.1 to compute the end bearing. Although no N values are located within a depth of $2B$ below the bottom of the shaft, it appears that $N_{60} = 22$ would be a reasonable value for design.

$$q_e{'} = 0.6\,\sigma_r N_{60} = (0.6)(100)(22) = 1320 \text{ kPa}$$

$$A_e = \frac{\pi(0.6)^2}{4} = 0.283 \text{ m}^2$$

Using a factor of safety of 2.5 (per Table 11.1) and Equation 11.4 in Chapter 11, the allowable downward capacity is:

$$P_a{'} = \frac{q_e{'} A_e + \sum f_s A_s}{F} = \frac{(1320)(0.283) + 1609}{2.5} = 793 \text{ kN} \qquad \Leftarrow \textbf{\textit{Answer}}$$

Cohesive Soils

End Bearing

Reese and O'Neill (1989) recommend the following function for end bearing in cohesive soils:

$$q_e' = N_c^* s_u \leq 80{,}000 \text{ lb/ft}^2 \ (4000 \text{ kPa}) \tag{15.6}$$

$$N_c^* = 6 \left[1 + 0.2 \left(D/B_b \right) \right] \leq 9 \tag{15.7}$$

Where:

N_c^* = bearing capacity factor

D = depth to the bottom of the shaft

B_b = diameter of shaft base

s_u = undrained shear strength in the soil between the base of the shaft and a distance $2B_b$ below the base

If the base diameter is more than 75 in (1900 mm), the value of q_e' from Equation 15.10 could produce settlements greater than 1 in (25 mm), which would be unacceptable for most buildings. To keep settlements within tolerable limits, reduce the value of q_e' to q_{er}', and use this value in Equation 11.4 or 11.5:

$$q_{er}' = F_r q_e' \tag{15.8}$$

$$F_r = \frac{2.5}{120 \, \psi_1 \, B_b/B_r + \psi_2} \leq 1.0 \tag{15.9}$$

$$\psi_1 = 0.0071 + 0.0021 \left(D/B_b \right) \leq 0.015 \tag{15.10}$$

$$\psi_2 = 1.59 \sqrt{\frac{s_u}{\sigma_r}} \qquad 0.5 \leq \psi_2 \leq 1.5 \tag{15.11}$$

Skin Friction

Evaluate the skin friction resistance in cohesive soils using either the α method or the β method.

The α Method

Most engineers use the α method to compute skin friction resistance in cohesive soils:

$$f_s = \alpha s_u \qquad (15.12)$$

Where:

α = adhesion factor

s_u = undrained shear strength of soil along the shaft

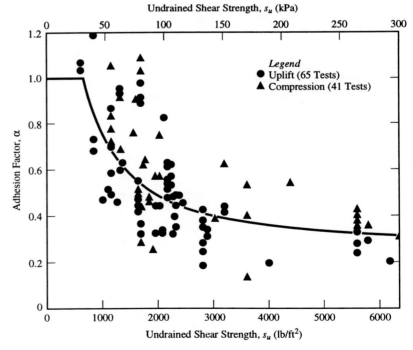

Figure 15.18 α function for drilled shafts (Adapted from Kulhawy and Jackson, 1989; Used with permission of ASCE).

Based on their analysis of load tests, Reese and O'Neill (1989) recommend using $\alpha = 0.55$ for soils with an undrained shear strength, s_u, of no more than 4000 lb/ft^2 (190 kPa). They also ignore the skin friction resistance in the upper 5 ft (1.5 m) of the shaft and along the bottom one diameter of straight shafts[2] because of interaction with the end bearing. For belled shafts, they recommend ignoring the skin friction along the surface of the bell and along the shaft for a distance of one shaft diameter above the top of the bell.

Kulhawy and Jackson (1989) gave an α function that varies with s_u, as shown in Figure 15.18. This function is based on the results of 106 load tests. Both α functions give smaller values that than those for driven piles (compare Figures 15.18 and 13.15).

Example 15.2

A 60 ft long drilled shaft is to be built using the open hole method in the soil shown in the figure to the right. It will be 24 inches in diameter and the bottom will be belled to a diameter of 60 inches. The structure is classified as "permanent" and the degree of control is "normal." Compute the allowable downward and uplift capacities.

Solution:

In stiff clays, the ratio of the as-built to design diameters is close to 1 (per Table 15.1). Therefore, use $B = 24$ in for design.

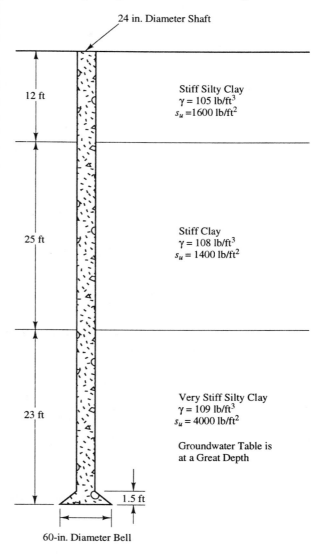

24 in. Diameter Shaft

12 ft

Stiff Silty Clay
$\gamma = 105$ lb/ft^3
$s_u = 1600$ lb/ft^2

25 ft

Stiff Clay
$\gamma = 108$ lb/ft^3
$s_u = 1400$ lb/ft^2

23 ft

Very Stiff Silty Clay
$\gamma = 109$ lb/ft^3
$s_u = 4000$ lb/ft^2

Groundwater Table is
at a Great Depth

1.5 ft

60-in. Diameter Bell

---

[2] Sometimes we may choose to ignore the end bearing resistance. In those cases, we can consider the skin friction resistance along the bottom one diameter.

Downward Capacity:

End bearing (using Equations 15.6 and 15.7):

$$D/B_b = 60/5 = 12$$

$$N_c^* = 6[1 + 0.2 D/B_b] = 6[1 + 0.2(12)] = 20 \qquad \therefore \text{Use } N_c^* = 9$$

$$q_e' = N_c^* s_u = 9(4000) = 36{,}000 \text{ lb/ft}^2$$

$$A_e = \frac{\pi B_b^2}{4} = \frac{\pi (5)^2}{4} = 19.6 \text{ ft}^2$$

$$P_e' = q_e' A_e = \frac{36000 (19.6)}{1000 \text{ lb/k}} = 706 \text{ k}$$

Skin friction (using Kulhawy and Jackson's method - Figure 15.18):

Layer No.	Depth (ft)	Thick-ness (ft)	s_u (lb/ft^2)	α	f_s (lb/ft^2)	A_s (ft^2)	P_s (k)
1	0 - 5	5.0	-	-	0	-	0
2	5 - 12	7.0	1600	0.53	848	44	37
3	12 - 37	25.0	1400	0.58	812	157	127
4	37 - 56.5	19.5	4000	0.34	1360	122	166
5	56.5 - 60	3.5	-	-	0	-	0
						$\Sigma =$	330

Use $F = 2.5$ (from Table 11.1) and Equation 11.3:

$$P_a = \frac{P_e' + P_s}{F} = \frac{706 + 330}{2.5} = \textbf{410 k} \qquad \Leftarrow \textit{Answer}$$

Uplift Capacity (using Equations 11.8 and 11.9):

$$D/B_b > 6 \qquad \therefore \; R = 1$$

Neglect skin friction below a point $2B_b$ above the bottom of the bell.

$$2 B_b = 2(5) = 10 \text{ ft}$$

Thus, layer 4 changes to: Depth = 37 - 50 ft; $A_s = 82$ ft²; $P_s = 112$ k

$$\sum P_s = 37 + 127 + 112 = 276 \text{ k}$$

$$N_u = 2\,(D/B_b - 0.5) = 2\,(60/5 - 0.5) = 23 \qquad \therefore \text{ use } N_u = 9$$

$$\sigma_D = \sum \gamma h = 105(12) + 108(25) + 109(23) = 6470 \text{ lb/ft}^2$$

$$P_{ub} = (s_u N_u + \sigma_D)(\pi/4)(B_b^2 - B_s^2)$$

$$= \frac{[(4000)(9) + 6470](\pi/4)(5^2 - 2^2)}{1000 \text{ lb/k}}$$

$$= 700 \text{ k}$$

$$W_f = \left[\frac{\pi(2)^2}{4}(58.5) + \frac{\pi[(2+5)/2]^2}{4}(1.5) \right](0.150 \text{ k/ft}^3) = 30 \text{ k}$$

(note no adjustment for buoyancy is necessary because the groundwater table is below the bottom of the shaft).

Use $F = (2.5)(1.5) = 3.7$ (per Chapter 11) and Equation 11.10:

$$P_{au} = 0.9 \, W_f + \frac{P_{ub} + R \sum f_s A_s}{F}$$

$$= (0.9)(30) + \frac{700 + 1(276)}{3.7}$$

$$= 291 \text{ k} \qquad \Leftarrow Answer$$

Comment: This fairly long foundation is required because of the moderate shear strength of the upper soils.

The β Method

As discussed in Chapter 13, the β method is theoretically superior to the α method because it is based on the effective stresses in the soil. However, it is difficult to properly evaluate β in cohesive soils, so this method is not used as widely.

As discussed earlier, β is usually larger near the ground surface because of the greater overconsolidation there, and gradually becomes smaller with depth, eventually reaching a constant value. However, we need additional instrumented load tests to develop reliable plots of β vs. depth. As an interim measure, Kulhawy and Jackson (1989) compiled a plot of the *average* β (known as $\bar{\beta}$) vs. depth of embedment, D, as shown in Figure 15.19. When using this data, determine only one $\bar{\beta}$ value for the entire shaft. This plot is based on load tests performed on 106 drilled shafts in stiff to hard cohesive soils. The undrained shear strength of most of these soils was greater than 1000 lb/ft^2 (50 kPa), so this plot may not be valid for weaker soils.

The average unit skin friction resistance is:

$$\bar{f}_s = \bar{\beta} \, \bar{\sigma}'_v \qquad (15.13)$$

Where:

$\bar{f}_s$ = average unit skin friction resistance

$\bar{\sigma}'_v$ = average vertical effective stress along length of shaft

Use this value in Equation 11.5 in Chapter 11.

Although the β method needs additional refinement, when used in stiff or hard soils it probably provides load predictions that are as good or better than those from the α method.

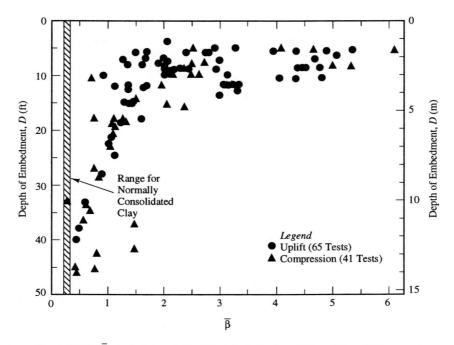

Figure 15.19 $\overline{\beta}$ vs. shaft length for drilled shafts in clays (Adapted from Kulhawy and Jackson, 1989; Used with permission of ASCE).

The λ Method

Stas and Kulhawy (1984) considered the use of the λ method[3] to predict skin friction resistance for drilled shafts in clay. They found that the Vijayvergiya and Focht (1972) λ function (which was developed for piles) overestimated the capacity of drilled shafts by a large margin. In addition, there was much scatter in the data, so it was not possible to develop a new λ function specifically for shafts.

Although this method is useful for pile design, it does not seem to be suitable for drilled shafts.

Rock

A drilled shaft socketed into rock can be an excellent foundation with a large load capacity and small settlements. However, for reasons discussed in Chapter 13, rock is a much more difficult material to evaluate.

---

[3] See discussion of the λ method in Chapter 13.

It also is difficult to determine if certain earth materials are soft rocks or hard clays. These are sometimes known as *intermediate geomaterials*. Reese and O'Neill (1988) suggest classifying them, as shown in Table 15.4. We can evaluate intermediate geomaterials by interpolating between methods intended for clay and those intended for rock.

TABLE 15.4 CLASSIFICATION OF CLAYS, INTERMEDIATE GEOMATERIALS, AND ROCK FOR DRILLED SHAFT DESIGN

Undrained Shear Strength, s_u		Unconfined Compressive Strength, q_{uc}		Classification
(lb/ft^2)	(kPa)	(lb/ft^2)	(kPa)	
< 4,000	< 190	< 8,000	< 380	Clay
4,000 - 18,000	190 - 860	8,000 - 36,000	380 - 1720	Intermediate geomaterial
> 18,000	> 860	> 36,000	> 1720	Rock

Adapted from Reese and O'Neill (1988).

Until the early 1990s, full-scale load tests on shafts in rock were very expensive because of the large loads required, and thus were not cost-effective for routine design problems. Therefore, engineers only had methods based on rock properties, such as those described below. These methods are conservative - perhaps too conservative (Osterberg, 1992) - so alternative methods would be helpful. Fortunately, the Osterberg load test, shown in Figure 15.17, and the high strain dynamic impact tests, described in Section 15.3, seem to be accurate, yet inexpensive ways to measure load capacity and may be especially cost-effective in rock sockets.

The skin friction resistance in rock may have a brittle load-settlement curve. This means that the concrete/rock bond breaks when the shaft reaches its capacity. This can occur at a settlement much less than that required to mobilize the full end bearing. Therefore, for design purposes, consider either skin friction along the sides of the rock socket or end bearing at the bottom, but not both (Reese and O'Neill, 1988).

End Bearing

Engineers often use presumptive end bearing values obtained from building codes. This and other methods of computing q_e' are discussed in Section 6.11 in Chapter 6.

For intermediate geomaterials, evaluating end bearing as follows:

- If no clear discontinuities exist, analyze as a clay (Equation 15.6).
- If clear discontinuities do exist, analyze as a rock (Section 6.11).

Skin Friction

The roughness of the drilled hole has a considerable effect on the skin friction resistance. For example, auger-drilled holes through sandstones probably will be rough enough to achieve a good bond, whereas a cored hole through hard rock may be too smooth. However, it is possible to make grooves in cored holes, thus improving their skin friction capacity.

Osterberg (1992) noted the approximate relationship between skin friction and the unconfined compressive strength of the rock, as shown in Table 15.5.

TABLE 15.5 APPROXIMATE RELATIONSHIP BETWEEN UNIT SKIN FRICTION RESISTANCE, f_s, AND UNCONFINED COMPRESSIVE STRENGTH, q_{uc}, IN ROCK

Rock Classification	Unconfined Compressive Strength, q_{uc}		f_s/q_{uc}
	(lb/in^2)	(MPa)	
Weak (shale, mudstone, etc.)	50 - 500	0.35 - 3.5	0.3 - 0.5
Intermediate[a]	500 - 2000	3.5 - 14	0.1 - 0.3
Strong (limestone, granite, hard sandstone, etc.)	2000 - 8000	14 - 55	0.03 - 0.1

Adapted from Osterberg (1992).
[a] This is not the same as the intermediate geomaterials described earlier.

Reese and O'Neill (1989) suggested the following empirical formula for skin friction in rock (as defined in Table 15.4):

$$f_s = 0.67 \, \sigma_r \sqrt{\frac{q_{uc}}{\sigma_r}} \leq 0.15 \, q_{uc} \qquad (15.14)$$

Where:

q_{uc} = unconfined compressive strength of NW size or larger rock core or of the drilled shaft concrete, whichever is smaller

σ_r = reference stress = 2000 lb/ft^2 = 100 kPa = 14 lb/in^2 = 0.10 MPa

This formula assumes that there will not be any special effort to roughen or groove the hole.

For intermediate geomaterials, compute f_s using both the clay method (Equation 15.12 or 15.13) and the rock method (Table 15.5 or Equation 15.14) and linearly interpolate based on the unconfined compressive strength.

Accuracy

Engineers should always strive to understand the degree of uncertainty in the analyses we perform. For example, predictions of the deflection of steel beams subjected to flexural loading are usually very reliable, whereas predictions of the rate of consolidation of natural soils are generally poor.

Empirical analysis methods often appear to be quite good when compared to the same load test data from which they were derived. Although such comparisons are useful, they tend to represent a type of circular reasoning. Comparisons with independent load tests are more informative, and this was done as a part of a load capacity prediction symposium (Finno, 1989).

Two drilled shafts were built at the Northwestern University campus in Evanston, Illinois, one using casing and the other using a bentonite slurry. The university provided soil test information (more than the usual amount) and the design shaft dimensions to anyone who wished to predict the ultimate capacities. This was an interesting exercise because the predictors did not have prior access to the load test information. The sponsors received 19 predictions of the capacity of the shaft built with casing and 20 of the shaft built with slurry. Their predictions, which were based on a variety of methods, are shown in Figures 15.20 and 15.21 along with the measured capacity of each shaft.

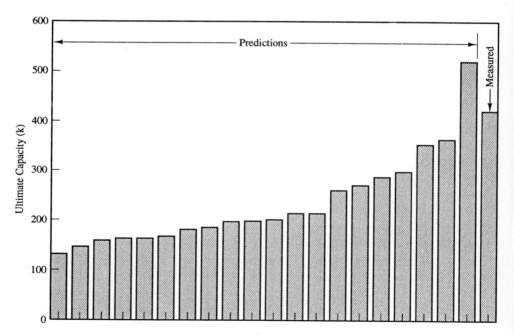

Figure 15.20 Predicted ultimate load capacities of drilled shaft built using bentonite slurry compared to load test results (Adapted from Finno, 1989; Used with permission of ASCE).

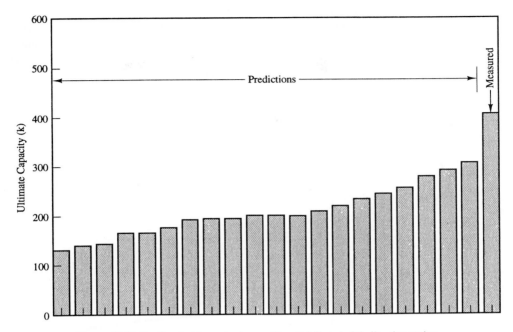

Figure 15.21 Predicted ultimate load capacities of drilled shaft built using casing compared to load test results (Adapted from Finno, 1989; Used with permission of ASCE).

All but one of the predictions underestimated the ultimate capacity, and the average predicted capacity was about half of the actual capacity. This is especially interesting because the predictors had more than the usual amount of subsurface information and probably conducted their analyses with much more than the usual degree of care. Part of this discrepancy seems to be due to differences between the planned dimensions and the as-built dimensions (which were slightly larger), but much of it simply demonstrates how little we really know about the behavior of drilled shafts.

The scatter between the predictions would have been much greater if only the usual amount of soil data was available and the analyses had been performed as part of a routine design. However, when divided by a normal factor of safety, all would have produced safe designs.

The Northwestern tests demonstrate the need for additional research and development on this topic. They also suggest that current analysis methods are probably conservative, which is appropriate, and provide safe designs while we are waiting for more advanced methods.

Site-Specific and Region-Specific Functions

One way to improve the accuracy of these analyses is to develop site-specific or region-specific functions for $q_e{}'$ and f_s from load tests and use them in lieu of the generic functions described earlier. For example, Chang and Broms (1991) developed design criteria specifically for soils in Singapore, and Price et al. (1992) developed custom criteria for a series of highway bridges in Utah. These custom functions should produce more reliable and economical foundations. However, they are very expensive to develop, and thus may be justified only for very large projects or as a cooperative effort between many projects.

15.5 SETTLEMENT

The settlement of drilled shafts under service loads usually is less than 1.0 in (25 mm), which is acceptable for many structures. However, if a large portion of the capacity is due to end bearing, the settlement may be greater, as discussed in Section 15.4.

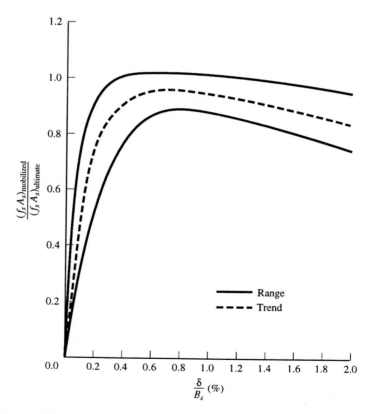

Figure 15.22 Normalized curves showing load transfer in skin friction vs. settlement for drilled shafts in cohesive soils (Reese and O'Neill, 1988).

Reese and O'Neill (1988) developed the charts in Figures 15.22 - 15.25 to estimate the settlement of drilled shafts under service loads. These charts express the settlement in terms of the ratio of the mobilized resistance to the actual resistance. If the computed settlement is too large, use these charts to modify the design accordingly.

Drilled shafts in hard or dense soils tend to have load-settlement curves toward the upper end of the ranges shown in Figures 15.22 - 15.25 (i.e., less settlement is required to reach their ultimate capacities). Conversely, those in soft or loose soils tend toward the lower end of these ranges, and require greater settlement.

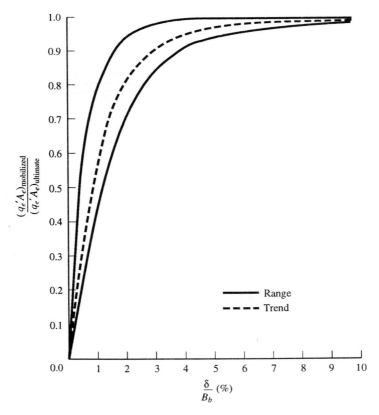

Figure 15.23 Normalized curves showing load transfer in end bearing vs. settlement for drilled shafts in cohesive soils (Reese and O'Neill, 1988).

   If a highly compressible soil strata is present below the drilled shaft as shown in Figure 15.26, add its consolidation settlement to that computed from Figures 15.22 - 15.25. Compute this additional settlement using the methods described in Section 14.2 of Chapter 14.

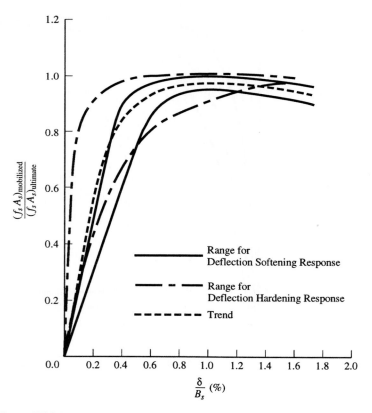

Figure 15.24 Normalized curves showing load transfer in skin friction vs. settlement for drilled shafts in cohesionless soils (Reese and O'Neill, 1988).

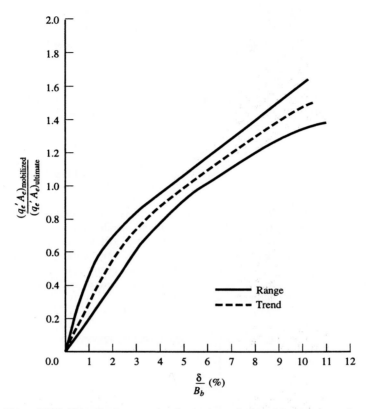

Figure 15.25 Normalized curves showing load transfer in end bearing vs. settlement for drilled shafts in cohesionless soils (Reese and O'Neill, 1988).

Example 15.3

Compute the settlement of the shaft in Example 15.2 when it supports the allowable downward load.

Solution:

From Example 15.2:

$$B = 24 \text{ in}$$
$$B_b = 60 \text{ in}$$
$$P_s = 330 \text{ k}$$
$$P_e' = 706 \text{ k}$$
$$P_a = 410 \text{ k}$$

Try $\delta = 0.2$ in

   $\delta/B = 0.2/24 = 0.8\%$.   From Fig. 15.22 $\rightarrow$ working $P_s = (1.0)(330)$ $= 330$ k
   $\delta/B_b = 0.2/60 = 0.3\%$.   From Fig. 15.23 $\rightarrow$ working $P_e' = (0.25)(706) = 176$ k

   _____

                                                                      506 k > 410

Try $\delta = 0.1$ in

   $\delta/B = 0.1/24 = 0.4\%$.   From Fig. 15.22 $\rightarrow$ working $P_s = (0.92)(330)$ $= 304$ k
   $\delta/B_b = 0.1/60 = 0.2\%$.   From Fig. 15.23 $\rightarrow$ working $P_e' = (0.1)(706)$ $= 71$ k

   _____

                                                                      375 k < 410

$$\delta = 0.1 + 0.1\left(\frac{410 - 375}{506 - 375}\right) = \mathbf{0.13\ in} \qquad\qquad \Leftarrow \textit{Answer}$$

Note how the skin friction resistance provides only 32% of the ultimate capacity, yet it carries 77% of the working load. This is because the shaft will mobilize the full skin friction after only a small settlement (in this case, about 0.2 in), whereas mobilization of the full end bearing requires much more settlement (in this case, about 3 in). Therefore, under working loads, this shaft has mobilized nearly all of the skin friction, but only a small portion of the end bearing. Thus, the remaining end bearing capacity becomes the reserve that forms the factor of safety. It would need to settle about 3 inches to attain the ultimate capacity.

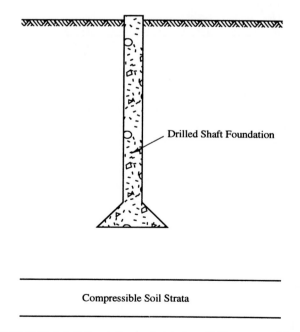

Figure 15.26 Drilled shaft foundations underlain by a highly compressible soil strata.

15.6 DESIGN CHARTS

Usually it is necessary to develop general design criteria for a specific site and express it using a design chart, such as the one in Figure 15.27. Normally, the geotechnical engineer develops this chart and presents it to the structural engineer, who then sizes the individual foundations.

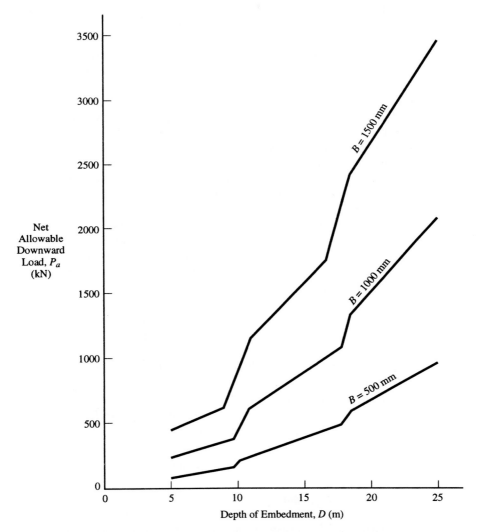

Figure 15.27 Typical design chart for drilled shaft foundations at a particular site. The jogs at 10 and 18 m are the result of increased end bearing capacity when the tip reaches a new soil strata. Design charts also may include a family of settlement curves developed using Figures 15.22 - 15.25.

15.7 QUALITY CONTROL AND INTEGRITY TESTING

As discussed earlier in this chapter, the successful performance of a drilled shaft foundation is very dependent on the contractor's workmanship. A well-designed and well-constructed shaft provides an excellent foundation, but careless construction may produce a failure, even in the most carefully designed shaft.

The best defense against construction defects is to use only experienced and conscientious contractors. In addition, the prudent owner will have representatives on site to observe the contractor's work. The traditional modes of observation include watching the contractor's operations, viewing the open shaft from the ground surface, and in some cases physically entering and inspecting the hole (ADSC, 1989). Unfortunately, even diligent observations may not detect a defective shaft. The possible consequences of such a defect are magnified because each column is generally supported by only one shaft and therefore does not have the redundancy offered by a group of driven piles.

With end bearing shafts, the contractor must confirm that the bottom is clean and free of debris before placing the concrete. In dry holes, it may be possible to shine a light down the hole and observe the bottom from the ground surface. A variety of remote techniques have also been used, especially down-hole video cameras. Holden (1988) describes a down-hole camera equipped with water and air jets that is capable of observing the bottom of holes drilled under bentonite slurry.

Engineers also should be concerned about the structural integrity of shafts, especially those constructed through caving soils and those constructed under water because these are most likely to have undetected problems. These concerns about integrity have prompted the development of various methods of nondestructive integrity testing to evaluate the quality of a completed shaft without damaging it (Litke, 1986; Hertlein, 1992). These methods include the following:

- *Sonic logging*: Install two vertical tubes (usually PVC pipe) in the shaft before placing the concrete. Once the shaft is completed, lower a compression wave source down one tube and a receiver down the other while taking readings of the wave propagation through the shaft. Any voids will show up as anomalies in the wave propagation pattern.

- *Nuclear logging*: This method, also known as gamma-gamma logging, is similar to sonic logging except that it uses radiation instead of compression waves. Christopher, Baker, and Wellington (1989) describe a case history.

- *Vibration analysis*: This consists of placing a mechanical vibrator at the top of the shaft and varying the frequency to determine the natural frequencies and impedance of the shaft.

- *Stress-wave propagation*: This may be the most common method of integrity testing because it is reasonably reliable (except in marginal cases) and relatively inexpensive. It consists of tapping the top of the shaft with a large mallet and using geophones to monitor the stress waves it creates. The geophones can be located at the top of the shaft or embedded within it (Hearne, Stokoe, and Reese,

1981 and Olson and Wright, 1989).

This method is able to detect large defects, say, those that are on the order of half of the shaft diameter. However, it will probably not reveal smaller defects because they do not generate a sufficiently strong reflection.

- *Tomography*: This technique is being used for medical purposes to produce three-dimensional images of the human body. Some researchers have suggested this method might be applied to drilled shafts, thus producing a three-dimensional image of the entire shaft and showing even the smallest defects. However, major technical hurdles must be overcome before this method becomes usable.

Other methods of testing are also possible, such as full scale load tests (too expensive) or coring the shaft top-to-bottom (expensive and easy to miss defects).

Integrity tests have been used primarily when the engineer or contractor suspects that the shaft has a problem. However, these tests may become a routine quality control measure in the future.

SUMMARY OF MAJOR POINTS

1. Drilled shafts are the most common alternative to a driven pile foundation. They are often less expensive, especially on small to medium-size projects where the lower mobilization/demobilization costs are particularly important. They also are appropriate for many large structures.

2. A variety of drill rigs and drilling tools are available to accommodate most site conditions. The construction methods, as well as the diameter and length of the shaft, can be easily changed to accommodate unanticipated soil conditions.

3. Caving or squeezing soils usually can be handled using either the casing method or the drilling fluid method.

4. Conventional full-scale load tests are not cost-effective except on extremely large projects. However, the Osterberg load test and high strain dynamic impact tests may be cost-effective alternatives to conventional tests.

5. Engineers usually design drilled shafts using analyses based on soil properties. Unfortunately, we do not fully understand the interactions between the shaft and soil and the mechanisms that develop the load capacity. Therefore, these methods include a fair degree of empiricism. These methods will probably become more rational and more accurate as in-situ measurements of lateral stress in the soil become a more common part of engineering practice.

6. The settlement of drilled shafts under working loads is usually small (i.e., less than 0.5 in (12 mm). However, it may be much larger if most of the working load is carried by end bearing or if the bottom of the shaft is underlain by compressible soils.

7. Less displacement is required to develop the skin friction resistance than is required

to develop end bearing. Therefore, under working loads, much of the resistance will probably be due to skin friction. (However, this will not be the case when the shaft penetrates through weak soils and is bottomed in a very strong soil. End bearing will probably dominate in that case.)

8. Quality control is an important part of successful drilled shaft construction. As a minimum, this should consist of field observations by an independent inspector. In the future, we will probably see more use of nondestructive testing as a routine quality control measure.

COMPUTER SOFTWARE – Program SHAFT

A computer program that computes the axial load capacity of drilled shaft foundations, SHAFT, is included with this book. It uses Reese and O'Neill's method as described in this chapter. To access this program, install the software in accordance with the instructions in Appendix C, and then type **SHAFT** at the DOS prompt (If you encounter an out-of-memory error, see the instructions in Appendix C). A brief introductory screen will appear, followed by the worksheet. Then, follow this procedure:

1. Choose the units of measurement (English or SI) by pressing **ALT-U** and selecting from the menu. English units are the default.

2. Choose whether or not to consider end bearing by pressing **ALT-E** and selecting from the menu. The default is Yes.

3. Use the arrow keys to move the cursor and input the title, shaft geometry, uplift reduction factor (R in Equation 11.6) and soil properties. Replace all of the **XXXXX** indicators with the appropriate data.

4. Hold the down-arrow key until the lower portion of the worksheet scrolls onto the screen. Notice that the computer has divided the soil into many very thin layers. The depths in the first two columns are measured from the ground surface. Move the curser to the soil type column for first layer. Then press **ALT-S**, select the soil type from the menu, and press the Enter key.

5. Leave the cursor at its present position and copy the soil type to lower layers by pressing **ALT-M** and selecting **COPY** from the menu. This will copy it down to the bottom of the worksheet. If the soil type changes with depth, move the cursor to the shallowest layer with the new type, press **ALT-M**, select the soils type, and then use the copy function as before. Thus, it is possible to input any distribution of soil type with depth. The worksheet extends to a depth of 100 ft or 33.3 m.

6. Move the cursor to the unit weight column for the first layer and input the unit weight, γ. Use the copy function to create the desired distribution of unit weight with depth.

7. Use the same procedure to fill in the soil strength column. Note that the required soil strength parameter differs depending on the soil classification.

Example 15.4

Solve Example 15.1 program SHAFT.

Partial output (full output extends to a depth of 33.3 m):

```
        SHAFT   Version 1.0    (c) 1994 by Prentice Hall, Inc.
              Axial Load Capacity of Drilled Shaft Foundations

Title: Example 15.4
                                       Date:23-May-93
Units of Measurement:      SI          Time: 04:55 PM
    (Press ALT-U to change)
                                   * *************************
Consider End Bearing?:     Yes     *                         *
    (Press ALT-E to change)        *         RESULTS         *
                                   *                         *
                                   * ULTIMATE LOAD CAPACITIES *
                                   *                         *
                                   *  Skin Friction =  1727 kN*
                                   *  End Bearing   =   373 kN*
Shaft Diameter          =   600  mm  *                         *
Shaft Length            =  14.0  m   *                         *
Depth to Groundwater    =   2.0  m   * ALLOWABLE LOAD CAPACITIES *
Uplift Reduction Factor =  1.00      *                         *
Factor of Safety                     *  Downward     =   840 kN*
         Downward       =  2.50      *  Uplift       =   486 kN*
         Uplift         =  4.00      * *************************
```

```
: Layer Depth : Soil Type : Unit  :            : Unit  : Skin   :
:    (m)      :           :Weight :  Strength  : Skin  :Friction :
:------------:Press ALT-S :       :            : Friction :        :
: Top :Bottom:  to set    :(kN/m3):            : (kPa) :  (kN)   :
:-----+------+------------+-------+------------+---------+---------+
: 0.0 : 0.3 :Cohesionless: 17.0 :N =    10   :    3  :    2.1 :
: 0.3 : 0.7 :Cohesionless: 17.0 :N =    10   :   10  :    6.4 :
: 0.7 : 1.0 :Cohesionless: 17.0 :N =    11   :   17  :   10.7 :
: 1.0 : 1.3 :Cohesionless: 17.0 :N =    11   :   24  :   15.0 :
: 1.3 : 1.7 :Cohesionless: 17.0 :N =    11   :   31  :   19.2 :
: 1.7 : 2.0 :Cohesionless: 17.0 :N =    12   :   36  :   22.9 :
: 2.0 : 2.3 :Cohesionless: 20.0 :N =    12   :   41  :   25.6 :
: 2.3 : 2.7 :Cohesionless: 20.0 :N =    12   :   43  :   27.3 :
: 2.7 : 3.0 :Cohesionless: 20.0 :N =    12   :   46  :   29.0 :
: 3.0 : 3.3 :Cohesionless: 20.0 :N =    13   :   49  :   30.7 :
: 3.3 : 3.7 :Cohesionless: 20.0 :N =    14   :   51  :   32.3 :
: 3.7 : 4.0 :Cohesionless: 20.0 :N =    15   :   54  :   33.8 :
: 4.0 : 4.3 :Cohesionless: 20.0 :N =    15   :   56  :   35.2 :
: 4.3 : 4.7 :Cohesionless: 20.0 :N =    16   :   58  :   36.6 :
: 4.7 : 5.0 :Cohesionless: 20.0 :N =    17   :   60  :   38.0 :
: 5.0 : 5.3 :Cohesionless: 20.0 :N =    18   :   62  :   39.3 :
: 5.3 : 5.7 :Cohesionless: 20.0 :N =    19   :   64  :   40.5 :
: 5.7 : 6.0 :Cohesionless: 20.0 :N =    19   :   66  :   41.7 :
: 6.0 : 6.3 :Cohesionless: 20.0 :N =    20   :   68  :   42.8 :
: 6.3 : 6.7 :Cohesionless: 20.0 :N =    21   :   70  :   43.9 :
: 6.7 : 7.0 :Cohesionless: 20.0 :N =    22   :   72  :   45.0 :
: 7.0 : 7.3 :Cohesionless: 20.0 :N =    22   :   73  :   46.0 :
: 7.3 : 7.7 :Cohesionless: 20.0 :N =    21   :   75  :   46.9 :
: 7.7 : 8.0 :Cohesionless: 20.0 :N =    21   :   76  :   47.8 :
: 8.0 : 8.3 :Cohesionless: 20.0 :N =    21   :   77  :   48.7 :
: 8.3 : 8.7 :Cohesionless: 20.0 :N =    20   :   79  :   49.5 :
: 8.7 : 9.0 :Cohesionless: 20.0 :N =    20   :   80  :   50.3 :
: 9.0 : 9.3 :Cohesionless: 20.0 :N =    20   :   81  :   51.0 :
: 9.3 : 9.7 :Cohesionless: 20.0 :N =    20   :   82  :   51.7 :
```

8. Review the output using the arrow keys or the Home key. The computed skin friction and end bearing capacities, along with the allowable downward and uplift load capacities are shown in the upper right portion of the worksheet.

9. If a printout is needed, press **ALT-M** to access the menu and select one of the following:

> **PRINT** - Prints the worksheet on a printer
> **PRINT TO DISK** - Prints the worksheet to a disk file

> If you select **PRINT TO DISK**, the program will ask for a disk file name. You must select a new file name; the program will not write over an old file.

10. To erase the user input data, press **ALT-M** and select **REFRESH** from the menu.

11. To exit to DOS, press **ALT-M** and select **QUIT** from the menu.

When evaluating load capacities in cohesionless soils, the results from the SHAFT program may differ slightly from the results of hand computations. This difference occurs because the program divides the soil into very thin layers, whereas hand computations use larger layers. The β factor varies non-linearly with depth, and is computed at the midpoint of each layer. Therefore, these two computations use slightly different distributions of β.

QUESTIONS AND PRACTICE PROBLEMS

15.1 Describe two situations where a drilled shaft would be preferable over a driven pile. Then describe two situations where the reverse would be true.

15.2 What are typical dimensions of drilled shafts?

15.3 When is it important to clean out the loose soils in the bottom of a drilled shaft excavation?

15.4 When would you expect caving or squeezing conditions to be a problem? What construction methods could the contractor use to overcome these problems?

15.5 Why are most drilled shaft designs based exclusively on soil properties without the benefit of any on-site load tests?

15.6 An office building is to be supported on a series of 700 mm diameter, 12.0 m long drilled shafts that will be built using the open hole method. The soil profile at this site is as follows:

Depth (m)	Soil Classification	Undrained Shear Strength, s_u (kPa)
0 - 2.2	Stiff clayey silt (CL)	70
2.2 - 6.1	Stiff silty clay (CL)	85
6.1 - 11.5	Very stiff sandy clay (CL)	120
11.5 - 20.0	Very stiff sandy clay (CL)	180

The groundwater table is at a depth of 50 m and the degree of control is "normal" and the design life is 50 years. Using hand computations, compute the allowable downward and uplift capacities of each shaft.

15.7 Solve Problem 15.6 using Program SHAFT. Note that the unit weight is not needed for this analysis.

15.8 Compute the settlement of the shaft in Problem 15.6.

15.9 A "monumental" structure is to be supported on a series of belled drilled shaft foundations as shown in the figure below. The degree of control is "normal" and the groundwater table is at a great depth. The shaft will be drilled using the open-hole method. Using hand computations, compute the allowable downward and uplift capacities of each shaft.

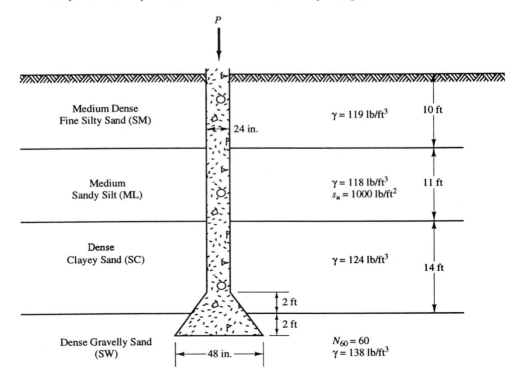

15.10 Use the data given in Problem 15.9 and Program SHAFT to design a straight drilled shaft that will support a downward load of 250 k. Use a factor of safety of 2.8.

15.11 Compute the settlement of the shaft in Problem 15.10.

15.12 A 1000 mm diameter drilled shaft penetrates through 8.5 m of silty sand ($\gamma = 19.9$ kN/m^3), and then 0.5 m into sandstone ($q_{uc} = 13$ MPa). A perched groundwater table exists between depths of 6.0 and 8.5 m. This shaft will be built using casing. Compute the allowable downward load capacity using a factor of safety of 2.8.

15.13 Using the CPT data in Figure 4.22 of Chapter 4 and the LCPC Method described in Chapter 13, compute the allowable downward capacity of a 24 inch diameter, 42 ft deep drilled shaft. Assume this shaft will be built using casing, and use a factor of safety of 3.0.

15.14 Using the data from Problem 15.6 and program SHAFT, develop a design chart for allowable downward loads on 500, 1000 and 1500 mm diameter drilled shafts. Plot all three curves on the same chart and consider depths of embedment between 5.0 and 15.0 m. Assume these shafts will be built using the open hole method, and use a factor of safety of 3.0.

15.15 A five-story office building with a design life of 50 years is to be built on the sandy soil described in the case study at the end of Chapter 8. The downward column loads will be between 100 and 450 k. Because the sandy soil at this site is not saturated and has some fines, it appears that shafts could be built using the open hole method. However, the groundwater table is 30 ft below the ground surface, so caving probably would be a problem below that depth. Therefore, consider shaft lengths of up to 28 ft. Use $\gamma = 120$ lb/ft^3.

Using Program SHAFT, develop a design chart for straight shafts in this soil. Plot the shaft length (embedment depth) on the horizontal axis and allowable downward capacity on the vertical axis. Generate three or four curves, each for a different diameter.

15.16 A full-scale load test has been conducted on a 24 in diameter, 40 ft long instrumented drilled shaft similar to the one shown in Figure 15.17. The test crew maintained records of the load-settlement data and the forces in each of the five load cells. The applied load at failure (using Davisson's method as described in Chapter 13) was 739,600 lb. The corresponding forces in the load cells were as follows:

Load Cell Number	Depth (ft)	Force (lb)
1	3.0	719,360
2	12.0	636,120
3	21.0	487,500
4	30.0	304,320
5	39.0	135,400

There are two soil strata: the first extends from the ground surface to a depth of 15 ft and has a unit weight of 117 lb/ft^3; the second extends from 15 ft to 60 ft and has a unit weight of 120 lb/ft^3 above the groundwater table and 127 lb/ft^3 below. The groundwater table is at a depth of 17 ft.

Compute the average β factor in each of the two soil strata, and the net unit end bearing resistance, $q_e{}'$.

Note: Once these site-specific β and $q_e{}'$ values have been computed, they could be used to design shafts of other diameters or lengths at this site.

16

Other Types of Deep Foundations

> *Let the foundations of those works be dug from a solid site and*
> *to a solid base if it can be found. But if a solid foundation is*
> *not found, and the site is loose earth right down, or marshy, then*
> *it is to be excavated and cleared and remade with piles of alder*
> *or of olive or charred oak, and the piles are to be driven close*
> *together by machinery, and the intervals between are to be*
> *filled with charcoal ... even the heaviest foundations may be*
> *laid on such a base.*
>
> Marcus Vitruvius, Roman Architect and Engineer,
> 1st Century BC
> (translated by Morgan, 1914)

Several special types of deep foundations also are available. These types are neither piles nor drilled shafts. Although these special types are not as common, they do offer special features that can be very useful in certain circumstances.

16.1 MANDREL-DRIVEN THIN-SHELLS FILLED WITH CONCRETE

One method of combining some of the best features of driven piles and cast-in-place drilled shafts is to use a mandrel-driven thin shell, as shown in Figure 16.1. Alfred Raymond developed an early version of this method in 1897 and later refined it to create the *step-taper pile*.

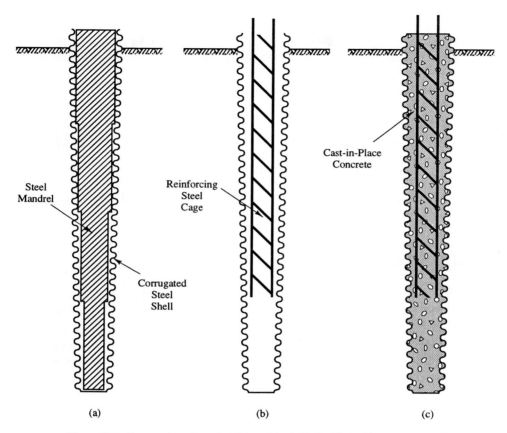

(a)   (b)   (c)

Figure 16.1 Construction of mandrel-driven thin-shell pile filled with concrete: (a) drive the thin shell into the ground using a steel mandrel; (b) Remove the mandrel and insert a cage of reinforcing steel; and (c) fill the shell with concrete.

This type of foundation is built as follows:

1. Hold a steel mandrel in the leads of the pile driver. This mandrel is a cylindrical and matches the inside of the thin shell. The purpose of the mandrel is to transmit the driving stresses from the hammer to the sides and bottom of the shell.
2. Pull the thin shell (which resembles corrugated steel drain pipe) onto the mandrel.
3. Drive the mandrel and shell into the ground using a pile hammer.
4. Remove the mandrel and inspect the shell.
5. (optional) Place a cage of reinforcing steel into the shell.
6. Fill the shell with concrete.

The advantages of this design include:

- The shell provides a clean place to cast the concrete, so the structural integrity may be better than a drilled shaft.
- The displacement developed during driving and the corrugations on the shell produce high skin friction resistance.
- The shells and mandrel can be shipped to the job site in pieces and assembled there, so it is possible to build long piles.

However, it also has disadvantages:

- It is necessary to mobilize a pile driver and other equipment, so the cost per pile will be at least as high as that for conventional driven piles.
- They cannot be spliced, so the total length is limited by the height of the pile driver.

16.2 AUGER-CAST PILES

The *auger-cast pile* is a type of cast-in-place pile developed in the United States during the late 1940s and early 1950s (Neate, 1989). They are known by many names (DFI, 1990), including:

- Augered pressure grouted (APG) pile
- Augered cast-in-place pile
- Continuous flight auger pile
- Intruded mortar piles
- Augerpress pile
- AugerPile
- Grouted bored pile
- Augered grout-injected pile

Specialty contractors build auger-cast piles as described below and as shown in Figure 16.2:

1. Using a hollow-stem auger with a temporary bottom plug, drill to the required depth. In the United States, 12, 14 or 16 inch (300, 350 or 400 mm) diameter augers are most common. Japanese contractors have built piles as large as 39 inches (1 m) in diameter. This equipment is similar to, but larger than, the hollow-stem augers used for soil exploration purposes. These augers are suspended from a crane and driven by a pneumatic or hydraulic motor. The depth of the pile may be as great as 90 ft (27 m), but lengths of 20 - 50 ft (6 - 15 m) are more typical.

2. Inject cement grout (sand, portland cement and water) under high pressure through the middle of the auger. This grout forces the bottom plug out and then begins to flow out of the bottom of the auger.

3. While the grout is being injected, slowly and smoothly raise and rotate the auger to form the pile while bringing the soil cuttings to the ground surface.

4. Upon reaching the ground surface, remove the auger and insert reinforcing steel into the grouted hole. This may consist of a single centrally located bar or a prefabricated steel cage. Because the grout has a very high slump and no gravel, it is possible to insert the steel directly into the newly grouted pile.

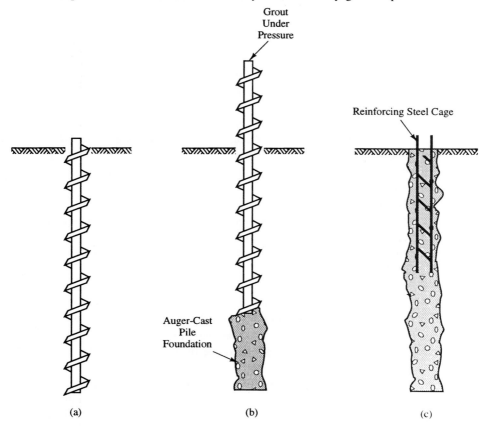

Figure 16.2 Construction of an auger-cast pile: (a) Drill to the required depth using a hollow-stem auger; (b) Withdraw the auger while injecting cement grout; (c) Install steel rebars (optional).

The advantages of this method include:

• The cost of construction is low, partly because the crane can be rented locally, thus reducing mobilization costs.

- The noise and vibration levels are much lower than those with driven piles.
- The auger protects the hole from caving, thus reducing the potential for ground movements during drilling.
- The grout is injected under pressure, so it penetrates the soil and provides and good bond. The pressure also provides some compaction of the soil.
- The technique is usable in a wide variety of soils, including some of those that might cause difficulties with driven piles or drilled shafts.

However, it also has disadvantages (Massarsch et al., 1988; Brons and Kool, 1988):

- The quality and integrity of the completed pile are very dependent on the skills of the contractor. For example, if the auger is raised too quickly or not rotated sufficiently, the concrete may become contaminated with soil. It is also difficult for the equipment operator to judge the correct grout pressure to use.
- In certain conditions, the augering process can draw up too much soil, thus causing a reduction in the lateral stresses in the ground. For example, if an auger passes through a loose sand, and then encounters a stronger strata, the auger will not advance as quickly, and continued rotation may bring too much of the sand to the surface.
- The construction process is very sensitive to equipment breakdowns. Once grouting has begun, any significant breakdown becomes cause for abandoning the foundation.
- Placement of reinforcing steel cages can be difficult, especially if heavy reinforcement is required. This may limit the pile's ability to resist lateral loads (resist uplift loads by placing a single large bar in the middle of the auger before grouting).
- These piles can not be used in soils that contain cobbles or boulders (the auger will not excavate them) or in thick deposits of highly organic soils (they compress under the grout pressure and thus require excessively large grout volumes).
- Unlike driven piles, there is no hammer blow count to use as an independent check on the pile capacity (the auger torque does not appear to be a good indicator).

Although auger-cast piles typically do not have as great a load capacity as conventional driven piles of comparable dimensions, they are often much less expensive. Because the equipment mobilization costs are much less than those for driven piles, and the technique works well in caving soils, auger-cast piles are most often used on small to medium-size structures on sandy soils.

Axial Load Capacity

Determine the axial load capacity of an auger-cast pile from a full-scale load test or from the soil properties. Neely (1991) developed empirical formulas for computing the unit end bearing and skin friction resistance of auger-cast piles in sands. These formulas are based on the results of 66 load tests on piles with diameters between 12 and 24 in (300-600 mm) and lengths between 15 and 85 ft (4.5 - 26 m). Grout consumption data were not always available, so he based his analyses on the auger diameter even though the actual pile diameter is somewhat larger.

Neely computes the net unit end bearing resistance, q_e', based on the SPT N value in the soil immediately beneath the pile tip:

$$q_e' = 1.9 \, \sigma_r \, N_{60} \leq 150,000 \text{ lb/ft}^2 \text{ (7500 kPa)} \tag{16.1}$$

Where:

σ_r = reference stress = 2000 lb/ft^2 = 100 kPa

N_{60} = SPT N_{60} value without overburden correction

He computes the average unit skin friction resistance using the β method:

$$\bar{f}_s = \bar{\beta} \, \bar{\sigma}_v' \leq 2800 \text{ lb/ft}^2 \text{ (140kPa)} \tag{16.2}$$

Where:

$\bar{f}_s$ = average unit skin friction resistance

$\bar{\beta}$ = average skin friction factor (from Figure 16.3)

$\bar{\sigma}_v'$ = average vertical effective stress along length of pile

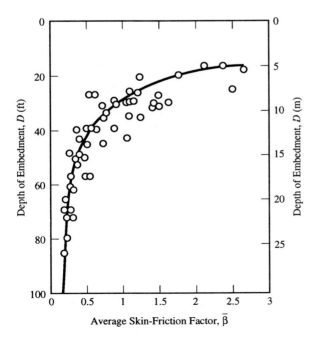

Figure 16.3 Average skin friction factor, $\bar{\beta}$ for use in Equation 16.2 (Adapted from Neely, 1991; Used by permission of ASCE).

The skin friction factor β varies nonlinearly with depth. Figure 16.3 gives the average value, $\bar{\beta}$, for various pile lengths. Therefore, when using this chart, determine only one $\bar{\beta}$ value based on the length of the pile and use it in Equation 16.2. Do not divide the soil into layers to determine the β value for each layer.

Use these q_e and f_s values in Equations 11.5 and 11.6 in Chapter 11 to compute the allowable load capacity. Base the end bearing and skin friction areas, A_e and A_s, on the diameter of the auger, even though the actual diameter will be larger.

Example 16.1

Compute the allowable vertical downward load capacity of a 400 mm diameter, 18 m long auger-cast pile to be built in the soil described in Example 13.2. Use a factor of safety of 3.5.

Solution:

End Bearing (using Equation 16.1):

$$N_{60} = \frac{23 + 32}{2} = 27$$

$$q'_e = 1.9\,\sigma_r\,N_{60} = (1.9)(100)(27) = 5130 \text{ kPa}$$

$$A_e = \frac{\pi\,(0.400)^2}{4} = 0.126 \text{ m}^2$$

$$P'_e = q'_e\,A_e = (5130)(0.126) = 646 \text{ kN}$$

Skin Friction (using Equation 16.2):

$$\bar{\beta} = 0.25 \quad \text{(from Figure 16.3)}$$

$$\bar{\sigma}'_v = \sum \gamma h - \gamma_w h_w = (18)(3) + (21)(6) - (9.8)(6) = 121 \text{ kPa}$$

$$f_s = \bar{\beta}\,\bar{\sigma}'_v = (0.25)(121) = 30.2 \text{ kPa}$$

$$A_s = \pi(0.400)(18) = 22.6\ m^2$$

$$P_s = f_s A_s = (30.2)(22.6) = 682\ kN$$

Capacity (using Equation 11.3):

$$P_a = \frac{P_e' + P_s}{F} = \frac{646 + 682}{3.5} = \textbf{380 kN} \qquad \Leftarrow Answer$$

16.3 PRESSURE-INJECTED FOOTINGS

Edgard Frankignoul developed the pressure-injected footing (PIF) foundation in Belgium before the First World War. This technique uses cast-in-place concrete that is rammed into the soil using a drop hammer. This ramming effect compacts the surrounding soil, thus increasing its load bearing capacity. PIF foundations are often called *Franki piles*. Other names include *bulb pile, expanded base pile, compacted concrete pile*, and *compacto pile*.

The construction techniques used to build PIFs are described below and shown in Figure 16.4.

Phase 1: Driving

The process begins by temporarily inserting the *drive tube* into the ground. This tube is a specially built 12 to 24 in (300 - 600 mm) diameter steel pipe. The contractor does so using one of the following methods:

- *Top driving method*: Install a temporary bottom plate on the drive tube, and then drive the tube to the required depth using a diesel pile hammer.
- *Bottom driving method*: Place a plug of low-slump concrete in the bottom of the tube and pack it in using the drop hammer. Then, continue to strike this plug, thus pulling the tube into the ground.

Phase 2: Forming the Base

Once the drive tube reaches the required depth, hold it in place using cables, place small charges of concrete inside the tube, and drive them into the ground with repeated blows of the drop hammer. This hammer has a weight of 3 to 10 kips

(1400 - 4500 kg) and typically drops from a height of 20 ft (6 m). If the top driving method was used, this process will expel the temporary bottom plate. Thus, a bulb of concrete is formed in the soil, which increases the end-bearing area and compacts the surrounding soil. This process continues until a specified number of hammer blows is required to drive out a certain volume of concrete.

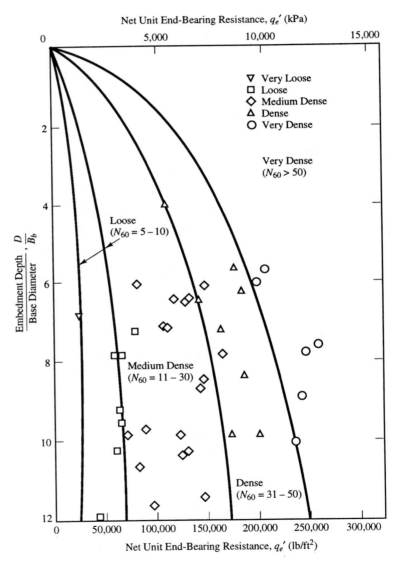

Figure 16.4 Construction of a PIF foundation: (a) top driving; (b) bottom driving; (c) finished base; (d) uncased shaft; and (e) cased shaft (Adapted from brochure by Franki Northwest Co.)

Phase 3: Building the Shaft

The shaft extends the PIF base to the ground surface. Two types of shafts are commonly used:

- To build a *compacted shaft*, raise the drive tube in increments while simultaneously driving in additional charges of concrete. This technique compacts the surrounding soil, thus increasing the skin friction resistance. It also increases the end bearing resistance by providing a stronger soil over the base.
- To build a *cased shaft*, insert a corrugated steel shell into the drive tube, place and compact a zero-slump concrete plug, and withdraw the tube. Then fill the shell with conventional concrete. Although this method does not develop as much load capacity, it is often more economical for piles that are longer than about 30 ft (9 m). A cased shaft may be mandatory if very soft soils, such as peat, are encountered because these soils do not provide the lateral support required for the compacted shaft method.

The contractor can reinforce either type of shaft to resists uplift or lateral loads. For the compacted shaft, the reinforcing cage fits between the drop hammer and the drive tube, thus allowing the hammer to fall freely.

PIF foundations may be installed individually or in a group of two or more and connected with a pile cap. Table 16.1 gives typical dimensions and capacities of PIF foundations. Figure 16.5 shows a "mini" PIF that was extracted out of the ground.

TABLE 16.1 TYPICAL PIF DIMENSIONS AND CAPACITIES

PIF Type	Allowable Downward Capacity		Base Diameter[a]		Nominal Shaft Diameter			
					Compacted		Cased	
	(k)	(kN)	(in)	(mm)	(in)	(mm)	(in)	(mm)
Mini	100	450	24 - 30	600 - 750	n/a	n/a	10.6 - 11.1	270 - 280
Medium	200	900	34 - 40	850 - 1000	17	430	12.2 - 14	300 - 360
Standard	400	1800	34 - 40	850 - 1000	22	560	16 - 17.6	400 - 450
Large	500	2200	34 - 40	850 - 1000	23	580	19	480
Maxi	600	2700	34 - 40	850 - 1000	25	630	22	560

Adapted from a brochure by Franki Northwest Company.

[a] In very loose soils, the base diameter may be larger than listed here. Conversely, when PIFs are installed in groups, it may be slightly smaller.

Figure 16.5 This "mini" PIF was extracted from the ground. It had a base diameter of 24 in (600 mm). (Photo courtesy of William J. Neely.)

The advantages of PIF foundations include:

- The construction process compacts the soil, thus increasing its strength and load-bearing capacity. This benefit is most pronounced in sandy and gravelly soils with less than about 15% passing the #200 sieve, so PIFs are best suited for these kinds of soils.
- When compacted shafts are used, the construction process produces a rough interface between the shaft and the soil, thus further improving the skin friction resistance.
- It is possible to build PIFs with large bases (thus gaining the additional end bearing area) in soils such as loose sands where belled drilled shafts would be difficult or impossible to build.

Disadvantages include:

- The unit skin friction resistance for cased PIFs is unreliable because of the annular space between the casing and the soil. Although this space is filled with sand after the drive tube is lifted, we cannot be certain about the integrity of the connection between the shaft and the soil.

- The construction process generates large ground vibrations and thus may not be possible near sensitive structures. These vibrations also can damage wet concrete in nearby PIFs.
- The construction equipment is bulky and cumbersome, and thus requires large work areas.
- Compacted shafts cannot include large amounts of reinforcing steel.
- Although each PIF will have a higher load capacity than a pile or drilled shaft of comparable dimensions, it also is more expensive to build. Therefore, the engineer must evaluate the alternatives for each project individually to determine which type is most economical.
- They are generally economical only when the length is less than about 30 ft (9 m) for compacted PIFs or about 70 ft (21 m) for cased PIFs.

Axial Load Capacity

The methods of predicting axial load capacity described in Chapters 13 and 15 underestimate the capacity of PIFs because they do not consider the soil compaction that occurs during construction. Therefore, a load capacity analysis must consider this effect, either explicitly or implicitly.

End Bearing

Neely (1989, 1990a, 1990b) developed empirical methods of computing the net unit end bearing resistance, q_e', from SPT N values. This method is based on the results of load tests conducted 93 PIFs with cased shafts and 41 PIFs with compacted shafts. All of these piles were bottomed in sands or gravels. His formula for cased PIFs bottomed in clean cohesionless soils is:

$$q_e' = 0.28\ \sigma_r\ N_{60}' \frac{D}{B_b} \leq 2.8\ \sigma_r\ N_{60}' \tag{16.3}$$

Where:

q_e' = net unit end bearing resistance

σ_r = reference stress = 2000 lb/ft^2 = 100 kPa

N_{60}' = corrected SPT N value beneath pile tip

D = depth to bottom of PIF base

B_b = diameter of PIF base

For compacted shaft PIFs, Neely presented his design end bearing recommendations in graphical form, as shown in Figure 16.6.

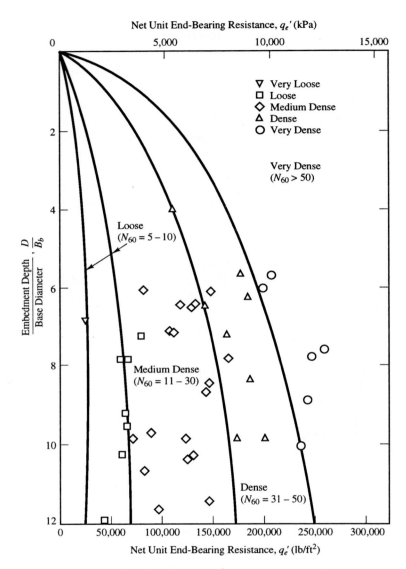

Figure 16.6 Unit end bearing resistance, q_e', for compacted shaft PIFs in clean cohesionless soils (Adapted from Neely, 1990b; Used with permission of ASCE).

Neely's data suggest that silty sands or clayey sands will have less end bearing resistance than a clean sand with a comparable N value. Soils with a fines content of 25% may have an end bearing resistance of about half that predicted by Equation 16.3 or Figure 16.6.

The end bearing capacity of a compacted shaft PIF is much greater than that of a drilled shaft of comparable dimensions, especially in clean sands (compare Equation 15.1 in Chapter 15 with Figure 16.6).

Skin Friction

As discussed earlier, the unit skin friction resistance for cased PIFs is unreliable because of the annular space between the casing and the soil. Therefore, for design purposes we neglect it.

Conversely, a compacted shaft PIF will have excellent contact between the pile and the soil. The construction process for compacted shafts makes them a high displacement pile with a rough interface, so both the coefficient of lateral earth pressure, K, and the coefficient of friction, tan ϕ_s, will be high and significant skin friction resistance can develop.

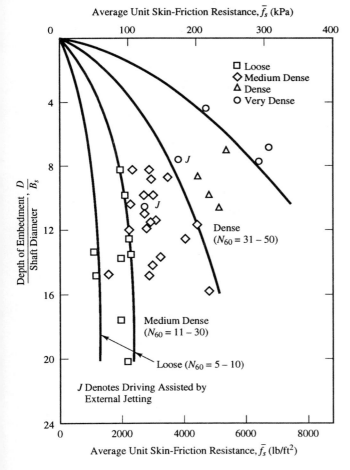

Figure 16.7 Average unit skin friction resistance for compacted shaft PIFs in cohesionless soils (Adapted from Neely, 1990b; Used with permission of ASCE). These values are appropriate only when the drive tube is driven into the ground in the conventional fashion. The use of jetting to assist this process seems to reduce the skin friction capacity.

Neely (1990b) used the results from 41 load tests to develop an empirical relationship between the average unit skin friction resistance, f_s , in cohesionless soils and SPT N values, as shown in Figure 16.7. These are three to five times that produced by driven piles in the same soil conditions.

Use Equation 11.5 in Chapter 11 to combine the design skin friction (Figure 16.7) and end bearing (Figure 16.6) resistances and compute the allowable load capacity. Do not attempt to combine Neely's skin friction resistance values with the unit end bearing resistance obtained from any other method.

Example 16.2

An 18 m deep "standard" PIF is to be built in the soil profile described in Example 13.2 in Chapter 13. The shaft will be cased. Compute the allowable downward load capacity using a factor of safety of 3.5.

Solution:

End Bearing:

$$N_{60} \text{ at tip} = 23 \quad \text{(from Example 13.2)}$$

$$\sigma' \text{ at tip} = 222 \text{ kPa (from Example 13.2)}$$

Using Equations 4.4 and 16.3:

$$C_N = \frac{2}{1 + \sigma_v'/\sigma_r} = \frac{2}{1 + 222/200} = 0.95$$

$$N_{60}' = C_N N_{60} = (0.95)(23) = 22$$

$$q_e' = 2.8 \, \sigma_r N_{60}' = (2.8)(100)(22) = 6160 \text{ kPa}$$

$$B_b \approx 0.925 \text{ m} \quad \text{(from Table 16.1)}$$

$$A_e = \frac{\pi(0.925)^2}{4} = 0.672 \text{ m}^2$$

$$P_e' = q_e' A_e = (6160)(0.672) = 4140 \text{ kN}$$

Skin Friction:

Because the PIF is compacted, we will neglect any skin friction resistance.

$$P_a = \frac{P_e' + P_s}{F} = \frac{4140 + 0}{3.5} = \textbf{1180 kN} \qquad \Leftarrow \textit{Answer}$$

Notice that this is three times the capacity of an auger-cast pile (Example 16.1).

Example 16.3

A 20 ft long "large" PIF with a compacted shaft is to be built in a silty sand, as shown in the figure below. Compute its allowable downward load capacity.

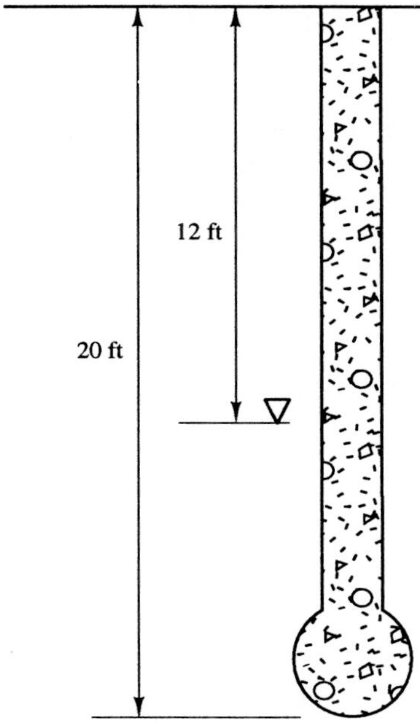

Depth (ft)	N_{60}
4	12
7	15
10	18
13	14
17	21
21	28
25	25
30	24

Solution:

End Bearing:

$$B_b \approx 37 \text{ in} = 3.1 \text{ ft} \quad \text{(Table 16.1)}$$

$$D/B_b = 20/3.1 = 6.5$$

$$N_{60} = 23 \quad \text{(The } N_{60} = 28 \text{ at } 21 \text{ ft seems to be an anomaly)}$$

$$q_e' = 40(2000) = 80,000 \text{ lb/ft}^2$$

$$A_e = \frac{\pi(3.1)^2}{4} = 7.55 \text{ ft}^2$$

$$P_e' = q_e' A_e = \frac{(80,000)(7.55)}{1000 \text{ lb/k}} = 604 \text{ k}$$

As discussed earlier, the fines will reduce the end bearing capacity by up to 50%. This soil has some fines, so we will subtract 30% from the computed capacity:

$$\therefore \text{ Use } P_e' = 604 \times 0.70 = 423 \text{ k}$$

Skin Friction:

$$B_s \approx 22 \text{ in} = 1.8 \text{ ft} \quad \text{(Table 16.1)}$$

$$\frac{D}{B_s} = \frac{20 - 3.1}{1.8} = 9.4$$

$$\bar{N}_{60} = \frac{12 + 15 + 18 + 14 + 21}{5} = 16$$

$$\bar{f}_s = 1.2(2000) = 2400 \text{ lb/ft}^2$$

$$A_s = \pi(1.8)(20 - 3.1) = 95.6 \text{ ft}^2$$

$$P_s = \bar{f}_s A_s = \frac{(2400)(95.6)}{1000 \text{ lb/k}} = 229 \text{ k}$$

$$P_a = \frac{P_e' + P_s}{F} = \frac{423 + 229}{3.5} = \textbf{190 k} \qquad \Leftarrow \textbf{\textit{Answer}}$$

Construction Control

Nordlund (1982) offered an alternative method of computing load capacities based on the energy delivered by the drop hammer. This method is similar in many ways to the pile driving formulas discussed in Chapter 13. It is most often used for construction control, especially if it has been calibrated with an on-site load test.

PIFs With Auger-Cast Shafts

Some contractors have combined a PIF base with an auger-cast shaft (Massarsch et al., 1988). The skin friction resistance of this type of foundation will be much greater and more reliable than that of a conventional cased PIF, but not as large as that of a compacted shaft PIF. However, an auger-cast shaft could be built more quickly and at less cost than a compacted shaft PIF, thus providing reasonably high capacity at a moderate cost. Very few of these foundations have been built.

16.4 ANCHORS

The term *anchor* generally refers to a foundation designed primarily to resist uplift (tensile) loads. Although most foundations are able to resist some uplift, anchors are designed specifically for this task and are often able to do so more efficiently and at a lower cost.

Lightweight structures often require anchors because the lateral loads are often large compared to the downward loads. These include power transmission towers, radio antennas, and mobile homes. Anchors are also used to support pipelines, retaining walls, dams, and other structures that require tensile restraint.

Kulhawy (1985) divided anchors into three categories, as shown in Figure 16.8. These include:

- *Spread anchors* - Wide structural members embedded in the ground.
- *Helical anchors* - Steel shafts with helices that are screwed into the ground.
- *Grouted anchors* - Grouted holes with steel tendons.

Some of these designs are proprietary. The manufacturers can provide specific design guidelines.

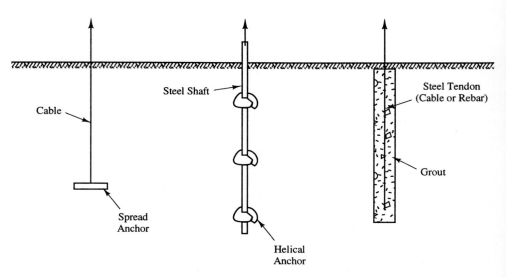

Figure 16.8 Types of anchors (Adapted from Kulhawy, 1985; Used with permission of ASCE).

SUMMARY OF MAJOR POINTS

1. Special types of deep foundations are available that are neither piles nor drilled shafts. These are less common, but are appropriate in certain situations.
2. Mandrel driven thin shells filled with concrete provide the high displacement associated with piles with cast-in-place concrete.
3. Auger-cast piles consist of grout pumped through a hollow stem auger. They are often less expensive than some other types of deep foundations.
4. Pressure-injected footings (PIFs) are built by ramming concrete into the ground. The construction process densifies loose soils, thus improving their load bearing capacity.
5. Special foundations designed primarily to resist uplift (tension) are called anchors.

QUESTIONS AND PRACTICE PROBLEMS

16.1 Step 4 of the construction procedure for mandrel-driven thin shell piles says to "remove the mandrel and inspect the shell." What should an inspector look for?

16.2 How could auger-cast piles become contaminated with soil during construction?

16.3 Why do pressure injected footings have a higher net unit end bearing pressure, $q_e{}'$, than other foundations with an equal base diameter?

16.4 An 18 inch diameter auger-cast pile is to be built in a well graded silty sand and must support a compressive load of 150 k. The soil has a unit weight of 120 lb/ft^3 above the groundwater table, 128 lb/ft^3 below and $N_{60} = 25$. The groundwater table is at a depth of 45 ft. Compute the required depth of embedment using a factor of safety of 3.0.

16.5 Using the data in Problem 16.4 and a 30 ft deep cased-shaft pressure injected footing instead of an auger-cast pile, determine the required base diameter.

16.6 A 25 ft deep "large" pressure-injected footing with a compacted shaft is to be built in the soil described on page 258. Using a factor of safety of 3.5, compute the allowable downward load capacity. Discuss the sources and potential magnitudes of error in this analysis.

17

Laterally Loaded Deep Foundations

> *I read some of the papers last night where some of these pile driving formulas were derived, and the result was that my sleep was very much disturbed.*
>
> Pioneer foundation engineer Lazarus White (1936)

The discussions in Chapters 11 to 16 considered only axial loads. However, many deep foundations also must support *lateral loads*. These include shear and/or moment loads as shown in Figure 17.1. Sources of these loads include the following:

- Earth pressures on the back of retaining walls
- Wind loads
- Seismic loads
- Berthing loads from ships as they contact piers or other harbor structures
- Downhill movements of earth slopes
- Vehicle acceleration and braking forces on bridges
- Eccentric vertical loads on columns
- Ocean wave forces on offshore structures
- River current forces on bridge piers
- Cable forces from electrical transmission towers
- Structural loads on abutments for arch or suspension bridges

These loads are often large and may exceed the axial loads.   Therefore, they are an important aspect of foundation analysis and design.

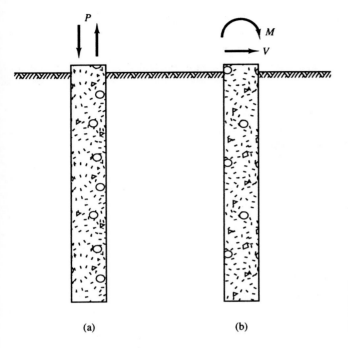

Figure 17.1 Loading on a deep foundation: (a) axial loads-downward (compression) or uplift (tension); (b) lateral loads-moment and/or shear.

17.1 BATTER PILES

Until the middle twentieth century, engineers did not know how to evaluate lateral loads on piles, so they assumed deep foundations were only able to resist axial loads. Therefore, when horizontal loads were present, engineers installed some of the piles at an angle to the vertical, as shown in Figure 17.2. Thus, in theory, each pile would be subjected to axial loads only.

Although it is possible to build most types of deep foundations on an angle, driven piles are the most common choice. Piles driven in this fashion are known as *batter piles* or *raker piles*. Contractors can usually install them at a batter of up to 4 vertical to 1 horizontal by simply tilting the leads on the pile driver.

Although batter piles have generally performed well, there are at least two situations that cause problems:

- Large offshore structures, such as drilling platforms, are subjected to very large lateral loads, but it is not practical to drive piles at a sufficiently large angle to accommodate these loads.
- Batter piles form a very stiff foundation system. This is suitable when only static

loads are present, but can cause problems when dynamic loads are applied, such as during an earthquake, as shown in Figure 17.3. A more flexible system would be better.

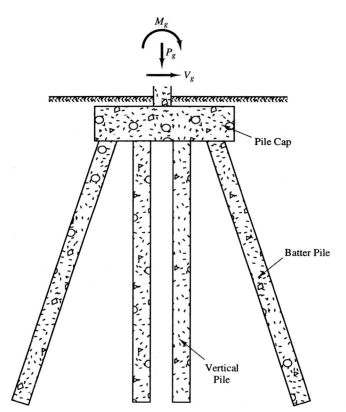

Figure 17.2 The use of batter piles to resist combined vertical and horizontal loads.

Engineers also recognized that groups of batter piles, such as shown in Figure 17.2, are not subject to axial loads only. In reality, shear and moment loads also form because:

- Only certain combinations of applied vertical and horizontal loads on the pile cap produce only axial loads in all of the piles. Because the horizontal loads often are temporary (i.e., wind or seismic loads), the actual working loads often are not properly combined. For example, if the horizontal load is sometimes zero, then the vertical load generates flexural stresses in the battered piles.

- The soil may consolidate after the piles are driven (see discussion of downdrag loads in Section 19.1 of Chapter 19). This downward movement of the soil produces lateral loads on the battered piles.

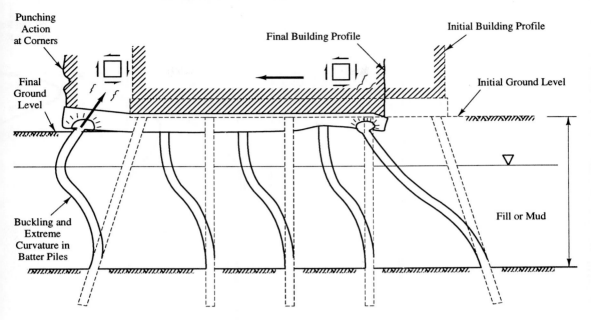

Figure 17.3 Behavior of batter piles during seismic loading (Adapted from Dowrick, 1987; Used with permission of John Wiley and Sons).

17.2 RESPONSE TO SHEAR AND MOMENT LOADS

Because of the difficulties with batter piles and a desire to produce more economical designs, engineers reconsidered the assumption that deep foundations are only able to resist axial loads. A foundation system that consists only of vertical members able to resist both axial and lateral loads is more flexible and thus resists dynamic loads more effectively. It also is less expensive to build, especially on offshore structures.

Applied shear and/or moment loads induce flexural stresses in the foundation, as shown in Figure 17.4. In response to these stresses, the foundation moves laterally and thus mobilizes resisting forces in the soil. The lateral soil resistance per unit length of the foundation is p and has units of force per length. The magnitude of p increases as the lateral deflection increases and eventually reaches a peak value, p_{ult}, as shown in Figure 17.5.

The behavior of a laterally loaded foundation depends on the properties of both the foundation and the soil. Relevant foundation properties include:

- Diameter, B
- Length, D
- Modulus of elasticity, E
- Moment of inertia, I

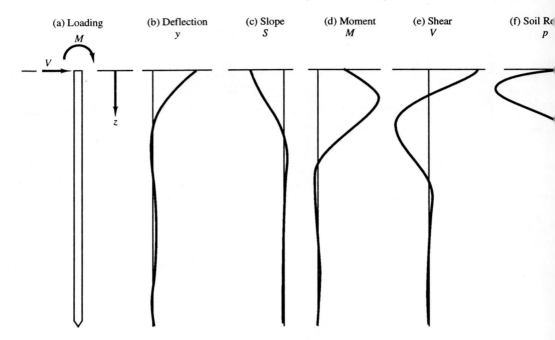

Figure 17.4 Forces and deflections in a vertical deep foundation subjected to shear and/or moment loads (Adapted from Matlock and Reese, 1960; Used with permission of ASCE).

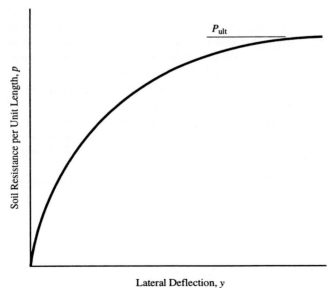

Figure 17.5 Soil resistance per unit length, p, as a function of lateral deflection, y.

The soil properties are implicit in the p-y function shown in Figure 17.5 and include the soil type, confining stress, and other characteristics.

The type of connection between the pile and the structure is also important because it determines the kinds of restraint, if any, acting on the pile. Engineers usually assume that one of the following conditions prevails:

- The *free-head condition*, shown in Figure 17.6a, means that the top of the pile may freely move laterally and rotate when subjected to shear and/or moment loads.
- The *restrained-head condition* (also known as the *fixed-head condition*), shown in Figure 17.6b, means that the top of the pile may move laterally, but is not permitted to rotate. Piles connected to a very stiff pile cap closely approximate this condition.
- The *pure moment condition*, shown in Figure 17.6c, occurs when there is an applied moment load, but no applied shear load. It results in rotation of the top of the pile, but no lateral movement.

Selecting one of these alternatives defines the boundary conditions at the top of the pile and thus permits the analysis to continue. Although real piles often have connections that are intermediate between these three ideal conditions, one of these assumptions is usually reasonable for design purposes.

Both the shear and moment at the bottom of a deep foundation are zero.

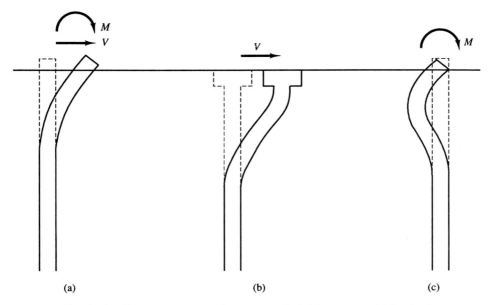

Figure 17.6 Types of connections between the pile and the structure: (a) free-head; (b) restrained-head; and (c) pure moment.

The foundation transmits most of the applied lateral loads to the upper soils, so the soil properties within about 10 diameters of the ground surface are most important. Therefore, be especially careful to consider scour or other phenomena that might eliminate some of the upper soils.

17.3 ULTIMATE SOIL CAPACITY

The ultimate soil capacity is the greatest lateral load the soil can sustain, regardless of the lateral deflection. Think of it as an ultimate lateral bearing capacity. It depends on the diameter and length of the foundation, the strength of the soil, and other factors.

To compute the ultimate soil capacity, assume that the foundation has sufficient flexural strength that the soil fails under lateral load before a flexural failure can occur. If a free-head connection is present, failure occurs when the foundation rotates enough to mobilize the ultimate soil capacity, as shown in Figure 17.7a. A restrained-head connection causes a translational failure, as shown in Figure 17.7b.

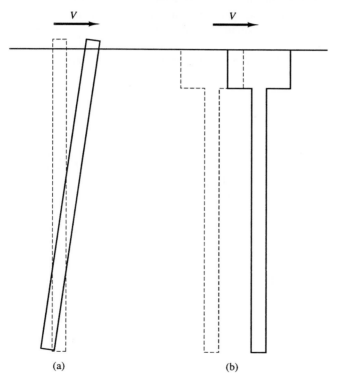

(a)   (b)

Figure 17.7 Modes of failure that control the ultimate soil capacity: (a) rotational failure of a free-head foundation; (b) translational failure of a fixed-head foundation.

In most design problems, the applied shear and moment loads are given, so we must determine the minimum required embedment depth, D_{min}, to provide a sufficient factor of safety against a soil failure. The design embedment depth, D, must be at least equal to D_{min}. If a significant axial load also is present, D will probably need to be much greater than D_{min} to provide sufficient axial load capacity.

Brinch Hansen (1961a) and Broms (1964a, 1964b, 1965) developed methods of computing the ultimate lateral capacity of deep foundations. We will consider only Broms' method.

Broms' Method

Broms considered both the free-head and restrained-head conditions. However, his method is limited to homogeneous soil deposits that are either purely cohesive ($\phi = 0$) or purely cohesionless ($c = 0$). He did not consider the pure moment loading condition.

The applied lateral load, V, acts at a distance e above the ground surface, as shown in Figure 17.8. Thus, the applied moment M equals Ve. In most problems, V and M are given and e is computed.

Cohesive Soils

The ultimate static lateral resistance, p_{ult}, of cohesive soils is approximately $2s_u B$ at the ground surface (where s_u = the undrained shear strength of the soil, and B = the diameter of the foundation), and increases to 8 to 12 $s_u B$ at a depth of about three diameters. For analysis purposes, Broms used a simplified distribution, as shown in Figures 17.8 and 17.9. It uses $p_{ult} = 0$ from the ground surface to a depth of $1.5 B$ and $p_{ult} = 9s_u B$ below that depth. He then used the principles of structural analysis to compute the ultimate lateral load capacity.

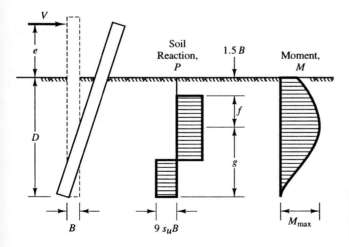

Figure 17.8 Soil reactions and foundation moments for free-head foundations in cohesive soil (Adapted from Broms, 1964a; Used with permission of ASCE).

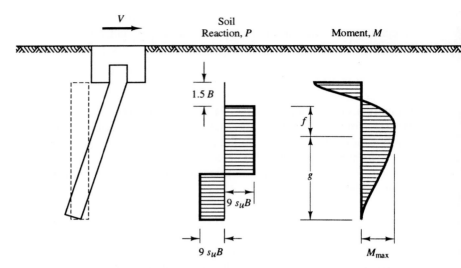

Figure 17.9 Soil reactions and foundation moments for restrained-head foundations in cohesive soil (Adapted from Broms, 1964a; Used with permission of ASCE).

For a free-head foundation, D_{min} is:

$$D_{min} = \sqrt{\frac{FV(e + 1.5B + 0.5f)}{2.25Bs_u}} + 1.5B + f \tag{17.1}$$

$$f = \frac{FV}{9s_uB} \tag{17.2}$$

For a restrained-head foundation, use:

$$D_{min} = \frac{FV}{9s_uB} + 1.5B \tag{17.3}$$

Where:

D_{min} = minimum required foundation embedment
F = factor of safety (typically 3.0)
V = applied shear load at top of foundation
$e = M/V$

M = applied moment load at top of foundation

B = foundation diameter

s_u = undrained shear strength of soil

Cohesionless Soils

For cohesionless soils, Broms used $p_{ult} = 3\,\sigma_v'\,K_p\,B$ where σ_v' is the vertical effective stress, K_p is the Rankine coefficient of passive earth pressure (described in Chapter 23), and B is the foundation diameter.   In a uniform soil, this produces a triangular distribution, as shown in Figures 17.10 and 17.11.

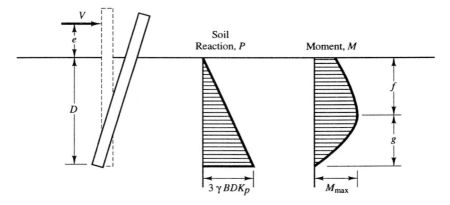

Figure 17.10 Soil reactions and foundation moments for free-head foundations in cohesionless soil (Adapted from Broms, 1964b; Used with permission of ASCE).

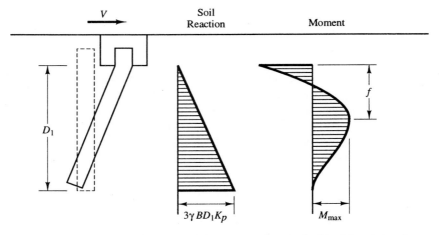

Figure 17.11 Soil reactions and foundation moments for restrained-head foundations in cohesionless soil (Adapted from Broms, 1964b; Used with permission of ASCE).

For free-head foundations, D_{min} is the smallest positive root of:

$$F = \frac{0.5\,\gamma\,B\,D_{min}^3\,K_p}{V\,(D_{min} + e)}$$ (17.4)

For restrained-head foundations, use:

$$D_{min} = \sqrt{\frac{F\,V}{1.5\gamma B K_p}}$$ (17.5)

Where:

F = factor of Safety (typically 3.0)

γ = unit weight of soil

B = foundation diameter

D_{min} = minimum required foundation embedment

K_p = coefficient of passive earth pressure = $\tan^2(45 + \phi/2)$

ϕ = friction angle of soil

V = applied shear load at top of foundation

$e = M/V$

M = applied moment load at top of foundation

If the groundwater table is at or above the ground surface, use the submerged unit weight $(\gamma - \gamma_w)$ instead of γ. If the groundwater table is between the ground surface and a depth D_{min}, use a weighted average of γ above the groundwater table and $(\gamma - \gamma_w)$ below.

17.4 ULTIMATE STRUCTURAL CAPACITY

Sometimes we can design laterally loaded deep foundations without considering the lateral deflections that occur during loading. A foundation that supports a large highway sign would be an example. We can base these designs on the *ultimate structural capacity*, which is the greatest lateral load the foundation can sustain without inducing a flexural failure. This capacity depends on the magnitude and distribution of the lateral soil resistance and on the flexural strength of the foundation.

For most design problems, the applied shear and moment loads are given and we need to compute the corresponding maximum moment, M_{max}, induced in the foundation. Then, using the methods described in Chapter 18, we provide a structural design with enough flexural strength to safely resist this moment. This forms the basis for the structural design when there are no limitations on lateral deflection.

For free-head foundations, M_{max} occurs at some depth below the ground surface, as shown in Figure 17.12a. The same is true for relatively long restrained-head foundations. However, the maximum moment in shorter restrained-head foundations occurs immediately below the pile cap, as shown in Figure 17.12b. The exact length that separates "short" from "long" depends on the properties of both the soil and the foundation.

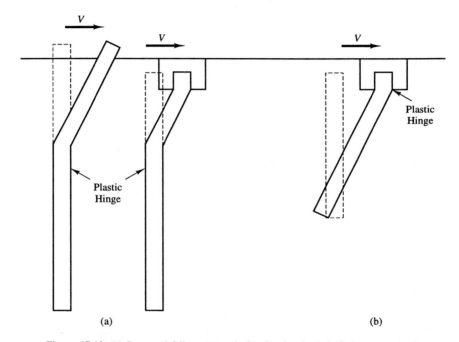

Figure 17.12 (a) Structural failure occurs in free-head and relatively long restrained-head foundations when a plastic hinge forms at some depth below the ground surface; (b) Shorter restrained-head foundations fail when a plastic hinge forms immediately below the pile cap (Adapted from Broms, 1965; Used with permission of ASCE).

Broms' Method

Broms also provided methods of computing M_{max} using the ultimate soil resistance distributions described earlier.

Cohesive Soils

For free-head foundations in cohesive soils, M_{max} is as follows (where V, e, B and f are as defined for Equations 17.1 - 17.3.):

$$M_{max} = V(e + 1.5B + 0.5f) \qquad (17.6)$$

For restrained-head foundations, compute the moment immediately below the pile cap and the moment at a depth of $1.5B + f$. These are M_1 and M_2, respectively.

$$M_1 = 9s_u Bf(1.5B + 0.5f) - 2.25s_u Bg^2 \geq 0 \qquad (17.7)$$

$$g = D_{min} - 1.5B - f \qquad (17.8)$$

$$M_2 = \frac{V(1.5B + 0.5f)}{2} \qquad (17.9)$$

Design the foundation using the greater of these two moments.

Cohesionless Soils

For free-head foundations in cohesionless soils, M_{max} is:

$$M_{max} = V(e + 0.67f) \qquad (17.10)$$

$$f = 0.82 \sqrt{\frac{FV}{BK_p\gamma}} \qquad (17.11)$$

For restrained-head foundations, compute M_1 and M_2:

$$M_1 = VD - (0.5\gamma BD^3 K_p) \geq 0 \qquad (17.12)$$

$$M_2 = 0.67VD \qquad (17.13)$$

M_1 is the moment just below the pile cap; M_2 is at a depth f below the ground surface. Use the greater of these two moments for design.

Example 17.1

An outdoor billboard is to be supported by a steel pipe column that will rest on a single 1000 mm diameter drilled shaft. The loads acting on the foundation will be: downward normal load $= 30.5$ kN, shear load $= 90.9$ kN, and moment load $= 955$ kN-m. The soil is a clay with $s_u = 175$ kPa. Compute the required embedment to support the axial and lateral loads. Then, compute the maximum moment, M_{max}. Consider this to be a free-head foundation with no limits on lateral deflection.

Solution:

1. Use the techniques in Chapter 15 to compute the required embedment for axial loads.

$$q_e' = N_c^* s_u = (9)(175) = 1570 \text{ kPa}$$

$$A_e = \frac{\pi (1.000)^2}{4} = 0.785 \text{ m}^2$$

$$q_e' A_e = (1570)(0.785) = 1230 \text{ kN}$$

For $F = 2.5 \rightarrow$ required ultimate load $= (2.5)(30.5) = 76.2$ kN

$\therefore$ End bearing is more than adequate to resist axial loads; need only minimal embedment.

2. Compute D_{min} for lateral loads:

$$e = \frac{M}{V} = \frac{955}{90.9} = 10.5 \text{ m}$$

Using Equations 17.2 and 17.1:

$$f = \frac{FV}{9 s_u B} = \frac{(3)(90.9)}{(9)(175)(1)} = 0.17 \text{ m}$$

$$D_{min} = \sqrt{\frac{FV(e + 1.5B + 0.5f)}{2.25 B s_u}} + 1.5B + f$$

$$= \sqrt{\frac{(3)(90.9)[10.5 + (1.5)(1) + (0.5)(0.17)]}{(2.25)(1)(175)}} + (1.5)(1) + 0.17$$

$$= 4.6 \text{ m}$$

$\therefore$ Lateral load controls minimum required embedment; use $D = \textbf{4.6 m}$ $\Leftarrow$ *Answer*

3. Compute M_{max} using Equation 17.6

$$M_{max} = V(e + 1.5B + 0.5f)$$
$$= (90.9)[10.5 + (1.5)(1) + (0.5)(0.17)]$$
$$= \mathbf{1100\ kN\text{-}m} \qquad \qquad \Leftarrow \textit{Answer}$$

Therefore, provide sufficient reinforcement to safely resist a moment of 1100 kN-m.

Accuracy

Figure 17.13 shows 90% confidence intervals for ultimate soil capacities and maximum moments computed using Broms' method. Although much of this data were obtained from model piles, it suggests that the moment computations are good, whereas the soil capacity computations are usually conservative.

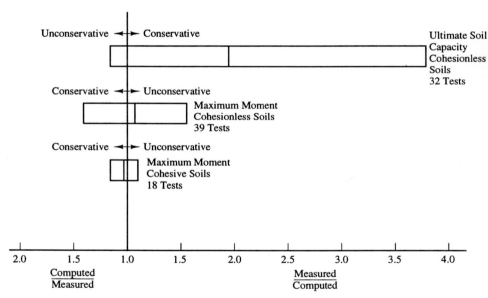

Figure 17.13 90% confidence intervals for ultimate soil capacities and ultimate moments using Broms' method (based on data from Broms 1964a, 1964b). The lines in the middle of the bars represent the average prediction, and the numbers to right indicate the number of data points.

17.5 LOAD-DEFORMATION BEHAVIOR

Occasionally, the analysis described in Sections 17.3 and 17.4 is sufficient to design laterally loaded deep foundations. However, many structures cannot tolerate the large lateral deflection required to mobilize the lateral soil capacity. For example, buildings, bridges, and other similar structures typically can tolerate no more than 0.25 to 0.75 in (6 - 18 mm) of lateral movement. Therefore, when there are limits on the allowable lateral deflections, we must conduct a *load-deformation analysis* to determine the lateral load that corresponds to a certain allowable deflection. This analysis also evaluates the maximum moment, M_{max}, that corresponds to this deflection.

Analysis Methods

Load-deformation analyses must consider both the flexural stiffness of the foundation and the lateral resistance from the soil. The flexural stiffness depends on the modulus of elasticity, E, and moment of inertia, I, of the foundation and is easily assessed using the principles of structural analysis. However, the soil resistance is much more difficult to evaluate.

Early attempts to analyze lateral deflections, such as Matlock and Reese (1960), considered the soil to be a linear elastic material where the ratio of the lateral soil force to the lateral deflection was defined as the stiffness, k. These analyses use the beam-on-elastic-foundation technique described in Chapter 9. Some methods used a k that was constant with depth, and others used a k that increased with depth according to some function.

During the 1960s and 1970s, engineers often expressed the results of these analyses using a *depth to fixity* (Davisson, 1970), as shown in Figure 17.14. This model ignored the soil between the ground surface and the depth to fixity and considered the soil below that depth to be infinitely strong. Thus, the structural engineer simply computed the flexural stresses and deflections in the foundation as if there were a rigid connection at the depth of fixity and no soil resistance above that point. This method was especially convenient for the structural engineer because it easily fit into conventional structural analyses.

These early methods always considered k to be constant at any particular depth. In other words, they assumed the lateral resistance was proportional to the lateral deflection. In reality, this function is clearly nonlinear. McClelland and Focht (1958) were the first to recognize the importance of this nonlinearity. They were also the first to report experimental data from an instrumented full-scale lateral load test.

The *p-y* Method

Analyses that consider the nonlinear relationship between lateral resistance and deflection more accurately model the true behavior of laterally loaded deep foundations. However, the implementation of such methods required test data to define the nature of this function and digital computers to perform the necessary computations. Fortunately, both of these obstacles were overcome during the 1970s and 1980s. Engineers now have access to enough experimental data and computers to conduct nonlinear analyses on routine design projects.

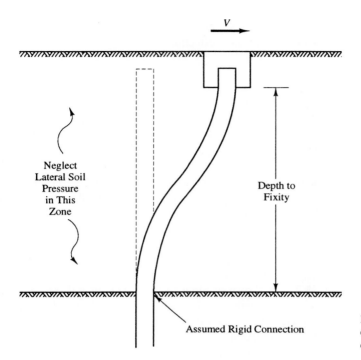

Figure 17.14 Depth to fixity method of expressing the lateral load-deformation characteristics of a deep foundation.

The most widely used nonlinear analysis is the *p-y method*, which is based on the work by McClelland and Focht (1958). It models the soil resistance using a series of nonlinear springs, as shown in Figure 17.15. Much of the research and development for this method was conducted at the University of Texas at Austin by Reese and his colleagues.

The *p-y* method has been well received because it was calibrated from full scale load tests and because it is able to consider many variables, including:

- Any nonlinear load-deflection curve
- Variations of the load-deflection curve with depth
- Variations in the foundation stiffness (*EI*) with depth
- Elastic-plastic (i.e., nonlinear) flexural behavior in the foundation
- Any defined head constraint condition, including free, restrained, pure moment, and others

When the lateral load approaches the ultimate lateral capacity, the nonlinear *p-y* curves generate large deflections, so the computed displacement of the foundation also becomes larger. Therefore, the *p-y* method implicitly includes an ultimate lateral capacity analysis. Thus, there is no need to conduct a separate analysis as described in Section 17.4.

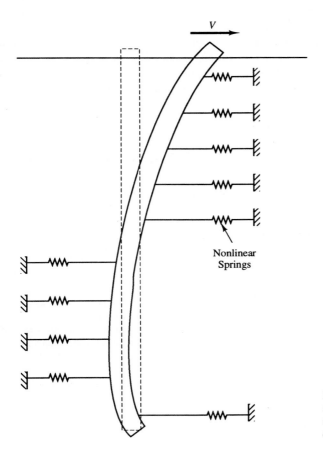

Figure 17.15 The *p-y* method models the soil resistance using a series of nonlinear springs. The load-displacement function of each spring is defined by its *p-y* curve.

p-y Curves

The heart of the *p-y* method is the definition of the lateral load-deflection relationships between the foundation and the soil. These are expressed in the form of *p-y* *curves*, where *p* is the lateral soil resistance per unit length of the foundation (expressed in units of force per length), and *y* is the lateral deflection. Typical *p-y* curves are shown in Figure 17.17.

The *p-y* relationship might first appear to be a nonlinear extension of the Winkler beam-on-elastic-foundation concept described in Chapter 9. However, there is an important difference between the two: The Winkler model considers only compressive forces between the foundation and the soil, whereas the lateral soil load acting on a deep foundation is the result of compression on the leading side, shear friction on the two adjacent sides, and possibly some small compression on the back side. These components are shown in Figure 17.16. Thus, it is misleading to think of the *p-y* curve as a compression phenomenon only (Smith, 1989).

The ultimate compression resistance will probably be much greater than the ultimate side shear resistance. However, mobilization of the side shear requires much less deflection, so it may be an important part of the total resistance at the small deflections generally associated with the working loads. We need additional research to more fully understand this behavior.

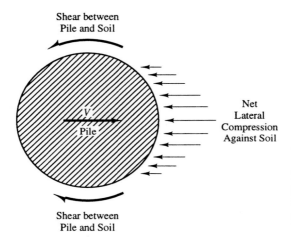

Figure 17.16 The soil resistance to lateral pile movement has both compression and shear components. The sum of them is the *p* in *p-y* curves (Adapted from Smith, 1989; Used with permission of ASCE).

The *p-y* curve for a particular point on a foundation depends on many factors, including the following:

- Soil type
- Type of loading (i.e., short term static, sustained static, repeated, or dynamic)
- Foundation diameter and cross-sectional shape
- Coefficient of friction between foundation and soil
- Depth below the ground surface
- Foundation construction methods
- Group interaction effects

The influence of these factors is not well established, so it has been necessary to develop *p-y* curves empirically from full-scale load tests. Most of this data were obtained from 10 to 24 in (250 - 600 mm) diameter steel pipe piles. Reese (1984, 1986) summarizes many of the tests conducted thus far and provides recommended *p-y* curves for analysis and design.

Some curves are *ductile*, as shown in curve A in Figure 17.17. These curves reach the maximum resistance, p_{max}, at a certain deflection, and then maintain this resistance at greater deflections. Other curves are *brittle*, as shown in curve B in Figure 17.17, and have a decreasing *p* at large deflections. Brittle curves can occur in some clays, especially if they are stiff or if the loading is repeated or dynamic. Soft clays under static loading and sands appear to have ductile curves. Brittle curves are potentially more troublesome because of their potential for producing large foundation movements.

Although the empirical *p-y* data collected thus far have been an essential part of making the *p-y* method a practical engineering tool, we continue to need more data. Additional instrumented load tests would further our understanding of this relationship and thus provide more accurate data for analysis and design. For large projects, it may be appropriate to conduct full-scale lateral load tests and develop site-specific *p-y* curves, as described later in this chapter.

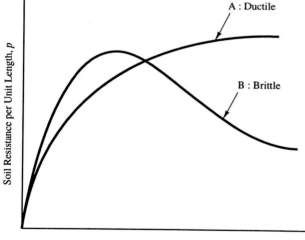

Figure 17.17 Typical *p-y* curves.

There have been some efforts to develop *p-y* curves from in-situ pressuremeter or dilatometer tests (Baguelin et al., 1978; Robertson et al., 1989). Although these tests directly measure something similar to the compression component of lateral pile resistance, they do not address the side shear component. This may be at least a partial explanation of Baguelin's assessment that this approach produces pessimistic results. There also are scale effects to consider. Further research may improve the reliability of these methods.

Numerical Modeling

Practical p-y analyses must consider changes in the p-y relationship with depth, as shown in Figure 17.18a. We can accomplish this using a *finite difference analysis* that divides the foundation into n intervals, as shown in Figure 17.18b. Assign a p-y curve and foundation stiffness EI to each interval, thus forming a mathematical model of the foundation and its lateral interaction with the soil.

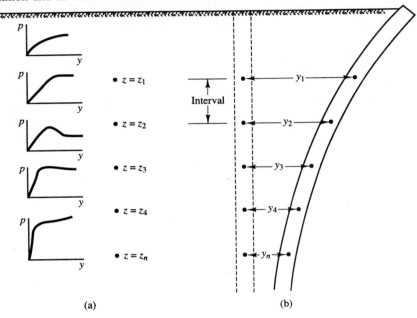

z = depth below ground surface

Figure 17.18 (a) Changes in the p-y relationship with depth; (b) A finite difference model.

It also is necessary to apply appropriate boundary conditions. There are two known conditions at the bottom of the foundation: both the shear and the moment are zero ($V_b = M_b = 0$). The boundary conditions at the top depend on the type of restraint:

- For the free-head condition, the applied shear and moment loads, V and M, respectively, are known. There is both rotation and deflection at the top ($S_t \neq 0$, $y_t \neq 0$).
- For the restrained-head condition, the applied shear load, V, and the slope, S_t, are known. Normally S_t is taken to be zero, but it could have any fixed value.
- For the pure moment condition, the applied moment, M, is known, the applied shear, V, is zero, and the lateral deflection at the top, y_t, is zero. However, there is a rotation at the top, so S_t is not zero.

The following equations define the foundation's behavior across each interval:

$$S_z = \frac{dy}{dz} \tag{17.14}$$

$$M_z = EI\frac{d^2y}{dz^2} \tag{17.15}$$

$$V_z = EI\frac{d^3y}{dz^3} \tag{17.16}$$

$$p_z = EI\frac{d^4y}{dz^4} \tag{17.17}$$

Where:
  S_z = slope of foundation at depth z
  M_z = bending moment in foundation at depth z
  V_z = shear force in foundation at depth z
  p_z = lateral soil resistance per unit length of the foundation at depth z
  E = modulus of elasticity
  I = moment of inertia in the direction of bending
  y = lateral deflection
  z = depth below ground surface

Using this information with an iterative solution, a computer program finds the equilibrium condition between the soil reaction and the stresses in the foundation and thus computes the shear, moment, and lateral deflection at each interval. Both public domain and proprietary programs are available.

Accuracy

Reese and Wang (1986) compared the results of 22 static full-scale lateral load tests with the predicted deflections. They also compared the measured and predicted maximum moments for 12 of these tests. These data, expressed in Figure 17.19 using confidence intervals, suggest that this method provides good predictions of deflections and very good predictions of moments.

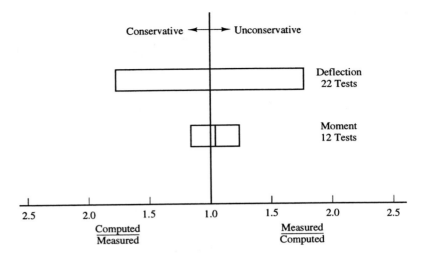

Figure 17.19 90% confidence intervals for computed lateral deflection and bending moment predictions from *p-y* analysis (based on data from Reese and Wang, 1986). The line in the middle of each bar represents the average prediction, and the number to the right is the number of data points.

Evans and Duncan's Method

Evans and Duncan (1982) developed a convenient method of expressing the lateral load-deflection behavior in chart form. They compiled these charts from a series of p-y method computer analyses using the computer program COM624.

The advantages of these charts include the following:

- The analyses can be performed more quickly, and they do not require the use of a computer.
- The load corresponding to a given pile deflection can be determined directly, rather than by trial.
- The load corresponding to a given maximum moment in the pile can be determined directly, rather than by trial.

These charts are also a useful way to check computer output from more sophisticated analyses.

The charts presented here are a subset of the original method and apply only to deep foundations that satisfy the following criteria:

- The stiffness, *EI*, is constant over the length of the pile.
- The shear strength of the soil, expressed either as s_u or ϕ, and the unit weight, γ, are constant with depth.

• The pile is long enough to be considered fixed at the bottom. For relatively flexible piles, such as timber piles, this corresponds to a length of at least 20 diameters. For relatively stiff piles, such as those made of steel or concrete, the length must be at least 35 diameters.

We can idealize deep foundations that deviate slightly from these criteria, such as tapered piles, by averaging the EI, s_u, ϕ, or γ values from the ground surface to a depth of 8 pile diameters.

Characteristic Load and Moment

Evans and Duncan defined the *characteristic shear load*, V_c, and *characteristic moment load*, M_c, as follows:

$$V_c = \lambda B^2 E R_I \left(\frac{\sigma_p}{E R_I} \right)^m (\varepsilon_{50})^n \tag{17.18}$$

$$M_c = \lambda B^3 E R_I \left(\frac{\sigma_p}{E R_I} \right)^m (\varepsilon_{50})^n \tag{17.19}$$

$$R_I = \frac{I}{\pi B^4 / 64} \tag{17.20}$$

$$= 1.00 \text{ for solid circular cross sections}$$
$$= 1.70 \text{ for square cross sections}$$

For plastic clay and sand:

$$\lambda = 1.00 \tag{17.21}$$

For brittle clay[1]:

$$\lambda = (0.14)^n \tag{17.22}$$

---

[1] A brittle clay is one with a residual strength that is much less than the peak strength.

For cohesive soils:

$$\sigma_p = 4.2\, s_u \qquad (17.23)$$

For cohesionless soils:

$$\sigma_p = 2\, C_{p\phi}\, \gamma B \tan^2(45 + \phi/2) \qquad (17.24)$$

Where:

V_c = characteristic shear load

M_c = characteristic moment load

λ = a dimensionless parameter dependent on the soil's stress-strain behavior

B = diameter of foundation

E = modulus of elasticity of foundation

= 29,000,000 lb/in^2 (200,000 MPa) for steel

= 15,200 $\sigma_r\, (f_c'/\sigma_r)^{0.5}$ for concrete

= 1,600,000 lb/in^2 (11,000 MPa) for Southern Pine or Douglas Fir

f_c' = 28-day compressive strength of concrete

σ_r = reference stress = 14 lb/in^2 = 0.10 MPa

R_I = moment of inertia ratio (dimensionless)

σ_p = representative passive pressure of soil

ε_{50} = strain at which 50% of the soil strength is mobilized

m, n = exponents from Table 17.1

I = moment of inertia of foundation

= $\pi B^4/64$ for solid circular cross-sections

= $B^4/12$ for square cross-sections

Also see tabulated values in Chapter 18

s_u = undrained shear strength of soil from the ground surface to a depth of 8 pile diameters

ϕ = friction angle of soil (deg) from ground surface to a depth of 8 pile diameters

$C_{p\phi}$ = passive pressure factor = $\phi/10$

γ = unit weight of soil from ground surface to a depth of 8 pile diameters. If the groundwater table is in this zone, use a weighted average of γ and γ_b, where γ_b is the buoyant unit weight in the zone below the groundwater table.

The value of ε_{50} could be determined from triaxial compression tests. Typically, $\varepsilon_{50} \approx 0.01$ for clays and $\varepsilon_{50} \approx 0.002$ for medium dense sands containing little or no mica.

TABLE 17.1 VALUES OF EXPONENTS m AND n FOR USE IN EQUATIONS 17.18 and 17.19

Soil Type	For V_c		For M_c	
	m	n	m	n
Cohesive	0.683	-0.22	0.46	-0.15
Cohesionless	0.57	-0.22	0.40	-0.15

Evans and Duncan (1982).

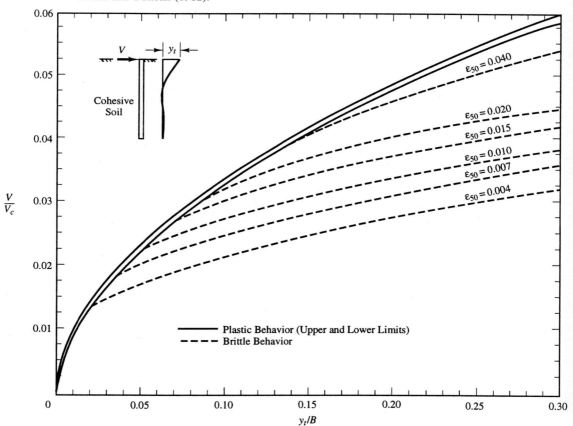

Figure 17.20 Shear load vs. lateral deflection curves for free-head conditions in cohesive soil (Evans and Duncan, 1982).

Using the Charts

The charts in Figures 17.20 to 17.29 express the relationships between the actual shear, moment and deflection, where:

V = applied shear at top of foundation

M = applied moment at top of foundation

M_{max} = maximum moment in foundation

y_t = lateral deflection at top of foundation

Some foundations are subjected to both shear and moment loads. As a first approximation, compute the lateral deflections and moments from each component separately and add them. Alternatively, use the nonlinear superposition procedure described in Evans and Duncan (1982).

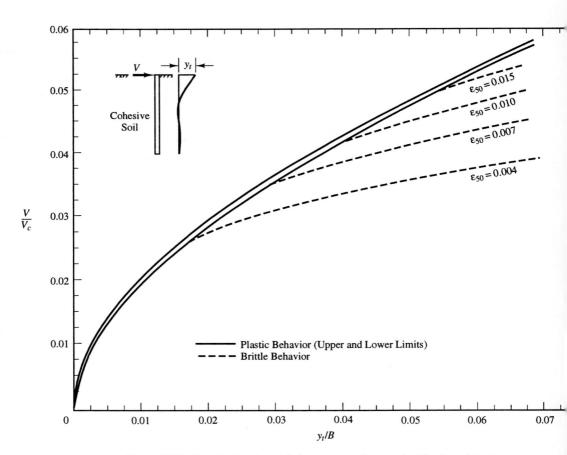

Figure 17.21 Shear load vs. lateral deflection curves for restrained-head condition in cohesive soil (Evans and Duncan, 1982).

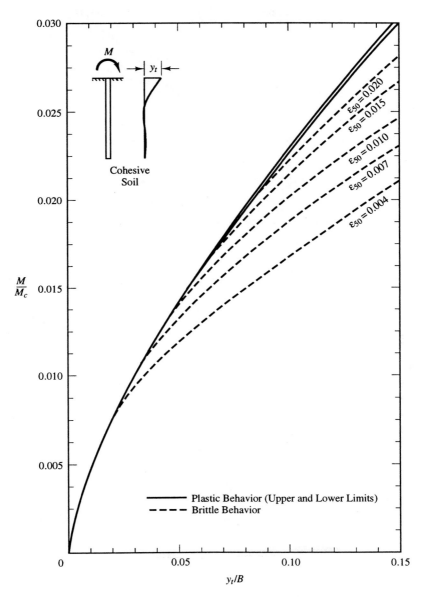

Figure 17.22 Moment load vs. lateral deflection curves for free-head condition in cohesive soil (Evans and Duncan, 1982).

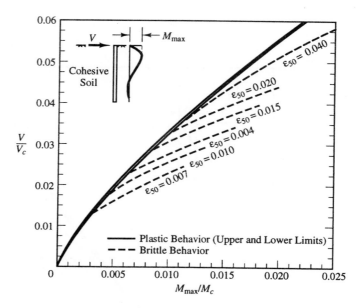

Figure 17.23 Shear load vs. maximum moment curves for free-head condition in cohesive soil (Evans and Duncan, 1982).

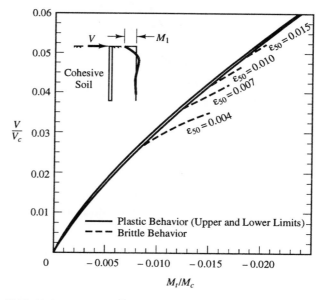

Figure 17.24 Shear load vs. maximum moment curves for restrained-head condition in cohesive soil (Evans and Duncan, 1982).

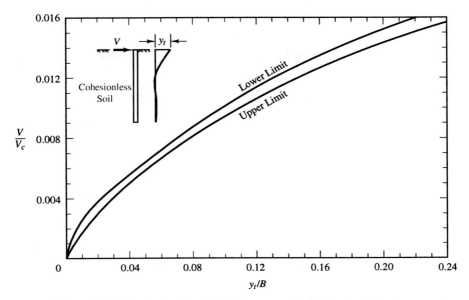

Figure 17.25 Shear load vs. lateral deflection curves for free-head condition in cohesionless soil (Evans and Duncan, 1982).

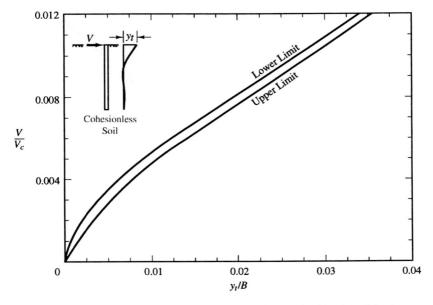

Figure 17.26 Shear load vs. lateral deflection curves for restrained-head condition in cohesionless soil (Evans and Duncan, 1982).

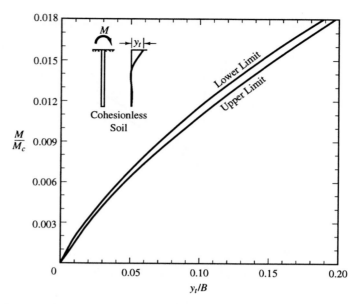

Figure 17.27 Moment load vs. lateral deflection curves for free-head condition in cohesionless soil (Evans and Duncan, 1982).

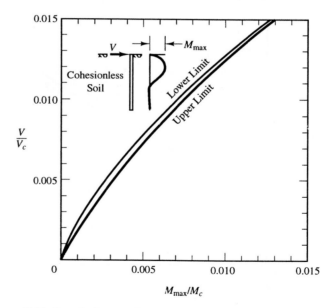

Figure 17.28 Shear load vs. maximum moment curves for free-head condition in cohesionless soils (Evans and Duncan, 1982).

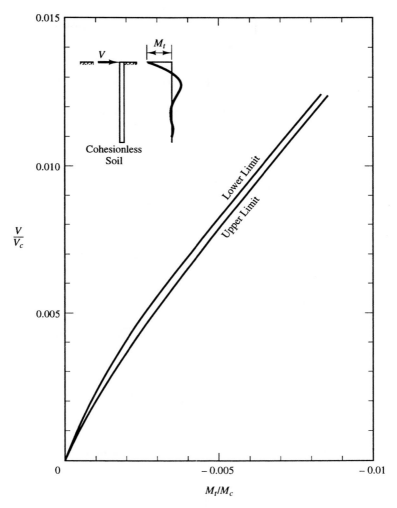

Figure 17.29 Shear load vs. maximum moment curves for restrained-head condition in cohesionless soils (Evans and Duncan, 1982).

Example 17.2

A 20 k shear load will be applied to a 12 inch square, 60 ft long restrained-head concrete pile. The soil is a sand with $\phi = 36°$ and $\gamma = 120$ lb/ft^3. The groundwater table is at a depth of 40 ft. The pile is made of concrete with a 28-day compressive strength of 6000 lb/in^2. Compute the lateral deflection at the top of this pile and the maximum moment.

Solution:

Use units of pounds and inches. $\lambda = 1.00$; $R_I = 1.70$; $\varepsilon_{50} = 0.002$:

$$C_{p\phi} = \frac{\phi}{10} = \frac{36}{10} = 3.6$$

$$\begin{aligned}
\sigma_p &= 2\,C_{p\phi}\,\gamma B \tan^2(45 + \phi/2) \\
&= (2)(3.6)\left(\frac{120}{12^3}\right)(12)\tan^2(45 + 36/2) \\
&= 23.1 \text{ lb/in}^2
\end{aligned}$$

$$\begin{aligned}
E &= 15{,}200\,\sigma_r (f_c'/\sigma_r)^{0.5} \\
&= (15{,}200)(14)(6000/14)^{0.5} \\
&= 4{,}400{,}000 \text{ lb/in}^2
\end{aligned}$$

Using Equation 17.18:

$$\begin{aligned}
V_c &= \lambda B^2 E R_I \left(\frac{\sigma_p}{ER_I}\right)^m (\varepsilon_{50})^n \\
&= (1.00)(12)^2(4{,}400{,}000)(1.70)\left(\frac{23.1}{(4{,}400{,}000)(1.70)}\right)^{0.57}(0.002)^{-0.22} \\
&= 3{,}056{,}000 \text{ lb}
\end{aligned}$$

$$V/V_c = 20{,}000/3{,}056{,}000 = 0.0065$$

From Figure 17.26: $y_t/B = 0.0150$

$$y_t = (0.0150)(12) = \textbf{0.18 in} \qquad \Leftarrow \textit{Answer}$$

Using Equation 17.19:

$$\begin{aligned}
M_c &= \lambda B^3 E R_I \left(\frac{\sigma_p}{ER_I}\right)^m (\varepsilon_{50})^n \\
&= (1.00)(12)^3(4{,}400{,}000)(1.70)\left(\frac{23.1}{(4{,}400{,}000)(1.70)}\right)^{0.40}(0.002)^{-0.15} \\
&= 205{,}200{,}000 \text{ in-lb}
\end{aligned}$$

From Figure 17.28: $M_{max}/M_c = 0.0041$

$M_{max} = (0.0041)(205,200,000) = $ **841,000 in-lb**   ⇐ *Answer*

17.6 GROUP EFFECTS

The analysis of lateral loads becomes more complex when we consider pile groups. There are two basic questions:

- How are the applied loads distributed among the piles in the group?
- How does the ultimate capacity and load-deflection behavior of the group compare to that of a single isolated pile?

Unfortunately, both are difficult to evaluate, largely because of the many factors that influence group behavior (O'Neill, 1983). These factors include the following:

- The number, size, spacing, orientation, and arrangement of the piles
- The soil type
- The type of connection at the top (fortunately, group piles are connected with a pile cap, so only the restrained-head condition need be considered)
- The interaction between the cap and the piles
- The vertical contact force between the cap and the soil
- The lateral resistance developed between the side of the cap and the soil
- The differences between the *p-y* curves for the inner piles and those for the leading row of piles
- The method and sequence of pile installation
- The as-built inclination of the piles (although the design drawings may show them perfectly plumb, in reality they will have some accidental batter)

Theories have been developed that consider some of these factors (e.g., Poulos, 1979), but no comprehensive method has yet been proposed. The limited load test data reported in the literature also have not produced a clear picture.

An important characteristic to define is *pile - soil - pile interaction (PSPI)* (O'Neill, 1983). This mechanism works as follows: The lateral movement of a pile relieves some of the stress on the soil behind it. This soil, in turn, provides less resistance to lateral movement of the next pile. Thus, different piles may have different *p-y* curves. This has also been called the *shadow effect*.

PSPI is most pronounced when the piles are closely spaced. Because of this effect, the leading row of piles carries more than a proportionate share of the load. Some load test data confirm this behavior (Holloway et al., 1982). The net result is that the lateral deflection of a pile group will be greater than that of a single isolated pile subjected to proportionate share of the group load. For conventionally spaced onshore pile groups, this ratio may be on the order of 2 to 3. Bogard and Matlock (1983) and Brown and Shie (1990) have suggested methods of computing group deflections that consider this effect, but both are in preliminary form.

Response to Applied Shear Loads

Because of PSPI, the leading row of piles will carry more than its share of the applied shear load. For design purposes, it may be reasonable to assume that each pile in the leading row carries twice its share of the applied shear load:

$$V \approx \frac{2 V_g}{N} \tag{17.25}$$

Where:

V = shear load on a leading row pile

V_g = applied shear load on pile group

N = number of piles in the group

A complete analysis also might consider the lateral resistance on the side of a buried pile cap. It may be appropriate to consider a portion of the passive earth pressure acting on the leading face less the active pressure acting on the trailing face. Methods for computing these pressures are described in Chapter 23. However, the lateral deflection required to mobilize the full passive pressure may be greater than the allowable lateral deflection (see Table 23.2). It would be prudent to ignore this resistance when scour, settlement, or some other mechanism might threaten the integrity of the cap-soil contact. It is always appropriate to ignore the shear resistance along the bottom of the cap.

Response to Applied Moment Loads

A moment load applied to a pile group induces uplift and downward loads on the individual piles. To compute these loads, assume pinned connections exist between the piles and the cap and that the incremental load in each pile is proportional to its incremental axial displacement, as defined in Equations 17.26 and 17.27 and Figure 17.30. Add this incremental axial load to that from the applied axial load.

$$M_g = \sum P_i r_i \tag{17.26}$$

$$P_i = P_{max}\left(\frac{r_i}{r_{max}}\right) \qquad (17.27)$$

Where:

M_g = applied moment on pile group

P_i = incremental axial load in pile i due to applied moment

r_i = distance from centerline of cap to pile i

P_{max} = incremental axial load in the piles farthest from centerline

r_{max} = distance from centerline of cap to farthest pile

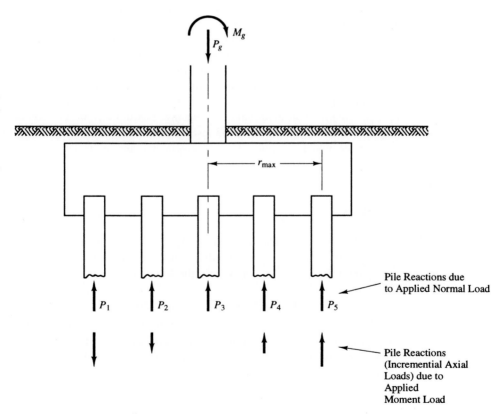

Figure 17.30 Incremental axial load in piles to resist an applied moment.

17.7 FULL-SCALE LATERAL LOAD TESTS

The sophisticated computer analyses often used to evaluate laterally loaded deep foundations give the impression that this topic is well defined and well understood. In reality, there is much we do not understand. The *p-y* curves are highly empirical and our insight into group effects is very limited. Therefore, full-scale load tests (Reese, 1984) may be warranted for some projects. Although such tests are less common than full-scale axial tests, they are performed on occasion. Standard test procedures are described in ASTM D3966.

As discussed earlier, most of the published *p-y* data were obtained from 10 to 24 in (250 - 600 mm) diameter steel pipe piles and on drilled shafts. Therefore, we introduce additional uncertainties when extrapolating this data to other diameters or foundation types. Full-scale tests would be especially useful in those situations.

An easy way to conduct lateral load tests is to build two deep foundations near each other and install a jacking system or a pulley system between them. Thus, each acts as the reaction for the other. A jacking system would spread them apart; a pulley system would draw them closer together. Measure absolute movements using reference beams anchored well away from the movement.

The interpretation of lateral load test results are generally more difficult than those for axial tests. This is because the test foundation probably does not replicate the head constraint conditions of the production foundation. We usually conduct the test under free-head conditions, but the proposed structure will impose some restraint on the production foundation. Therefore, it is necessary to analytically adjust the results.

We can obtain additional data by placing instruments in the foundation and monitoring its subsurface response to the test loads. This instrumentation might consist of an *inclinometer*, which measures lateral movements, and/or strain gages that measure stresses. Then, use this data to develop site-specific *p-y* curves (Kramer, 1991).

17.8 SEISMIC DESIGN

Section 14.3 of Chapter 14 discusses the modes of deep foundation failures observed during earthquakes. It appears that virtually all of these failure modes can be avoided by incorporating the following precautions into the foundation analysis and design:

- Identifying and resolving any soil liquefaction problems.
- Avoiding the use of batter piles, thus providing a more ductile foundation system.
- Using grade beams to tie the foundations together, as discussed in Section 18.6 of Chapter 18, thus reducing the potential for differential lateral movements.
- Providing ductile connections between the foundations and the pile caps, and ductility in the foundations, as discussed in Section 18.6.

Some research has been conducted to develop methods of predicting the lateral displacements and associated moments in deep foundations subjected to dynamic loads. Most of this work has focused on machine foundations, and may not be appropriate for developing earthquake-resistant designs. Although the analytical procedures can be quite sophisticated, our understanding of the soil response parameters is very limited. Therefore, the vast majority of seismic designs currently are based on pseudostatic analyses, as described in Section 14.3 of Chapter 14.

The proper selection of p-y curves for use in pseudostatic analyses is another difficult matter. The curves for seismic loads are softer than those for static loads. However, we usually can accept a lower factor of safety. See the discussion in Section 14.3 on the effects of cyclic loads on soil strength.

Unfortunately, the use of ductile foundation systems can increase the lateral movements in a structure. These movements could cause rupture of utility lines as they enter the structure, especially if soft soils are present. The use of flexible-extendible pipe couplers, as shown in Figure 2.3 in Chapter 2, should eliminate this problem.

17.9 IMPROVING LATERAL CAPACITY

If the computed lateral resistance of a pile is not satisfactory, we could improve it by adjusting the factors described earlier (i.e., diameter, moment of inertia, etc.). However, Broms (1972) described other methods that may be more cost-effective. These are shown in Figure 17.31.

SUMMARY OF MAJOR POINTS

1. A *lateral load* is any load that acts perpendicular to the foundation axis. Thus, shear or moment loads are lateral loads, but axial compression or tension or torsional loads are not.

2. Until the middle of the twentieth century, engineers assumed that deep foundations were only able to resist axial loads, so they used batter piles to resist horizontal loads. More recently, we have reconsidered that assumption and now rely on both axial and lateral capacities.

3. The utilization of lateral capacities in design often produces foundations that are more economical to build and more resistant to seismic loads.

4. When conducting lateral load analyses, engineers usually assume one of the following boundary conditions at the top of the foundation: The *free-head* condition, the *restrained-head* condition, or the *pure moment* condition.

5. The ultimate soil capacity is the greatest lateral load the soil can sustain. This capacity is most easily improved by increasing the foundation embedment, D. Therefore, use a D at least equal to the minimum required embedment, D_{min}.

6. If lateral deflections are of no concern, base the lateral load design on the *ultimate structural capacity*. This is the greatest lateral load the foundation can sustain

without inducing a flexural failure. Compute the maximum moment, M_{max}, using Broms' Method and develop a structural design to resist this moment.

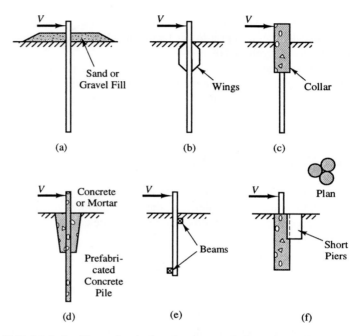

Figure 17.31 Methods of increasing the lateral resistance of piles (Adapted from Broms, 1972).

7. If the structure will tolerate only a limited lateral deflection, then the allowable lateral capacity may be less than that computed using Broms' method. In these situations, conduct a load-deformation analysis to determine the lateral load that corresponds to a certain deflection. Then, design the foundation based on the M_{max} that corresponds to that load.

8. The current state of practice is to use the nonlinear *p-y method* to evaluate the load-deflection behavior. This method uses nonlinear *p-y curves* to describe the lateral soil pressure acting on the foundation and a finite difference analysis to compute the deflections, shears, and moments. Design *p-y curves* are based on empirical data obtained from full-scale lateral load tests.

9. Evans and Duncan have developed chart solutions based on *p-y* method analyses. These charts are not as flexible as a computer solution, but are useful when a computer is not available or to check computer output.

10. The behavior of pile groups subjected to lateral loads is only partially understood. With typical pile spacings (i.e., about 3 diameters on center), the lateral deflections will be greater than that of a single isolated pile subjected to a proportionate share of the applied load.

11. Full-scale lateral load tests are sometimes an appropriate way to determine the design lateral load capacity, either directly or through the development of site-specific p-y curves.

QUESTIONS AND PRACTICE PROBLEMS

17.1 What are the advantages of using laterally loaded vertical piles instead of batter piles?

17.2 Draw typical shear, moment, and lateral deflection diagrams for a laterally loaded free-head deep foundation.

17.3 Draw typical shear, moment, and lateral deflection diagrams for a laterally loaded restrained-head deep foundation.

17.4 A PP16×0.500 restrained-head pile (see p. 564) is subjected to a shear load of 50 k. The soil is a plastic clay with $s_u = 1500$ lb/ft^2. Compute the required embedment, D_{min}.

17.5 A 500 mm diameter, free-head drilled shaft is subjected to a shear load of 150 kN and a moment load of 1300 kN-m. The soil is a clay with $s_u = 100$ kPa. Compute the minimum required embedment, D_{min}.

17.6 A 14 inch square, free-head reinforced concrete pile is subjected to a shear load of 35 k and a moment load of 1000 ft-k. The soil is a sand with $\phi = 35°$ and $\gamma = 126$ lb/ft^3. Compute the minimum embedment depth, D_{min}.

17.7 The pile in Problem 17.4 has no restrictions on lateral deflection. Compute the maximum moment in this pile due the applied shear load.

17.8 The pile described in Problem 17.6 has no restrictions on lateral deflection. Compute the maximum moment in this pile due to the applied shear and moment loads.

17.9 What are the benefits of using a computer program to conduct a p-y analysis instead of Evans and Duncan's charts?

17.10 The pile described in Problem 17.4 is 60 ft long. Use Evans and Duncan's charts to compute the lateral deflection at the top of this pile due to the applied shear load.

17.11 A 75 ft long restrained-head H-pile must resist an applied shear load of 36 k while deflecting no more than 0.40 in. The soil is a silty sand with $c = 0$, $\phi = 34°$ and $\gamma = 127$ lb/ft^3, and the groundwater table is at a great depth. Determine the required H-pile section (see p. 563). Ignore the flexural strength in the pile, even though in reality it often controls the design.

18

Deep Foundations – Structural Design

The final solution to a problem is that which is in hand when time and money run out.

Geotechnical and construction considerations, as described in Chapters 11 to 17, generally control the required diameter and length of deep foundations. Once these have been set, the next step is to develop a structural design that provides enough strength to resist the internal stresses produced by the applied loads. For piles, the structural design also must accommodate handling and driving stresses.

Usually, we can provide sufficient structural integrity within the diameter obtained from the geotechnical design. For example, for steel pipe piles, geotechnical considerations control the overall diameter, and structural considerations control the wall thickness. However, it sometimes is necessary to increase the diameter to provide sufficient integrity, especially when large lateral loads are present.

18.1 DESIGN PHILOSOPHY

The technical literature gives very little information on structural design of deep foundations. Some building codes present design criteria, but they often are incomplete or ambiguous. Therefore, there are no universally accepted standards.

Buckling

Even the softest soils provide enough lateral support to prevent buckling in axially loaded deep foundations, especially when a pile cap is present. Bethlehem Steel Corporation (1979) quotes several load tests on steel H-piles, one of which penetrated through soft soils with no hammer blows (i.e., it sank under its own weight). None of these piles buckled.

Piles subjected to both axial and lateral loads are more likely to buckle. Therefore, when large lateral loads are likely and the soils are very soft, check for buckling using a *p-y* analysis, as described in Chapter 17.

Buckling is a greater concern during pile driving (Fleming et al., 1985), especially in long, slender piles driven through water. Contractors can handle these cases by limiting the hammer size or by providing temporary lateral support.

Comparison with Superstructure Design

Because buckling is generally not a concern for the design engineer, the structural design of deep foundations is similar to that for short columns in the superstructure. However, there are some important differences:

- The construction tolerances for foundations are much wider and quality control is more difficult.
- Piles can be damaged during driving, so the as-built capacity may be less than anticipated.
- Residual stresses may be locked into piles during driving, so the actual stresses in the piles after the structure is completed may be greater than those generated by the applied structural loads.
- Concrete in drilled shafts and other cast-in-place foundations is not placed under ideal conditions, and thus may experience segregation of the aggregates, contamination from the soil, and other problems.

Therefore, most engineers use more conservative design criteria for deep foundations. This extra conservatism often appears in the form of lower allowable stresses and conservative simplifications in the analysis methods.

Working Stress vs. Ultimate Strength Design

Structural engineers use both ultimate strength methods (also known as LRFD: Load and Resistance Factor Design) and working stress methods (also known as ASD: Allowable Stress Design) to design superstructures. Ultimate strength methods dominate with concrete structures, whereas working stress methods dominate with timber. Both are commonly used for steel.

In principle, we could use either method to design deep foundations, as long as it incorporates an appropriate level of conservatism. In practice, most engineers use ultimate strength analyses for drilled shafts and working stress analyses for piles.

18.2 COMPUTING LOADS AND STRESSES

The structural design must consider both axial (compression or tension) and lateral (shear and moment) loads in the foundation. Torsion loads are usually negligible.

The axial tension or compression stress at a depth z in a foundation subjected to axial and/or lateral loads is:

$$\sigma = \frac{P}{A} \pm \frac{Mc}{I} \qquad (18.1)$$

Where:

σ = axial normal stress (tension or compression) in the foundation at depth z

P = axial force in the foundation at depth z

A = cross-sectional area of the foundation at depth z

M = moment in the foundation at depth z

c = distance from neutral axis

I = moment of inertia

Moments of inertia for steel H-piles and common steel pipe piles are tabulated in Tables 18.2 and 18.3. For foundations with solid circular cross sections of diameter B, such as drilled shafts, use:

$$I = \frac{\pi B^4}{64} \qquad (18.2)$$

For square cross-sections with side width B, such as concrete piles, use:

$$I = \frac{B^4}{12} \qquad (18.3)$$

The largest and smallest normal stresses occur at the edges of the foundation farthest from the neutral axis. These are the *extreme fiber stresses*. Compute them using $c = B/2$ in Equation 18.1. Depending on the magnitudes of P and M, these stresses can be either compressive or tensile.

Structural engineers often consider axial and flexural loads separately using the following formulas:

$$f_a = \frac{P}{A} \tag{18.4}$$

$$f_b = \frac{M\,c}{I} = \frac{M\,B}{2\,I} \tag{18.5}$$

Sometimes it is convenient to rewrite Equation 18.5 as:

$$f_b = \frac{M}{S} \tag{18.6}$$

$$S = \frac{2\,I}{B} \tag{18.7}$$

Where:

f_a = average normal stress due to axial load
f_b = normal stress in extreme fiber due to flexural load
B = foundation diameter
S = elastic section modulus

Elastic section moduli for standard pile sections are presented later in this chapter.

For foundations with a constant diameter, the stress f_a becomes smaller with depth as some of the axial load shifts to the soil through skin friction. The stress f_b can increase or decrease with depth according to the moment. Chapter 17 discusses methods of computing the moment as a function of depth.

When using a working stress design method, the allowable axial and flexural stresses are F_a and F_b, respectively, and the design must satisfy the following equation:

$$\frac{f_a}{F_a} + \frac{f_b}{F_b} \le 1 \tag{18.8}$$

Therefore, the presence of moment loads in a foundation reduces its axial load capacity.

It also is possible to develop an *interaction diagram* that shows all possible combinations of axial and moment loads for a given cross section.

For analysis purposes, neglect any interaction between the shear loads and the axial or moment loads. For a working stress analysis, the shear stress, f_v, must not exceed the shear capacity, F_v:

$$f_v = \frac{V}{A} \tag{18.9}$$

$$f_v \leq F_v \tag{18.10}$$

Where:

f_v = shear stress in foundation at depth z
V = shear force in foundation at depth z
A = cross-sectional area of foundation at depth z
F_v = allowable shear capacity of foundation at depth z

18.3 PILES

The structural design of piles must consider each of the following loading conditions:

- *Handling loads* are those imposed on the pile between the time it is fabricated and the time it is in the pile driver leads and ready to be driven. They are generated by cranes, forklifts, and other construction equipment.
- *Driving loads* are produced by the pile hammer during driving.
- *Service loads* are the design loads from the completed structure.

The most critical handling loads often occur when the pile is suspended in a nearly horizontal position from a few support points. This can produce flexural stresses that are greater than those from the service loads. Concrete piles are especially prone to damage from these loads because of their greater weight and smaller tensile strength. Accommodate handling loads by specifying pickup points along the length of the pile, as shown in Figure 18.1.

Driving loads also are important, especially if the contractor uses a large hammer. Timber and concrete piles are especially prone to such damage. Driving stresses are primarily compressive, but tensile stresses can develop in some circumstances. Use a wave equation analysis to predict these driving stresses, and thus guide the selection of an appropriate hammer and pile driving appurtenances.

Obtain the service loads from an analysis of the superstructure. These are the same loads we have been using throughout this book. For working stress analyses, use the dead load plus the live load; for ultimate strength analyses, use the factored load as described in Section 10.1 of Chapter 10. It is often necessary to consider various combinations of service loads to determine the most severe combination.

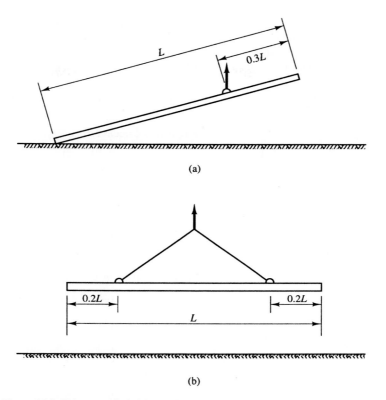

Figure 18.1 Using specified pickup points to keep handling stresses within tolerable limits: (a) single-point pickup; (b) double-point pickup.

Timber Piles

Because wood is a natural material, not a manufactured product, it is difficult to assign allowable design stresses. Design criteria for wood piles must consider many factors, including the following:

- The species of tree
- The quality of the wood (i.e., knots, straightness, etc.)
- The moisture content of the wood
- The extent of any damage incurred during driving
- The type and method of treatment (normally reduces strength)
- The number of piles in a pile group (redundancy if one pile is weak)

The vast majority of timber piles used in North America are either Southern Pine or Douglas Fir, and most building codes give an allowable axial stress under service loads, F_a, of about 1200 lb/in^2 (8.3 MPa) for either type. However, many engineers believe this allowable stress is too high (Armstrong, 1978), especially when there is significant end bearing, and some have advocated values as low as 700 lb/in^2 (4.8 MPa).

TABLE 18.1 MINIMUM REQUIRED TIP CIRCUMFERENCES FOR TIMBER PILES (EXCEPT SOUTHERN PINE) (in)

Pile Length (ft)	Minimum Required Circumference (in), 3 ft from Butt										
	22	25	28	31	35	38	41	44	47	50	57
20	16.0	16.0	16.0	18.0	22.0	25.0	28.0				
25	16.0	16.0	16.0	17.0	20.5	23.5	26.5	29.5			
30	16.0	16.0	16.0	16.0	19.0	22.0	25.0	28.0			
35				16.0	18.0	21.0	24.0	27.0	30.0		
40				16.0	17.0	20.0	23.0	26.0	29.0		
45				16.5	18.5	21.0	24.0	27.0	30.0		
50					16.0	17.0	19.0	22.0	25.0	28.0	
55						16.5	17.5	20.3	23.3	26.3	31.3
60						16.0	16.0	18.6	21.6	24.6	31.6
65						16.0	16.0	17.3	18.9	21.9	28.9
70						16.0	16.0	16.0	16.2	19.2	26.2
75							16.0	16.0	16.1	17.6	24.0
80							16.0	16.0	16.0	16.0	21.8
85							16.0	16.0	16.0	16.0	20.6
90							16.0	16.0	16.0	16.0	19.5
95							16.0	16.0	16.0	16.0	18.8
100							16.0	16.0	16.0	16.0	18.0
105										16.0	17.0
110										16.0	16.0
115											16.0
120											16.0

Per ASTM D25-91; Used with permission of the American Society for Testing and Materials.

Davisson (1989) recommends using $F_a = 800$ lb/in^2 (5.5 MPa). However, a higher value might be appropriate if the design load is of short duration, and a lower value would be necessary if the pile is not part of a group (i.e., no redundancy) or if it might encounter obstructions during driving. Graham (1985) suggests a method of determining the design strength based on several such factors.

The allowable extreme fiber stress due to flexure (bending), F_b, is typically about $2 F_a$. The allowable shear stress, F_v, is typically $0.09 F_a$ to $0.10 F_a$. Because timber piles are tapered, they are not well suited to resist uplift loads. Therefore, limit the allowable uplift capacity to the 90% of the weight of the pile. Therefore, the tensile stresses will be small.

Timber piles are generally not subject to structural damage during handling, but the contractor must avoid large abrasions that might remove the preservative treatment and expose untreated wood. However, these piles can easily be damaged during driving, as described in Chapter 12. To avoid such damage, the maximum driving stresses should not exceed $3 F_a$. This means that timber piles should only be driven with lightweight hammers and they should not be used at sites with hard driving conditions.

Table 18.1 gives ASTM specifications for diameter and length of timber piles.

Example 18.1

A certain column load is to be supported on a group of 16 timber piles. Each pile is 30 ft long and has a 12 in diameter butt and a 7 in diameter tip. The top of each pile in the leading row will be subjected to an axial load of 40 k a shear load of 3.9 k. The maximum moment of 7.0 ft-k occurs at a depth of 8 ft. Is the structural design satisfactory?

Solution:

1. Check the axial and flexural stresses at depth of maximum moment:

$$B \text{ at } 8 \text{ ft} = 12 - (8/30)(12 - 7) = 10.7 \text{ in}$$

$$A = \frac{\pi B^2}{4} = \frac{\pi (10.7)^2}{4} = 89.9 \text{ in}^2$$

Assume P at 8 ft = P at top (conservative).
Using Equations 18.4, 18.2, 18.5, and 18.8:

$$f_a = \frac{P}{A} = \frac{40,000}{89.9} = 445 \text{ lb/in}^2$$

$$I = \frac{\pi B^4}{64} = \frac{\pi (10.7)^4}{64} = 643 \text{ in}^4$$

$$f_b = \frac{MB}{2I} = \frac{(7.0 \text{ ft-k})(12 \text{ in/ft})(1000 \text{ lb/k})(10.7 \text{ in})}{2 (643)} = 699 \text{ lb/in}^2$$

$$\frac{f_a}{F_a} + \frac{f_b}{F_b} = \frac{445}{800} + \frac{699}{1600} = 0.99 \leq 1 \quad OK$$

2. Check shear stress at the top of the pile (point of maximum shear):

$$F_v = 0.09 \, F_a = (0.09)(800) = 72 \text{ lb/in}^2$$

$$A = \frac{\pi B^2}{4} = \frac{\pi (12)^2}{4} = 113 \text{ in}^2$$

Using Equation 18.9:

$$f_v = \frac{V}{A} = \frac{3900}{113} = 34.5 \text{ lb/in}^2 < 72 \quad OK$$

$$\therefore \textbf{ Design is acceptable} \qquad\qquad \Leftarrow \textit{Answer}$$

Steel Piles

Steel piles are usually made of mild steel that conforms with ASTM standard A36. This material has a yield strength, F_y, of 36,000 lb/in² (250 MPa), and is adequate for most projects. Piles with F_y as high as 65,000 lb/in² (450 MPa) are available, but they are not used as often because most applications do not need the higher strength, and because they are more difficult to weld. In addition, some building codes do not permit the use of higher yield strengths in piles.

The allowable axial stress in steel piles, F_a, is typically $0.35 \, F_y$ to $0.50 \, F_y$ (12,600 18,000 lb/in² or 87 - 124 MPa) for either tension or compression (Rempe, 1979). The higher value is appropriate only in the most favorable conditions where the pile will drive straight and not be deflected by boulders or other obstructions. In most cases, use $F_a = 0.35 \, F_y$. For comparison, engineers use an allowable stress of $0.60 \, F_y$ to $0.66 \, F_y$ for A36 steel in the superstructure.

Steel design methods often use a different allowable stress, F_b, for flexural (bending) loads. However, for piles use $F_b = F_a$. Thus, the extreme fiber stress under combined axial and flexural loads must not exceed F_a.

Structural engineers use an allowable shear stress of $F_v = 0.40 \, F_y$ in the superstructure. The greatest shear stress in piles occurs at the top, so the differences

TABLE 18.2 STANDARD STEEL H-PILE SECTIONS USED IN THE UNITED STATES[b]

Designation [a]	Area A (in^2) Total	Area A (in^2) Web Only	Depth B_1 (in)	Width B_2 (in)	$F_a A$ (k) c	X-X Axis I (in^4)	X-X Axis S (in^3)	Y-Y Axis I (in^4)	Y-Y Axis S (in^3)
HP 8×36	10.6	3.56	8.02	8.16	134	119	29.8	40	9.9
HP 10×42	12.4	4.05	9.70	10.07	156	210	43.4	72	14.2
HP 10×57	16.8	5.65	9.99	10.22	212	294	58.8	101	19.7
HP 12×53	15.5	5.11	11.78	12.04	195	393	66.8	127	21.1
HP 12×63	18.4	6.18	11.94	12.12	232	472	79.1	153	25.3
HP 12×74	21.8	7.34	12.13	12.21	275	569	93.8	186	30.4
HP 12×84	24.6	8.39	12.28	12.29	310	650	106	213	34.6
HP 13×60	17.5	5.75	12.54	12.90	221	503	80.3	165	25.5
HP 13×73	21.6	7.20	12.75	13.00	272	630	98.8	207	31.9
HP 13×87	25.5	8.65	12.95	13.10	321	755	117	250	38.1
HP 13×100	29.4	10.04	13.15	13.20	370	886	135	294	44.5
HP 14×73	21.4	6.88	13.61	14.58	270	729	107	261	35.8
HP 14×89	26.1	8.53	13.83	14.69	329	904	131	326	44.3
HP 14×102	30.0	9.87	14.01	14.78	378	1050	150	380	51.4
HP 14×117	34.4	11.47	14.21	14.88	433	1220	172	443	59.5

From *AISC Manual of Steel Construction*, American Institute of Steel Construction; Used with permission.

a. The first number in the pile designation is the nominal section depth in inches. The second number is the weight in pounds per foot.

b. Soft metric HP sections are the same size as those listed here.

c. The allowable axial load when no moment is present ($F_a A$) is based on $F_a = 0.35 f_y$ and $f_y = 36$ k/in^2.

TABLE 18.3 COMMON STEEL PIPE PILE SECTIONS USED IN THE UNITED STATES

Designation	Area A (in^2)	Weight (lb/ft)	I (in^4)	S (in^3)	$F_a A$ (k)
PP 12.75×0.250	9.82	33	192	30.1	124
PP 12.75×0.375	14.58	50	279	43.8	184
PP 12.75×0.500	19.24	65	362	56.7	242
PP 12.75×0.625	23.81	81	439	68.8	300
PP 12.75×0.750	28.27	96	511	80.1	356
PP 14×0.250	10.80	37	255	36.5	136
PP 14×0.375	16.05	55	373	53.3	202
PP 14×0.500	21.21	72	484	69.1	267
PP 14×0.625	26.26	89	589	84.1	331
PP 14×0.750	31.22	106	687	98.2	393
PP 14×1.000	40.84	139	868	124.0	515
PP 16×0.250	12.37	42	384	48.0	156
PP 16×0.375	18.41	63	562	70.3	232
PP 16×0.500	24.35	83	732	91.5	307
PP 16×0.625	30.19	103	894	111.7	380
PP 16×0.750	35.93	122	1047	130.9	453
PP 16×1.000	47.12	160	1331	166.4	594
PP 18×0.250	13.94	47	549	61.0	176
PP 18×0.375	20.76	71	807	89.6	262
PP 18×0.500	27.49	94	1053	117.0	346
PP 18×0.625	34.12	116	1289	143.2	430
PP 18×0.750	40.64	138	1515	168.3	512
PP 18×1.000	53.41	182	1936	215.1	673
PP 20×0.375	23.12	79	1113	111.3	291
PP 20×0.500	30.63	104	1457	145.7	386
PP 20×0.625	38.04	129	1787	178.7	479

TABLE 18.3 (cont'd) COMMON STEEL PIPE PILE SECTIONS USED IN THE UNITED STATES

Designation	Area A (in^2)	Weight (lb/ft)	I (in^4)	S (in^3)	$F_a A$ (k)
PP 20×0.750	45.36	154	2104	210.4	571
PP 20×1.000	59.69	203	2701	270.1	752
PP 24×0.500	36.91	126	2549	212.4	465
PP 24×0.750	54.78	186	3705	308.8	690
PP 24×1.000	72.26	246	4787	398.9	910
PP 24×1.250	89.34	304	5797	483.1	1126
PP 24×1.500	106.03	361	6739	561.6	1336
PP 30×0.500	46.34	158	5042	336.1	584
PP 30×0.750	68.92	235	7375	491.7	868
PP 30×1.000	91.11	310	9589	639.3	1148
PP 30×1.250	112.90	384	11687	779.1	1423
PP 30×1.500	134.30	457	13674	911.6	1692

a. The first number in the pile designation is the outside diameter; the second number is the wall thickness.

b. The allowable axial load when no moment is present ($F_a A$) is based on $F_a = 0.35 f_y$ and $f_y = 36$ k/in^2.

between the pile and the superstructure listed earlier are not as significant. Therefore, use the same allowable shear stress for piles. However, do not use the entire cross-sectional area for shear resistance. For H-piles, use only the area of the web; for pipe piles, use half of the total cross-sectional area.

Standard H-pile sections are listed in Table 18.2. Pipe piles are available in a wide variety of diameters and wall thicknesses; some of the more common sizes are listed in Table 18.3. Steel pipe piles have a constant moment of inertia, I, regardless of the direction of the lateral load. However, I for H-piles depends on the direction of the load relative to the web. Generally, the designer has no control over the as-built web orientation, so we must use I of the weak (Y-Y) axis.

For short piles, say, those with less than about 30 ft (10 m) of embedment, engineers usually design for the largest normal, moment, and shear loads at any point in the pile. However, these loads may become significantly smaller with depth, as illustrated in Figure 18.2. Therefore, we can obtain a more economical design by using progressively thinner members with depth. For example, a 16 inch diameter steel pipe pile might have a 0.500 inch wall thickness near the top, 0.375 inch in the middle, and 0.250 inch near the bottom.

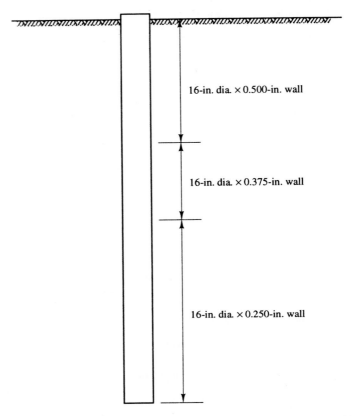

Figure 18.2 Using progressively thinner pile sections to match the diminishing stresses with depth.

Because of their high strength and ductility, steel piles are normally not subject to damage during handling. However, they might be damaged during driving, especially if the contractor uses a large hammer. Davisson (1979) recommends limiting driving stresses to $1.1 F_y$.

Example 18.2

A large sign is to be supported on a single, 400 mm diameter steel pipe pile with 10 mm wall thickness. The sign will impose a vertical downward load of 20 kN, a shear load of 12 kN, and an overturning moment of 95 kN-m onto the top of the pile. Because it is a free-head pile, the greatest loads are at the ground surface. The pile is made of A36 steel. Is this design adequate?

Solution:

Check axial and flexural stresses at the top of the pile.

$$A = \frac{\pi \, 0.400^2}{4} - \frac{\pi \, 0.380^2}{4} = 0.0123 \text{ m}^2$$

$$f_a = \frac{20}{0.0123} = 1{,}600 \text{ kPa} = 1.6 \text{ MPa}$$

$$I = \frac{\pi \, 0.400^4}{64} - \frac{\pi \, 0.380^4}{64} = 2.33 \times 10^{-4} \text{ m}^4$$

$$f_b = \frac{M\,B}{2\,I} = \frac{(95)(0.400)}{(2)(2.33 \times 10^{-4})} = 81{,}500 \, kPa = 81.5 \text{ MPa}$$

$$F_y = 250 \text{ MPa}$$

$$F_a = F_b = 0.35 \, F_y = (0.35)(250) = 87 \text{ MPa}$$

$$\frac{f_a}{F_a} + \frac{f_b}{F_b} = \frac{1.6}{87} + \frac{81.5}{87} = 0.96 < 1.0 \qquad \text{OK}$$

Check shear stresses using half of the cross-sectional area.

$$f_v = \frac{12}{0.0123/2} = 1950 \text{ kPa} = 1.95 \text{ MPa}$$

$$F_v = 0.4 \, F_y = (0.4)(250) = 100 \text{ MPa}$$

$$f_v \ll F_v \qquad \text{OK}$$

The design is satisfactory.

Note how the flexural stresses dominate this design.

Concrete-Filled Steel Pipe Piles

When empty steel pipe piles do not provide sufficient structural capacity, engineers sometimes fill them with concrete. The weight of the concrete also provides additional uplift capacity.

Design concrete-filled steel pipe piles using a working stress analysis:

For axial compression loads:

$$F_a = \frac{0.35 F_y A_s + 0.33 f_c' A_c}{A} \tag{18.11}$$

For axial tension loads:

$$F_a = \frac{0.35 F_y A_s}{A} \tag{18.12}$$

For flexural loads:

$$F_b = 0.35 F_y \tag{18.13}$$

For shear loads:

$$F_v = \frac{0.40 F_y A_s + 4 A_c \sqrt{f_c'/\sigma_r}}{A} \tag{18.14}$$

Where:

F_a = allowable axial stress
F_b = allowable flexural stress
F_v = allowable shear stress
F_y = yield stress of steel (usually 36,000 lb/in^2 (250 MPa))
f_c' = 28-day compressive strength of concrete
σ_r = reference stress = 14 lb/in^2 = 0.10 MPa
A_s = cross-sectional area of steel
A_c = cross-sectional area of concrete
A = total cross-sectional area = $A_s + A_c$

The coefficient of the first term in Equations 18.11 - 18.14 may be increased to 0.50 if the driving conditions are especially favorable, as described earlier.

Prestressed Concrete Piles

The Prestressed Concrete Institute (PCI, 1993) uses a working stress analysis to design fully-embedded prestressed concrete piles. They suggest using concrete with a 28-day compressive strength, f_c', of 5000 - 8000 lb/in² (34 - 55 MPa) with an effective prestress, f_{pc}, of 400 - 1200 lb/in² (2.8 - 8.3 MPa). The allowable stresses under service loads for fully-embedded piles are given in Table 18.4, where:

F_a = allowable axial stress in concrete

F_b = allowable flexural stress in concrete

f_c' = 28-day compressive strength of concrete

f_{pc} = effective prestress stress on the gross section

σ_r = reference stress = 14 lb/in² = 0.10 MPa

TABLE 18.4 ALLOWABLE STRESSES IN FULLY-EMBEDDED PRESTRESSED CONCRETE PILES UNDER SERVICE LOADS

Loading Condition	Allowable Stress
Axial compression	$F_a = 0.33 f_c' - 0.27 f_{pc}$
Axial tension	
Permanent or repetitive loads	$F_a = f_{pc}$
Transient loads	$F_a = 22 (f_c'/\sigma_r)^{0.5} + f_{pc}$
Compression due to flexure[a]	
Normal	$F_b = 0.45 f_c' - f_{pc}$
Bridges and marine structures	$F_b = 0.40 f_c' - f_{pc}$
Tension due to flexure[a]	
Permanent or repetitive loads	
Noncorrosive environments	$F_b = 15 (f_c'/\sigma_r)^{0.5} + f_{pc}$
Corrosive environments	$F_b = f_{pc}$
Transient loads	$F_b = 22 (f_c'/\sigma_r)^{0.5} + f_{pc}$

Adapted from PCI, 1993; Used with permission.

[a] When flexural stresses are present, compute F_b for both tension and compression, and then use the lower value for design.

[b] Shear stresses are accommodated by providing sufficient reinforcement in the pile.

The design criteria in Table 18.4 are conservative. Therefore, the design of heavily loaded piles, especially those with combined axial and lateral loads, may benefit from an ultimate strength analysis as described in PCI (1993).

PCI also recommends computing handling stresses using 1.5 times the pile weight and a maximum tensile stress of $22 \, (f_c' / \sigma_r)^{0.5} + f_{pc}$; and driving stresses of no more than $0.85 f_c' - f_{pc}$ (compression) and $22 \, (f_c' / \sigma_r)^{0.5} + f_{pc}$ (tension).

Standard prestressed concrete pile designs are shown in Tables 18.5 and 18.6 and in Figure 18.3.

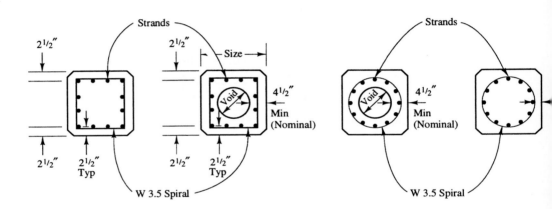

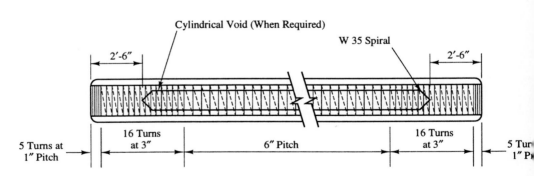

Figure 18.3 Standard prestressed pile designs (PCI, 1984).

Prestressed concrete piles subjected to seismic loads may be prone to a brittle failure at the connection with the pile cap. To avoid this problem, many engineers embed mild steel bars into the top of the pile, and use them to connect it with the pile cap. These bars are more ductile than the prestressing strands.

TABLE 18.5 STANDARD SQUARE PRESTRESSED CONCRETE PILES

Pile Size (in)	Core Diameter (in)	Area A (in²)	Weight (lb/ft)	Moment of Inertia I (in⁴)	Section Modulus S (in³)	Perimeter (ft)	$F_a A$ in Compression (k) $(M=0; f_{pc}=700 \text{ lb/in}^2)$ f_c' (lb/in²)			
							5000	6000	7000	8000
10	Solid	100	104	833	167	3.33	146	178	212	244
12	Solid	144	150	1,728	288	4.00	210	258	304	352
14	Solid	196	204	3,201	457	4.67	286	350	416	480
16	Solid	256	267	5,461	683	5.33	374	458	542	628
18	Solid	324	338	8,748	972	6.00	472	580	688	794
20	Solid	400	417	13,333	1,333	6.67	584	716	848	980
20	11	305	318	12,615	1,262	6.67	444	546	646	746
24	Solid	576	600	27,648	2,304	8.00	840	1,030	1,220	1,410
24	12	463	482	26,630	2,219	8.00	676	828	982	1,134
24	14	422	439	25,762	2,147	8.00	616	754	894	1,034
24	15	399	415	25,163	2,097	8.00	582	714	846	976
30	18	646	672	62,347	4,157	10.00	942	1,156	1,370	1,582
36	18	1,042	1,085	134,815	7,490	12.00	1,522	1,866	2,210	2,552

Adapted from PCI (1993).

TABLE 18.6 STANDARD OCTAGONAL AND ROUND PRESTRESSED CONCRETE PILES

Pile Size (in)	Pile Shape/ Core Diameter (in)	Area A (in²)	Weight (lb/ft)	Moment of Inertia I (in⁴)	Section Modulus S (in³)	Perimeter (ft)	$F_a A$ in Compression (k) $(M=0; f_{pc} = 700$ lb/in²) f'_c			
							5000	6000	7000	8000
10	Oct/Solid	83	85	555	111	2.76	120	148	176	202
12	Oct/Solid	119	125	1,134	189	3.31	172	212	252	290
14	Oct/Solid	162	169	2,105	301	3.87	236	290	344	396
16	Oct/Solid	212	220	3,592	449	4.42	308	378	448	518
18	Oct/Solid	268	280	5,705	639	4.97	390	480	568	656
20	Oct/Solid	331	345	8,770	877	5.52	482	592	702	810
20	Oct/11	236	245	8,050	805	5.52	342	422	500	578
22	Oct/Solid	401	420	12,837	1,167	6.08	584	718	850	982
22	Oct/13	268	280	11,440	1,040	6.08	390	480	566	656
24	Oct/Solid	477	495	18,180	1,515	6.63	696	854	1012	1168
24	Oct/15	300	315	15,696	1,318	6.63	438	536	636	736
36	Round/26	487	507	60,007	3,334	9.43	710	872	1032	1192
48	Round/38	675	703	158,222	6,592	12.57	986	1208	1430	1654

Adapted from PCI (1993).

18.4 DRILLED SHAFTS

There are no handling or driving loads with drilled shafts, so we only need to consider the service loads. Drilled shafts also are free from the residual stresses found in piles.

Ultimate Strength Design

Design drilled shafts using ultimate strength design methods per the American Concrete Institute.[1] Although drilled shafts are similar to concrete columns, ACI [1.1.5] treats them separately.

Compute the ultimate (or factored) axial load, P_u, shear load, V_u, and moment load, M_u, using Equations 10.1 - 10.6 in Chapter 10. Use positive values regardless of the directions of these loads.

Because the quality of the concrete in drilled shafts may not be as good as that in the superstructure, it is appropriate to use a more conservative design. Therefore, the compressive strength, f_c', used in the analysis should be 20 to 30% smaller than that quoted in the construction specifications. For example, we might specify 3000 lb/in^2, but use only 2200 lb/in^2 in the analysis. Additional conservatism might be warranted when using the slurry method of construction; less when drilling an open hole in firm soils above the groundwater table.

Designing for Axial and Flexural Loads

The shaft design must satisfy the following criteria:

$$\frac{P_u}{\phi P_n} + \frac{M_u}{\phi M_n} \leq 1 \tag{18.15}$$

Where:

P_u = ultimate axial load

P_n = nominal axial load capacity

M_u = ultimate moment load

M_n = nominal moment load capacity

ϕ = strength reduction factor

= 0.75 when using spiral reinforcement

= 0.70 when using tied reinforcement

---

[1] See Chapter 10 for a summary of the ACI ultimate strength design method. See MacGregor (1992) for more detailed discussions.

Although ACI permits the use of ϕ values as high as 0.9 when the axial load is small, it is best to use the values listed after Equation 18.15 because of the uncertainties in the concrete quality near the shaft circumference. Use the same ϕ value for both terms in Equation 18.15.

The nominal compressive axial load capacity, P_n, for compressive loads is:

$$P_n = \beta [0.85 f_c' (A_g - A_s) + A_s f_y] \tag{18.16}$$

For tensile (uplift) axial loads, consider only the steel:

$$P_n = A_s f_y \tag{18.17}$$

Where:

P_n = nominal axial load capacity

β = reduction factor to account for possible eccentricities

= 0.85 when using spiral reinforcement

= 0.80 when using tied reinforcement

f_c' = 28-day compressive strength of concrete

A_g = gross cross-sectional area (i.e., concrete plus steel)

A_s = cross-sectional area of reinforcing steel

f_y = yield strength of reinforcing steel

Methods of evaluating the flexural capacity, M_n, of drilled shafts are beyond the scope of this text. See MacGregor (1992) for more information on flexural capacity of circular "columns" and use these methods with drilled shafts.

The moment in a deep foundation diminishes rapidly with depth, as shown in Figure 17.4 in Chapter 17, so the required steel area also diminishes. Thus, it is possible to economize the design by progressively cutting some of the bars at various depths. Some engineers always carry a nominal amount of steel, perhaps using $\rho = 0.005$ to 0.010, to the bottom, whereas others use unreinforced concrete in the lower portion.

However, be much more cautious when large uplift loads are present. Concrete will not reliably carry tension, and uplift failures are more hazardous, so be sure the steel penetrates deep enough to transmit all of the uplift to the soil through skin friction.

Designing for Shear Loads

Consider only the shear strength of the concrete when computing the nominal shear capacity, V_n. When compressive axial loads are present, use:

$$V_n = 0.54\,\sigma_r A_v \left(1 + \frac{P_u}{144\,A_g\,\sigma_r}\right)\sqrt{\frac{f_c'}{\sigma_r}} \qquad (18.18)$$

When tensile axial loads are present, use:

$$V_n = 0.54\,\sigma_r A_v \left(1 - \frac{P_u}{36\,A_g\,\sigma_r}\right)\sqrt{\frac{f_c'}{\sigma_r}} \geq 0 \qquad (18.19)$$

Then design the shaft so the following statement is true:

$$V_u \leq \phi\,V_n \qquad (18.20)$$

Where:

V_u = ultimate shear load

V_n = nominal shear load capacity

P_u = axial load (always positive)

A_g = gross cross-sectional area (concrete + steel)

A_v = cross-sectional area available for shear resistance = $0.95\,A_g$

σ_r = reference stress = 2000 lb/ft^2 = 100 kPa = 14 lb/in^2 = 0.10 MPa

ϕ = strength reduction factor = 0.85 for shear

18.5 PILE CAPS

The design of pile caps is very similar to the design of spread footings. Both must distribute a concentrated load from the column across the bottom of the footing or cap. The primary differences are:

- The loads are larger
- The applied load is distributed over a small portion of the bottom (i.e., the piles) instead of being uniformly distributed

Pile caps are nearly always made of reinforced concrete and designed using ultimate strength methods. See MacGregor (1992) for a discussion of the potential modes of failure and methods of analysis and design.

18.6 SEISMIC DESIGN

Structures subjected to significant seismic loads may have problems with differential lateral displacements between the foundations, especially if the upper soils are weak. Therefore, engineers often tie the foundations together using grade beams as shown in Figure 12.18a in Chapter 12. These grade beams should be designed to resist tension or compression equal to 10% of the vertical load acting on the column (ICBO, 1991a).

Building codes normally allow the engineer to design structural members using material strengths one-third higher than the static values. This increase reflects the short-term nature of the seismic loads, allows for a lower factor of safety, and provides additional ductility. Designing a ductile connection between the foundation and the pile cap is especially important. Brittle foundation members, especially those made of unreinforced concrete, must be avoided.

SUMMARY OF MAJOR POINTS

1. Usually, the diameter and length of a deep foundation are controlled by geotechnical and structural considerations, and we accommodate the structural design within these overall dimensions. However, structural requirements may dictate the use of a greater diameter, especially when the lateral loads are large.

2. The structural design of piles must consider handling loads, driving loads, and service loads. Drilled shaft designs need consider only service loads.

3. Buckling is not a problem for axially loaded foundations under service loads, but may be a concern during driving or when subjected to large lateral loads acting against soft soils.

4. The structural design must consider the interaction between axial and flexural loads. Thus, a foundation that carries a moment has less axial load capacity.

5. Because of the more difficult construction environment, residual stresses, and other factors, the structural design of deep foundations is more conservative than that of members in the superstructure.

QUESTIONS AND PRACTICE PROBLEMS

18.1 Why is it appropriate to use more conservative structural designs for foundations than for comparable superstructure members?

18.2 What type of pile is most susceptible to damage during handling and driving?

18.3 A certain 250 mm diameter Douglas Fir pile has an allowable compressive load capacity (based on geotechnical considerations) of 400 kN. There will be no applied shear or moment loads. The soil is sandy with occasional cobbles. Is this design satisfactory from a structural point-of-view?

18.4 A PP 18×0.500 steel pipe pile made of A36 steel carries a compressive load of 200 k. Compute the maximum allowable moment and shear forces that keep the pile stresses within acceptable limits.

18.5 Develop an interaction diagram (a plot of allowable moment load vs. allowable compressive load) for an HP 12×84 pile made of A36 steel. Assume bending occurs along the weak axis.

18.6 A 14 inch square, prestressed concrete pile with $f_c' = 5000$ lb/in^2 and $f_{pc} = 700$ lb/in^2 is embedded 50 ft into the ground. The net unit end bearing resistance is 180,000 lb/ft^2 and the unit skin friction resistance is 1400 lb/ft^2. The geotechnical factor of safety is 2.8. Compute the allowable compressive load (using Equation 11.4 in Chapter 11) and determine whether the pile has sufficient structural capacity to carry this load.

18.7 Select the most inexpensive A36 steel pipe pile that will carry a compressive load of 375 k, a shear load 20 k, and a moment load of 200 ft-k simultaneously. Assume the price of steel per pound is the same for every pile size.

18.8 A 24 in diameter drilled shaft must carry an unfactored compressive dead load of 400 k and an unfactored compressive live load of 320 k. There are no moment loads. Determine the minimum required 28-day compressive strength of the concrete. Assume a steel ratio of 1.0%, $f_y = 40$ k/in^2, and spiral reinforcement.

18.9 Caesar needs to drive a group of VI piles to support part of a new edifice. Each pile must carry a compressive load of CCXL kips and a moment of CCL ft-k. The unit skin friction resistance, f_s, is MC lb/ft^2, and the net unit end bearing resistance, q_e', is L x 10^3 lb/ft^2. Each pile must have a geotechnical factor of safety of at least 3.00.

    Prestressed concrete piles ($f_c' = $ MMMMM lb/in^2; $f_{sc} = $ DCC lb/in^2) are available for a cost of XXV x 10^{-3} Denarii/lb, and A36 steel piles are available for LXX x 10^{-3} Denarii/lb. Because of the ready availability of slave labor, piles up to C ft long can be transported to the site and installed at no cost. Unfortunately, the slaves are not able to splice piles.

    Develop the most inexpensive design, and compute the total cost for the six piles.

    Hint: Use the information in this chapter to find the smallest structural section required to accommodate the loads, then use the unit skin friction and end bearing capacities to determine the required embedment depth. Repeat for each pile type and determine the costs.

    For those who have not recently used the Caesar's math, I = 1, V = 5, X = 10, L = 50, C = 100, D = 500, and M = 1000.

18.10 A 75 ft long, restrained-head A36 steel pipe pile is to be used to support a 36 k shear load and a 300 k compressive load. The soil properties are described in Problem 17.11 on page 553. Determine the diameter and wall thickness required to keep the stresses inside the pile within tolerable limits.

    Hint: Use Evans and Duncan's method in Chapter 17 to compute the moment load in the pile.

    Note: A complete design would need to consider both the lateral deflection, as described in Problem 17.11, and the stresses in the pile, as in this problem.

Part D

Foundations on Difficult Soils

19

Foundations on Weak
and Compressible Soils

*As the correct solution of any problem depends primarily on a
true understanding of what the problem really is, and wherein
lies its difficulty, we may profitably pause upon the threshold
of our subject to consider first, in a more general way, its
real nature; the causes which impede sound practice;
the conditions on which success or failure depends; the
directions in which error is most to be feared.*

A. M. Wellington, 1887

As cities grow and land becomes more scarce, it often becomes necessary to erect buildings and other structures on sites underlain by poor soils. These sites are potentially troublesome, so the work of the foundation engineer becomes even more important.

The most common of these problematic soils are the soft, saturated clays and silts often found near the mouths of rivers, along the perimeter of bays, and beneath wetlands. These soils are very weak and compressible, and thus are subject to bearing capacity and settlement problems. They frequently include organic material, which further aggravates these problems.

The areas underlain by these soft soils frequently are subject to flooding, so it often becomes necessary to raise the ground surface by placing fill. Unfortunately, the weight of these fills causes large settlements. For example, Scheil (1979) described a building constructed on fill underlain by varved clay in the Hackensack Meadowlands of New Jersey. About 10 in (250 mm) of settlement occurred during placement of the fill, 0.5

in (12 mm) during construction of the building, and an additional 4 in (100 mm) over the following 10 years.

In seismic areas, loose saturated sands also can become weak through the process of *liquefaction*. Moderate to strong ground shaking can create large excess pore water pressures in these soils, which temporarily decrease the effective stress and shear strength. One of the most dramatic illustrations of this phenomenon occurred in Niigata, Japan, during the 1964 earthquake. Many buildings suddenly settled more than 3 ft (1 m), and these settlements were often accompanied by severe tilting (Seed, 1970). One apartment building tilted to an angle of 80 degrees from the vertical!

Fortunately, engineers and contractors have developed methods of coping with these problematic soils, and have successfully built many large structures on very poor sites. These methods include the following, either individually or in combination:

- Support the structure on deep foundations that penetrate through the weak soils.
- Support the structure on shallow foundations and design them to accommodate the weak soils.
- Use a floating foundation, either deep or shallow.
- Improve the engineering properties of the soils.

19.1 DEEP FOUNDATIONS

One of the most common methods of dealing with poor soils is to use deep foundations that penetrate down to stronger soils or bedrock, as shown in Figure 19.1. Thus, the structural loads bypass the troublesome soils.

Although deep foundations often perform well, watch for a special problem: downdrag loads. If the upper soils are consolidating, perhaps in response to the weight of a fill, they are moving downward with respect to the foundation. Therefore, the skin friction force in this zone now acts downward instead of upward, and thus becomes a load instead of a resistance. This load is known as *downdrag*, or *negative skin friction*. It can be very large, and often causes excessive settlements (Bozozuk, 1972, 1981).

Sometimes, downdrag loads are greater on some piles and less on others. This can cause differential settlements in the structure, and, in severe cases, the soil may pull some of the piles out of their caps.

Although downdrag loads are most pronounced in soft soils, they can occur whenever the soils are moving downward in relation to the pile. Thus, end bearing piles bottomed in bedrock may be subjected to downdrag loads, even if the upper soils are moderately stiff.

It is difficult to accurately predict the magnitude of downdrag loads. Some engineers use the β method (Equation 13.7 in Chapter 13) with a $\bar{\beta}$ of 0.25 - 0.30 to estimate the downdrag load on isolated piles in soft clays (Burland, 1973; Meyerhof, 1976). However, the total downdrag load on the inner piles in a group seems to be less than that on an isolated pile (Kuwabara and Poulos, 1989).

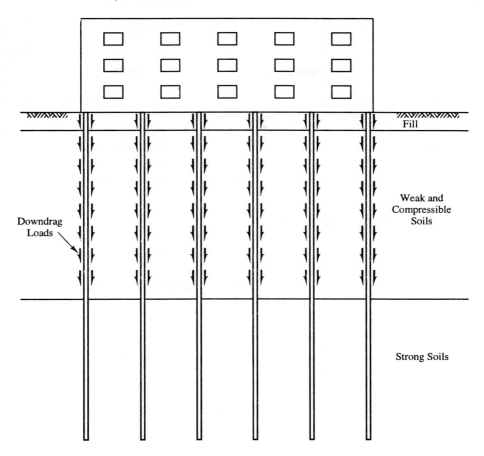

Figure 19.1 Bypassing weak and compressible soils using deep foundations; formation of downdrag loads.

Fortunately, several methods are available to reduce downdrag loads, including the following:

- Coat the pile with bitumen, thus reducing ϕ_s (Bjerrum et al., 1969).
- Use a large diameter predrill hole, possibly filled with bentonite, thus reducing K).
- Use a pile tip larger in diameter than the pile, thus making a larger hole as the pile is driven.
- Drive an open-end steel pipe pile through the consolidating soils, remove the soil plug, and then drive a smaller diameter load-bearing pile through the pipe and into the lower bearing strata. This isolates the inner pile from the downdrag loads.
- Accelerate the settlement using surcharge fills or other techniques, and then drive the piles after the settlement is complete.

However, even if they are designed to accommodate downdrag loads, deep foundations built at sites that are settling also can pose another problem: As the ground settles, the structure remains at the same elevation. Thus, a gap forms between the structure and the ground. Settlements of 2 ft (0.5 m) or more are not unusual at a soft ground site, so the structure could eventually be suspended well above the ground surface. This creates serious serviceability problems, including access difficulties, stretched and broken utility lines, and poor aesthetics.

Thus, properly designed deep foundations can safely support a structure on a soft ground site, but they also introduce new problems of downdrag loads and structure/ground separation. Such foundations can also be very expensive, especially for lightweight structures, because the downdrag loads may be greater than the structural loads.

19.2 SHALLOW FOUNDATIONS

To avoid the problems with downdrag loads and structure/ground separation, some engineers have used shallow foundation systems (either spread footings or mats), as shown in Figure 19.2. These are especially suitable for lightweight structures.

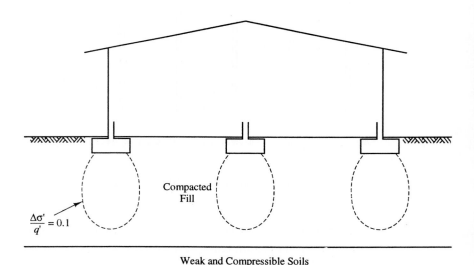

Figure 19.2 Use of spread footings to support structures on fills underlain by weak soils.

Lightly loaded foundations located in a fill probably will not have bearing capacity problems. However, they may be subjected to large total and differential settlements. Therefore, the engineer must provide a way to accommodate or avoid these settlement problems.

Coping with Settlement Problems

One way to cope with settlement problems is to place the fill, and then delay construction of the structure until most of the settlement has occurred. This method is reliable and inexpensive, but may require many years of waiting. However, it also is possible to accelerate the settlement process using surcharge fills, as described in Section 19.4 of this chapter.

Another method is to design the structure to accommodate large settlements. An airline terminal building at LaGuardia Airport in New York City (originally built to serve Eastern Airlines) is a good example (York and Suros, 1989). The building site is underlain by an 80 ft (24 m) deep deposit of soft organic clay that was covered with 20 to 38 ft (6 - 12 m) of incinerated refuse fill in the late 1930s. The organic clay has a compressibility, C, of 0.29 - 0.33. When the construction of this building began in 1979, the ground surface had already settled more than 7 ft (2 m) and was expected to settle an additional 18 in (450 mm) during the following 20 years.

Other buildings at the airport are supported on pile foundations and require continual maintenance to preserve access for aircraft, motor vehicles, and people. Therefore, the Eastern Airlines terminal building was built on a floating foundation with spread footings. It was also designed to be very tolerant of differential settlements and included provisions for leveling jacks between the footings and the building.

By 1988, the building had settled as much as 12.4 in (315 mm), with differential settlements of up to 2.2 in (56 mm). However, because of the tolerant design, the structure was performing well. The first leveling operation was planned for 1989.

In this case, a pile foundation would have been much more expensive, yet it would not have performed as well.

Buildings with heavily loaded slab-on-grade floors, such as warehouses, also can have problems because the floor may settle more than the footings. Thus, we must either structurally support the floor or provide construction details that permit this differential movement.

Using Lightweight Fills

The weight of the fill is usually the primary cause of settlement, so it is helpful to use fill materials that have a very low unit weight. These can consist of natural materials, such as lightweight aggregates, or synthetics. The most common synthetic is *expanded polystyrene* (EPS), a plastic foam with a unit weight of 1.2 lb/ft^3 (0.19 kN/m^3). This is the same material used to make disposable coffee cups. In the United States, it is available in $2 \times 4 \times 8$ ft ($610 \times 1220 \times 2440$ mm) blocks that can be stacked to form a fill (Horvath, 1992).

Some projects have involved excavating some of the upper soft soils, placing EPS, and then covering the foam with a thin soil layer. Thus, the net increase in stress on the natural soils can be very small. The foam can support light to moderately loaded spread footings.

19.3 FLOATING FOUNDATIONS

Another way to reduce the settlement of a structure is to build a *floating foundation* (also known as a *compensated foundation*) (Golder, 1975). This consists of excavating a volume of soil with a weight nearly equal to that of the proposed structure, and then building the structure in this excavation, as shown in Figure 19.3. Thus, the increase in vertical effective stress in the soil, $\Delta\sigma_v'$, is very small. Nearly all floating foundations are mats or pile supported mats.

The earliest documented floating foundation was for the Albion Mill, which was built on a soft soil in London around 1780. It had about 50% flotation (i.e., the excavation reduced $\Delta\sigma_v'$ by 50%) and, according to Farley (1827), "the whole building would have floated upon it, as a ship floats in water." In spite of this pioneering effort, floating foundations did not become common until the early twentieth century. Early examples include the Empress Hotel in Victoria, British Columbia, 1912; the Ohio Bell Telephone Company Building in Cleveland, 1925; and the Albany Telephone Building in Albany, New York, 1929.

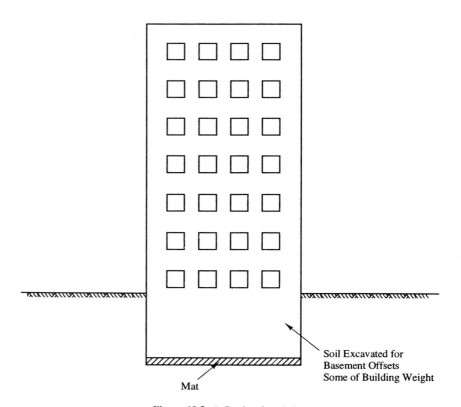

Figure 19.3 A floating foundation.

The Tower Latino Americana

The construction of the 43-story Tower Latino Americana in Mexico City was an important milestone in floating foundation technology (Zeevaert, 1957). It is significant because of the exceptionally difficult soils there.

The soil profile is generally as follows:

0 - 18 ft (0 - 5.5 m) depth
> Old fill that includes Aztec artifacts. Groundwater table at 7 ft (2 m).

18 - 30 ft (5.5 - 9.1 m) depth
> *Becarra sediments*—Interbedded sands, silts, and clays.

30 - 110 ft (9.1 - 33.5 m) depth
> *Tacubaya clays*—Soft volcanic clay; moisture content = 100 - 400%, $C \approx 0.80$; $s_u = 700 - 1400$ lb/ft^2 (35 - 70 kPa).

110 - 230 ft (33.5 - 70.0 m) depth
> *Tarango sands and clays*—Harder and stronger deposits; much less compressible than the Tacubaya clays.

The highly compressible Tacubaya clays have caused dramatic settlements in other structures. For example, the Palace of Fine Arts, located across the street from the Tower, settled over 10 ft (3 m) from 1904 to 1962 and continues to settle at a rate of 0.5 in (12 mm) per year (White, 1962)!

To avoid these large settlements, engineers excavated to a depth of 43 ft (13.0 m) and built a mat supported on piles driven to the Tarango sands. The removal of this soil compensated for about half the building weight. Thus, the building does load the deeper soils and has settled. However, this is by design, and the settlement has approximately matched that of the surrounding ground.

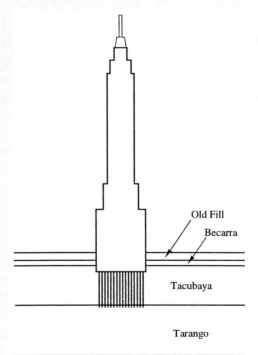

Old Fill
Becarra
Tacubaya
Tarango

Figure 19.4 The Tower Latino Americana.

If the dead and live loads from the structure were constant, we theoretically could select a depth of excavation that would produce $\Delta\sigma_v' = 0$. However, the live load varies over time, so the weights are not always perfectly balanced. Fluctuations in the groundwater table elevation are also important because the groundwater produces an upward buoyancy force on submerged and dewatered basements, thus further reducing $\Delta\sigma_v'$. Therefore, we must determine the worst possible loading condition and select a depth of excavation such that $\Delta\sigma_v'$ is always greater than zero.

When considering floating foundations, it is often useful to conduct parametric studies that evaluate the effect of various parameters on $\Delta\sigma_v'$. Sometimes small changes in loads, groundwater level, or other factors can produce a much greater change in $\Delta\sigma_v'$.

The greatest problems with floating foundations usually occur during construction. The most critical time is when the excavation is open, but construction of the structure has not yet begun. These excavations may not be stable, and thus may require special precautions.

19.4 SOIL IMPROVEMENT

Another way of dealing with weak or compressible soils is to improve them. Some soil improvement techniques have been used for many years, whereas others are relatively new (Mitchell et al., 1978; Mitchell and Katti, 1981). This section discusses some of the more common methods.

Removal and Replacement

Sometimes poor soils can simply be removed and replaced with good quality compacted fill. This alternative is especially attractive if the thickness of the deposit is small, the groundwater table is deep, and good quality fill material is readily available.

Surcharge Fills

Covering poor soils with a temporary surcharge fill, as shown in Figure 19.5, causes them to consolidate more rapidly. When the temporary fill is removed, some or all of the soil is now overconsolidated, and thus stronger and less compressible. This process is known as *preloading* or *precompression* (Stamatopoulos and Kotzias, 1985).

Engineers have primarily used preloading to improve saturated silts and clays because these soils are most conducive to consolidation under static loads. Sandy and gravelly soils respond better to vibratory loads.

If the soil is saturated, the time required for it to consolidate depends on the ability of the excess pore water to move out of the soil voids (see the discussion of consolidation theory in Chapter 3). This depends on the thickness of the soil deposit, its coefficient of permeability, and other factors, and can be estimated using the principles of soil mechanics. The time required could range from only a few weeks to 30 years or more.

The consolidation process can be accelerated by an order of magnitude or more by installing *vertical drains* in the natural soil, as shown in Figure 19.5. These drains provide a pathway for the excess water to escape more easily. The most common design is a *wick drain*, which is inserted into the ground using a device that resembles a giant sewing machine.

Preloading is less expensive than some other soil improvement techniques, especially when the surcharge soils can be moved from place to place, thus preloading the site in sections. Vertical drains, if needed, may double the cost.

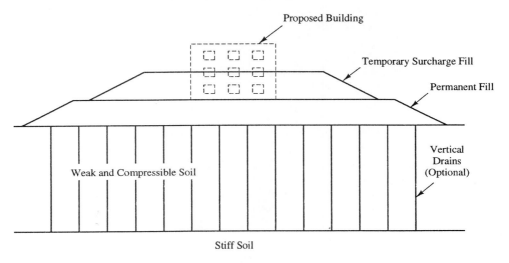

Figure 19.5 Soil improvement by preloading.

Vibro-Compaction and Vibro-Replacement

Sandy and gravelly soils consolidate most effectively when subjected to vibratory loads. This is especially true at depths of about 10 ft (3 m) or more because of the greater overburden stress.

Engineers in Germany recognized this behavior and developed depth vibrators in the 1930s. These methods were refined during the 1970s and now are frequently used throughout the world.

Some of these depth vibrators are essentially vibratory pile hammers attached to a length of steel pipe pile (Brown and Glenn, 1976). Known as a *terra probe*, this device is vibrated into the ground, densifying the adjacent soils, and then retracted. This process is repeated on a grid pattern across the site until all of the soil has been densified.

Another technique uses a probe with a built-in vibrator that is lowered into the ground, also on a grid pattern, as shown in Figure 19.6. The probe is known as a *vibroflot* and the technique is called *vibroflotation*. The vibroflots are also equipped with water jets to aid penetration and improve soil compaction.

Both of these techniques may be classified as *vibro-compaction* methods because they compact the soils in-situ using vibration. In either case, additional sandy soil is added to assist the compaction process. Another closely related method is *vibro-replacement*, which uses the vibrator to create a shaft, and then the shaft is backfilled with gravel to form a *stone column* (Mitchell and Huber, 1985). This technique may be used in cohesive soils and is primarily intended to provide load bearing members that extend through the weak soils. The stone column also acts as a vertical drain, thus helping to accelerate consolidation settlements.

Figure 19.6 This crane is lowering a vibroflot into the ground (Photo courtesy of GKN Hayward Baker, Inc.).

Dynamic Consolidation

The Soviets tried dropping heavy weights to stabilize loess in the 1930s. However, this method did not become widely accepted until the French developed *dynamic consolidation* (also known as *heavy tamping*) in 1970. This technique consists of dropping 5 to 40 ton (4 - 36 Mg) weights, called *pounders*, from heights of 20 to 100 ft (6 - 30 m) (Mitchell et al., 1978). This equipment is shown in Figure 19.7. The impact of the falling weight compacts the soil to significant depths. This tamping process is repeated on a grid pattern, and then the upper soils are smoothed and compacted using conventional earthmoving and compaction equipment.

     Dynamic consolidation has been effectively used with a wide range of soil types. It is relatively inexpensive, but it also generates large shock waves and therefore cannot be used close to existing structures.

Figure 19.7 This crane has just dropped a large weight, thus producing dynamic consolidation in the ground below (Photo courtesy of GKN Hayward Baker, Inc.).

Reinforcement Using Geosynthetics

Soil is similar to concrete in that both materials are strong in compression but weak in tension. Fortunately, both can be improved by introducing tensile members to form a composite material. Steel bars are used to form reinforced concrete, which is vastly superior to plain concrete. Likewise, various metallic and nonmetallic materials can be used to reinforce soil.

Engineers most frequently use materials known as *geosynthetics* to reinforce soils (Koerner, 1990). Although most of these applications have been in the context of retaining structures and earth slopes, strategically placed tensile reinforcement could also provide flexural strength in soil beneath spread footings.

SUMMARY OF MAJOR POINTS

1. Many areas have a scarcity of good building sites, thus forcing development on sites with poor soil conditions. These soils are often weak and compressible, and create special problems for the foundation engineer.

2. Structures can be supported on deep foundations that extend through the weak soils and into deeper, stronger strata. However, these foundations are often subjected to

large downdrag loads. In addition, a gap may form between the ground and the structure.

3. Shallow foundations avoid the downdrag load and structure/ground separation problems, but can have problems with excessive settlements. These can be accommodated by delaying construction or by providing jacks between the structure and the foundation.

4. Lightweight fills can be used to achieve the desired ground surface elevation without excessively loading the soil, thus reducing settlements.

5. Floating foundations compensate for some of the structural load by excavating a certain weight of soil. These can be a very effective way of controlling settlements.

6. Some poor soil conditions can be remedied by improving the soil. Many techniques are available, and they are often a cost-effective solution.

QUESTIONS AND PRACTICE PROBLEMS

19.1 A 300 mm diameter steel pipe pile must penetrate through 2 m of recently placed compacted fill and 16 m of soft clay before reaching firm bearing soils. The unit weights of the fill and clay are 20.0 kN/m^3 and 13.0 kN/m^3, respectively, and the groundwater table is at a depth of 2 m below the top of the fill. Estimate the downdrag load.

19.2 Two soft clay deposits are identical except that one has a higher coefficient of lateral earth pressure, K_0. A 12 inch square reinforced concrete pile is to be driven into each deposit. Will the downdrag load be the same for both piles? Explain.

19.3 A certain structure is supported on a floating foundation located below the groundwater table. What would happen if $\Delta\sigma_v'$ in the soil below this foundation became less than zero?

19.4 The most common remediation for soils prone to seismic liquefaction is to densify them. Which soil improvement methods might be suitable for this task?

19.5 A proposed construction site is underlain by 50 ft of soft clays that have a unit weight of 80 lb/ft^3. The ground surface elevation is +2.0 ft and the groundwater table is at elevation +1.0 ft.

   The proposed site development will consist of excavating some of the soft clay (with temporary dewatering), placing expanded polystyrene blocks to elevation +6.0 ft, and then covering the blocks with 2.0 ft of 120 lb/ft^3 compacted fill. The proposed building will be supported on spread footings in the compacted fill. The total weight of this building divided by its footprint area will be 100 lb/ft^2.

   Compute the required elevation of the bottom of the temporary excavation so that the net increase in effective stress in the soft clay will be 75 lb/ft^2. Be sure to consider buoyancy effects in your computations.

20

Foundations on Expansive Soils

Question: *What causes more property damage in the United States than all the earthquakes, floods, tornados, and hurricanes combined?*

Answer: *Expansive soils!*

Each year, expansive soils in the United States inflict about $9 billion in damages to buildings, roads, airports, pipelines, and other facilities—more than twice the combined damage from the disasters listed above (Jones and Holtz, 1973; Jones and Jones, 1987). The distribution of these damages is approximately as shown in Table 20.1. Many other countries also suffer from expansive soils. Although it is difficult to estimate the global losses, this is clearly an international problem.

Sometimes the damages from expansive soils are minor maintenance and aesthetic concerns, but often they are much worse, even causing major structural distress, as illustrated in Figures 20.1 through 20.3. According to Holtz and Hart (1978), 60% of the 250,000 new homes built on expansive soils each year in the United States experience minor damage and 10% experience significant damage, some beyond repair.

In spite of these facts, we do not expect to see newspaper headlines "Expansive Soils Waste Billions" and certainly not "Expansive Soil Kills Local Family." Expansive soils are not as dramatic as hurricanes or earthquakes and they cause only property damage, not loss of life. In addition, they act more slowly and the damage is spread over wide areas rather than being concentrated in a small locality. Nevertheless, the economic loss is large and much of it could be avoided by proper recognition of the problem and incorporating appropriate preventive measures into the design, construction, and maintenance of new facilities.

TABLE 20.1 ANNUAL DAMAGE IN THE UNITED STATES FROM EXPANSIVE SOILS

Category	Annual Damage
Highways and streets	$4,550,000,000
Commercial buildings	1,440,000,000
Single family homes	1,200,000,000
Walks, drives and parking areas	440,000,000
Buried utilities and services	400,000,000
Multi-story buildings	320,000,000
Airport installations	160,000,000
Involved in urban landslides	100,000,000
Other	390,000,000
Total annual damages (1987)	$9,000,000,000

Adapted from Jones and Holtz 1973; Jones and Jones, 1987;
Used with permission of ASCE.

Figure 20.1 Heaving of an expansive soil caused this brick wall to crack. The $490,000 spent to repair this and other walls, ceilings, doors and windows represented nearly one-third of the original cost of the 6 year old buildings (Photo courtesy of the Colorado Geological Survey).

Figure 20.2   Heaving of expansive soils caused this 0.1-inch wide crack in the ceiling of a one-story wood-frame house.

Figure 20.3   Expansive soil caused this brick building to crack (Photo courtesy of the Colorado Geological Survey).

Foundation engineers must be aware of this potential problem and be ready to take appropriate action when encountering such soils.

20.1 THE NATURE, ORIGIN, AND OCCURRENCE OF EXPANSIVE SOILS

When geotechnical engineers refer to expansive soils, we usually are thinking about clays or sedimentary rocks derived from clays and the volume changes that occur as a result of changes in moisture content. This is the most common expansion phenomenon, and thus is the primary focus of this chapter[1]. Other less common mechanisms of soil expansion are discussed in Section 20.6.

Clays are fundamentally very different from gravels, sands, and silts. All of the latter consist of relatively inert bulky particles and their engineering properties depend primarily on the size, shape, and texture of these particles. In contrast, clays are made up of very small particles that are usually plate shaped. The engineering properties of clays are strongly influenced by the very small size and large surface area of these particles and their inherent electrical charges.

What Causes a Clay to Expand?

Several different *clay minerals* occur in nature, the differences being defined by their chemical makeup and structural configuration. Three of the most common clay minerals are *kaolinite*, *illite*, and *montmorillonite* (part of the smectite group). The different chemical compositions and crystalline structures of these minerals give each a different susceptibility to swelling, as shown in Table 20.2.

Swelling occurs when water infiltrates between and within the clay particles, causing them to separate. Kaolinite is essentially nonexpansive because of the presence of strong hydrogen bonds that hold the individual clay particles together. Illite contains weaker potassium bonds that allow limited expansion, and montmorillonite particles are only weakly linked. Thus, water can easily flow into montmorillonite clays and separate the particles. Field observations have confirmed that the greatest problems occur in soils with a high montmorillonite content.

TABLE 20.2 SWELL POTENTIAL OF PURE CLAY MINERALS

Surcharge Load		Swell Potential (%)		
(lb/ft^2)	(kPa)	Kaolinite	Illite	Montmorillonite
200	9.6	Negligible	350	1500
400	19.1	Negligible	150	350

Adapted from Budge et al. (1964).

Several other forces also act on clay particles, including the following:

- Surface tension in the menisci of water contained between the particles (tends to pull the particles together, compressing the soil).
- Osmotic pressures (tend to bring water in, thus pressing the particles further apart

---

[1] We are considering frost heave, as discussed in Chapter 2, to be a different phenomenon. It is not an expansive soil, but heave due to the formation of ice lenses in the soil.

and expanding the soil).

• Pressures in entrapped air bubbles (tend to expand the soil).

• Effective stresses due to external loads (tend to compress the soil).

• London-Van Der Waals intermolecular forces (tend to compress the soil).

Expansive clays swell or shrink in response to changes in these forces. For example, consider the effects of changes in surface tension and osmotic forces by imagining a montmorillonite clay that is initially saturated, as shown in Figure 20.4a. If this soil dries, the remaining moisture will congregate near the particle interfaces, forming menisci, as shown in Figure 20.4b, and the resulting surface tension forces pull the particles closer together causing the soil to shrink. We could compare the soil in this stage to a compressed spring; both would expand if it were not for forces keeping them compressed.

The soil in Figure 20.4b has a great affinity for water and will draw in available water using osmosis. We would say that it has a very high *soil suction* at this stage. If water becomes available, the suction will draw it into the spaces between the particles and the soil will swell, as shown in Figure 20.4c. Returning to our analogy, the spring has been released and perhaps is now being forced outward.

What Factors Control the Amount of Expansion?

The portion of a soil's potential expansion that will actually occur in the field depends on many factors. One of these factors is the percentage of expansive clays in the soil. For example, a pure montmorillonite could swell more than 15 times its original volume (definitely with disastrous results!), but it is rarely found in such a pure form. Usually, the expansive clay minerals are mixed with more stable clays and with sands or silts. A typical "montmorillonite" (really a mixed soil) would probably not expand more than 35 to 50%, even under the worst laboratory conditions, much less in the field.

There are two types of montmorillonite clay, calcium montmorillonite and sodium montmorillonite (also known as bentonite). The latter is much more expansive, but less common.

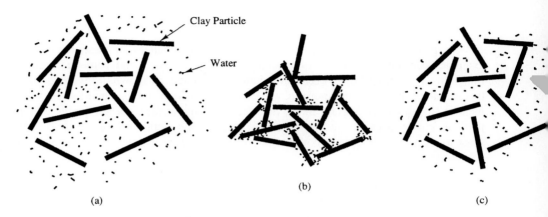

Figure 20.4 Shrinkage and swelling of an expansive clay.

Two of the most important variables to consider are the initial moisture content and the surcharge pressure. If the soil is initially moist, then there is much less potential for additional expansion than if it were dry. Likewise, even a moderate surcharge load restrains much of the swell potential (although large loads are typically required to completely restrain the soil). Figure 20.5 illustrates a typical relationship between swell potential, initial moisture content, and surcharge load.

This relationship demonstrates why pavements and slabs-on-grade are so susceptible to damage from expansive soils (see Table 20.1). They provide such a small surcharge load that there is little to resist the soil expansion. However, it also demonstrates how even a modest increase in surcharge, such as 12 in (300 mm) or so of subbase, would significantly decrease the potential heave.[2]

Remolding a soil into a compacted fill may make it more expansive (O'Neill and Poormoayed, 1980), probably because this process breaks up cementation in the soil and produces high negative pore water pressures that later dissipate. Many other factors also affect the expansive properties of fills, especially the methods used to compact the fill (kneading vs. static) and the as-compacted moisture content and dry unit weight (Seed and Chan, 1959).

Figure 20.6 illustrates how compacting a soil wet of the optimum moisture content reduces its potential for expansion. It also illustrates that compacting the soil to a lower

---

[2] The use of a subbase for this purpose is somewhat controversial. Although it provides additional surcharge load, which is good, it can also become an avenue for additional water to enter the expansive soils, which is bad. Many engineers therefore feel that the risk of additional water infiltration is too great and therefore do not use this method.

dry unit weight also reduces its swell potential (although this also will have detrimental effects, such as reduced shear strength and increased compressibility).

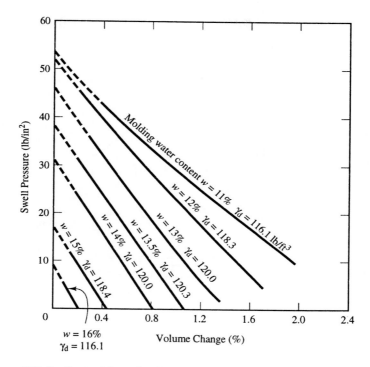

Figure 20.5 Swell potential as a function of initial moisture content and surcharge load (typical). (Adapted from Seed, Mitchell, and Chan, 1962).

Although laboratory tests are useful, they may not accurately predict the behavior of expansive soils in the field. This is partly because the soil in the lab is generally inundated with water, whereas the soil in the field may have only limited access to water. The flow of water into a soil in the field depends on many factors, including the following:

- The supply of water (depends on rainfall, irrigation, and surface drainage).
- Evaporation and transpiration (depends on climate and vegetation; large trees can extract large quantities of water from the soil through their roots).
- The presence of fissures in the soil (water will flow through the fissures much more easily than through the soil).
- The presence of sand or gravel lenses (helps water penetrate the soil).
- The soil's affinity for water (its suction).

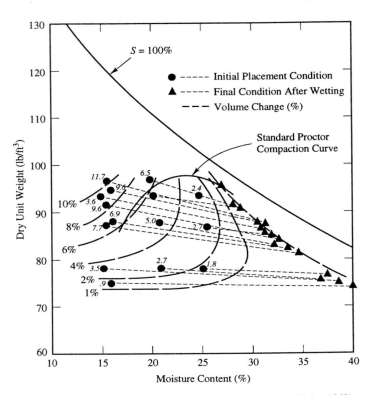

Figure 20.6 Swell potential of compacted clays (Adapted from Holtz, 1969).

Because of these factors, Jones and Jones (1987) suggested that soils in the field typically swell between 10 and 80% of the total possible swell.

Occurrence of Expansive Clays

Chemical weathering of materials such as feldspars, micas, and limestones can form clay minerals. The particular mineral formed depends on the makeup of the parent rock, topography, climate, neighboring vegetation, duration of weathering, and other factors.

Montmorillonite clays often form as a result of the weathering of ferromagnesian minerals, calcic feldspars, and volcanic materials. They are most likely to form in an alkaline environment with a supply of magnesium ions and a lack of leaching. Such conditions would most likely be present in semi-arid regions. *Bentonite* (sodium montmorillonite) is formed by chemical weathering of volcanic ash. Figure 20.7 shows the approximate geographical distribution of major montmorillonite deposits in the United States.

Figure 20.7 Approximate distribution of major montmorillonite clay deposits in the United States (Adapted from Tourtelot, 1973). Erosion, glacial action, and other geologic processes have carried some of these soils outside the zones shown here. Thus, the unshaded areas are not immune to expansive soils problems.

Expansive clays are also common in the Canadian prairie provinces, Israel, South Africa, Australia, Morocco, India, Sudan, Peru, Spain, and many other places in the world.

Figure 20.7 illustrates regional trends only. Not all the soils in the shaded areas are expansive and not all the soils outside are nonexpansive. The local occurrence of expansive soils can vary widely, as illustrated in Figure 20.8. However, even maps of this scale are only an aid, not a substitute for site-specific field investigations.

Influence of Climate on Expansion Potential

As discussed earlier, any expansive soil could potentially shrink and swell, but in practice this occurs only if its moisture content changes. The likelihood of such changes depends on the balance between water entering a soil (such as by precipitation or irrigation) and water leaving the soil (often by evaporation and transpiration).

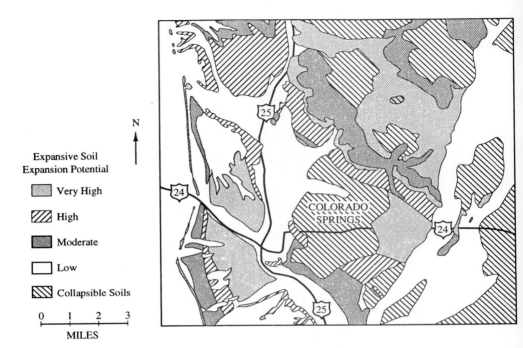

Expansive Soil
Expansion Potential

Very High

High

Moderate

Low

Collapsible Soils

0 1 2 3
MILES

N

COLORADO
SPRINGS

Figure 20.8 Distribution of expansive soils in the Colorado Springs, Colorado area (Adapted from Hart, 1974).

In humid climates, the soil is moist or wet and tends to remain so throughout the year. This is because the periods of greatest evaporation and transpiration (the summer months) also coincide with the greatest rainfall. The climate in North Carolina, as shown in Figure 20.9a, is typical of this pattern. Because the variations in moisture content are small, very little shrinkage or swelling will occur. However, some problems have been reported during periods of extended drought when the soil dries and shrinks (Hodgkinson, 1986; Sowers and Kennedy, 1967).

Most of the problems with expansive soils occur in arid, semi-arid, and monsoonal areas because the seasonal distribution of precipitation and evaporation/transpiration causes wide fluctuations in the soil's moisture content. Most of the precipitation in arid and semi-arid areas occurs during the winter and spring when evaporation and transpiration rates are low. Thus, the moisture content of the soil increases. Then, during the summer, precipitation is minimal and evaporation/transpiration is greatest, so the soil dries. Thus, the soil expands in the winter and shrinks in the summer. The climate in Los Angeles, shown in Figure 20.9b, displays this pattern.

A useful measure of precipitation and evaporation/transpiration as it affects expansive soil problems is the Thornthwaite Moisture Index (TMI) (Thornthwaite, 1948). This index is a function of the difference between the mean annual precipitation and the amount of water that could be returned to the atmosphere by evaporation and transpiration. A positive value indicates a net surplus of soil moisture whereas a negative value indicates a net deficit. Using this index, Thornthwaite classified climates as shown in Table 20.3.

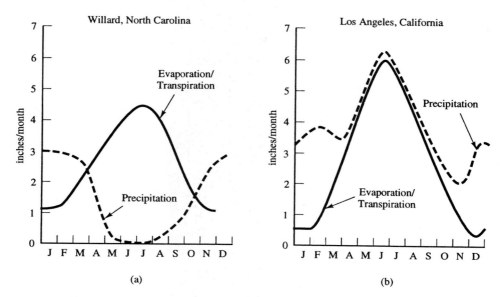

Figure 20.9 Annual distribution of precipitation and potential evaporation/transpiration in (a) Willard, North Carolina and (b) Los Angeles, California (Adapted from Thornthwaite, 1948). Note how the wet Los Angeles winters are followed by very dry summers. In contrast, the total annual precipitation in Willard is much higher and most of it occurs during the summer. Used with permission of the American Geographical Society.

TABLE 20.3 CLASSIFICATION OF CLIMATE BASED ON THORNTHWAITE MOISTURE INDEX (TMI)

TMI	Climate Type
-60 to -40	Arid
-40 to -20	Semiarid
-20 to 0	Dry subhumid
0 to 20	Moist subhumid
20 to 100	Humid
> 100	Perhumid

Adapted from Thornthwaite (1948); Used with permission of the American Geographical Society.

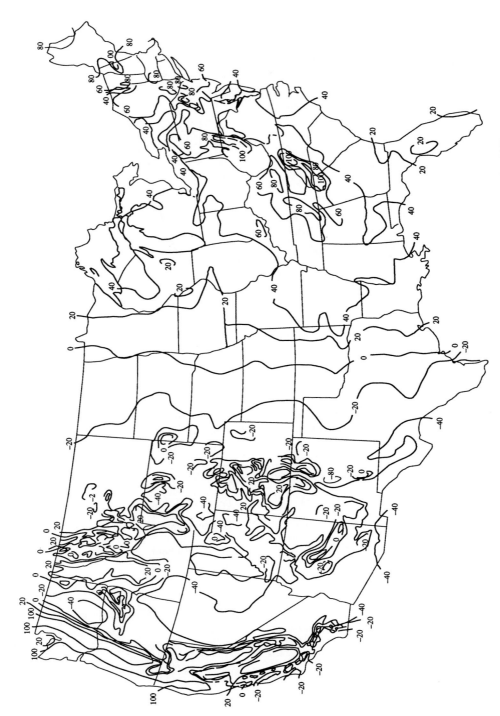

Figure 20.10 Thornthwaite Moisture Index distribution in the United States (Adapted from Thornthwaite, 1948, and PTI, 1980). Used with permission of the American Geographical Society and the Post-Tensioning Institute.

Because expansive soils are most troublesome in areas where there the moisture content varies during the year, and this is most likely to occur in arid climates, regions with the lowest TMI values should have the greatest potential for problems. Researchers have observed that expansive soils are most prone to cause problems in areas where the TMI is no greater[3] than +20.

Figure 20.10 shows Thornthwaite contours for the United States. Combining this information with Figure 20.7 shows that the areas most likely to have expansive soils problems include the following:

- Central and southern Texas
- Portions of Colorado outside the Rocky Mountains
- Much of California south of Sacramento
- The northern plains states
- Portions of the great basin area (Arizona, Nevada, Utah)

Man-made improvements can change the TMI at a given site, as discussed later in this section.

Depth of the Active Zone

In its natural state, the moisture content of a soil fluctuates more near the ground surface than at depth. This is because these upper soils respond more rapidly to variations in precipitation and evaporation/transpiration.

An important criterion when evaluating expansive soils problems is the *depth of the active zone*, which is the greatest depth of moisture content fluctuations (see Figure 20.11). Presumably, the moisture content is reasonably constant below that depth, so no expansion occurs there. From Figure 20.9, the soil moisture content in Los Angeles varies dramatically from summer to winter, so we would expect the active zone there to extend much deeper than in Willard, where the moisture content is less variable. As a result, expansive soils should be more of a problem in Los Angeles (and they are!).

O'Neill and Poormoayed (1980) presented active zone depths for selected cities as shown in Table 20.4.

According to Table 20.4, a soil profile in San Antonio would generate more heave than an identical profile 200 miles (320 km) away in Houston. As a result, the appropriate preventive design measures for identical structures would also be different.

Clays that are heavily fissured will typically have deeper active zones because the fissures transmit the water to greater depths. For example, field studies conducted in Colorado indicate the active zone in some locations may extend as deep as 50 ft below

---

[3] This is not an absolute upper limit. For example, some expansive soil problems have occurred in Alabama and Mississippi (TMI ≈ 40), but such problems do not have the magnitude or frequency of those in areas with drier climates.

the ground surface (Holtz, 1969).

The depth of the active zone at a given site is difficult to determine, and this is a major source of uncertainty in heave analyses.

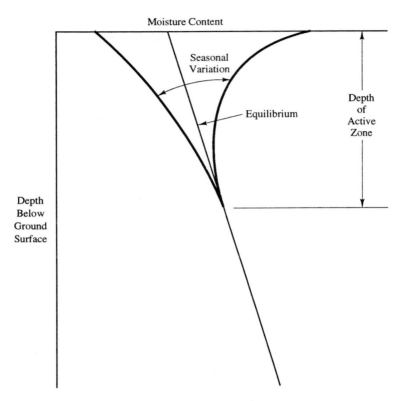

Figure 20.11 The active zone is the layer of soil that has a fluctuating moisture content.

TABLE 20.4 APPROXIMATE DEPTH OF THE ACTIVE ZONE IN SELECTED CITIES

City	Depth of the Active Zone		TMI (from Fig. 20.10)
	(ft)	(m)	
Houston	5 - 10	1.5 - 3.0	18
Dallas	7 - 15	2.1 - 4.2	-1
Denver	10 - 15	3.0 - 4.2	-15
San Antonio	10 - 30	3.0 - 9.0	-14

Adapted from O'Neill and Poormoayed (1980).

Influence of Human Activities

Engineers also need to consider how human activities, especially new construction, change the moisture conditions at a particular site. For example:

- Removal of vegetation brings an end to transpiration.
- Placement of slab-on-grade floors, pavements, or other impervious materials on the ground stops both evaporation and the direct infiltration of rain water.
- Irrigation of landscaping introduces much more water into the ground. In southern California, some have estimated that irrigation in residential areas is the equivalent of raising the annual precipitation from a natural 15 in (380 mm) to an inflated 60 in (1500 mm) or more.
- Placement of aggressive trees or heated buildings can enhance desiccation.

These changes are difficult to quantify, but the net effect in arid and semi-arid areas is normally to increase the moisture content under structures. This results in more swelling and more structural damage.

20.2 IDENTIFYING, TESTING, AND EVALUATING EXPANSIVE CLAYS

When working in an area where expansive soils can cause problems, geotechnical engineers must have a systematic method of identifying, testing, and evaluating the swelling potential of troublesome soils. The ultimate goal is to determine which preventive design measures, if any, are needed to successfully complete a proposed project.

An experienced geotechnical engineer is usually able to visually identify potentially expansive soils. To be expansive, a soil must have a significant clay content, probably falling within the unified symbols CL or CH (although some ML, MH, and SC soils also can be expansive). A dry expansive soil will often have fissures, slickensides, or shattering, all of which are signs of previous swelling and shrinking. When dry, these soils will have cracks at the ground surface. However, any such visual identification is only a first step; we must obtain more information before we can develop specific design recommendations.

The next stage of the process—determining the degree of expansiveness—is more difficult. A wide variety of testing and evaluation methods have been proposed, but none of them is universally or even widely accepted. Engineers who work in certain geographical areas often use similar techniques, which may be quite different from those used elsewhere. This lack of consistency continues to be a stumbling block.

We can classify these methods into three groups. The first group consists of purely *qualitative methods* that classify the expansiveness of the soil with terms such as "low," "medium," or "high" and form the basis for empirically based preventive measures. The second group includes *semiquantitative methods*. They generate numerical results, but engineers consider them to be an *index* of expansiveness, not a fundamental physical property. The implication here is that the design methods will also be empirically based. The final group includes methods that provide *quantitative* results that are measurements of fundamental physical properties and become the basis for a rational or semirational design procedure.

TABLE 20.5 CORRELATIONS WITH COMMON SOIL TESTS

Percent Colloids	Plasticity Index	Shrinkage Limit	Liquid Limit	Swelling Potential
< 15	< 18	< 15	< 39	Low
13 - 23	15 - 28	10 - 16	39 - 50	Medium
20 - 31	25 - 41	7 - 12	50 - 63	High
> 28	> 35	> 11	> 63	Very high

Adapted from Holtz (1969) and Gibbs (1969).

TABLE 20.6 CORRELATIONS WITH COMMON SOIL TESTS

Laboratory and Field Data			Degree of Expansiveness			
Percent Passing #200 Sieve	Liquid Limit	SPT N Value	Probable Expansion (%) [a]	Swelling Pressure		Swelling Potential
				(k/ft^2)	(kPa)	
< 30	< 30	< 10	< 1	1	50	Low
30 - 60	30 - 40	10 - 20	1 - 5	3 - 5	150 - 250	Medium
60 - 95	40 - 60	20 - 30	3 - 10	5 - 20	250 - 1000	High
> 95	> 60	> 30	> 10	> 20	> 1000	Very High

Adapted from Chen (1988); Used with permission of Elsevier Science Publishers.
[a] Percent volume change when subjected to a total stress of 1000 lb/ft^2 (50 kPa).

Qualitative Evaluations

This category of evaluations is usually based on correlations with common soil tests, such as the Atterberg limits or the percent colloids.[4] Such correlations are approximate, but they can be useful, especially for preliminary evaluations.

---

[4] Colloids are usually defined as all particles smaller than 0.001 mm; clay-size particles are sometimes defined as those smaller than 0.002 mm. A hydrometer test is an easy way to measure the percentage of colloids or clay-size particles in a soil, which can be a rough indicator of its potential expansiveness.

The U.S. Bureau of Reclamation developed the correlations in Table 20.5. An engineer could use any or all of them to classify the swelling potential of a soil, but the plasticity index and liquid limit correlations are probably the most reliable. Montmorillonite particles are generally smaller than illite or kaolinite, so expansiveness roughly correlates with the percent colloids. Engineers rarely perform the shrinkage limit test, and some have questioned the validity of its correlation with expansiveness.

Chen (1988) proposed the correlations in Table 20.6 based on his experience in the Rocky Mountain area.

Semiquantitative Evaluations

Loaded Swell Tests

The most common semiquantitative method of describing expansive soils is in terms of its *swell potential*, which engineers usually measure in some kind of *loaded swell test*. Unfortunately, these are very ambiguous terms because there are many different definitions of swell potential and an even wider range of test methods.

Loaded swell tests usually utilize a laterally confined cylindrical sample, as shown in Figure 20.12. The initially dry sample is loaded with a surcharge, and then soaked. The sample swells vertically and this displacement divided by the initial height is the swell potential, usually expressed as a percentage.

This methodology is attractive because it measures the desired characteristics directly, is relatively easy to perform, and does not require exotic test equipment (the test can be performed in a conventional consolidometer). However, because there is no universally accepted standard test procedure, the specifics of the test vary and results from different tests are not always comparable. The typical ranges of test criteria are as follows:

- **Sample size**: What is its diameter and height? Typically, the sample is 2.0 to 4.5 in (50 - 112 mm) in diameter and 0.5 to 1.5 in (12 - 37 mm) tall. Larger diameter samples are less susceptible to side friction and therefore tend to swell more.

- **Method of preparation**: Is the sample undisturbed or remolded? If it is undisturbed, how was it sampled and prepared? If it is remolded, how was it compacted, to what density and moisture content, and what curing, if any, was allowed?

- **Initial moisture content**: What is the moisture content at the beginning of the test? Some possibilities include:
  - In-situ moisture content
  - Optimum moisture content
  - Air dried moisture content
  Other options are also possible.

- **Surcharge load**: How large is the surcharge load? It is usually between 60 and

1500 lb/ft^2 (2.9 - 71.8 kPa). Some prefer to use a surcharge equal to the in-situ or anticipated as-built overburden stress.

- **Duration**: Expansive soils do not swell immediately upon application of water. It takes time for the water to seep into the soil. This raises the question of how long to allow the test to run. Some conduct the test for a specified time (i.e., 24 hours) whereas others continue until a specified rate of expansion is reached (such as no more than 0.001 in/hr). The latter could take several days in some soils.

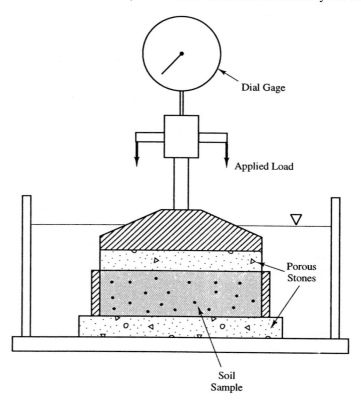

Figure 20.12 Typical loaded swell test.

Snethen (1984) suggested the following definition of potential swell:

> Potential swell is the equilibrium vertical volume change or deformation from an oedometer-type[5] test (i.e., total lateral confinement), expressed as a percent of original height, of an undisturbed specimen from its natural water content and density to a state of saturation under an applied load equivalent to the in-situ overburden pressure.

---

[5] The terms *oedometer* and *consolidometer* are synonymous.

Snethen also suggested that the applied load should consider any applied external loads, such as those from foundations.

Using Snethen's test criteria, we could classify the expansiveness of the soil, as shown in Table 20.7.

TABLE 20.7 TYPICAL CLASSIFICATION OF SOIL EXPANSIVENESS BASED ON LOADED SWELL TEST RESULTS AT IN-SITU OVERBURDEN STRESS

Swell Potential (%)	Swell Classification
< 0.5	Low
0.5 - 1.5	Marginal
> 1.5	High

Adapted from Snethen (1984).

Expansion Index Test

The *expansion index test* [ASTM D4829] (ICBO, 1991b; Anderson and Lade, 1981) is an attempt to standardize the loaded swell test. In this test a soil sample is remolded into a standard 4.01 in (102 mm) diameter, 1 in (25 mm) tall ring at a degree of saturation of about 50%. A surcharge load of 1 lb/in^2 (6.9 kPa) is applied, and then the sample is saturated and allowed to stand until the rate of swelling reaches a certain value or 24 hours, whichever is longer. The amount of swell is expressed in terms of the *expansion index*, or *EI*, which is defined as follows:

$$EI = 1000\,h\,F \tag{20.1}$$

Where:

EI = expansion index

h = expansion of the soil (in)

F = percentage of the sample by weight that passes through a #4 sieve

Table 20.8 gives the interpretation of *EI* test results.

If the *EI* varies with depth, the test procedure also includes a series of weighting factors that emphasize the shallow soils and deemphasize the deeper soils. Although this concept is sound in principle, the stated factors imply an active zone depth of only 4 ft (1.2 m), which is far too shallow. Therefore, it may be best to ignore this portion of the test procedure.

Because the expansion index test is conducted on a remolded sample, it may mask certain soil fabric effects that may be present in the field.

TABLE 20.8 INTERPRETATION OF EXPANSION INDEX TEST RESULTS

EI	Potential Expansion
0 - 20	Very Low
21 - 50	Low
51 - 90	Medium
91 - 130	High
> 130	Very High

Reproduced from the 1991 Edition of the *Uniform Building Code*, © 1991, with permission of the publisher, the International Conference of Building Officials.

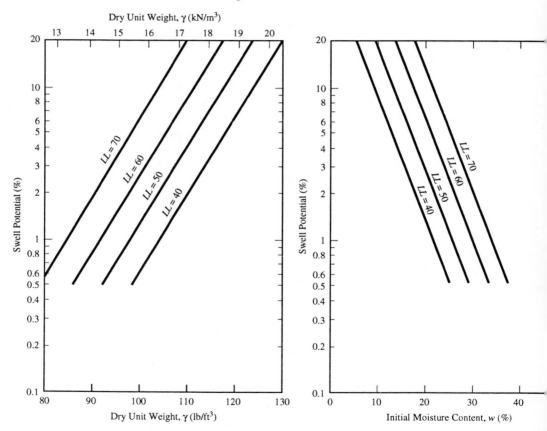

Figure 20.13 Correlations between swell potential, liquid limit, LL; initial moisture content, w; and dry unit weight, γ_d. Adapted from Vijayvergiya and Ghazzaly (1973), used with permission of the Israel Geotechnical Society.

Correlations

Several researchers have developed empirical correlations between swell potential and basic engineering properties. Vijayvergiya and Ghazzaly (1973) developed relationships for undisturbed soils, as shown in Figure 20.13. They use moisture content, liquid limit, and dry unit weight as independent variables and define the swell potential at a surcharge load of 200 lb/ft^2 (9.6 kPa) and an initial moisture content equal to the in-situ moisture content.

Expansion Pressure Tests

As discussed earlier in this chapter, the surcharge pressure affects the swelling of a soil. Higher pressures provide more restraint, and a certain pressure, called the *expansion pressure*, or *swell pressure*, σ_s, prevents all swell. Some engineers elect to measure σ_s and use it as a measure of expansiveness.

Engineers can measure the expansion pressure at the end of a loaded swell test by simply reloading the sample in increments until it returns to its original volume, as shown in Figure 20.14. Another method is to use a modified consolidometer that allows no vertical strain, as shown in Figure 20.15. The first method tends to produce larger swelling pressure. However, neither test precisely duplicates the actual sequence of loading and wetting in the field.

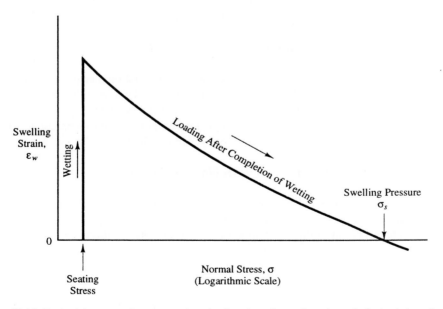

Figure 20.14 Determining expansion pressure by reloading the soil sample at the end of a loaded swell test.

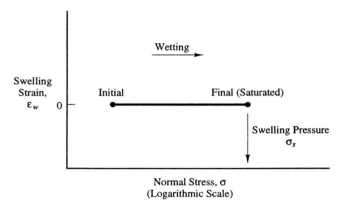

Figure 20.15 Determining expansion pressure using a zero-strain consolidometer.

Some engineers believe the expansion pressure is independent of the initial moisture content, initial degree of saturation, and strata thickness of the soil and varies only with the dry unit weight and is therefore a fundamental physical property of an expansive soil (Chen, 1988). Others disagree with this evaluation and claim that it varies.

When testing undisturbed samples, Chen (1988) recommends defining the swelling pressure as that required to keep the soil at its in-situ dry unit weight. He also recommends testing remolded samples at 100% relative compaction and defining the swelling pressure as that required to maintain this dry unit weight.

Quantitative Evaluations

There is increasing emphasis on developing more rational analysis techniques to deal with expansive soils problems, a concept discussed later in this chapter. These methods require tests that evaluate the soil on a more fundamental basis. This approach has not yet been developed in detail and is not commonly used in practice. However, it will probably become much more dominant in the future.

Variation of Swell with Normal Stress

The swell potential varies with the normal stress acting on the soil. Shallow, lightly loaded soils will swell more than those that are deeper and more heavily loaded. Therefore, any rational method must consider this relationship.

One way to assess this relationship is to obtain undisturbed soil samples at various depths and perform a loaded swell test on each. The *constant volume swell* (CVS) test

(Johnson and Stroman, 1976) is one such test. The procedure is generally as follows[6] and as shown in Figure 20.16:

1. Place an undisturbed soil sample in a consolidometer and apply a normal load equal to the in-situ overburden stress.
2. Inundate the sample and begin increasing the normal load in increments as necessary to restrain any swelling. Continue until the swelling pressure is fully developed.
3. Unload the soil in increments to obtain the swell curve. Continue until the load is less than the in-situ overburden stress.

Another similar procedure is the *modified swell overburden* (MSO) test (Johnson and Stroman, 1976), which follows and is shown in Figure 20.17:

1. Place an undisturbed soil sample in a consolidometer and apply a normal load equal to the design overburden stress (i.e., the stress at the sample location after the foundation is in place).
2. Inundate the sample and allow it to swell under the design overburden stress.
3. After the swelling is complete, load the sample in increments until the soil returns to its original volume. The pressure that corresponds to the original volume is the swelling pressure.
4. Unload the sample in increments to a stress less than the in-situ overburden.

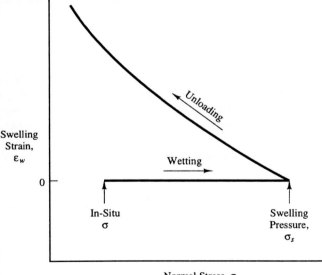

Figure 20.16 Constant volume swell (CVS) test results.

---

[6] See the original reference for complete details on the CVS and MSO test procedures.

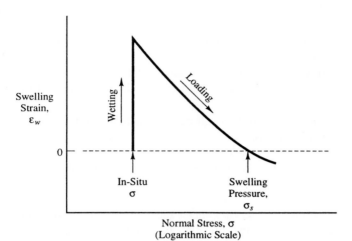

Figure 20.17 Modified swell overburden (MSO) test results.

Swelling Strain, ε_w

Wetting

Loading

0

In-Situ σ

Swelling Pressure, σ_s

Normal Stress, σ
(Logarithmic Scale)

Johnson and Stroman (1976) recommend the MSO test when the design overburden pressures are known in advance, and the CVS test when they are not known in advance.

The strain measured in these tests is the *potential swell strain*, ε_w, for each normal stress. Note that the test results are expressed in terms of total stress, σ, not effective stress, σ'. Thus, this information will be used in a *total stress analysis*, not an *effective stress analysis*. The slope of these plots is C_e, the *expansion coefficient*. This coefficient varies with effective stress.

Wetting Processes

The swell strain that will occur in the field will not necessarily be equal to that measured in the laboratory. This is because the soil in the field may not become completely saturated. The ratio of the actual swell strain to the potential swell strain is the *wetting coefficient*, α. If the soil remains at its in-situ moisture content, then $\alpha = 0$; if it becomes completely saturated, $\alpha = 1$.

Chen (1988) suggests that α is approximately proportional to the change in the degree of saturation. Thus:

$$\alpha \approx \frac{S - S_0}{1 - S_0} \tag{20.2}$$

Where:

   S_0 = degree of saturation before wetting (in decimal form)

   S = degree of saturation after wetting (in decimal form)

Unfortunately, it is very difficult to predict the degree of saturation that will occur in the field. It depends on many factors, including the following:

- The rate and duration of water inflow (due to wetting) and outflow (primarily due to evaporation and transpiration).
- The rate at which water flows through the soil.
- Stratification of the soil.

Flow in unsaturated soils is driven by soil suction, not by conventional hydraulic heads, so some engineers have attempted to use measurements of soil suction to predict the degree and depth of wetting. See Nelson and Miller (1992) for more details.

20.3 ESTIMATING POTENTIAL HEAVE

The current state-of-practice in most areas is to move directly from laboratory test results to recommended design measures with no quantitative analyses to connect the two. Such leaps are possible only when the engineer is able to rely on local experience obtained from trial-and-error. For example, we may know that in a certain geologic formation, slab-on-grade floors have performed adequately only if the expansion index is less than some certain value. If the EI at a new project site in that formation is less than the specified value, then the engineer will recommend using a slab-on-grade floor; if not, then some other floor must be used.

This kind of methodology implicitly incorporates such factors as climate, depth of the active zone, permeability (especially the presence of fissures), and structural tolerance of differential heave, so they are limited only to the geologic formations, geographic areas and types of structures that correspond to those from which the method was derived. They generally work well as long as these restrictions are observed, but can be disastrous when extrapolated to new conditions.

We would prefer to have a more rational method of designing structures to resist the effects of expansive soils; one that *explicitly* considers these factors. Ideally, such a method would predict potential heave and differential heave. Just as engineers design spread footings based on their potential for settlement, it would be reasonable to design structures on expansive soils based on the potential for heave.

Laboratory Testing

Heave analyses are normally based on laboratory swell tests, such as the MSO or CVS tests described earlier. Conduct these tests on undisturbed samples from different depths within the active zone to establish the expansive properties of each strata. Typically, the moisture content of the soil at the beginning of each test is equal to the in-situ moisture content. Thus, the laboratory tests represent the swelling that would occur if the soil becomes wetter than the in-situ moisture. Sometimes, engineers will first dry the samples to a lower moisture content, thus modeling a worse condition.

Because the laboratory swell tests are laterally confined, they model a field condition in which the swell occurs only in the vertical direction. This may be a suitable model when the ground surface is level, but a poor one when it is sloped or when a retaining wall is present. In the latter cases, the horizontal swell is often very important.

In the field, some of the swell may be consumed by the filling of fissures in the clay. This is not reflected in the laboratory tests because the samples normally will not include fissures. However, this error should be small and conservative, and thus can be ignored.

Analysis

The heave due to soil expansion is:

$$\delta_w = \sum \alpha_i H_i \varepsilon_{wi} \tag{20.3}$$

Where:

δ_w = heave due to wetting

α_i = wetting coefficient in layer i

H_i = thickness of layer i

ε_{wi} = potential swell strain in layer i

Implement this analysis as follows:

Step 1.   Divide the active zone of soil beneath the foundation into layers in a fashion similar to that used for settlement analyses. These layers should be relatively thin near the bottom of the footing (perhaps 1 ft or 0.25 m thick) and become thicker with depth. The bottom of the last layer should coincide with the bottom of the active zone.

Step 2.   Compute the vertical total stress, σ_v, at the midpoint of each layer. This stress must consider both the weight of the soil and external loads acting on the footing.

Step 3.   Using the results of the laboratory swell tests, compute the potential swell strain, ε_w, at the midpoint of each layer.

Step 4.   Determine the initial profile of degree of saturation vs. depth. This would normally be based on the results of moisture content tests from soil samples recovered from an exploratory boring.

Step 5.   Estimate the final profile of degree of saturation vs. depth. As discussed earlier, this profile is difficult to predict. It is the greatest source of uncertainty in the analysis. Techniques for developing this profile include the following:

Option 1 Use empirical estimates based on observations of past projects. Studies of the equilibrium moisture conditions under covered areas, including buildings and pavements, suggest that the final moisture content is usually in the range of 1.1 to 1.3 times the plastic limit.

Option 2 Assume that the soil becomes saturated, but the pore water pressure above the original groundwater table remains equal to zero. This assumption is common for many foundation engineering problems and may be appropriate for many expansive soils problems, especially if extra water, such as that from irrigation or poor surface drainage, might enter the soil.

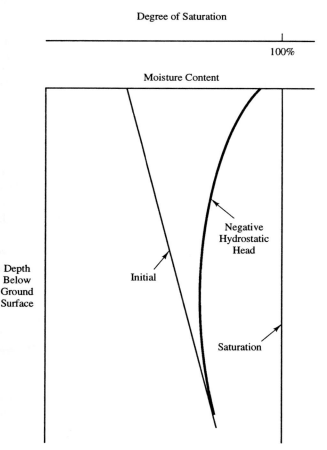

Figure 20.18   Final moisture content profiles.

Option 3 Assume that a suction profile will develop such that a negative hydrostatic head is present. This scenario is based on a soil suction that diminishes with depth at the rate of 62.4 lb/ft^2 per foot of depth.

Option 4 Assume $S = 100\%$ at the ground surface, and tapers to the natural S at the bottom of the active zone.

The second and third options are shown graphically in Figure 20.18.

Step 6. Compute the heave for each layer and sum them using Equation 20.3.

For additional information on heave estimates, see McDowell (1956), Lambe and Whitman (1959), Richards (1967), Lytton and Watt (1970), Johnson and Stroman (1976), Snethen (1980), Mitchell and Avalle (1984), and Nelson and Miller (1992).

Example 20.1

A compressive column load of 140 kN is to be supported on a 0.50 m deep spread footing. The net allowable bearing pressure is 150 kPa. The soils beneath this proposed footing are expansive clays that currently have a degree of saturation of 25%. This soil has a unit weight of 17.0 kN/m^3, and the depth of the active zone is 3.5 m. The results of laboratory swell tests are shown in the figure.

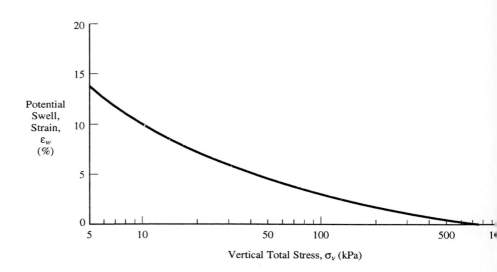

Compute the potential heave of this footing due to wetting of the expansive soils.

Solution:

$$B = \sqrt{\frac{P}{q_a{}''}}$$

$$= \sqrt{\frac{140}{150}}$$

$$= 0.97 \text{ m}$$

Use $B = 1.00$ m

$$q' = \frac{P}{A}$$
$$= \frac{140}{1.00^2}$$
$$= 140 \text{ kPa}$$

Assume S after wetting varies from 100% at the ground surface to 25% at the bottom of the active zone. Compute $\Delta\sigma_v$ using Equation 7.6 and add it to σ_{v0} to compute σ_v. Find ε_s using the lab data, α using Equation 20.2, and δ_w using Equation 20.3.

Depth (m)	H_i (mm)	At Midpoint of Soil Layer								δ_{wi} (mm)
		z_f (m)	σ_{v0} (kPa)	$\Delta\sigma_v$ (kPa)	σ_v (kPa)	ε_{wi} (%)	S_0 (%)	S (%)	α	
0.50 - 0.75	250	0.12	11	140	151	2.0	25	90	0.87	4.3
0.75 - 1.00	250	0.32	15	130	145	2.1	25	80	0.73	3.8
1.00 - 1.50	500	0.75	21	71	92	3.0	25	70	0.60	9.0
1.50 - 2.00	500	1.25	30	40	70	3.5	25	50	0.33	5.8
2.00 - 3.00	1000	2.00	42	19	61	3.8	25	30	0.07	2.7
									Total	26

The estimated heave is **26 mm**.   ⇐ *Answer*

Differential Heave

Just as differential settlements often control the design of normal building foundations, differential heaves control the design of foundations on expansive soils. The differential heave can range from zero to the total heave, but is typically between one-quarter and one-half of the total heave (Johnson and Stroman, 1976). The greatest differential heaves are most likely to occur when swelling is due to such extraneous influences as broken water lines, poor surface drainage, or aggressive tree roots. Soil profiles with numerous fissures are also more likely to have higher differential heaves because of their higher permeability.

Donaldson (1973) recommended designing for the following heave factors (where heave factor = differential heave/total heave):

Profile with high permeability (i.e., fissures) in upper 10 ft (3 m)
    Without extraneous influences   0.50
    With extraneous influences   0.75

Profile with low permeability in upper 10 ft (3 m)
    Without extraneous influences   0.25
    With extraneous influences   0.40

20.4 TYPICAL STRUCTURAL DISTRESS PATTERNS

It is difficult to describe a typical distress pattern in buildings on expansive soils because the exact pattern of heaving depends on so many factors. However, in broad terms, buildings in arid areas tend to experience an *edge lift*, as shown in Figure 20.19, that causes them to distort in a concave-up fashion (Simons, 1991). Conversely, in humid climates, the expansive soil may shrink when it dries, causing the edges to depress, as shown in Figure 20.19. Heated buildings with slab-on-grade floors in colder climates sometimes experience a center depression due to drying and shrinkage of the underlying clay soils.

Special local conditions will often modify this pattern. For example, poor surface drainage or a leaky water line near one corner of the building will probably cause additional local heave. Conversely, aggressive tree roots at another location may locally dry the soil and cause local shrinkage (Byrn, 1991).

In the Dallas area, heaves of 5 to 6 in (125 - 150 mm) are common, and heaves of 8 to 12 in (200 - 300 mm) have been measured (Greenfield and Shen, 1992).

In arid climates, the heaving usually responds to seasonal moisture changes, producing annual shrink-swell cycles, as shown in Figure 20.20. However, during the first 4 to 6 years, the cumulative heave will usually exceed the cumulative shrinkage, so there will be a general heaving trend, as shown in Figure 20.20. As a result, expansive soil problems in dry climates will usually become evident during the first 6 years after construction.

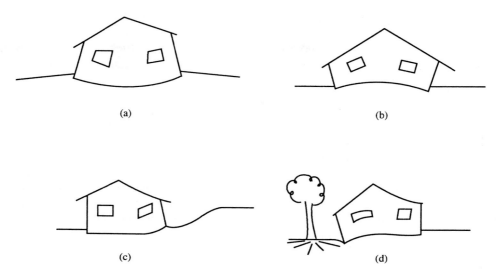

Figure 20.19 Typical distress patterns resulting from heave of expansive soils: (a) edge lift; (b) center lift; (c) localized heave due to drainage problems; and (d) localized shrinkage due to aggressive tree roots.

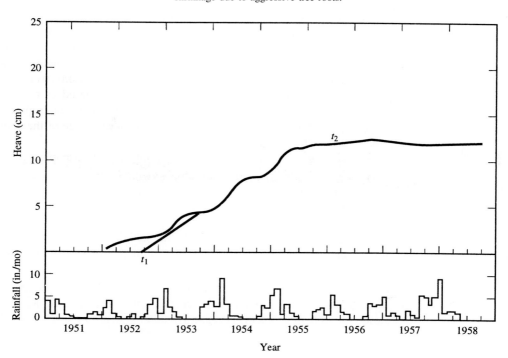

Figure 20.20 Heave record from single story brick house in South Africa (Adapted from Jennings, 1969).

Although this is a typical scenario, it does not mean that expansive soils will always behave this way. It is also very possible that unprecedented swelling may occur later in the life of the building. For example, an exceptionally wet year may invoke larger heaves. Likewise, changes in surface topography could cause ponding near the building and generate heave.

20.5 PREVENTIVE DESIGN AND CONSTRUCTION MEASURES

Having identified the potential for expansive soils problems at a project site, the geotechnical engineer will be concerned with methods of design and construction that will minimize (but not eliminate!) the potential for damage. As with most engineering problems, dealing with expansive soils ultimately boils down to a matter of risk vs. cost. A geotechnical engineer cannot guarantee that a structure will have no problems with expansive soils, but can recommend the use of certain preventive measures that seem to be an appropriate compromise between reducing the risk of damage (especially major damage) and keeping construction costs to a minimum.

Basic Preventive Measures

Any building site on an expansive soil should include at least the following features:

- **Surface Drainage**—Although good surface drainage is important at all building sites, it is especially critical where expansive soils are present. The ground surface should slope away from the structure, as shown in Figure 20.21. Bare or paved areas should have a slope of at least 2%, vegetated ground should have at least 5%. If possible, slope the ground within 10 feet of the structure at a 10% grade.

  It is also important to install gutters or other means of collecting rainwater from the roof and discharging it away from the foundation.

- **Basement Backfills**—If the structure has a basement, the backfill should consist of nonexpansive soils. Also, it should be well compacted to avoid subsequent settlement that would adversely affect the surface drainage patterns. A well compacted backfill will also be less pervious, so water will be less likely to infiltrate the soil.

  Install a drain pipe at the bottom of the backfill to capture any water that might enter and carry it outside or to a sump pump. Carefully design such drains to avoid having them act as a conduit to bring water *into* the backfill.

- **Landscaping**—Irrigation near the structure can introduce large quantities of water into the soil and is a common cause of swelling. This can be an especially troublesome source of problems because irrigation systems are usually installed by homeowners or others who are not sufficiently conscious of expansive soil concerns.

  Specific preventative measures include:

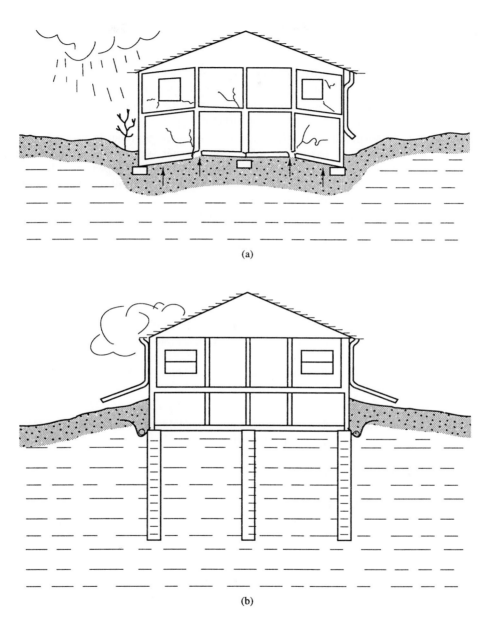

(a)

(b)

Figure 20.21 Surface drainage details: (a) poor drainage, wet expanded clay; and (b) good drainage, dry stable clay (Colorado Geological Survey).

- Avoid placing plants and irrigation systems immediately adjacent to the structure.
- Avoid placing irrigation pipes near the structure (to prevent problems from leaks).
- Direct all spray heads away from the structure.

As discussed earlier, large trees near the structure are often troublesome, especially those with shallow root systems. These trees can draw large quantities of water out of the soil, thus causing it to shrink. Therefore, it is best to avoid planting large trees near the structure.

- **Underground Utilities**—Utility lines often become distorted due to differential swelling of expansive soils. With water, sewer, or storm drain pipes, these distortions can case leaks that in turn cause more expansion. This progression is likely to occur where the pipes enter the building, and could cause large heaves and serious structural distress.

The risk of this potential problem can be reduced by:

- Using flexible pipe materials (i.e., PVC or ABS instead of clay or concrete pipe).
- Installing the pipe such that large shear or flexural stresses will not develop. In some cases, this may require the use of flexible joints.

Beyond these basics, it also is possible to incorporate more extensive measures. O'Neill and Poormoayed (1980) divided them into three basic categories:

- Alter the condition of the expansive clay to reduce or eliminate its swelling potential.
- Bypass the expansive clay by isolating the foundation from its effect.
- Provide a foundation capable of withstanding differential movements and mitigate their effect in the superstructure.

Each of these approaches includes several specific methodologies, as discussed below.

Altering the Expansive Clay

Replacement

Perhaps the most obvious method of overcoming expansive soil problems is to remove the soil and bring in a nonexpansive soil as a replacement. When done carefully, this can be a very effective, although expensive, method. However, be careful not to use a highly permeable soil that could provide an avenue for water to infiltrate the natural soils below (which are probably expansive), thus increasing the depth of the active zone.

Lime Treatment

When hydrated lime is mixed with an expansive clay, a chemical reaction occurs and the clay is improved in the following ways:

- Swelling potential is reduced
- Shear strength is increased
- Moisture content is decreased (helpful when working during the wet season because it increases the workability of the soil)

The lime can be mechanically mixed with the soil at a rate of about 2 to 8% by weight. Special equipment is needed to assure adequate mixing and the process generally limited to shallow depths (i.e., 12 in).

Another method of treating soil with lime is to inject it in a slurry form using a technique known as *pressure injected lime* (PIL). The lime slurry is forced into the soil under high pressure using equipment similar to that shown in Figure 20.22. This method is capable of treating soils to depths of up to about 8 ft.

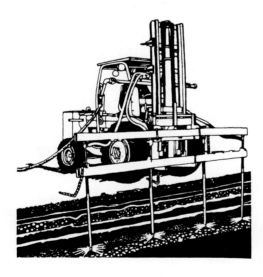

Figure 20.22 Pressure injected lime (PIL) system (Courtesy of GKN Hayward Baker, Inc.).

The PIL technique is most effective in highly fissured soils because the fissures provide pathways for the slurry to disperse. In addition to the chemical effects, the filling of the fissures with lime also retards moisture migration in the soil.

Lime treatment is most commonly used on canals, highways, and other projects that have no foundation. The pressure injected lime technique is also useful when making repairs to an existing structure that has suffered damage from expansive soils.

Prewetting

This technique, also known as *ponding*, *presoaking*, or *presaturation*, consists of covering the site with water before construction in an attempt to increase the moisture content of the soil, thus preswelling it. When used with a project that will include a slab-on-grade, the moisture will remain reasonably constant, especially if the perimeter of the site is landscaped and irrigated or if moisture barriers are installed around the structure. The idea here is to cause the soil to expand before building the structure, and then maintain it at a high moisture content.

In some areas, such as southern California, this technique works well and generally requires a soaking time of a few days or weeks. However, in other areas, the required soaking time is unacceptably long (i.e., many months). These differences in soaking times may be due to differing required depths of soaking (a function of the depth of the active zone) and the presence or absence of fissures in the clay.

The soaking process can be accelerated by first drilling a grid of vertical sand drains (borings filled with sand) to help the water percolate into the soil. The use of wetting agents in the water also can accelerate this process.

After the prewetting is completed, it is usually necessary to treat the ground surface to provide a working platform. This could consist of lime treating the upper soils or placing a 4 to 6 inch thick layer of sand or gravel on the surface.

Moisture Barriers

Impermeable moisture barriers, either horizontal or vertical, can be an effective means of stabilizing the moisture content of the soil under a structure. These barriers may be located on the ground surface in the form of sidewalks or other paved areas, or they can be buried. The latter could consist of underground polyethylene or asphalt membranes. Moisture barriers are especially helpful under irrigated landscape areas where they can take the form of sealed planter boxes.

The primary advantage of barriers is that they promote more uniform moisture conditions below the structure, thus reducing the differential heave. They may or may not affect the total heave.

Although moisture barriers can be very helpful, never consider them to be completely impervious. They are generally supplementary measures that work in conjunction with other techniques.

Bypassing the Expansive Clay

Because the moisture content of a soil will fluctuate more near the ground surface than it will at depth (the active zone concept described earlier), one method of mitigating swelling effects is to support the structure on deeper soils, thus bypassing some or all of the active zone. This method is also useful when the expansive soil strata is relatively thin and is underlain by a nonexpansive soil.

Deepened Footings

When working with mildly expansive soils it often is possible to retain a spread footing foundation system by simply deepening the footings, perhaps to 2 or 3 ft (0.5 m) below grade. This will generally be less than the depth of the active zone, so some heave would still be possible, but its magnitude will be much less.

This method also has the advantage of increasing the rigidity of the footing (which is usually supplemented by additional reinforcement—say, one or two #4 bars top and bottom), thus spreading any heave over a greater distance and improving the structure's tolerance of heave.

When spread footings are used, they should be designed using as high a bearing pressure as practicable to restrain the heaving. A bearing pressure equal to the swelling pressure would be ideal, but is generally possible only in very mildly expansive soils.

Drilled Shafts

In a highly expansive soil, deepened spread footings cease to be practical and a drilled shaft foundation often becomes the preferred system. In the Denver, Colorado area, shafts for lightweight structures are typically 10 - 12 in (250 - 300 mm) in diameter and 15 - 20 ft (4.5 - 6 m) deep (Greenfield and Shen, 1992). The shafts must extend well below the active zone. The individual shafts are connected with grade beams that are cast on top of collapsible cardboard or foam forms, as shown in Figure 20.23. The purpose of these forms is to permit the soil to freely expand without pressing against the grade beam. Another alternative is to use precast grade beams.

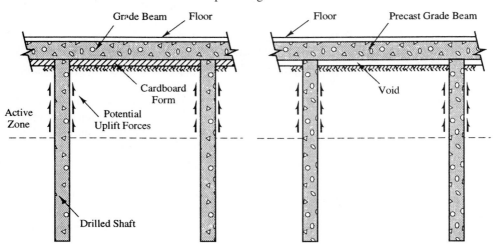

Figure 20.23 Typical lightweight building foundation consisting of drilled shafts and raised grade beams: (a) cast-in-place grade beam with cardboard forms are designed to collapse at pressures slightly greater than those from the wet concrete; (b) precast concrete grade beam. Note the uplift forces acting on the shaft due to the heave of the soil in the active zone.

This system also works well on larger structures, although much larger shafts are then required.

One of the special concerns of using drilled shafts in expansive soils is the development of uplift forces along the sides of the shafts within the active zone, as shown in Figure 20.23. The shafts must be designed to accommodate these forces, both in terms of load transfer and the need for tension steel extending through the active zone.

This matter is further aggravated because the soil also attempts to swell horizontally, which translates to an increased normal stress between the soil and the shaft. This, in turn, permits more skin friction to be mobilized and increases the uplift force.

Reese and O'Neill (1988) recommend computing the unit uplift skin friction in the active zone as follows:

$$f_s = a\,\sigma'_{hs}\,\tan\phi_r \tag{20.4}$$

Where:

f_s = unit uplift skin friction

a = an empirical coefficient

σ'_{hs} = horizontal swelling pressure

ϕ_r = effective residual friction angle of the clay

For design purposes, consider the horizontal swelling pressure, σ_{hs}, to be equal to the vertical swelling pressure obtained from a swelling pressure test.

Unfortunately, we cannot yet define the correlation coefficient, a, with any great precision. O'Neill and Poormoayed (1980) back-calculated a value of 1.3 from a single instrumented shaft in San Antonio. This value agrees very closely with Chen (1988) whose work was based on model tests and has been successfully used in practice. This is probably a reasonable number to use until we obtain more experimental data.

Obtain the effective residual friction angle from laboratory tests. Chen (1988) suggests that for stiff clay and clay shale, it is usually on the order of 5 to 10 degrees.

Consider the possibility that parts of the shaft might develop a net tensile force if the soil were to swell. Therefore, the design must include steel reinforcing bars and they should extend to the bottom of the shaft. In Denver, shafts that support lightweight structures typically have at least two full-length #5 grade 40 rebars, whereas in Dallas, two #6 bars are common (Greenfield and Shen, 1992).

An alternative to designing for uplift loads is to isolate the shaft from the soil in the active zone. One way to do so is by forming the shaft with a cardboard tube inside a permanent steel casing and filling the annular space with a weak but impervious material.

For design purposes, the resistance to either upward or downward movements is considered to begin at the bottom of the active zone. This resistance can be generated by a straight shaft, one with a bell, or one with shear rings. The latter two are commonly preferred when large uplift loads are anticipated.

Structurally Supported Floors

When the computed total heave exceeds about 1 to 2 in (25 - 50 mm), conventional slab-on-grade floors cease to perform well. In such cases, the most common design is to use a raised floor supported on a drilled shaft foundation that penetrates through the active zone, as shown in Figure 20.24. Not only does this design isolate the floor from direct heaving, it also provides ventilation of the ground surface while still shielding it from precipitation. This keeps the soil under the building much drier than it would be with a slab-on-grade floor or a mat floor.

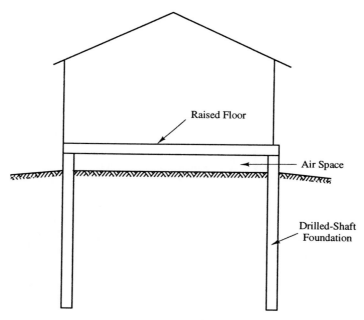

Figure 20.24 Bypassing an expansive clay with a raised floor and a drilled shaft foundation.

Mitigating Movements in the Structure

Another method of addressing differential settlement and heave problems is to make the structure more tolerant of these movements. There are many ways to accomplish this, and these measures can be used alone or in conjunction with the methods described earlier.

Flexible Construction

Some structures will tolerate a large differential movement and still perform acceptably. Lightweight industrial buildings with steel siding are an example of this type of construction. See the discussion in Chapter 2 for more details on this concept.

Another way of adding flexibility to a structure is to use *floating slabs*, as shown in Figure 20.25. Casting these slabs separate from the foundation and providing a slip joint between the slab and the wall allow it to move vertically when the soil swells.

Any construction resting on a floating slab must also be able to move. For example, furnaces would need a flexible plenum and vent pipe. Partition walls could be suspended from the ceiling with a gap at the bottom covered with flexible molding. Stairways also could be suspended from the ceiling and wall and not be connected to the slab. Each of these details is shown in Figure 20.25. Floating slabs are most commonly used in garages and basements because these design details are easier to implement.

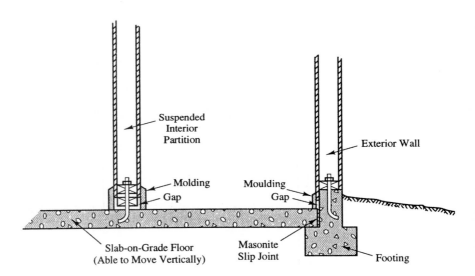

Figure 20.25 Floating slab and related design details.

Rigid Foundation System

The opposite philosophy is to provide a foundation system that is so rigid and strong that it moves as a unit. Differential heaves would then cause the structure to tilt without distorting. Conventionally reinforced mats have been used for this purpose. These mats are also known as *waffle slabs* because of the shape of their integral beams, as shown in Figure 20.26. These slabs are cast using collapsible cardboard forms to provide void spaces under the slab, thus allowing space for the soil to expand. Kantey (1980) noted routine success with brick buildings on this type of foundation in South Africa, where some had experienced heaves of up to 10 in (250 mm) and still performed adequately.

An alternative to conventional reinforcement is to use prestressed or post-tensioned slabs as a kind of mat foundation. PTI (1980) presents a complete design procedure for post-tensioned slabs on expansive soils. This method is gaining popularity, especially for residential projects, and has been used successfully on highly expansive soils in California, Texas, and elsewhere.

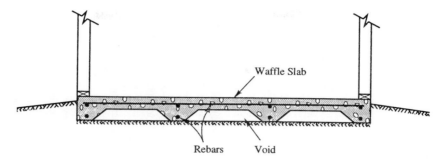

Figure 20.26 Conventionally reinforced mat foundation "waffle slab."

We can apply this concept more economically to mildly expansive soils by using conventional spread footings in such a way that no footing is isolated from the others. This can be accomplished by using continuous footings and/or grade beams, as shown in Figure 20.27. Such a system does not have the rigidity of a mat, but is much more rigid than isolated footings and will help to spread differential heaves over a longer distance.

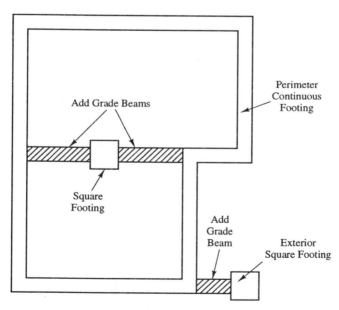

Figure 20.27 Use of continuous footings and grade beams to improve the rigidity of a spread footing system.

TABLE 20.9 PREVENTIVE DESIGN MEASURES BASED ON COMPUTED HEAVE

Total Predicted Heave (in)		Recommended Construction	Method	Remarks
$L/H = 1.25$	$L/H = 2.50$			
< 0.25	< 0.5	No precautions		
0.25 - 0.5	0.5 - 2	Rigid building tolerating movement (steel reinforcement as necessary)	Foundations: Spread footings, mats, or waffle	Footings should be narrow and deep, consistent with the soil bearing capacity.
			Floor slabs: Waffle, tile	Slabs should be designed to resist bending and should be independent of grade beams.
			Walls	Walls on a mat should be as flexible as the mat. No rigid connections vertically. Strengthen brickwork with tiebars or bands.
0.5 - 2	2 - 4	Building damping movement	Joints: Clear, flexible	Contacts between structural units should be avoided, or a flexible waterproof material may be inserted in the joint.
			Walls: Flexible, unit construction, steel frame	Walls or rectangular building units should heave as a unit.
			Foundations: Three-point, cellular or jacks	Cellular foundations allow slight soil expansion to reduce the swelling pressure. Three-point loading allows motion without duress.
> 2	> 4	Building independent of movement	Drilled shafts: Straight, belled	Use smallest diameter and most widely spaced shafts compatible with load. Allow clearance under grade beams.
			Suspended floor	Suspend floors on grade beams 12 - 18 in (300 - 450 mm) above the soil.

Adapted from Gromko (1974). Used with permission of ASCE.

Determining Which Methods to Use

Gromko (1974) suggested the criteria in Table 20.9 to guide the selection of preventive design measures based on the estimated heave and the length-to-height ratio, *L/H*, of the walls.

20.6 OTHER SOURCES OF HEAVE

Although expansive clays are the most common source of heave, other mechanisms also have been observed.

Expansive Rocks

Sedimentary rocks formed from clays, such as claystone, and shale, are often expansive (Lindner, 1976). The physical mechanisms are similar to those for clay soils, but the swelling pressures and potential heave are often very high because of the high unit weight of rock. However, these rocks do not transmit water as easily, so the potential heave may be more difficult to attain in the field.

Some other rocks can expand as a result of chemical processes, such as oxidation or carbonation. These processes often create by-products that occupy a larger volume than the original materials (Lindner, 1976).

Steelmaking Slag

The process of making steel from iron ore produces two principal types of solid wastes: *blast furnace slag* and *steelmaking slag* (Lankford et al., 1985). Blast furnace slag is the waste produced when iron ore and limestone are combined in a blast furnace to produce iron. This material has very favorable engineering properties and has been successfully used for concrete aggregate, base material beneath pavements, and many other applications (Lee, 1974). In contrast, steelmaking slag is produced when iron is made into steel, and it is a much more troublesome material.

The primary problem with steelmaking slag is that it can expand in volume after being put in place, thus producing problems similar to those caused by expansive clays. For example, Crawford and Burn (1968) described a case history of a floor slab built over sand-sized steelmaking slag that heaved up to 3 inches in 5 years. Uppot (1980) described another case history of an industrial building that experienced heaves of up to 8 inches in columns and up to 10 inches in floor slabs within 6 years of construction.

This expansion is the result of hydration of unslaked lime, magnesium oxides, and other materials, and may exceed 20% of the original volume of the slag. Spanovich and Fewell (1968) observed that the expansion potential is reduced by more than 50% if the slag is allowed to age for 30 days before being used. This aging process requires exposure to oxygen and water. However, they also observed that slag that had been buried for more than 30 years (and thus had not been cured) was still very expansive.

Some engineers have used cured steelmaking slag for applications that have a high tolerance of movements, such as open fills, unpaved roads, and railroad ballast. Engineers in Japan have used mixtures of steelmaking slag and blast furnace slag to produce aggregate base material for roadway pavements (Nagao et al., 1989). However, because of its expansive properties, steelmaking slag should not be used beneath structural foundations.

Salt Heave

Soils in arid areas sometimes contain high concentrations of water-soluble salts that can crystallize out of solution at night when the temperature is low and return to solution during the day as the temperature rises. The formation and dissipation of these salt crystals can cause daily cycles of heave and shrinkage in the soil, especially in the late fall and early spring when the difference between day and night air temperatures may be 60° F (33° C) or more. Although some moisture must be present for crystallization to occur, this process is driven by changes in temperature. This phenomenon, known as *salt heave*, or *chemical heave*, normally occurs only in the upper 1 - 2 ft (0.3 - 0.6 m) of the soil because the salt concentrations and temperature fluctuations are greatest in this zone.

Based on studies of soils in the Las Vegas, Nevada area, Cibor (1983) suggested that salt heave can generate swelling pressures of 200 - 300 lb/ft^2 (10 - 15 kPa). The bearing pressure beneath most spread footings is much larger than this swelling pressure, so these footings are able to resist the heaving forces without moving. In addition, the bottoms of footings are typically below the zone of greatest heave potential. However, heave has been observed in very shallow and lightly loaded footings.

Damage from salt heave is most often seen in slab-on-grade floors, sidewalks, and other shallow, very lightly loaded areas. For example, Blaser and Scherer (1969) observed heaves of 4 in (100 mm) in exterior concrete slabs.

Evaluate these soils by measuring the percentage of salts in solution. Typically, concentrations of greater than 0.5% soluble salts in soils with more than 15% fines may be of concern. The heave potential can be measured using a thermal swell test (Blaser and Scherer, 1969). This test is similar to the swell tests discussed earlier in this chapter, except that the soil expansion is induced by cooling the sample instead of adding water.

A common method of avoiding salt heave problems is to excavate the salt-laden natural soils to a depth of 1 - 2 ft (0.3 - 0.6 m) below the proposed ground surface and backfill with open graded gravel. Because the gravel provides thermal insulation, the temperature below will not vary as widely as on the ground surface, so salt heave will be less likely to occur. In addition, the weight of the gravel will resist any swelling pressures that might develop.

SUMMARY OF MAJOR POINTS

1. Expansive soils cause more property damage in the United States than all the earthquakes, floods, tornados, and hurricanes combined.

2. Most of this damage is inflicted on lightweight improvements, such as houses, small commercial buildings, and pavements.

3. Soil swelling is the result of the infiltration of water into certain clay minerals, most notably montmorillonite. Swelling will occur only if the moisture content of the soil changes.

4. Expansive soils are most likely to cause problems in subhumid, semiarid, and arid climates. The patterns of precipitation and evaporation/transpiration in these areas usually cause the moisture content of the soil to fluctuate during the year, which creates cycles of shrinking and swelling. New construction in these areas normally promotes an increased moisture content in the soil that further aggravates the problem.

5. The potential heave at a given site is a function of the soil profile, variations in soil moisture, overburden loads, and superimposed loads.

6. A wide variety of test methods are available to evaluate the degree of expansiveness. Some of these methods are primarily qualitative, and others provide quantitative results.

7. Preventive measures fall into three categories:

    a. Alter the condition of the expansive clay.

    b. Bypass the expansive clay by isolating the foundation from its effect.

    c. Provide a shallow foundation capable of withstanding differential movements and mitigating their effect in the superstructure.

8. The current state-of-practice in most places is to select the appropriate preventive measures based on qualitative or semiquantitative test results and local experience. These methods are generally applicable only to limited geographic areas and for certain types of structures.

9. The profession is moving toward the use of heave calculations as a basis for determining preventive measures. Such calculations can be based on either loaded swell tests or soil suction tests. Hopefully, these methods will help the engineer more accurately characterize the soil and soil-structure interaction.

QUESTIONS AND PRACTICE PROBLEMS

20.1 Why are lightweight structures usually more susceptible to damage from expansive soils?

20.2 What types of climates are most prone to cause problems with expansive clays?

20.3 How do human activities often aggravate expansive soil problems?

20.4 What is the "active zone"?

20.5 What is the "swelling pressure"?

20.6 What are the primary sources of uncertainty in heave analyses?

20.7 Why do the soils beneath buildings with raised floors tend to be drier than those beneath slab-on-grade floors?

20.8 The foundation described in Example 20.1 is to be redesigned using a net allowable bearing pressure of 75 kPa. Compute the potential heave and compare it with that computed in the example. Discuss.

20.9 An 8 inch diameter drilled shaft will penetrate through an expansive stiff clay to a depth well below the active zone. It will carry a downward load of 5200 lb. The horizontal swelling pressure in this soil is approximately 5000 lb/ft^2 and the active zone extends to a depth of 10 ft. No residual strength data are available. Determine the following:

    a. What is the uplift skin friction load?

    b. Is a tensile failure in the shaft is possible (do not forget to consider the weight of the shaft)?

    c. What is the required reinforcing steel to prevent a tensile failure (if necessary)? Use $f_y = 40$ k/in^2 and a load factor of 1.7.

21

Foundations on Collapsible Soils

*It is better to fail while attempting to do something worthwhile
than to succeed at doing something that is not.*

Foundation engineers who work in arid and semiarid areas of the world often encounter deposits of *collapsible soils*. These soils are dry and strong in their natural state and appear to provide good support for foundations. However, if they become wet, these soils quickly consolidate, thus generating unexpected settlements. Sometimes these settlements are quite dramatic, and many buildings and other improvements have been damaged as a result. These soils are stable only as long as they remain dry, so they are sometimes called *metastable soils*, and the process of collapse is sometimes called *hydroconsolidation*, *hydrocompression*, or *hydrocollapse*.

To avoid these kinds of settlements, the foundation engineer must be able to identify collapsible soils, assess the potential settlements, and employ appropriate mitigation measures when necessary.

21.1 ORIGIN AND OCCURRENCE OF COLLAPSIBLE SOILS

Collapsible soils consist predominantly of sand and silt size particles arranged in a loose "honeycomb" structure, as shown in Figure 21.1. Sometimes gravel is also present. This loose structure is held together by small amounts of water-softening cementing agents, such as clay or calcium carbonate (Barden et al., 1973). As long as the soil remains dry, these cements produce a strong soil that is able to carry large loads. However, if the soil becomes wet, the cementing agents soften and the honeycomb structure collapses.

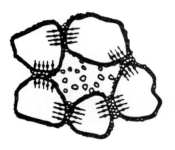

Loaded Soil Structure
Before Inundation

Loaded Soil Structure
After Inundation

Figure 21.1 Microscopic view of a collapsible soil. In their natural state, these soils have a honeycomb structure that is held together by water soluble bonds. However, if the soil becomes wet, these bonds soften and the soil consolidates (Adapted from Houston, et al., 1988; Used with permission of ASCE).

Various geologic processes can produce collapsible soils. By understanding their geologic origins, we are better prepared to anticipate where they might be found.

Collapsible Alluvial and Colluvial Soils

Some *alluvial soils* (i.e., soils transported by water) and some *colluvial soils* (i.e., soils transported by gravity) can be highly collapsible. These collapsible soils are frequently found in the southwestern United States as well as other regions of the world with similar climates. In these areas, short bursts of intense precipitation often induces rapid downslope movements of soil known as *flows*. While they are moving, these soils are nearly saturated and have a high void ratio. Upon reaching their destination, they dry quickly by evaporation, and capillary tension draws the pore water toward the particle contact points, bringing clay and silt particles and soluble salts with it, as shown in Figure 21.2. Once the soil becomes dry, these materials bond the sand particles together, thus forming the honeycomb structure.

When the next flow occurs, more honeycomb-structured soil forms. It, too, dries rapidly by evaporation, so the previously deposited soil remains dry. Thus, deep deposits of this soil can form. The remains of repeated flows are often evident from the topography of the desert, as shown in Figure 21.3. These deposits are often very erratic, and may include interbedded strata of collapsible and noncollapsible soils.

In some areas, only the upper few feet (1 - 2 m) of these soils is collapsible, whereas elsewhere the collapse-prone soils may extend 200 ft (60 m) or more below the ground surface. An example of the latter is the San Joaquin Valley in Central California. Irrigation canals are especially prone to damage in that area because even small leaks that persist for a long time may wet the soil to a great depth. Settlements of 2 to 3 ft (600 - 900 mm) are common, and some cases of up to 15 ft (4.7 m) have been reported (Dudley, 1970).

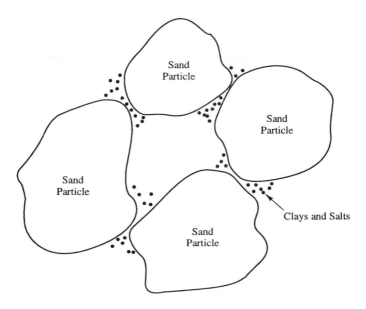

Figure 21.2 When flow deposits dry by evaporation, the retreating water draws the suspended clay particles and dissolved salts toward the particle contact points.

Figure 21.3 Aerial view of flow topography in the desert near Palm Springs, California. Both surface water and windstorms in this area move from the right side of the photo to the left side. Thus, the alluvial and aeolian soil deposits form long stripes on the ground. These soils are often highly collapsible, so settlement problems can occur in heavily irrigated developments, such as the golf course community shown in the bottom of this photograph.

Collapsible Aeolian Soils

Soils deposited by winds are known as *aeolian soils*. These include windblown sand dunes, loess, volcanic dust deposits, as well as other forms. *Loess* (an aeolian silt or sandy silt) is the most common aeolian soil and it covers much of the earth's surface. It is found in the United States, central Europe, China, Africa, Australia, the former Soviet Union, India, Argentina, and elsewhere (Pye, 1987). The locations of major loess deposits in the United States are shown in Figure 21.4 along with locations of other reported collapsible soil deposits.

Symbols

Major Loess Deposits

x   Reports of Collapse in Other Type Deposits

Figure 21.4 Locations of major loess deposits in the United States along with other sites of reported collapsible soils (Adapted from Dudley, 1970; Used with permission of ASCE).

Collapsible loess has a very high porosity (typically on the order of 50%) and a correspondingly low unit weight (typically $70 - 90$ lb/ft^3 ($11 - 14$ kN/m^3)). The individual particles are usually coated with clay, which acts as a cementing agent to maintain the loose structure. This cementation is often not as strong as that in many alluvial soils, so collapse can occur either by wetting under a moderate normal stress or by subjecting the soil to higher normal stresses without wetting it.

Loess deposits are generally much less erratic than other types of collapsible soils, but they are often much thicker. Deposits 200 ft (60 m) thick are not unusual.

Collapsible Residual Soils

Residual soils are those formed in-place by the weathering of rock. Sometimes this process involves decomposition of rock minerals into clay minerals that may be removed by leaching, leaving a honeycomb structure and a high void ratio. When this structure develops, the soil is prone to collapse. For example, Brink and Kantey (1961) reported that residual decomposed granites in South Africa often collapse upon wetting, leading to a 7 - 10% increase in unit weight. Residual soils derived from sandstones and basalts in Brazil also are collapsible (Hunt, 1984).

Dudley (1970) reported test results from a residual soil from Lancaster, California, that showed nearly zero consolidation when loaded dry to a stress of 14,000 lb/ft^2 (670 kPa) over the natural overburden stress, yet collapsed by 10% of its volume when soaked.

Residual soils are likely to have the greatest amount of spatial variation, thus making it more difficult to predict the collapse potential.

21.2 IDENTIFICATION, SAMPLING, AND TESTING

Engineers have used many different techniques to identify and evaluate collapsible soils. They may be divided into two categories: indirect methods and direct methods.

The indirect methods assess collapse potential by correlating it with other engineering properties such as unit weight, Atterberg limits, or percent clay particles. The result of such efforts is usually a qualitative classification of collapsibility, such as "highly collapsible." Although these classifications can be useful, they provide little, if any, quantitative estimates of the potential settlements (Lutenegger and Saber, 1988). In addition, most of them have been developed for certain types of soil, such as loess, and cannot necessarily be used for other types, such as alluvial soils.

As a result of these difficulties, many engineers prefer to use direct methods of evaluating potentially collapsible soils. These methods involve actually wetting the soil, either in the laboratory or in-situ, and measuring the corresponding strain. We can then extrapolate the results of such tests to the entire soil deposit and predict the potential settlements.

Obtaining Samples of Collapsible Soils

Laboratory tests are more useful than in-situ tests because of the greater degree of control that is possible in the lab. However, to perform such tests, we must obtain relatively undisturbed samples of the soil and bring them to the lab.

Collapsible soils that are moderately to well cemented and do not contain much gravel usually can be sampled for purposes of performing collapse tests without undue difficulty. This includes many of the alluvial collapsible soils. As with any sampling operation, strive to obtain samples that are as undisturbed as possible and representative of the soil deposit. Unfortunately, collapsible soils, especially those of alluvial or residual origin, are often very erratic. It is difficult to obtain representative samples of erratic soil deposits, so we must obtain many samples to accurately characterize the collapse potential. Considering the usual limitations of funding for soil sampling, it is probably wiser to obtain many good samples rather than only a couple of extremely high quality (but expensive) samples.

Sometimes, conventional thin-wall Shelby tube samplers can be pressed into the soil.   It is best to use short tubes to avoid compressing the samples.   Unfortunately, because collapsible soils are strong when dry (it is important not to artificially wet the soil during sampling), it often is necessary to hammer the tube in place.   Fortunately, studies by Houston and El-Ehwany (1991) suggest that hammering does not significantly alter the results of laboratory collapse tests for cemented soils.

Although Shelby tubes work well in silty and sandy soils, they are easily bent when used in soils that contain even a small quantity of gravel.   This is often the case in collapsible soils, so we may be forced to use a sampler with heavier walls, such as the ring-lined barrel sampler shown in Figure 4.5 in Chapter 4.   Although these thick-wall samplers generate more sample disturbance, their soil samples may still be suitable for laboratory collapse tests (Houston and El-Ehwany, 1991).

Lightly cemented collapsible soils, such as loess, are much more difficult to sample and require more careful sampling techniques.   Fortunately, loess is usually much more homogeneous, so fewer samples are needed.

Gravelly soils are more difficult to sample and therefore more difficult to evaluate. Special in-situ wetting tests are probably appropriate for these soils (Mahmoud, 1991).

Laboratory Soil Collapse Tests

Once samples have been obtained, they may be tested in the laboratory by conducting *collapse tests*.   These are conducted in a conventional oedometer (consolidometer) and directly measure the strain (collapse) that occurs upon wetting.   Although the details of the various test procedures vary, we can divide them into two groups: double oedometer tests and single oedometer tests.

Double Oedometer Method

Jennings and Knight (1956, 1957, 1975) developed the *double oedometer method* while investigating collapsible soils in South Africa.   This method uses two parallel consolidation (oedometer) tests on "identical" soil samples.   The first test is performed on a sample at its in-situ moisture content; the second on a soaked sample.   The test results are plotted together, as shown in Figure 21.5.   The vertical distance between the test results represents the *potential hydrocollapse strain*, ε_w, as a function of normal stress.

Single Oedometer Method

Many engineers prefer to use the single oedometer method to assess collapse potential. With this method, each test requires only one soil sample.

The single oedometer test is usually conducted as follows (Houston et al., 1988):

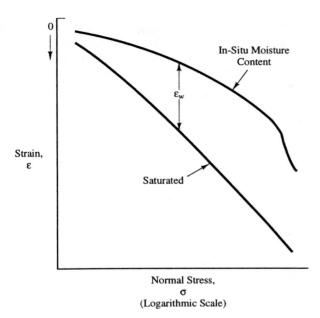

Figure 21.5 Typical results from double oedometer tests.

1. Place an undisturbed soil sample in an oedometer and maintain the in-situ moisture content.
2. Apply a seating load of 100 lb/ft^2 (5 kPa) and zero the dial gage.
3. Increase the vertical stress in increments, allowing the soil to consolidate each time. Normally the load may be changed when the rate of consolidation becomes less than 0.1% per hour. Continue this process until the vertical stress is equal to or slightly higher than that which will occur in the field.
4. Inundate the soil sample and monitor the resulting hydrocompression. This is the potential hydrocollapse strain, ε_w, for this overburden stress.
5. Once the hydroconsolidation has ceased, apply an additional stress increment and allow the soil to consolidate.

This procedure can normally be completed in 24 to 36 hours. The results of such a test are shown in Figure 21.6.

The single oedometer method is faster and easier and it more closely simulates the actual loading and wetting sequence that occurs in the field. It also overcomes the problem of obtaining the two identical samples needed for the double oedometer test. However, this method provides less information because it only gives the hydrocollapse strain at one normal stress. Therefore, the soil should be wetted at a normal stress that is as close as possible to that which will be present in the field.

Evaluation of Laboratory Collapse Test Results

The most common practice has been to evaluate collapse test results using criteria such as that shown in Table 21.1. When using this approach, engineers specify the required remedial measures, if any, based on a qualitative assessment of the test results. For example, soils with an ε_w greater than some specified value might always be excavated and compacted back in place.

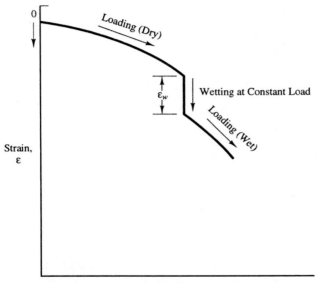

Figure 21.6 Typical results from a collapse test using the single oedometer method (Adapted from Houston et al., 1988; Used with permission of ASCE).

TABLE 21.1 CLASSIFICATION OF SOIL COLLAPSIBILITY

Potential Hydrocollapse Strain, ε_w	Severity of Problem
0 - 0.01	No problem
0.01 - 0.05	Moderate trouble
0.05 - 0.10	Trouble
0.10 - 0.20	Severe trouble
> 0.20	Very severe trouble

Adapted from Jennings and Knight, 1975; Proceedings 6th Regional Conference for Africa; ©A.A. Balkema Publishers, Brookfield, VT; Used with permission).

Although such methods often are sufficient, they do not provide an estimate of the potential settlement due to hydrocollapse. For example, a thick stratum of "moderate trouble" soil that becomes wetted to a great depth may cause more settlement than a "severe trouble" soil that is either thinner or does not become wetted to as great a depth. Therefore, it would be more useful to perform a *quantitative* evaluation of the test data to estimate the potential settlements due to collapse and develop any preventive design measures accordingly.

One of the soil characteristics to identify before conducting a settlement analysis is the relationship between the potential hydrocollapse strain, ε_w, and the normal stress, σ. Although additional research needs to be conducted to further understand this relationship, it appears that there is some *threshold collapse stress*, σ_t, below which no collapse will occur, and that ε_w becomes progressively larger at stresses above σ_t. This trend probably continues until the stress is large enough to break down the dry honeycomb structure, as shown in Figure 21.7. This latter stress is probably quite high in most collapsible soils, but in some lightly cemented soils, such as loess, it may be within the range of stresses that might be found beneath a foundation. For example, Peck, Hanson and Thornburn (1974) noted a sudden and dramatic collapse of certain dry loessial soils in Iowa when the normal stress was increased to about 5500 lb/ft^2 (260 kPa).

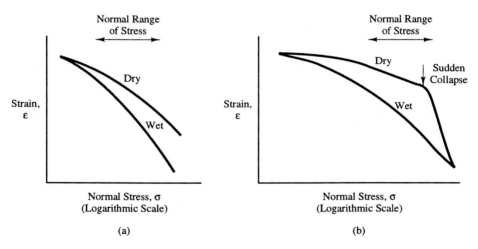

Figure 21.7 Relationship between hydrocollapse strain and normal stress: (a) for most collapsible soils; and (b) for loess.

For collapsible soils other than loess, it appears that the relationship between ε_w and σ_t within the typical range of stresses beneath foundations may be expressed by the following formula:

$$\varepsilon_w = C_w \left(\log \sigma - \log \sigma_t \right)$$

$$= C_w \log \left(\frac{\sigma}{\sigma_t} \right) \qquad (21.1)$$

$$C_w = \frac{d\,\varepsilon_w}{d\,(\log \sigma)} \qquad (21.2)$$

Where:

ε_w = potential hydrocollapse strain

C_w = hydroconsolidation coefficient

σ_t = threshold collapse stress

σ = normal stress at which hydroconsolidation occurs

The definition of C_w, as shown in Equation 21.2, also could be expressed as the additional hydrocollapse strain that occurs for every tenfold increase in normal stress.

We could determine C_w and σ_t by conducting one or more double oedometer tests. Alternatively, we could perform a series of single oedometer tests at a variety of normal stresses, and plot the results, as illustrated in Example 21.1. In either case, conduct the tests at stresses that are comparable to those in the field.

The magnitude of σ_t can often be taken to be about 100 lb/ft^2 (5 kPa) (Houston et al., 1988), although a higher value would be likely if the clay content or unit weight is high. Fortunately, any error in determining σ_t, and the corresponding error in computing C_w, does not significantly affect the settlement computations as long as the samples are inundated at normal stresses within the range of those that exist in the field.

In-Situ Soil Collapse Tests

Gravelly soils pose special problems because they are very difficult to sample and test; yet they still may be collapsible. To evaluate these soils, it may be necessary to conduct an in-situ collapse test. Some of these tests have consisted of large-scale artificial wetting with associated monitoring of settlements (Curtin, 1973), and others have consisted of small-scale wetting in the bottom of borings (Mahmoud, 1991).

21.3 WETTING PROCESSES

To assess potential settlements due to soil collapse, it is necessary to understand the processes by which the soil becomes wetted. We must identify the potential sources of water and understand how it infiltrates into the ground.

Usually, the water that generates the collapse comes from artificial sources, such as the following:

- Infiltration from irrigation of landscaping or crops
- Leakage from lined or unlined canals
- Leakage from pipelines and storage tanks
- Leakage from swimming pools
- Leakage from reservoirs
- Seepage from septic tank leach fields
- Infiltration of rainwater as a result of unfavorable changes in surface drainage

Although the flow rate from most of these sources may be slow, the duration is long. Therefore, the water often infiltrates to a great depth and wets soils that would otherwise remain dry.

As water penetrates the soil, a *wetting front* forms, as shown in Figure 21.8. This process is driven primarily by soil suction, so the wetting front will be very distinct. The distance it advances depends on the rate and duration of the water inflow as well as the hydraulic conductivity of the soil, the magnitude of the soil suction, and other factors. These are difficult to quantify, so this may be the greatest source of uncertainty in estimating the settlement due to collapse. Hopefully additional research and experience will generate more insight.

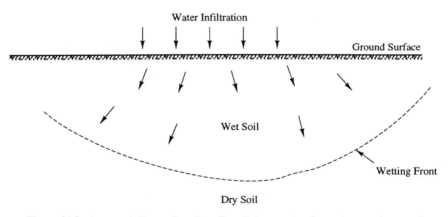

Figure 21.8 As water infiltrates into the soil, a distinct wetting front advances downward.

Curtin (1973) gives an interesting illustration of wetting processes in collapsible soils. He conducted large-scale wetting collapse tests in a deposit of collapsible alluvial soil in California's San Joaquin Valley. This soil is collapsible to a depth of at least 250 ft (75 m). After applying water continuously for 484 days, the wetting front advanced to a depth of at least 150 ft (45 m). The resulting collapse caused a settlement of 13.5 ft (4.1 m) at the ground surface.

In another case, irrigation of lawns and landscaping, and poor surface drainage around a building in New Mexico caused the wetting front to extend more than 100 ft (30 m) into the ground, which resulted in 1 - 2 in (25 - 50 mm) of settlement (Houston, 1991). In this case, the water inflow was slow and gradual, so the soil did not become completely saturated. If the rate of inflow had been greater, and the soil became wetter, the settlements would have been larger.

The New Mexico case illustrates the importance of defining the degree of saturation that might occur in the field. Laboratory collapse tests wet the soil to nearly 100% saturation, which may be a much worse situation than that in the field. This is why we refer to ε_w as the *potential* hydrocollapse strain; the actual strain depends on the degree of wetting.

Tests conducted on collapsible soils in Arizona and New Mexico indicate that these soils typically become only 40 to 60% saturated, even after extended periods of wetting (Houston, 1991). It appears that long-term intensive wetting, such as that obtained in Curtin's tests, is necessary to obtain greater degrees of saturation.

Obtain the relationship between the percentage of potential collapse and the degree of saturation for a given soil by conducting a series of single oedometer collapse tests with different degrees of wetting. Typical results from such tests are shown in Figure 21.9. The *wetting coefficient*, α, is the ratio between the collapse that occurs when the soil is partially wetted to that which would occur if it were completely saturated. Although very few such tests have been conducted, it appears that α values in the field will typically be on the order of 0.5 to 0.8. Once again, additional research and experience probably will give more insight on selecting α values for design.

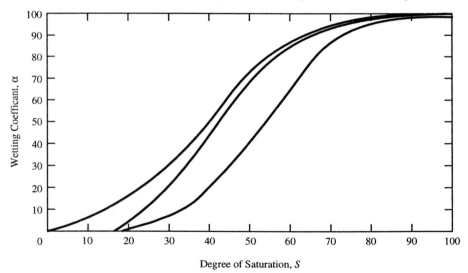

Figure 21.9 Experimental relationships between collapse and degree of saturation attained during wetting. These functions are for particular soils from Arizona, and must be determined experimentally for other soils (Adapted from Houston, 1992).

21.4 SETTLEMENT COMPUTATIONS

The settlement that results from collapse of the soil may be computed as follows:

$$\delta_w = \int \alpha C_w \log\left(\frac{\sigma_v}{\sigma_t}\right) dz \qquad (21.3)$$

Where:

δ_w = settlement due to hydroconsolidation

α = wetting coefficient

C_w = hydrocollapse coefficient

σ_v = vertical total stress

σ_t = threshold stress

z = depth

For practical analyses, evaluate this formula using several finite layers of soil as follows:

$$\delta_w = \sum \alpha\, C_w\, H_i \log\left(\frac{\sigma_v}{\sigma_t}\right) \qquad (21.4)$$

Where:

H_i = thickness of layer i

Compute the total stress, σ, at the midpoint of each layer using a unit weight that corresponds to about 50% saturation. The values of α, σ_t, and C_w may vary with depth and should be assigned to each layer accordingly. Continue the analysis down to the maximum anticipated depth of the wetted front or to the bottom of the collapsible soil, whichever comes first.

Notice that this is a *total stress analysis*, which means that both the laboratory and field data are evaluated in terms of total stress. This differs from the *effective stress analyses* used to evaluate consolidation settlement, as discussed in Chapter 7. An effective stress analysis would need to consider soil suction (negative pore water pressures).

Example 21.1

A 6 kip load is to be applied to a 2 ft square, 1.5 ft deep footing that will be supported on a collapsible soil. This soil extends to a depth of 12 ft below the ground surface and has a natural unit weight of 100 lb/ft^3. At 50% saturation, it has a unit weight of 120 lb/ft^3. Seven single oedometer collapse tests have been performed, the results of which are plotted in the following figure.

Solution:

For design purposes, the depth of wetting would probably be at least 12 ft. Therefore, base the analysis on the assumption that the entire depth of this collapsible soil will become wetted.

The line in the figure that follows is a conservative interpretation of the collapse test data. It represents $\sigma_t = 200$ lb/ft^2 and $C_w = 0.078$.

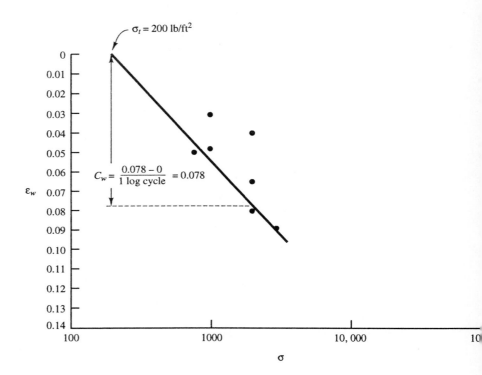

The limited data available suggests that α varies between 0.5 and 0.8. Therefore, it seems appropriate to use α values ranging from 0.9 near the bottom of the footing to 0.5 at a depth of 12 ft.

A tabular format, such as that which follows, is a convenient way to compute the settlement. The parameter σ_{v0} is the geostatic stress, $\Delta\sigma$ is the stress due to the footing load, and $\sigma_v = \sigma_{v0} + \Delta\sigma_v$. Compute $\Delta\sigma_v$ using Equation 7.6 and δ_w using Equation 21.4.

We also could compute the settlement that would occur if the footing was not present by using σ_{v0} instead of σ_v. Such an analysis produces a settlement of 3.3 in. Thus, even if the wetting and the soil were perfectly uniform, the differential settlement would be 4.8 - 3.3 = 1.5 in. In reality, both would be erratic, so the differential settlement could be even greater.

Depth Increment (ft)	H (ft)	z (ft)	σ_{v0} (lb/ft^2)	$\Delta\sigma_v$ (lb/ft^2)	σ_v (lb/ft^2)	σ_t	C_w	α	δ_w (ft)
1.5 - 2.0	0.5	0.25	210	1500	1710	200	0.078	0.9	0.033
2.0 - 3.0	1.0	1.00	300	1058	1358	200	0.078	0.8	0.052
3.0 - 5.0	2.0	2.50	480	428	908	200	0.078	0.8	0.082
5.0 - 7.0	2.0	4.50	720	166	886	200	0.078	0.7	0.071
7.0 - 9.0	2.0	6.50	960	66	1026	200	0.078	0.6	0.066
9.0 - 12.0	3.0	9.00	1260	9	1269	200	0.078	0.5	0.094
> 12.0							0		0

		Total	0.40 ft
Answer $\Rightarrow$		=	**4.8 in**

21.5 COLLAPSE IN DEEP COMPACTED FILLS

Although most collapsing soil problems are associated with natural soil deposits, engineers have observed a similar phenomenon in deep compacted fills (Lawton et al., 1992). Some deep fills can collapse even when they have been compacted to traditional standards (Lawton et al., 1989, 1991; Brandon et al., 1990). For example, Brandon reported settlements of as much as 18 in (450 mm) occurred in 100 ft (30 m) deep compacted fills near San Diego that became wet sometime after construction.

This suggests that the hydroconsolidation potential is dependent both on the void ratio and the normal stress. Very loose soils will collapse upon wetting even at low normal stresses, but denser soils will be collapsible only at higher stresses. In other words, these soils have a very high σ_t, but the normal stresses near the bottom of deep fills are greater than σ_t. At shallower depths (i.e. $\sigma_v < \sigma_t$), fills with a significant clay content may expand when wetted.

It appears that this phenomenon is most likely to occur in soils that are naturally dry and compacted at moisture contents equal to or less than the optimum moisture content. We can reduce the collapse potential by compacting the fill to a higher dry unit weight at a moisture content greater than the optimum moisture content.

21.6 PREVENTIVE AND REMEDIAL MEASURES

In general, collapsible soils are easier to deal with than are expansive soils because collapse is a one-way process, whereas expansive soils can shrink and swell again and again. Many mitigation measures are available, several of which consist of densifying the soil, thus forming a stable and strong material.

Houston and Houston (1989) identified the following methods of mitigating collapsible soil problems:

1. **Removal of the collapsible soil:** Sometimes, the collapsible soil can simply be excavated and the structure then may be supported directly on the exposed non-collapsible soils. We could accomplish this either by lowering the grade of the building site or by using a basement. This method would be most attractive when the collapsible soil extends only to a shallow depth.

2. **Avoidance or minimization of wetting:** Collapse will not occur if the soil is never wetted. Therefore, when working with collapsible soils, always take extra measures to minimize the infiltration of water into the ground. This should include maintaining excellent surface drainage, directing the outflow from roof drains and other sources of water away from the building, avoiding excessive irrigation of landscaping, and taking extra care to assure the water-tightness of underground pipelines.

   For some structures, such as electrical transmission towers, simple measures such as this will often be sufficient. If collapse-induced settlement did occur, the tower could be releveled without undue expense. However, the probability of success would be much less when dealing with foundations for buildings because there are many more opportunities for wetting and the consequences of settlement are more expensive to repair. Therefore, in most cases, we would probably combine these techniques with other preventive measures.

3. **Transfer of load through the collapsible soils to the stable soils below:** If the collapsible soil deposit is thin, it may be feasible to extend spread footing foundations through this stratum as shown in Figure 21.10. When the deposit is thick, we could use deep foundations for the same purpose. In either case, the ground floor would need to be structurally supported.

   When using this method, consider the possibility of negative skin friction acting on the upper part of the foundation.

4. **Injection of chemical stabilizers or grout:** Many types of soils, including collapsible soils, can be stabilized by injecting special chemicals or grout. These techniques strengthen the soil structure so future wetting will not cause it to collapse. These methods are generally too expensive to use over large volumes of soil, but they can be very useful to stabilize small areas or as a remedial measure beneath existing structures.

   A variation of this method, known as *compaction grouting*, involves injecting stiff grout that forms hard inclusions in the soil. This is often used to remediate settlement problems.

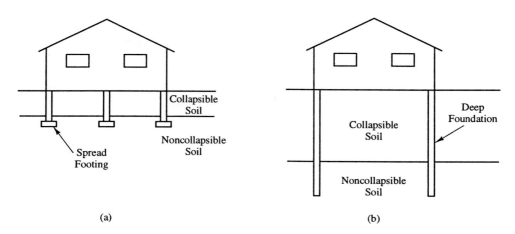

Figure 21.10 Transferring structural loads through collapsible soils to deeper, more stable soils: (a) with deepened spread footings; and (b) with deep foundations.

5. **Prewetting:** If the collapsible soils are identified before construction begins, they can often be remedied by artificially wetting the soils before construction (Knodel, 1981). This can be accomplished by sprinkling or ponding water at the ground surface, or by using trenches or wells. This method is especially effective when attempting to stabilize deep deposits, but may not be completely satisfactory for shallow soils where loads from the proposed foundations may significantly increase the normal stress.

   If the soil has strong horizontal stratification, as is the case with many alluvial soils, then the injected water may flow horizontally more than it does vertically. Therefore, be cautious when using this method near existing structures that are underlain by collapsible soils.

   It is important to monitor prewetting operations to confirm that the water penetrates to the required depth and lateral extent.

   This method can also be combined with a temporary surcharge load. The increased normal stress beneath such loads will intensify the collapse process and produce greater settlements.

6. **Compaction with rollers or vehicles:** Collapsible soils can be converted into excellent bearing material with little or no collapse potential by simply compacting them (Basma and Tuncer, 1992). Sometimes, this compaction may consist simply of passing heavy vibratory sheepsfoot rollers across the ground surface, preferably after first prewetting the soil.

   More frequently, this procedure includes excavating and stockpiling the soil, and then placing it back in layers using conventional earthmoving techniques. If the collapsible stratum is thin, say, less than 10 ft (3 m), this method can be used to completely eradicate the problem. It is often the preferred method when minimum risk is necessary and the collapsible soil deposit is shallow.

If the collapsible stratum is thick, then we may choose to estimate the depth of the wetting front and extend the removal and recompaction to that elevation. This method also reduces the likelihood that the lower soils will become wet because the recompacted soil has a reduced hydraulic conductivity (permeability). Also, if the lower soils should collapse, the compacted fill will spread any settlements over a larger area, thus reducing differential settlements.

Some collapsible soils have sufficient clay content to become slightly expansive when compacted to a higher unit weight.

7. **Compaction with displacement piles:** Large displacement piles, such as closed-end steel pipe piles, can be driven into the ground, compacting the soil around the pile. It may then be extracted and the hole backfilled with sandy gravel or some other soil. Repeating this process on a grid pattern across the site will reduce the collapse potential both by soil compaction and by the column action of the backfill.

8. **Compaction by heavy tamping:** This technique, which is discussed in more detail in Chapter 19, consists of dropping heavy weights (several tons) from large heights (several meters) to compact the soil. This process is continued on a grid pattern across the site, leaving craters that are later backfilled. This technique can be very effective, especially when combined with prewetting.

9. **Vibroflotation:** This technique, also discussed in more detail in Chapter 19, consists of penetrating the soil with a vibrating probe equipped with water jets (known as a vibroflot). The water softens the soil and the vibrations help the collapse process. The vertical hole formed by the vibroflot is also filled with gravel, thus reinforcing the soil and adding bearing capacity. This process is repeated on a grid pattern across the site.

10. **Deep blasting combined with prewetting:** Engineers in the former Soviet Union have experimented with stabilizing collapsible loess by detonating buried explosives. In some cases, the ground is first thoroughly prewetted and then the collapse is induced by vibrations from the detonations. In other cases, the explosives are detonated while the soil is still dry, and the voids created are filled first with water and then with sand and gravel.

11. **Controlled wetting:** This method is similar to method 5 in that it involves injecting water into the soil through trenches or wells. However, it differs in that the wetting is much more controlled and often concentrated in certain areas. This would be used most often as a remedial measure to correct differential settlements that have accidently occurred as a result of localized wetting. When used with careful monitoring, this method can be an inexpensive yet effective way of stabilizing soils below existing buildings that have already settled.

12. **Design structure to be tolerant of differential settlements:** As discussed in Chapter 2, some types of structures are much more tolerant of differential settlements than are others. Therefore, if the potential for collapse-induced settlement is not too large, we may be able to use a more tolerant structure. For example, a steel storage tank would be more tolerant than a concrete one.

The selection of the best method or methods to use at a given site depends on many factors, including the following:

- How deep do the collapsible soils extend?
- How deep would the wetting front extend if the soil was accidentally wetted?
- How much settlement is likely to occur if the soil is accidentally wetted?
- What portion of the total stress is due to overburden and what portion is due to applied loads?
- Is the building or other structure already in place?
- Has any artificial wetting already occurred?

SUMMARY OF MAJOR POINTS

1. Collapsible soils have a loose honeycomb structure and are dry in their natural state. If they are later wetted, this structure will collapse and settlements will occur.

2. Collapsible soils are usually encountered only in arid or semi-arid climates and are continually dry in their natural state. The water that causes them to collapse is normally introduced artificially (as compared to natural infiltration of rainfall). However, collapsible loess is also found in more humid climates.

3. Collapsible soils can be formed by various geologic processes. They may be alluvial soils (especially debris flow deposits), aeolian soils (especially loess), or residual soils.

4. Collapsible soils usually can be effectively sampled, as long as they are moderately to well cemented and do not contain too much gravel. Lightly cemented soils can be sampled with more difficulty, whereas gravelly soils are very difficult or impossible to sample.

5. Laboratory collapse tests may be conducted to measure the collapse potential as a function of overburden stress. Either the double oedometer test or the single oedometer test may be used.

6. In the field, collapsible soils are normally not wetted to 100% saturation and therefore do not strain as much as those tested in the laboratory. Typically, the strain in the field is 50 to 80% of that observed in a thoroughly wetted laboratory test specimen.

7. The settlement also depends on the depth of wetting, but this is difficult to evaluate in advance.

8. We can estimate the amount of settlement by projecting the laboratory collapse tests back to the field conditions while making appropriate corrections for overburden stress and degree of saturation.

9. Many remedial and preventive measures are available to prevent or repair structural damage caused by collapsible soils.

QUESTIONS AND PRACTICE PROBLEMS

21.1 What is a "honeycomb" structure?

21.2 Why are collapsing soils rarely found in areas with very wet climates?

21.3 What are the advantages of evaluating collapse potential using direct methods instead of indirect methods?

21.4 Why is it difficult to obtain samples of gravelly soils and use them in laboratory collapse tests?

21.5 The analysis method presented in this chapter includes the simplifying assumption that ε_w increases with the log of σ. Although this appears to be a reasonable assumption within the usual range of stresses beneath a spread footing, would you expect this relationship to hold true at very high stresses? Explain.

21.6 Why is σ_t usually larger in soils that have a greater clay content?

21.7 A proposed industrial building will include an 800 mm wide, 400 mm deep continuous footing that will carry a load of 90 kN/m. The soil beneath this proposed footing is collapsible to a depth of 2.5 m, and noncollapsible below that depth. The upper 2.5 m has a natural unit weight of 16.0 kN/m³, $\sigma_t = 7.2$ kPa and $C_w = 0.085$. At 50% saturation, this soil has a unit weight of 18.0 kN/m³. Compute the potential settlement due to wetting of this collapsible soil. Assume the depth of wetting is greater than 2.5 m.

21.8 What preventive measures would you use to avoid settlement problems in the footing described in Problem 21.7?

21.9 Assume you have several undisturbed samples of a collapsible soil and need to develop a plot of α vs. S similar to those in Figure 21.9. How would you obtain these data?

21.10 Is it possible for some soils to be expansive when the normal stress is low, and collapsible when it is high? Explain.

Part E

Earth Retaining Structure Analysis and Design

22

Earth Retaining Structures

It would be well if engineering were less generally thought
of ... as the art of constructing. In a certain important sense it
is the art of not constructing ... of doing that well with one
dollar which any bungler can do with two after a fashion.
Arthur M. Wellington (1887)

Foundation engineers often participate in the design of *earth retaining structures*. These are vertical or near-vertical walls that retain soil or rock, as shown in Figure 22.1. Many kinds of retaining structures are available, each best suited for particular applications.

O'Rourke and Jones (1990) classified earth retaining structures into two broad categories: *externally stabilized systems* and *internally stabilized systems*, as shown in Figure 22.2. Some hybrid methods combine features from both systems.

22.1 EXTERNALLY STABILIZED SYSTEMS

Externally stabilized systems are those that resist the applied earth loads by virtue of their weight and stiffness. This was the only type of retaining structure available before 1960, and they are still very common. O'Rourke and Jones subdivided these structures into two categories: *gravity walls* and *in-situ walls*.

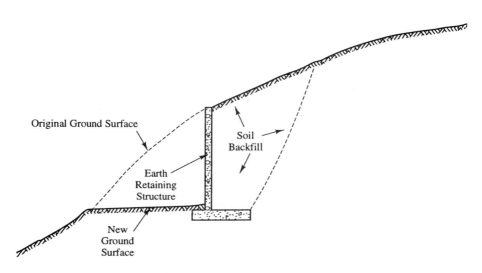

Figure 22.1 An earth retaining structure.

Gravity Walls

Massive Gravity Walls

The earliest retaining structures were *massive gravity walls*, as shown in Figure 22.3. They were often made of mortared stones, masonry, or unreinforced concrete and resisted the lateral forces from the backfill by virtue of their large mass. The construction of these walls is very labor-intensive and requires large quantities of materials, so they are rarely used today except for very short walls.

Cantilever Gravity Walls

The *cantilever gravity wall*, shown in Figure 22.4, is a refinement of the massive wall. These walls have a much smaller cross-section and thus require much less material. However, these walls have large flexural stresses, so they are typically made of reinforced concrete or reinforced masonry. Chapter 24 discusses these walls in more detail.

Crib Walls

A *crib wall*, shown in Figure 22.5, is another type of gravity retaining structure. It consists of precast concrete members linked together to form a crib. These members resemble a child's Lincoln Log toy. The zone between the members is filled with compacted soil.

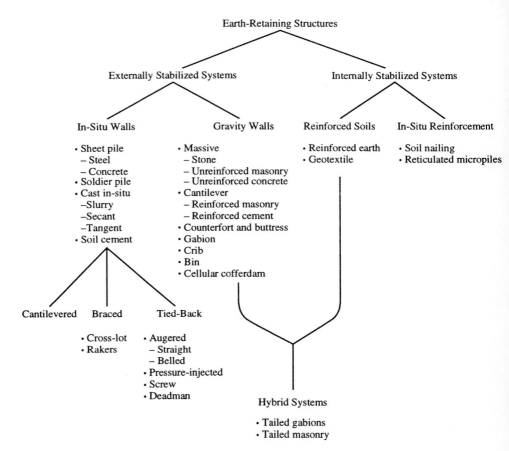

Figure 22.2 Classification of earth retaining structures (Adapted from O'Rourke and Jones, 1990; Used with permission of ASCE).

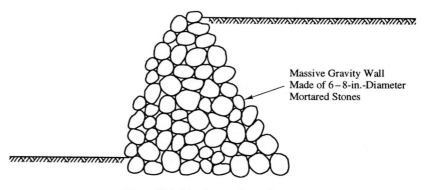

Figure 22.3 Massive gravity wall.

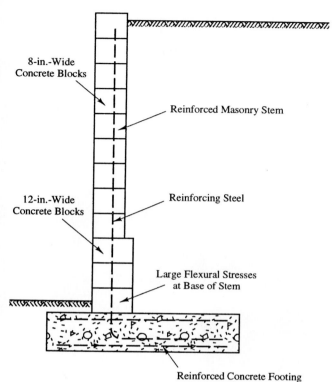

8-in.-Wide
Concrete Blocks

Reinforced Masonry Stem

Reinforcing Steel

12-in.-Wide
Concrete Blocks

Large Flexural Stresses
at Base of Stem

Reinforced Concrete Footing

Figure 22.4 A cantilever gravity wall.

Figure 22.5 A crib wall.

In-Situ Walls

In-situ walls differ from gravity walls in that they rely primarily on their flexural strength, not their mass.

Sheet Pile Walls

A *sheet pile* is a thin, wide pile driven into the ground using a pile hammer. A series of sheet piles in a row form a *sheet pile wall*, as shown in Figure 22.6. Most sheet piles are made of steel, but some are made of reinforced concrete.

Figure 22.6 A sheet pile wall.

It may be possible to simply cantilever a short sheet pile out of the ground, as shown in Figure 22.7. However, it is usually necessary to provide lateral support at one or more levels above the ground. This may be accomplished in either of two ways: by *internal braces* or by *tieback anchors*.

Internal braces are horizontal or diagonal compression members that support the wall, as shown in Figure 22.7. Tieback anchors are tension members drilled into the ground behind the wall. The most common type is a grouted anchor with a steel tendon.

Soldier Pile Walls

Soldier pile walls consist of vertical wide flange steel members with horizontal timber lagging. They are often used as temporary retaining structures for construction excavations.

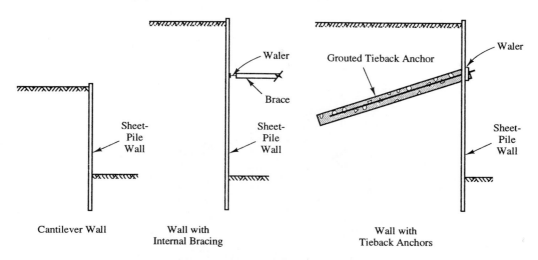

Figure 22.7 Short sheet pile walls can often cantilever; taller walls usually require internal bracing or tieback anchors.

Slurry Walls

Slurry walls are cast-in-place concrete walls built using bentonite slurry. The contractor digs a trench along the proposed wall alignment and keeps it open using the slurry. Then, the reinforcing steel is inserted and the concrete is placed using tremie pipes or pumps. As the concrete fills the trench, the slurry exits at the ground surface.

Slurry walls have been used as basement walls in large urban construction, and often eliminate the need for temporary walls.

22.2 INTERNALLY STABILIZED SYSTEMS

Internally stabilized systems reinforce the soil to provide the necessary stability. Various schemes are available, all of which have been developed since 1960. They can be subdivided into two categories: *reinforced soils* and *in-situ reinforcement.*

Reinforced Soils

The *reinforced soils* techniques embed tensile members into the backfill soils as the fill is being placed.

Reinforced Earth Walls

The *reinforced earth* method uses strips of metal or polymer embedded into the fill, as shown in Figure 22.8. These strips are connected to precast concrete face panels. However, the primary support is from the strips, not the panels.

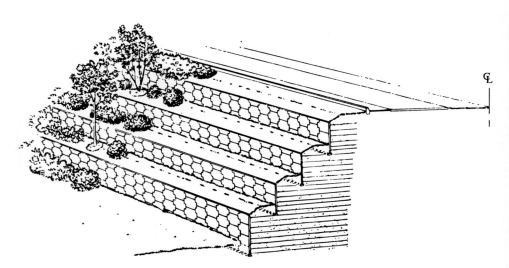

Figure 22.8 A series of reinforced earth walls (Walkinshaw, 1975).

In-Situ Reinforcement

In-situ reinforcement methods differ from reinforced soils in that the tensile members are inserted into a soil mass rather than being embedded during placement of fill.

Soil Nailing

Soil nailing consists of drilling near-horizontal holes into the ground, inserting steel tendons, and grouting. The face of the wall is typically covered with shotcrete, as shown in Figure 22.9.

These walls do not require a construction excavation, and thus are useful when space is limited.

Reticulated Micropiles

Reticulated micropile structures consist of large numbers of small diameter grouted piles tied together, as shown in Figure 22.10. They work together as a system to form a retaining structure.

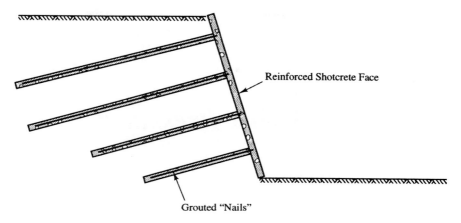

Reinforced Shotcrete Face

Grouted "Nails"

Figure 22.9 A soil nail wall.

Figure 22.10 Model of a reticulated micropile structure (Photo courtesy of Nicholson
Construction Co.).

23

Lateral Earth Pressures

Nature has no contract to agree with mathematics.
Pioneer foundation engineer Lazarus White (1936)
emphasizing the uncertainties in mathematical formulations
of natural processes

One of the first steps in designing any structural member is to determine the magnitude and direction of the applied loads. For retaining walls, the primary applied load is the resultant of the *lateral earth pressures* the soil imparts on the back of the wall. The primary function of the wall is to resist this load. Chapters 12 - 17 briefly discussed lateral earth pressures in the context of computing the skin friction resistance of deep foundations. This chapter examines them in much more detail.

The terms *pressure* and *stress* are nearly synonymous. Let us define *pressure* as the contact force per unit area between the wall and the soil or between the footing and the soil, and *stress* as the force per unit area within the soil or concrete.

23.1 HORIZONTAL STRESSES IN SOIL

The ratio of the horizontal effective stress to the vertical effective stress at any point in a soil was defined in Chapter 3 as the *coefficient of lateral earth pressure, K*:

666

$$K = \frac{\sigma_h'}{\sigma_v'} \qquad (23.1)$$

Where:

K = coefficient of lateral earth pressure

σ_h' = horizontal effective stress

σ_v' = vertical effective stress

In the context of this chapter, K is important because it is an indicator of the lateral earth pressures acting on a retaining wall.

For purposes of describing lateral earth pressures, engineers have defined three important soil conditions: the *at-rest condition*, the *active condition*, and the *passive condition*.

The At-Rest Condition

Let us assume a certain retaining wall is both *rigid* and *unyielding*. In this context, a rigid wall is one that does not experience any significant flexural movements. The opposite would be a *flexible* wall—one that has no resistance to flexure. The term *unyielding* means the wall does not translate or rotate, as compared to a *yielding* wall that can do either or both. Let us also assume this wall is built so that no lateral strains occur in the ground. Therefore, the lateral stresses in the ground are the same as they were in its natural undisturbed state.

The value of K in this situation is K_0, the *coefficient of lateral earth pressure at rest*. Chapter 3 discussed methods of measuring or computing K_0 when the ground surface is level. If the ground surface behind the wall is inclined upwards at an angle β from the horizontal, then we can rewrite Equation 3.13 from Chapter 3 as:

$$K_0 = (1 - \sin\phi)\,\text{OCR}^{\sin\phi}\,(1 + 0.5\tan\beta)^2 \qquad (23.2)$$

Where:

ϕ = friction angle of soil

OCR = overconsolidation ratio of soil

β = inclination of ground surface from horizontal

If no groundwater table is present, the horizontal pressure, σ_h', acting on this wall is, in theory, equal to the horizontal stress in the soil:

$$\sigma_h' = \sigma_v' K_0 \qquad\qquad (23.3)$$

In a homogeneous soil, K_0 is a constant and σ_v' varies linearly with depth. Therefore, in theory, σ_h' also varies linearly with depth, forming a triangular pressure distribution, as shown in Figure 23.1. Thus, if at-rest conditions are present, the horizontal force acting on a unit length of a vertical wall is:

$$P_0/b = \frac{\gamma H^2 K_0}{2} \qquad\qquad (23.4)$$

Where:

P_0/b = normal force acting between soil and wall per unit length of wall

b = unit length of wall (usually 1 ft or 1 m)

γ = unit weight of soil

H = height of wall

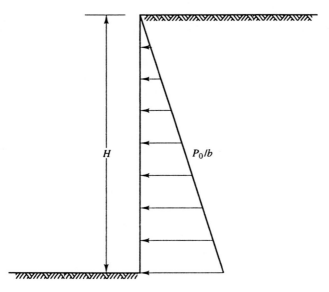

Figure 23.1 At-rest pressure acting on a retaining wall.

Example 23.1

An 8 ft tall basement wall retains a soil that has the following properties: $\phi = 35°$, $\gamma = 127$ lb/ft³, OCR = 2. The ground surface is horizontal and level with the top of the wall. The groundwater table is well below the bottom of the wall. Consider the soil to be in the at-rest condition and compute the force that acts between the wall and the soil.

Solution:

Using Equation 23.2

$$K_0 = (1 - \sin\phi)\ \text{OCR}^{\,\sin\phi}\ (1 + 0.5\tan\beta)^2$$
$$= (1 - \sin 35)\ 2^{\,\sin 35}\ (1 + 0.5\tan 0)^2$$
$$= 0.635$$

Using Equation 23.4

$$P_0/b = \frac{\gamma H^2 K_0}{2}$$

$$= \frac{(127)(8)^2(0.635)}{2}$$

$$= \textbf{2580 lb/ft} \qquad \Leftarrow\textit{Answer}$$

Because the theoretical pressure distribution is triangular, this resultant force acts at the lower third-point on the wall. However, as discussed in Chapter 24, the actual location of the resultant force is likely be higher than this.

The Active Condition

The at-rest condition is present only if the wall does not move. Although this may seem to be a criterion that all walls should meet, even very small movements alter the lateral earth pressure.

Suppose Mohr's circle **A** in Figure 23.2 represents the state of stress at a point in the soil behind the wall in Figure 23.3, and suppose this soil is in the at-rest condition. The inclined lines represent $\tau = s = c + \sigma'\tan\phi$ and are known as the *failure envelope*. Because the Mohr's circle does not touch the failure envelope, the shear stress, τ, is less than the shear strength, s.

Now, permit the wall to move outward a short distance. This relieves part of the horizontal stress, causing the Mohr's circle to expand to the left. Continue this process until the circle reaches the failure envelope and the soil fails in shear (circle **B**). This shear failure will occur along the planes shown in Figure 23.3, which are inclined at an angle of $45 + \phi/2$ degrees from the horizontal. A soil that has completed this process is said to be in the *active condition*. The value of K in a cohesionless soil in the active condition is known as K_a, the *coefficient of active earth pressure*.

Once the soil attains the active condition, the horizontal stress in the soil (and thus the pressure acting on the wall) will have reached its lower bound, as shown in Figure 23.4.

How far must the wall move to obtain the active condition? Figure 23.4 is a plot of K vs. wall movement for a typical retaining wall backfilled with dense sand. In this case, the active condition is achieved if the wall moves outward from the backfill a distance equal to only $0.003\,H$ (about 0.4 in for a 10 ft tall wall). Although basement walls, being braced at the top, cannot move even that distance, a cantilever wall not

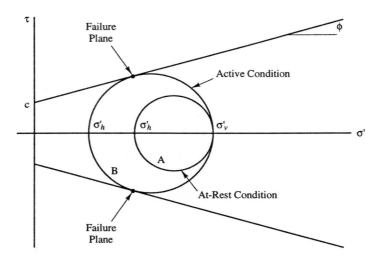

Figure 23.2 Changes in the stress condition in a soil as it transitions from the at-rest to the active condition.

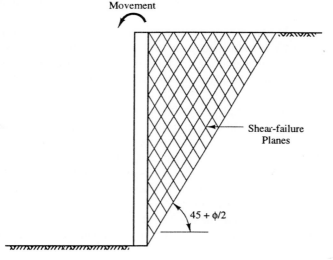

Figure 23.3 Development of shear failure planes in the soil behind a wall as it transitions from the at-rest to the active condition.

connected to a building or other structure could very easily move outward 0.4 in and such a movement would usually be acceptable. Thus, a basement wall may need to be designed to resist the at-rest pressure, whereas the design of a free-standing cantilever wall could use the active pressure, which is smaller.

This movement can be either translational or rotational about the bottom of the wall. The required amount of movement varies with the soil type, as shown in Table 23.1.

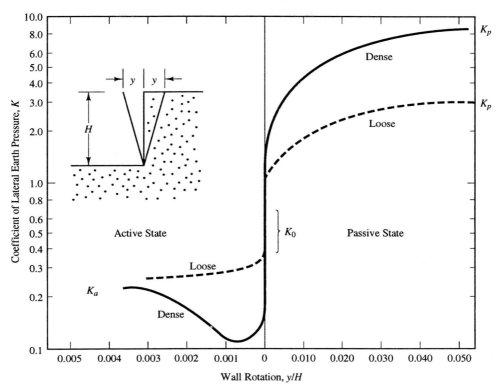

Figure 23.4 Effect of wall movement on lateral earth pressure in sand (Adapted from CGS, 1985).

TABLE 23.1 WALL MOVEMENT REQUIRED TO REACH THE ACTIVE CONDITION

Soil Type	Horizontal Movement Required to Reach the Active Condition
Dense cohesionless	$0.001 H$
Loose cohesionless	$0.004 H$
Stiff cohesive	$0.010 H$
Soft cohesive	$0.020 H$

Adapted from CGS (1985).
H = Wall height

The Passive Condition

The *passive condition* is the opposite of the active condition. In this case, the wall moves *into* the backfill, as shown in Figure 23.5, and the Mohr's circle changes, as shown in Figure 23.6. Notice how the vertical stress remains constant whereas the horizontal stress changes in response to the induced horizontal strains.

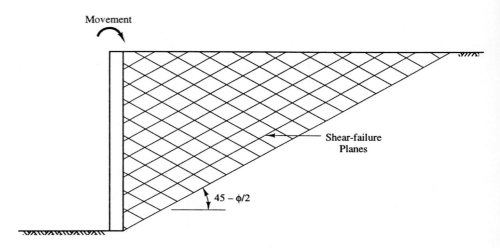

Figure 23.5 Development of shear failure planes in the soil behind a wall as it transitions from the at-rest to the passive condition.

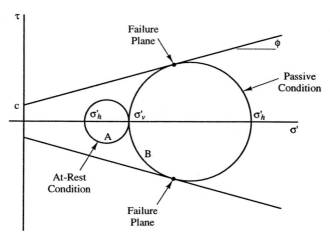

Figure 23.6 Changes in the stress condition in a soil as it transitions from the at-rest to passive condition.

In a homogeneous soil, the shear failure planes in the passive case are inclined at an angle of 45 - $\phi/2$ degrees from the horizontal. The value of K in a cohesionless soil in the passive condition is known as K_p, the *passive coefficient of lateral earth pressure*. This is the upper bound of K and produces the upper bound of pressure that can act on the wall.

Engineers often use the passive pressure that develops along the toe of a retaining wall footing to help resist sliding, as shown in Figure 23.7. In this case, the "wall" is the side of the footing.

More movement must occur to attain the passive condition than for the active condition. Typical required movements for various soils are shown in Table 23.2.

TABLE 23.2 WALL MOVEMENT REQUIRED TO REACH THE PASSIVE CONDITION

Soil Type	Horizontal Movement Required to Reach the Passive Condition
Dense cohesionless	0.020 H
Loose cohesionless	0.060 H
Stiff cohesive	0.020 H
Soft cohesive	0.040 H

Adapted from CGS (1985).
H = Wall height

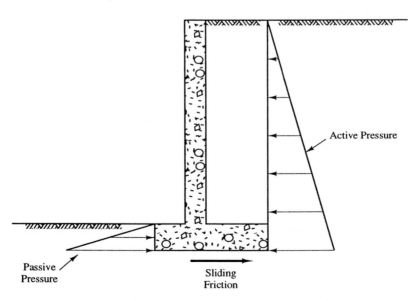

Figure 23.7 Active and passive pressures acting on a cantilever retaining wall.

Although movements on the order of those listed in Tables 23.1 and 23.2 are necessary to reach the full active and passive states, respectively, much smaller movements also will cause significant changes in the lateral earth pressure. While conducting a series of full-scale tests on retaining walls, Terzaghi (1934b) observed:

> With compacted sand backfill, a movement of the wall over an insignificant distance (equal to one-tenthousanth of the depth of the backfill) decreases the [coefficient of lateral earth pressure] to 0.20 or increases it up to 1.00.

Although this effect is not as dramatic in other soils, only the most rigid and unyielding structures will truly be subjected to the at-rest pressure.

23.2 CLASSICAL LATERAL EARTH PRESSURE THEORIES

The solution of lateral earth pressure problems was among the first applications of the scientific method to the design of structures (see the sidebar "Coulomb and the Scientific Method"). Two of the pioneers in this effort were the Frenchman Charles Augustin Coulomb and the Scotsman W. J. M. Rankine. Although many others have since made significant contributions to our knowledge of earth pressures, the contributions of these two men were so fundamental that they still form the basis for earth pressure calculations today. More than 50 earth pressure theories are now available; all of them have their roots in Coulomb and Rankine's work.

Coulomb presented his theory in 1773 and published it 3 years later (Coulomb, 1776). Rankine developed his theory more than 80 years after Coulomb (Rankine, 1857). In spite of this chronology, it is conceptually easier for us to discuss Rankine's theory first.

For clarity, we will begin our discussion of these theories by considering only cohesionless soils ($c = 0$). Once we have established the basic concepts, we will then expand the discussion to include soils with cohesion ($c \geq 0$).

Rankine's Theory for Cohesionless Soils

Assumptions

Rankine approached the lateral earth pressure problem with the following assumptions:

1. The soil is isotropic and homogeneous.
2. The most critical shear surface is a plane. In reality, it is slightly concave up, but this is a reasonable assumption (especially for the active case) and it simplifies the analysis.
3. The ground surface is a plane (although it does not necessarily need to be level).
4. The wall is infinitely long so that the problem may be analyzed in only two

dimensions. Geotechnical engineers refer to this as a *plane strain* condition.

5. The wall moves sufficiently to develop the active or passive condition.

6. The resultant of the normal and shear forces that act on the back of the wall is inclined at an angle parallel to the ground surface (Coulomb's theory provides a more accurate model of shear forces acting on the wall).

Active Condition

With these assumptions, we can treat the wedge of soil behind the wall as a free body and evaluate the problem using the principles of statics, as shown in Figure 23.8a. This is known as a *limit equilibrium analysis*, which means that we consider the conditions that would exist if the soil along the base of the failure wedge was about to fail in shear. This is very similar to the kinds of analyses geotechnical engineers perform to evaluate the stability of earth slopes.

Weak seams or other nonuniformities in the soil may control the inclination of the critical shear surface. However, if the soil is homogeneous, P_a/b is greatest when this surface is inclined at an angle of $45 + \phi/2$ degrees from the horizontal, as shown in the Mohr's circle in Figure 23.2. Thus, this is the most critical angle.

Solving this free body diagram for P_a/b and V_a/b gives:

$$P_a/b = \frac{\gamma H^2 K_a \cos\beta}{2} \tag{23.5}$$

$$V_a/b = \frac{\gamma H^2 K_a \sin\beta}{2} \tag{23.6}$$

$$K_a = \frac{\cos\beta - \sqrt{\cos^2\beta - \cos^2\phi}}{\cos\beta + \sqrt{\cos^2\beta - \cos^2\phi}} \qquad \beta \le \phi \tag{23.7}$$

If $\beta = 0$, Equation 23.7 reduces to:

$$K_a = \tan^2(45 - \phi/2) \tag{23.8}$$

A solution of P_a/b as a function of H would show that the theoretical pressure distribution is triangular. Therefore, the theoretical pressure and shear stress acting against the wall, $\sigma_h{}'$ and τ, respectively, are:

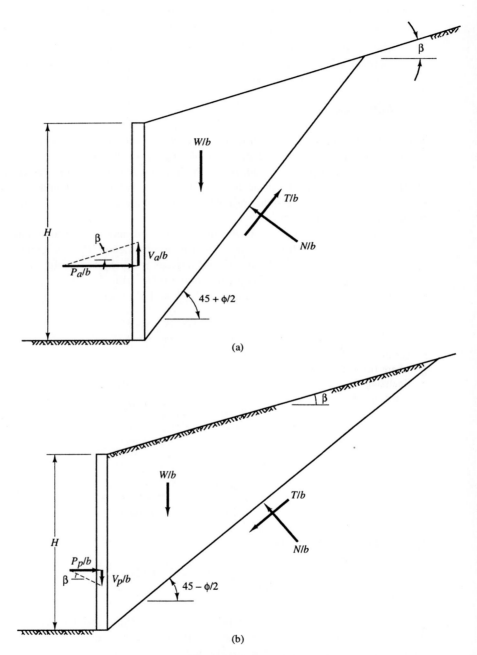

Figure 23.8 Free body diagram of soil behind a retaining wall using Rankine's solution: (a) active case; and (b) passive case.

Coulomb and the Scientific Method

Prior to the eighteenth century, there were virtually no rational methods available for designing structures. Builders based their designs on precedent and common sense that developed through years of trial-and-error. When something failed, they concluded that their design had overstepped some permissible limit and future designs carefully avoided this limit, thus setting a new precedent.

This method of developing design rules is particularly evident in the construction of the great cathedrals of Europe during the twelfth through sixteenth centuries. The builders of these magnificent structures developed and perfected design rules for many structural and architectural forms, such as arches and buttresses. Unfortunately, they suffered many dramatic failures in the process of developing these rules.

Such methods are, of course, very inefficient and wasteful, not only because of the high price of learning new precedents, but also because the precedents usually were not based on any real understanding of the underlying mechanisms. These difficulties prompted the beginning of the use of the scientific approach to building during the eighteenth century and Charles Augustin Coulomb was one of the pioneers in this effort. Although he is most commonly known for his work on electricity and magnetism, the earlier part of his career was concerned more with structures.

Coulomb graduated from the Mézières School of Military Engineers in France at the age of 26. Two years later, the young officer was sent to the Caribbean island of Martinique where he was placed in charge of building a fort to protect the harbor. In the process of finalizing the design of the fort, he became dissatisfied with the rules of thumb for sizing retaining walls because they dictated walls that were too large. He later wrote:

> I have often come across situations in which all the theories based on hypotheses or on small-scale experiments in a physics laboratory have proved inadequate in practice.

Therefore, he began a series of full-scale experiments to study the behavior of retaining walls. His theory of lateral earth pressure is based on the results of these experiments (Kerisel, 1987).

Coulomb was not the first to apply scientific methods to the design of structures, but his work was very influential, and helped lay the framework for modern engineering practice. Although precedent and common sense still play an important roles, the introduction of scientific methods has allowed us to build larger and more efficient structures than would otherwise have been possible.

$$\sigma_h' = \sigma_v' K_a \cos\beta \tag{23.9}$$

$$\tau = \sigma_v' K_a \sin\beta \tag{23.10}$$

Where:

P_a/b = normal force between soil and wall per unit length of wall

V_a/b = shear force between soil and wall per unit length of wall

   b = unit length of wall (usually 1 ft or 1 m)

  K_a = active coefficient of lateral earth pressure

  σ_v' = vertical effective stress

  β = inclination of ground surface above the wall

However, observations and measurements from real retaining structures indicate that the true pressure distribution, as shown in Figure 23.9, is not triangular. This difference is because of wall deflections, arching, and other factors. Chapter 24 provides recommendations for developing design earth pressures for gravity walls; Schnabel (1982) discusses earth pressures for other types of retaining structures.

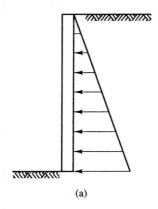

(a)

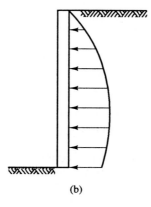

(b)

Figure 23.9   Comparison between (a) theoretical and (b) observed distributions of earth pressures acting behind retaining structures.

Example 23.2

A 6 m tall cantilever wall retains a soil that has the following properties: $c = 0$, $\phi = 30°$, and $\gamma = 19.2$ kN/m^3. The ground surface behind the wall is inclined at a slope of 3 horizontal to 1 vertical, and the wall has moved sufficiently to develop the active condition. Determine the normal and shear forces acting on the back of this wall using Rankine's theory.

Solution:

Use Equations 23.5 - 23.7:

$$\beta = \tan^{-1}(1/3)$$
$$= 18°$$

$$K_a = \frac{\cos\beta - \sqrt{\cos^2\beta - \cos^2\phi}}{\cos\beta + \sqrt{\cos^2\beta - \cos^2\phi}}$$

$$= \frac{\cos 18° - \sqrt{\cos^2 18° - \cos^2 30°}}{\cos 18° + \sqrt{\cos^2 18° - \cos^2 30°}}$$

$$= 0.415$$

$$P_a/b = \frac{\gamma H^2 K_a \cos\beta}{2} = \frac{(19.2)(6)^2(0.415)\cos 18°}{2} = \textbf{136 kN/m} \qquad \Leftarrow \textit{Answer}$$

$$V_a/b = \frac{\gamma H^2 K_a \sin\beta}{2} = \frac{(19.2)(6)^2(0.415)\sin 18°}{2} = \textbf{44 kN/m} \qquad \Leftarrow \textit{Answer}$$

These results are shown in the following figure.

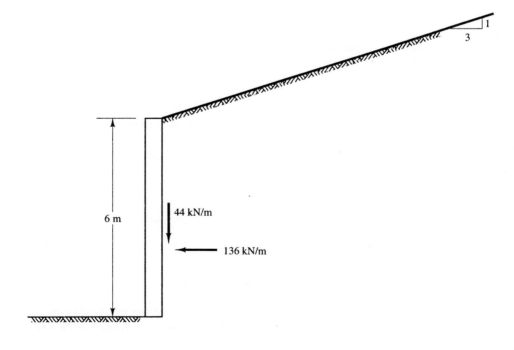

Passive Condition

Rankine analyzed the passive condition in a fashion similar to the active condition except that the shear force acting along the base of the wedge now acts in the opposite direction (it always opposes the movement of the wedge) and the free body diagram becomes as shown in Figure 23.8b. Notice that the failure wedge is much flatter than it was in the active case and the critical angle is now 45 - ϕ/2 degrees from the horizontal.

The normal and shear forces, P_p/b and V_p/b, respectively, acting on the wall in the passive case are:

$$P_p/b = \frac{\gamma H^2 K_p \cos\beta}{2} \tag{23.11}$$

$$V_p/b = \frac{\gamma H^2 K_p \sin\beta}{2} \tag{23.12}$$

$$K_p = \frac{\cos\beta + \sqrt{\cos^2\beta - \cos^2\phi}}{\cos\beta - \sqrt{\cos^2\beta - \cos^2\phi}} \qquad \beta \leq \phi \tag{23.13}$$

If $\beta = 0$, Equation 23.13 reduces to:

$$K_p = \tan^2(45 + \phi/2) \tag{23.14}$$

The theoretical pressure and shear acting against the wall, σ_h' and τ, respectively, are:

$$\sigma_h' = \sigma_v' K_p \cos\beta \tag{23.15}$$

$$\tau = \sigma_v' K_p \sin\beta \tag{23.16}$$

Coulomb's Theory for Cohesionless Soils

Coulomb's theory differs from Rankine's in that the resultant of the normal and shear forces acting on the wall is inclined at an angle ϕ_w from a perpendicular to the wall, where tan ϕ_w is the coefficient of friction between the wall and the soil, as shown in Figure 23.10.

Although Coulomb recognized the passive condition, he did not develop a numerical solution for it. However, others have expanded his work to include this condition, so we will discuss both active and passive pressures using Coulomb's theory.

Coulomb's formulas for lateral earth pressure are as follows and the parameters are as defined in Figure 23.10:

$$P_a/b = \frac{\gamma H^2 K_a \cos\phi_w}{2} \tag{23.17}$$

$$V_a/b = \frac{\gamma H^2 K_a \sin\phi_w}{2} \tag{23.18}$$

$$K_a = \frac{\cos^2(\phi-\alpha)}{\cos^2\alpha \cos(\phi_w+\alpha)\left[1 + \sqrt{\dfrac{\sin(\phi_w+\alpha)\ \sin(\phi-\beta)}{\cos(\phi_w+\alpha)\ \cos(\alpha-\beta)}}\right]^2} \tag{23.19}$$

$$P_p/b = \frac{\gamma H^2 K_p \cos\phi_w}{2} \tag{23.20}$$

$$V_p/b = \frac{\gamma H^2 K_p \sin\phi_w}{2} \tag{23.21}$$

$$K_p = \frac{\cos^2(\phi+\alpha)}{\cos^2\alpha \cos(\phi_w-\alpha)\left[1 - \sqrt{\dfrac{\sin(\phi_w+\alpha)\ \sin(\phi+\beta)}{\cos(\phi_w-\alpha)\ \cos(\alpha-\beta)}}\right]^2} \tag{23.22}$$

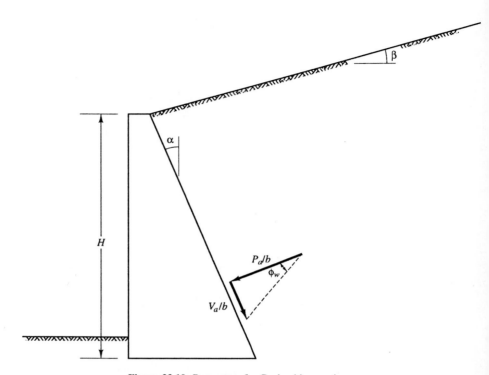

Figure 23.10 Parameters for Coulomb's equations.

These formulas are valid only for $\beta \leq \phi$. When designing concrete or masonry walls it is common practice to use $\phi_w = 0.67\,\phi$. Steel walls have less sliding friction, perhaps on the order of $\phi_w = 0.33\,\phi$.

Coulomb's values of K_p are often much higher than Rankine's. The difference ranges from about 10% at $\phi = 10°$ to 150% at $\phi = 40°$. This discrepancy occurs because the critical failure surface is not a plane, as both theories assume. In reality it is concave upward. Coulomb's theory seems to be especially sensitive to this and therefore produces erroneously high values of K_p. Therefore, for practical problems, it probably is best to use Rankine's theory to compute passive pressures.

23.3 LATERAL EARTH PRESSURES IN SOILS WITH COHESION

Rankine did not address lateral earth pressures in soils with cohesion ($c \geq 0$ and $\phi \geq 0$) and Coulomb did not address passive pressures. However, later investigators developed complete formulas for cohesive soils. Bell (1915) was among those who contributed.

Theoretical Behavior

Soils with cohesion can stand vertically to a height of no more than the critical height, H_c:

$$H_c = \frac{2c}{\gamma\sqrt{K_a}} \qquad (23.23)$$

In other words, if $H < H_c$ the earth will stand vertically without a wall. In practice we would apply some factor of safety to H_c (perhaps 1.5 to 2) before deciding not to build a wall. An engineer also would want to consider the potential for surface erosion and other modes of failure.

If $H > H_c$, the theoretical pressure distribution is as shown in Figure 23.11.

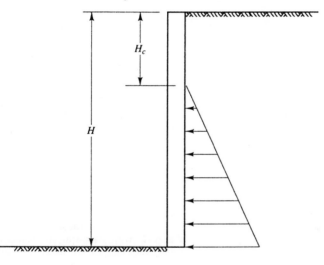

Figure 23.11 Theoretical active pressure distribution in soils with cohesion ($c \geq 0$, $\phi \geq 0$).

Equations 23.5 and 23.6 then become:

$$P_a/b = \left(\frac{\gamma H^2 K_a}{2} - 2cH\sqrt{K_a} + \frac{2c^2}{\gamma} \right)\cos\beta \geq 0 \qquad (23.24)$$

$$V_a/b = \left(\frac{\gamma H^2 K_a}{2} - 2cH\sqrt{K_a} + \frac{2c^2}{\gamma} \right)\sin\beta \geq 0 \qquad (23.25)$$

These formulas often are incorrectly stated without the $2c^2/\gamma$ term. This term must be present to account for the lack of tensile forces between the wall and the soil at depths shallower than H_c.

The Rankine equations for passive conditions in soils with cohesion are as follows:

$$P_p/b = \left(\frac{\gamma H^2 K_p}{2} + 2cH\sqrt{K_p}\right)\cos\beta \qquad (23.26)$$

$$V_p/b = \left(\frac{\gamma H^2 K_p}{2} + 2cH\sqrt{K_p}\right)\sin\beta \qquad (23.27)$$

The theoretical shape of the passive pressure distribution is shown in Figure 23.12.

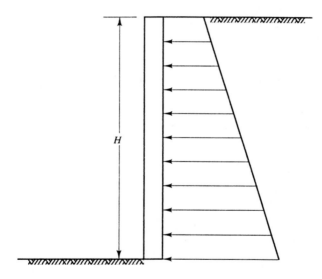

Figure 23.12 Theoretical distribution of passive earth pressure in soils with cohesion ($c \geq 0$, $\phi \geq 0$).

Actual Behavior

Unfortunately, lateral earth pressure computations in cohesive soils based on Rankine or Coulomb's theories are not very reliable, and may produce unconservative designs. This discrepancy occurs because the earth pressure theories do not consider the following aspects of clay behavior:

- Creep in these soils may change the earth pressures.
- The soil may be expansive.
- Clays obstruct drainage, and thus may trap groundwater behind the wall.

When the shear stresses in a clayey soil are a large percentage of the shear strength, a phenomenon known as *creep* can occur. The soil slowly shears and never reaches complete equilibrium unless the shear stresses become sufficiently small. When this occurs behind a retaining wall, the failure wedge slowly moves toward the wall. Therefore, it is impossible to maintain the active condition (which demands high shear stresses) for long periods. Therefore, design cantilever walls backfilled with clay using something higher than the active pressure.

Another potentially more significant problem is that clayey soils may be expansive (see Chapter 20). An expanding backfill places very large loads on the wall, far greater than the computed active or at-rest pressure. The exact magnitude of lateral pressures due to expansive soils is difficult to predict, but the passive pressure would be an upper bound.

The cyclic expansion and contraction cycles in expansive soils further aggravate this situation and could cause the following scenario:

1. The soil expands and the retaining wall moves outward.
2. The soil shrinks and moves slightly downhill, thus remaining in intimate contact with the wall. Cracks form.
3. Loose debris falls into the cracks, preventing them from closing.
4. The soil expands again and moves the wall farther out.

This process could continue indefinitely, moving the wall much farther than would a single cycle of expansion.

A slightly cohesive soil, such as an SC, would probably not pose any serious problems, but highly cohesive soils, such as a CH, could produce large movements or even failure.[1]

The best procedure is to avoid backfilling any wall with an expansive soil, especially in regions with adverse climates. It is often possible to bring in other soils from on-site or off-site to use for backfill material. However, if clayey soils must be used, the design must reflect their presence. Jennings (1973) described the South African practice of designing such walls using at-rest pressures with a K_0 of 0.8 to 1.0 and making them as flexible as possible.

Many of the preventive measures described in Chapter 20 are also appropriate for reducing the potential for expansion behind a retaining wall.

23.4 TERZAGHI AND PECK'S METHOD

Terzaghi and Peck (1967) offered an alternative method of computing design lateral earth pressures behind cantilever retaining walls. This is a semiempirical method based

---

[1] See Chapter 3 for an explanation of the Unified Soil Classification System symbols SC and CH.

partially on the observed performance of real walls. It is especially useful with clayey soils (because of the shortcomings of the theories) or when no soil test data are available.

This method classifies the soil into five types, as described in Table 23.3. The normal and shear forces acting on the wall are then computed using Equations 23.28 and 23.29 with Figures 23.13 and 23.14.

TABLE 23.3 CLASSIFICATION OF SOIL TYPES

Soil Type	Description
1	Coarse grained soil without admixture of fine soil particles, very permeable (i.e., clean sand or gravel).
2	Coarse-grained soil of low permeability due to admixture of particles of silt size.
3	Residual soil with stones, fine silty sand, and granular materials with conspicuous clay content.
4	Very soft clay, organic silts, or silty clays.
5	Medium or stiff clay, deposited in chunks and protected in such a way that a negligible amount of water enters the spaces between the chunks during floods or heavy rains. If this condition cannot be satisfied, the clay should not be used as backfill material. With increasing stiffness of the clay, danger to the wall due to infiltration of water increases rapidly.

Adapted from Terzaghi and Peck (1967).

$$P_a/b = \frac{G_h H^2}{2} \tag{23.28}$$

$$V_a/b = \frac{G_v H^2}{2} \tag{23.29}$$

Where:

P_a/b = normal force between soil and wall per unit length of wall

V_a/b = shear force between soil and wall per unit length of wall

b = unit length of wall (usually 1 ft or 1 m)

G_h, G_v = coefficients from Figure 23.13 or 23.14

H = height of wall measured from bottom of footing to ground surface, as shown in Figures 23.13 and 23.14.

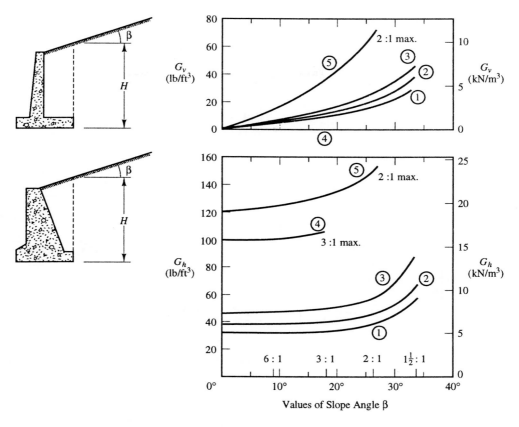

Figure 23.13 Charts for estimating the loads acting against a retaining wall beneath a planar ground surface (Adapted from Terzaghi and Peck, 1967).

This method probably is appropriate only for walls less than about 20 ft (6 m) in height.

23.5 EQUIVALENT FLUID METHOD

The coefficient G_h in Terzaghi and Peck's method is sometimes called the *equivalent fluid density*. Thus, we can imagine the wall is backfilled with a fluid that has a unit weight of G_h and compute the "earth" pressure using the principles of fluid statics.

For sandy soils, it is also possible to obtain G_h from lateral earth pressure theories, as illustrated in Example 23.3.

Example 23.3

A cantilever wall will retain a sandy soil with $c = 0$, $\phi = 35°$, and $\gamma = 124$ lb/ft³. The ground surface above the wall will be level ($\beta = 0$) and there will be no surcharge loads. Compute the active pressure and express it as the equivalent fluid density.

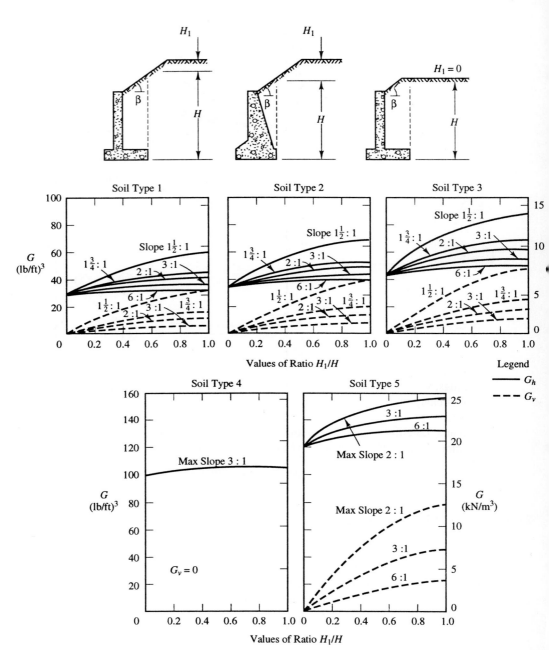

Figure 23.14 Charts for estimating the loads acting against a retaining wall below a ground surface that is sloped and then becomes level (Adapted from Terzaghi and Peck, 1967).

Solution:

$$P_a/b = \frac{\gamma H^2 K_a \cos\beta}{2} = \frac{G_h H^2}{2}$$

$\therefore\ G_h = \gamma K_a$ (This is true only for a cohesionless soil with $\beta = 0$)

$$K_a = \tan^2(45 - \phi/2)$$

$$= \tan^2(45 - 35°/2)$$

$$= 0.271$$

$$G_h = \gamma K_a$$

$$= (124)(0.271)$$

$$= \mathbf{34\ lb/ft^3} \qquad\qquad \Leftarrow Answer$$

Therefore, we would recommend that the structural engineer design the wall to retain a fluid that has a unit weight of 34 lb/ft^3.

23.6 SURCHARGE LOADS

Surcharge loads often are present along the ground surface above a retaining structure, as shown in Figure 23.15. These loads might be the result of structural loads on shallow footings, vehicles, above-ground storage, backfill compaction equipment, or other causes. If a surcharge load occurs within a horizontal distance of about H from the wall (where H equals the height of the wall), then it will impose additional lateral pressures on the wall and therefore is of interest. We can compute the magnitude and distribution of these additional pressures using elastic theory and superimpose them onto the conventional lateral earth pressure to obtain a pressure distribution suitable for design. U.S. Navy (1982b) includes several chart solutions.

An often neglected surcharge load is that imposed by the equipment that places and compacts the backfill. Although these loads are trivial if the contractor uses only lightweight hand-operated compaction equipment, they can be significant if larger equipment, such as loaders, dozers, or sheepsfoot rollers, operate behind the wall. Unfortunately, design engineers sometimes fail to consider construction procedures, so it is not unusual for walls to tilt excessively during backfilling because they were not designed to support heavy equipment loads.

One way to predict the effect of these loads on the wall is to combine a knowledge of the weight and position of the compaction equipment with the surcharge charts in this chapter. Another way is the method proposed by Ingold (1979).

Compacted backfills also induce large horizontal stresses in the ground (i.e., the K_0 value is large). This also increases the lateral earth pressure acting on the wall (Duncan et al., 1991).

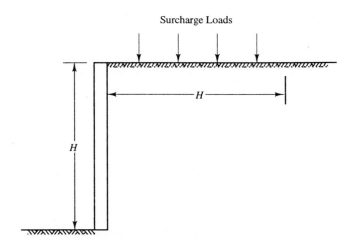

Figure 23.15 Surcharge loads on the ground surface above a retaining wall.

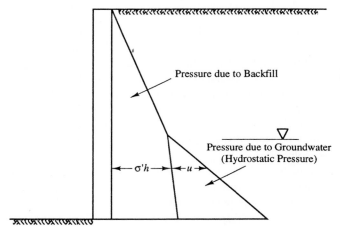

Figure 23.16 Theoretical lateral pressure distribution with shallow groundwater table.

23.7 GROUNDWATER EFFECTS

The discussions in this chapter have thus far assumed that the groundwater table is located below the base of the wall. If the groundwater table rises to a level above the base of the wall, three important changes occur:

- The effective stress in the soil below the groundwater table will decrease, which decreases the active, passive, and at-rest pressures.

- Horizontal hydrostatic pressures will develop against the wall and must be superimposed onto the lateral earth pressures.
- The effective stress between the bottom of the footing and the soil becomes smaller, so there is less sliding friction.

The net effect of the first two changes is a large increase in the total horizontal pressure acting on the wall (i.e., the increased hydrostatic pressures more than offset the decreased effective stress). The resulting pressure diagram is shown in Figure 23.16.

Example 23.4

This cantilever wall has moved sufficiently to create the active condition. Compute the lateral pressure distribution acting on this wall with the groundwater table at locations **a** and **b**, as shown in the figure on the next page.

The soil properties are: $c = 0$, $\phi = 30°$, $\gamma = 20.4$ kN/m^3, and $\gamma_{sat} = 22.0$ kN/m^3

Solution:

Use Rankine's Method (Equations 23.8 and 23.21):

$$K_a = \tan^2 (45 - \phi/2)$$

$$= \tan^2 (45 - 30/2)$$

$$= 0.333$$

With the groundwater table at **a** (Equation 23.9):

$$\sigma_h' = \sigma_v' K_a \cos \beta$$

$$= \gamma z K_a \cos \beta$$

$$= 20.4 \, z \, (0.333) \cos 0$$

$$= 6.79 z$$

Where:

z = depth below the top of the wall

$$\beta = 0 \quad \therefore \ V_a = 0 \ \text{(per Rankine)}$$

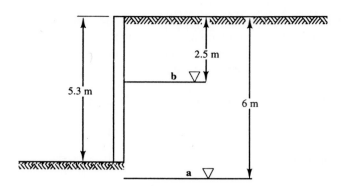

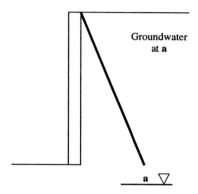

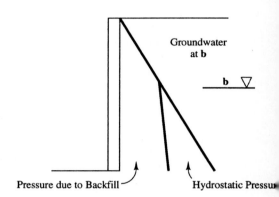

With the groundwater table at **b**:

$$\sigma_h' @ z \geq 2.5\,\text{m} = \sigma_v' K_a \cos \beta$$

$$= \left(\sum \gamma z - u \right) K_a \cos \beta$$

$$= \left((20.4)(2.5) + (22.0)(z - 2.5) - u \right) (0.333) \cos 0$$

$$= 7.33z - 0.33u - 1.33$$

$$u = 9.80\,(z - 2.5) \geq 0$$

Total horizontal pressure on wall $= \sigma_h' + u$

z (m)	Groundwater at **a**	Groundwater at **b**		
	$\sigma_h{}'$ (kPa)	u (kPa)	$\sigma_h{}'$ (kPa)	Total Pressure (kPa)
0.0	0.00	0.00	0.00	0.00
0.5	3.40	0.00	3.40	3.40
1.0	6.79	0.00	6.79	6.79
1.5	10.19	0.00	10.19	10.19
2.0	13.58	0.00	13.58	13.58
2.5	16.98	0.00	16.98	16.98
3.0	20.37	4.90	19.04	23.94
3.5	23.77	9.80	21.09	30.89
4.0	27.16	14.70	23.14	37.84
4.5	30.56	19.60	25.19	44.79
5.0	33.95	24.50	27.24	51.74
5.3	35.99	27.44	28.46	55.90

Example 23.4 demonstrates the profound impact of groundwater on retaining walls. If the groundwater table rises from **a** to **b**, the total horizontal force acting on the wall increases by about 30%. Therefore, the factor of safety against sliding and overturning could drop from 1.5 to about 1.0, and the flexural stresses in the stem would be about 30% larger than anticipated. There are two ways to avoid these problems:

1. Design the wall for the highest probable groundwater table. This can be very expensive, but it may be the only available option.
2. Install drains to prevent the groundwater from rising above a certain level. These could consist of *weep holes* drilled in the face of the wall or a *perforated pipe drain* installed behind the wall. Drains such as these are the most common method of designing for groundwater. Chapter 24 discusses drainage methods in more detail.

23.8 PRACTICAL APPLICATION

Even the most thorough analytical method is a simplification of the truth. Traditional lateral earth pressure calculations ignore many real-world effects, such as temperature fluctuations, readjustment of soil particles due to creep, and vibrations from traffic. Add the ever-present uncertainties in defining soil profiles and obtaining truly representative samples for testing and we should not be surprised when measured earth pressures are often quite different from those that the equations predict. See Gould (1970) for interesting comparisons of predicted and measured earth pressures on real structures.

This does not mean that the analyses are not useful. However, be careful to apply them carefully and with judgment. Specific practical guidelines include the following:

1. Use Extra Caution in Cohesive Soils

Although these methods model cohesionless soils reasonably well, they are less reliable when used with cohesive soils because of their more complex behavior. Therefore, an extra note of caution is appropriate when working with cohesive soils.

One important characteristic of cohesive soils is their tendency to creep, which is a continued deformation under a constant applied load. We normally expect materials subjected to a given stress to develop a corresponding deformation and then stabilize, but materials that creep will continue to deform and not reach equilibrium until the stress is relieved. Although creep can occur in cohesive soils at any stress level, it will usually be most pronounced when the shear stress is at least half of the shear strength.

In the context of lateral earth pressures, creep can be a problem when we design cohesive soils for the active or passive condition, which, by definition, stress the soil to a level near its shear strength. This means that once a wall has moved sufficiently to develop the active or passive condition, the soil will begin to creep and the lateral earth pressures will begin to migrate back toward the at-rest condition.

Therefore, do not design cantilever retaining walls backfilled with clay using the active pressure. A design pressure between the active and at-rest values would be most appropriate. Likewise, it is best not to rely on the full passive pressure, at least not for long-term loads. Terzaghi and Peck's method could be used to estimate the active pressure, and Rankine's method (with a high factor of safety) for the passive.

2. Select Appropriate Strength Parameters

Retaining wall designs should nearly always be based on the saturated strength of the soil. Even though the soil may be relatively dry during construction, the designer must assume that it may one day become wet.

If the design incorporates a drainage system, such as weep holes, then designers can safely assume that the groundwater will not rise above the level of the drain. In such cases, use zero pore water pressure in the soil and zero hydrostatic pressure against the wall.

It is also important to remember that walls are usually built with construction excavations that are smaller than the active wedge. Any analysis of such walls should be based on the soil strength along the base of the wedge, not those from the narrow backfill. In all cases, use plane strain strengths.

3. Use an Appropriate Method of Analysis

Morgenstern and Eisenstein (1970) compared the earth pressure coefficients from Coulomb and four other theories with those of Rankine. They found that the computed values of K_a are within 10% of Rankine's K_a for all values of ϕ, which we would consider to be very close agreement. Coulomb's theory produces K_a values that are slightly more accurate than Rankine's, but the difference is so small that for practical problems an engineer could achieve satisfactory results with either method. However, as discussed earlier, passive pressures are best computed using Rankine's K_p.

Chapter 24 includes additional recommendations for cantilever retaining walls.

23.9 LATERAL CAPACITY OF SPREAD FOOTINGS

Lateral earth pressures also act on the sides of spread footings and may be used to resist applied shear loads.

If no shear load is present, the at-rest pressure acts on both sides of the footing, as shown in Figure 23.17. However, an applied shear load will cause a lateral displacement, eventually producing the active condition on one side and the passive on the other. Therefore, the difference between the passive and active pressures is available to resist the applied shear load.

Another source of lateral resistance is sliding friction force, V_f, along the bottom of the footing. Its magnitude is often greater than the passive resistance, so it is usually the primary source of lateral resistance in spread footings.

Compute V_f using the principles of sliding friction:

$$V_f = P_f \tan\phi_f \tag{23.30}$$

$$\phi_f = \tan^{-1}\mu \tag{23.31}$$

Where:

V_f = ultimate shear resistance along bottom of footing

P_f = normal force between footing and soil = $P + W_f$

P = applied normal load acting on footing

W_f = weight of footing

ϕ_f = footing-soil interface friction angle

μ = coefficient of friction between footing and soil

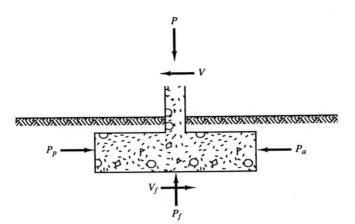

Figure 23.17 Lateral earth pressures and sliding friction acting on a footing subjected to an applied shear load.

TABLE 23.4 DESIGN VALUES OF ϕ_f FOR CAST-IN-PLACE CONCRETE FOOTINGS

Soil or Rock Classification	ϕ_f (deg)
Clean sound rock	35
Clean gravel, gravel-sand mixtures, coarse sand	29 - 31
Clean fine to medium sand, silty medium to coarse sand, silty or clayey gravel	24 - 29
Clean fine sand, silty or clayey fine to medium sand	19 - 24
Fine sandy silt, nonplastic silt	17 - 19
Very stiff and hard residual or overconsolidated clay	22 - 26
Medium stiff and stiff clay and silty clay	17 - 19

Adapted from U.S. Navy (1982b).

Potyondy (1961) and others have measured the footing-soil interface friction angle, ϕ_f, for various soils. The values in Table 23.4 are usually suitable for design. Alternatively, use $\phi_f \approx 0.7\,\phi$, where ϕ is the friction angle of the soil beneath the footing. Thus, the allowable shear load on a spread footing, V_{af}, is:

$$V_{af} = \frac{V_f + P_p - P_a}{F} \qquad (23.32)$$

Where:

V_{af} = allowable shear load on spread footing
V_f = sliding friction along base of footing
P_p = passive resistance along side of footing
P_a = active resistance along side of footing
F = factor of safety = 2 to 3

Unfortunately, the lateral displacement required to mobilize the passive pressure may be more than the structure can tolerate. Therefore, it may be necessary to reduce the computed value of P_p using the guidelines in Table 23.2 and assuming the resistance is proportional to the displacement. If the footing is to be built using wooden forms, then backfilled, neglect P_p unless the project specifications call for the backfill to be compacted.

Only about 0.5 in (12 mm) shear displacement is required to mobilize the sliding friction along the base of the footing, so there is no need to reduce V_f prior to using it in Equation 23.32.

Some engineers also consider sliding friction along the bottom of slab-on-grade floors. This is acceptable, as long as the floor is suitably connected to the rest of the structure.

SUMMARY OF MAJOR POINTS

1. The lateral earth pressures acting on a structure can vary from a lower bound, known as the active condition, to an upper bound, known as the passive condition. The magnitude of the active pressure could be as low as zero, whereas the passive pressure could be as high as several times the vertical stress.

2. The lateral stress condition in a soil depends on the lateral strain. If the soil expands laterally a sufficient distance, then the active condition will exist. If it contracts laterally a sufficient distance, then the passive condition will exist. If no lateral strains occur, then the at-rest condition exists. Intermediate states are also possible.

3. The ratio of the horizontal to vertical effective stress at a point is known as the coefficient of lateral earth pressure, K. The value of K in the active, passive, and at-rest conditions is represented by K_a, K_p, and K_0, respectively.

4. The two classical theories for computing active and passive pressures are Coulomb's theory and Rankine's theory. Either theory is acceptable for computing active pressures, but only Rankine's theory accurately predicts passive pressures.

5. Use special care when computing lateral earth pressures in clayey soils because of the potential for creep and the potential for expansion upon wetting.

6. Surcharge loads that act within a distance H behind the wall, where H is the height of the wall, will increase the lateral pressures.

7. The presence of groundwater behind the wall can significantly increase the lateral earth pressure.

8. Lateral earth pressure and sliding friction can be used to resist applied shear loads on spread footings.

QUESTIONS AND PRACTICE PROBLEMS

23.1 Explain the difference between the active, at-rest, and passive earth pressure conditions.

23.2 Which of the three earth pressure conditions should be used to design a rigid basement wall? Why?

23.3 A 4 m tall cantilever wall is to be backfilled with a stiff cohesive soil. How far must this wall move to attain the active condition in the soil behind it? Is it appropriate to use the active pressure for design? Explain.

23.4 A 3 m tall cantilever retaining wall with a vertical back is to be backfilled with a soil that has an equivalent fluid density of 6.0 kN/m^3. Compute the lateral force per meter acting on the back of this wall.

23.5 A 10 ft tall concrete wall with a vertical back is to be backfilled with a silty sand that has a unit weight of 122 lb/ft^3, cohesion of 0, and friction angle of 32°. The ground behind the wall will be level. Using Rankine's method, compute the normal force per foot acting on the back of the wall. Assume the wall moves sufficiently to develop the active condition in the soil.

23.6 A 12 ft tall concrete wall with a vertical back is to be backfilled with a clean sand that has a unit weight of 126 lb/ft^3, cohesion of 0, and friction angle of 36°. The ground behind the wall will be inclined at a slope of 2 horizontal to 1 vertical. Using Rankine's method, compute the normal and shear forces per foot acting on the back of the wall. Assume the wall moves sufficiently to develop the active condition in the soil.

23.7 Repeat Problem 23.6 using Coulomb's method.

23.8 Repeat Problem 23.6 using Terzaghi and Peck's method.

23.9 A 5 ft wide, 2 ft deep continuous footing carries a vertical load of 12,000 lb/ft. The soil surrounding this footing is a clean fine-to-medium sand with $c = 0$, $\phi = 37°$ and $\gamma = 129$ lb/ft^3. Compute the allowable shear load per foot using a factor of safety of 3.

23.10 A 1.2 m square, 0.5 m deep spread footing carries a vertical compressive load of 150 kN, and a shear load of 16 kN. The soil is a find sandy silt with $c = 0$, $\phi = 28°$ and $\gamma = 17.5$ kN/m^3. The factor of safety on the shear load must be at least 3. Is this design satisfactory?

24

Cantilever Retaining Walls

Theories require assumptions which often are true only to a limited degree. Thus any new theoretical ideas have questionable points which can be removed only by checks under actual conditions. This statement holds especially true for theories pertaining to soil mechanics because assumptions relative to soil action are always more or less questionable. Such theories may sometimes be checked to a limited degree by laboratory tests on small samples, but often the only final and satisfactory verification requires observations under actual field conditions.

MIT Professor Donald W. Taylor (1948)

The most common type of retaining structure is a concrete or masonry cantilever retaining wall. It is often the most economical design, especially when the wall height is less than about 15 ft (5 m).

The design of these walls must satisfy two major requirements: First, the wall must have adequate *stability*, which means it must remain fixed in the desired location (except for small movements required to mobilize the active or passive pressures). Second, it must have sufficient *structural integrity* so it is able to carry the necessary internal stresses without rupturing. Both are shown in Figure 24.1. These are two separate requirements, and each must be satisfied independently. Extra effort in one does not compensate for a shortcoming in the other. For example, adding more rebars (improving the structural integrity) does not compensate for a footing that is too short (a deficiency in stability).

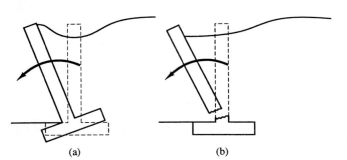

Figure 24.1 (a) A wall that lacks sufficient stability moves away from its desired location; (b) A wall with inadequate structural integrity is unable to carry the necessary internal stresses and will rupture.

(a) (b)

24.1 DESIGN EARTH PRESSURES

The stability and structural integrity of a wall depend on the forces acting between the wall and the soil. These forces, shown in Figure 24.2, are part of a soil-structure interaction problem because their magnitude depends on the movement of the wall, whereas the movement of the wall depends on the magnitude of the forces.

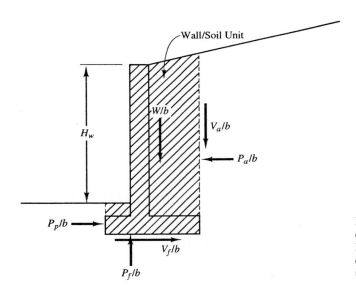

Figure 24.2 Forces acting between a cantilever retaining wall and the ground. Note how the soil above the footing is combined with the wall to form a wall-soil unit.

Measurements and observations of retaining walls have shown that actual in-service earth pressures are often quite different from those predicted by the classical earth pressure theories described in Chapter 23. This is especially true for walls that retain clayey soils.

Duncan et al. (1990) recommended the following methods for developing design earth pressures for gravity earth retaining structures (which includes cantilever retaining walls):

- **For walls with clayey soils in the backfill or below the footing:** Since classical earth pressure theories do not account for creep, they do not accurately predict the pressures acting on the wall. Therefore, use semiempirical methods such as that proposed by Terzaghi and Peck (Figures 23.13 and 23.14 in Chapter 23).

- **For walls with sandy or gravelly soils both in the backfill and below the footing:** These walls have historically performed well, and the classical earth pressure theories provide more accurate estimates of the true earth pressures. However, we must consider the vertical component of the active pressure, V_a/b, to avoid an overly conservative design. Therefore, use Coulomb's method to compute active earth pressures. When using the wall-soil unit shown in Figure 24.2, set the value of ϕ_w in Equation 23.17 equal to 0.8 ϕ, where ϕ is the friction angle of the backfill soil. Another option is to use Terzaghi and Peck's method.

- **For walls with sandy or gravelly backfills and foundations founded on rock:** Engineers often design these walls using more conservatism than necessary or appropriate. This is primarily because the vertical component of the active pressure, V_a/b, is needlessly ignored. A more reasonable approach is to use the following design pressures:

For walls with level backfills:

Compute the horizontal and vertical components of the earth pressure using the following formulas:

$$P/b = \frac{\gamma H^2 K_h}{2} \tag{24.1}$$

$$V/b = \frac{\gamma H^2 K_v}{2} \tag{24.2}$$

Where:

P/b = normal force acting between soil and wall per unit length of wall

V/b = shear force acting between soil and wall per unit length of wall

b = unit length of wall (usually 1 ft or 1 m)

γ = unit weight of soil behind wall

H = height of wall

K_h = horizontal earth pressure coefficient

   = 0.45 for compacted backfill

   = 0.55 for uncompacted backfill

K_v = vertical earth pressure coefficient

   = 0.1

In some cases, a much higher K_v may be appropriate and a finite element analysis might be used to determine it.

For walls with inclined backfills:

Duncan et al. (1990) did not provide recommendations for inclined backfills. However, the following K_h and K_v values should be suitable for design when the ground surface above the wall is inclined at an angle β from the horizontal:

$$K_v = 0.1 \, (1 + 0.5 \tan \beta)^2 \qquad (24.3)$$

For compacted backfills:

$$K_h = 0.45 \, (1 + 0.5 \tan \beta)^2 \qquad (24.4)$$

For uncompacted backfills:

$$K_h = 0.55 \, (1 + 0.5 \tan \beta)^2 \qquad (24.5)$$

• **For all walls:** Although classical earth pressure theories indicate that the resultant of the earth pressure on the back of the wall acts at the lower one-third point, field measurements suggest it is actually higher. For design, place it at a point 40% of the wall height above the bottom, as shown in Figure 24.3. This behavior may be due to a combination of soil arching and differences in soil strain between the top and bottom of the wall.

In some circumstances, such as walls with little or no heel extension, the vertical shear force acting on the back of the wall-soil unit may appear to be the primary source of wall stability. Extra conservatism is appropriate in such cases because the wall may not remain in intimate contact with the backfill, so this shear force may be unreliable.

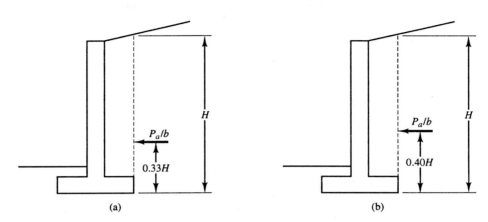

Figure 24.3 Location of resultant earth pressure acting on back of wall: (a) theoretical, and (b) recommended for design (Adapted from Duncan et al., 1990).

The passive resistance, P_p/b, acting on the front of the footing is much smaller than the force acting behind the wall. Use Rankine's method to compute it. Engineers often use a design value less than the theoretical passive resistance because of the following:

- The horizontal displacement required to mobilize the full passive resistance may be larger than the tolerable movement of the wall.
- The soil below the wall is often disturbed by landscaping or other activities and thus may not be as strong as anticipated.

Another important force between the wall and the soil is the shear force V_f along the bottom of the footing, as discussed in Section 23.10 in Chapter 23. It usually forms the bulk of the sliding resistance. Because retaining walls are supported on continuous footings, rewrite Equation 23.30 as :

$$V_f/b = (P_f/b)\ \tan\phi_f \tag{24.6}$$

Where:

V_f/b = shear resistance along bottom of footing per unit length of wall

P_f/b = normal force between footing and soil per unit length of wall

ϕ_f = footing-soil interface friction angle

Section 23.10 gives guidelines for determining ϕ_f.

24.2 STABILITY

A cantilever retaining wall must be stable in all the following ways:

- It must not *slide* horizontally, as shown in Figure 24.4a.
- It must not *overturn*, as shown in Figure 24.4b and Figure 24.5.
- The normal force that acts on the base of the footing must be within the middle third of the footing, as shown in Figure 24.4c.
- The foundation must not experience a *bearing capacity failure*, as shown in Figure 24.4d.
- It must not undergo a *deep-seated shear failure*, as shown in Figure 24.4e.

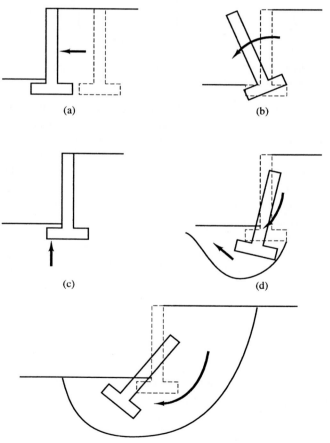

Figure 24.4 Potential stability problems in a cantilever retaining wall: (a) sliding failure, (b) overturning failure, (c) normal force acting on the base of the footing not within the middle third, (d) bearing capacity failure, and (e) deep-seated shear failure.

The stability of a wall in each of these modes is dependent on its dimensions and on the forces between the wall and the ground. Figure 24.5 shows a wall that does not satisfy the overturning criteria and is slowly rotating outward.

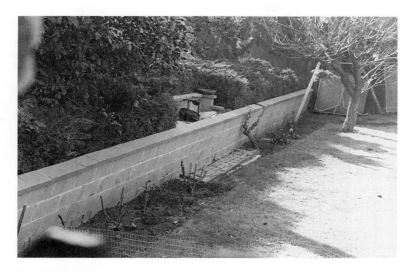

Figure 24.5 This retaining wall, which was backfilled with an expansive clay, is slowly failing by overturning and will eventually collapse. This photograph was taken approximately 30 years after construction.

When evaluating stability, engineers consider the wall and the soil above the footing as a unit, as shown in Figure 24.2. We will refer to it as the *wall-soil unit* and evaluate its stability using the principles of statics.

We can evaluate the stability of a wall-soil unit only after its dimensions are known. Therefore, first develop a trial design using the guidelines in Figure 24.6, then check its stability, and progressively refine the design. Continue this converging trial-and-error process until an optimal design is obtained (one that minimizes costs while satisfying all stability criteria).

Sliding

Evaluate the sliding stability using a *limit equilibrium* approach by considering the forces acting on the wall-soil unit if it were about to fail. The factor of safety is the ratio of the forces required to cause the wall to fail to those that actually act on it.

The forces tending to cause sliding (known as the *driving forces*) are as follows:

- The horizontal component of the lateral earth pressures acting on the back of the wall-soil unit.
- Hydrostatic forces, if any, acting on the back of the wall-soil unit.

These are countered by the following *resisting forces*:

- Lateral earth pressures acting on the front of the wall-soil unit.
- Sliding friction along the bottom of the footing.
- Hydrostatic forces, if any, acting on the front of the wall-soil unit.

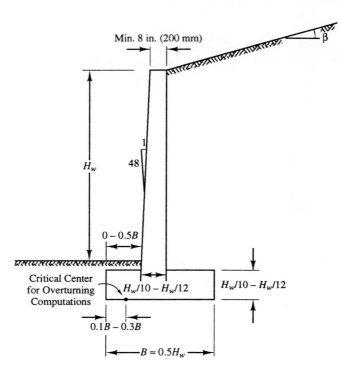

Figure 24.6 Suggested first trial dimensions for cantilever retaining walls backfilled with cohesionless soils. For strong soils and level backfill, the toe extension will be about $0.5B$ and the critical center will be about $0.10B$ from the toe. For weaker soils or inclined backfill, the toe extension will be less (with a corresponding increase in the heel extension) and the critical center will be further from the toe.

A more convenient way to compute the factor of safety, F, is to divide the horizontal resisting forces, $\Sigma(P_R/b)$ by the horizontal driving forces, $\Sigma(P_D/b)$:

$$F = \frac{\Sigma(P_R/b)}{\Sigma(P_D/b)} \tag{24.7}$$

The unit length, b, is usually 1 ft or 1 m. This formula is numerically equivalent to the factor of safety as defined earlier.

In cohesionless soils, design the wall such that the factor of safety against sliding is at least 1.5. In cohesive soils, it should be at least 2.0. Alternatively, some engineers use a pseudo ultimate-strength criteria, and design the wall so that $0.9 \Sigma(P_R/b) \geq 1.7 \Sigma(P_D/b)$, which is the same as using a factor of safety of 1.89.

The earth pressures and sliding friction coefficients used in stability analyses should be ultimate values. Unfortunately, some engineers unintentionally use allowable values (i.e., those that already include a factor of safety), and then compound the factor of safety when conducting the stability analyses.

Example 24.1 — Part A

A 12 ft tall retaining wall supports a backfill inclined at a slope of 4:1 as shown in the figure. The soil behind the wall is a fine to medium sand with the following properties: $c = 0$, $\phi = 35°$, and $\gamma = 122$ lb/ft^3. The soil below the footing is also a fine to medium sand: Its properties are $c = 0$, $\phi = 38°$, $\gamma = 125$ lb/ft^3. Determine the required footing dimensions to satisfy the stability criteria.

Solution:

1. Select the trial dimensions using Figure 24.6.

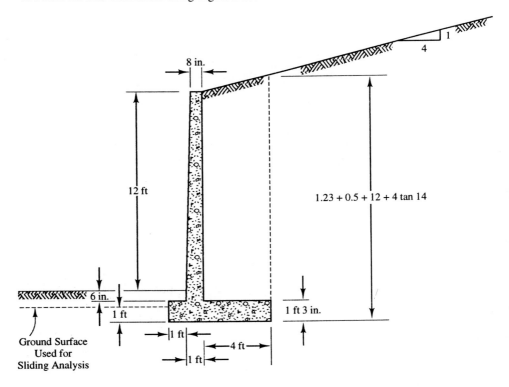

2. Analyze the sliding stability of the trial design.

Active earth pressure - use Coulomb's method (Equation 23.17 in Chapter 23).

$$\beta = \tan^{-1}(1/4) = 14°$$

$$\phi_w = (2/3)(35°) = 23°$$

$$K_a = \frac{\cos^2(\phi - \alpha)}{\cos^2\alpha \cos(\phi_w + \alpha)\left[1 + \sqrt{\dfrac{\sin(\phi_w + \alpha)\ \sin(\phi - \beta)}{\cos(\phi_w + \alpha)\ \cos(\alpha - \beta)}}\right]^2}$$

$$= \frac{\cos^2(35 - 0)}{\cos^2 0 \cos(23 + 0)\left[1 + \sqrt{\dfrac{\sin(23 + 0)\ \sin(35 - 14)}{\cos(23 + 0)\ \cos(0 - 14)}}\right]^2}$$

$$= 0.374$$

$$H = 1.25 + 0.5 + 12 + 4\tan 14°$$
$$= 14.75 \text{ ft}$$

$$P_a/b = \frac{\gamma H^2 K_a}{2}\cos\phi_w = \frac{122(14.75)^2(0.374)}{2}\cos 23° = 4570 \text{ lb/ft}$$

$$V_a/b = \frac{\gamma H^2 K_a}{2}\sin\phi_w = \frac{122(14.75)^2(0.374)}{2}\sin 23° = 1940 \text{ lb/ft}$$

Passive earth pressure - use Rankine method (Equation 23.11 in Chapter 23). For purposes of computing the passive earth pressure, ignore the upper 0.75 ft of soil because:

- The ground may not be brought to finish grade until after the wall is backfilled
- The upper soils may become loosened with time or excavated later

$$K_p = \tan^2(45 + \phi/2)$$
$$= \tan^2(45 + 38/2)$$
$$= 4.20$$

$$P_p/b = \frac{\gamma H^2 K_p}{2}\cos\beta$$

$$= \frac{122(1)^2(4.20)}{2}\cos 0$$

$$= 260 \text{ lb/ft}$$

Sliding friction

Weight on footing:

Stem: $[(8/12 + 12/12)/2](12.5)(150 \text{ lb/ft}^3)$	= 1562 lb/ft
Footing: $(1.25)(6)(150)$	= 1125 lb/ft
Soil behind wall: $(4)[12.5 + (1/2)(4)\tan 14](122 \text{ lb/ft}^3)$	= 6343 lb/ft
V_a/b	= 1940 lb/ft
P_f/b	= 10970 lb/ft

From Table 23.4, use $\phi_f = 26°$.

$$V_f/b = (P_f/b) \tan \phi_f = 10{,}970 \tan 26 = 5350 \text{ lb/ft}$$

$$F = \frac{\Sigma(P_R/b)}{\Sigma(P_D/b)} = \frac{V_f/b + P_p/b}{P_a/b} = \frac{5350 + 260}{4570} = 1.23 < 1.5 \quad \text{NG}$$

If the trial design does not satisfy the sliding requirements, as in Example 24.1, select one or more of the following modifications (see Figure 24.7):

- **Extend the heel of the footing:** This increases the weight acting on the footing, thus increasing the sliding resistance. Unfortunately, it also increases construction costs because it requires a larger construction excavation. If the wall is near a property line or some other limit to construction, this method may not be feasible.

- **Add a key beneath the footing:** This improves the sliding stability by increasing the passive pressure. Unfortunately, the active pressure also increases. However, the increase in the passive is greater than the increase in the active, so there is a net gain. This method is most effective in soils with a relatively high friction angle because the ratio K_p/K_a is greatest in such soils and the net increase in resisting force is greatest.

  Some engineers do not favor the use of keys, especially in clean cohesionless soils (where they are theoretically most effective) because they feel that the benefits of the key are more than offset by the soil disturbance during construction.

- **Use a stronger backfill soil:** This soil must extend at least to the line shown in Figure 24.7 to assure that the critical failure surface passes through it. Usually this requires a larger construction excavation and often requires the use of imported soil. Therefore, this method is usually expensive.

- **Install tiedown anchors:** This increases the normal force acting on the footing, thus increasing the sliding friction. This method is most effective when ϕ_f is large.

- **Install a tieback anchor:** This increases the total resisting force, $\Sigma(P_R/b)$. Tieback anchors might be in the form of a *deadman* as shown here, or they may be as shown in Figure 22.7 in Chapter 22.

Conversely, if the sliding factor of safety is excessive, reduce the heel extension and/or remove or shorten the key. Adjusting the toe extension has very little effect on the sliding stability.

Express the final dimensions of the footing and the key (if used) as a multiple of 3 in or 0.1 m. Therefore, the trial-and-error design process ends when the required dimensions are known within this tolerance.

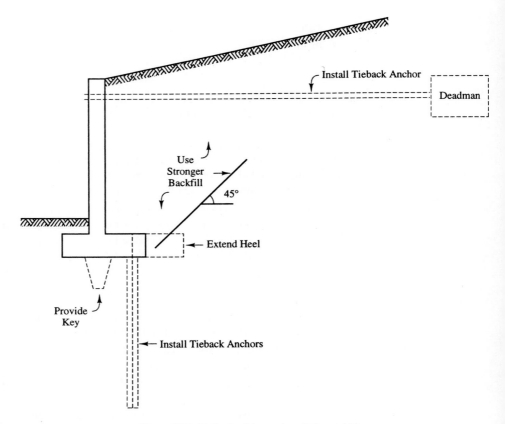

Figure 24.7 Methods of improving sliding stability.

Example 24.1 — Part B

3. Try adding 1.75 ft to the heel extension (total heel extension = 5.75 ft)

$$H = 1.25 + 0.5 + 12 + 5.75 \tan 14° = 15.18 \text{ ft}$$

$$P_a/b = \frac{\gamma H^2 K_a}{2} \cos\phi_w = \frac{122\,(15.18)^2\,(0.3741)}{2} \cos 23° = 4840 \text{ lb/ft}$$

$$V_a/b = \frac{\gamma H^2 K_a}{2} \sin\phi_w = \frac{122\,(15.18)^2\,(0.3741)}{2} \sin 23 = 2055 \text{ lb/ft}$$

Sliding friction

Weight on footing:

Stem: $[(8/12 + 12/12)/2](12.5)(150 \text{ lb/ft}^3)$ = 1562 lb/ft
Footing: $(1.25)(7.75)(150)$ = 1453 lb/ft
Soil behind wall: $(5.75)[12.5 + (1/2)(4)\tan 14](122 \text{ lb/ft}^3)$ = 9272 lb/ft
V_a/b = 2055 lb/ft
                                                              P_f/b = 14342 lb/ft

$$V_f/b = (P_f/b)\tan\phi_f = 14342 \tan 26° = 6990 \text{ lb/ft}$$

$$F = \frac{\Sigma(P_R/b)}{\Sigma(P_D/b)} = \frac{6990 + 260}{4840} = 1.50 \geq 1.5 \quad \text{OK}$$

Additional trials would demonstrate that this is the optimal solution.

Example 24.1 — Part C

4.   Alternative solution: Use original heel extension (4.00 ft) and add a 2 ft deep key, as shown in the figure on the next page.

$$H = 14.75 + 2 = 16.75 \text{ ft}$$

$$P_a/b = \frac{\gamma H^2 K_a}{2}\cos\phi_w = \frac{122(16.75)^2(0.374)}{2}\cos 23 = 5890 \text{ lb/ft}$$

$$V_a/b = \frac{\gamma H^2 K_a}{2}\sin\phi_w = \frac{122(16.75)^2(0.374)}{2}\sin 23 = 2500 \text{ lb/ft}$$

$$P_p/b = \frac{\gamma H^2 K_p}{2}\cos\beta = \frac{122(1+2)^2(4.20)}{2}\cos 0 = 2360 \text{ lb/ft}$$

Sliding friction

Weight on footing:

Stem   = 1562 lb/ft
Footing   = 1125 lb/ft
Soil behind wall   = 6343 lb/ft
V_a/b = 2500 lb/ft
Key $(2)(1)(150)$ = 300 lb/ft
Soil in front of key $(2)(1)(125)$ = 250 lb/ft
                                        P_f/b = 12080 lb/ft

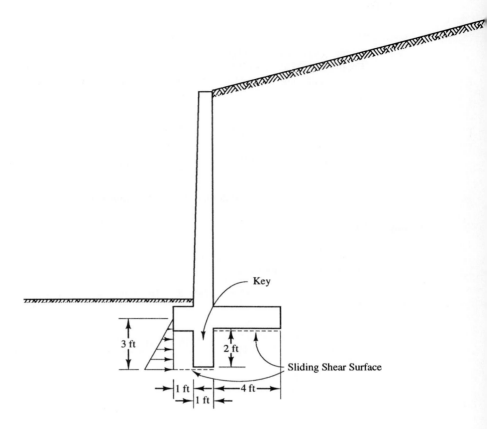

Use a weighted average coefficient of friction, μ: 1 ft of shear through the soil ($\mu = \tan 38$) and 5 ft along the soil-concrete interface ($\mu = \tan 26$) as shown in the earlier sketch.

$$\tan \phi_f = \mu_{avg} = \frac{1 \tan 38 + 5 \tan 26}{6} = 0.537$$

$$V_f/b = (P_f/b) \tan \phi_f = (12080)(0.537) = 6480 \text{ lb/ft}$$

$$F = \frac{\Sigma(P_R/b)}{\Sigma(P_D/b)} = \frac{6480 + 2360}{5890} = 1.50 \geq 1.5 \quad \text{OK}$$

The required factor of safety has been met exactly, so no further trials are necessary.

Overturning

Once the trial design satisfies the sliding stability requirements, begin evaluating its overturning stability. Continue to use a limit equilibrium approach with the wall-soil unit shown in Figure 24.8. However, the factor of safety is now defined in terms of moments:

$$ F = \frac{\Sigma(M_R/b)}{\Sigma(M_D/b)} \tag{24.8} $$

Where:

F = factor of safety against overturning

$\Sigma M_R/b$ = sum of the resisting moments per unit length of wall

$\Sigma M_D/b$ = sum of the driving moments (also known as overturning moments) per unit length of wall

b = unit length (usually 1 ft or 1 m)

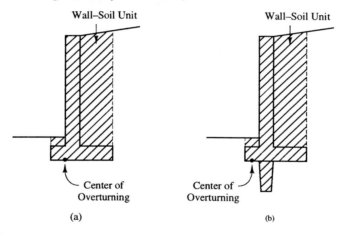

Figure 24.8 Wall-soil unit for overturning analysis: (a) for walls without a key; and (b) for walls with a key.

This is not a $\Sigma M = 0$ analysis. We are taking the sum of all moments acting in one direction (the resisting moments) and dividing them by the sum of all moments acting in the other direction (the driving moments). Thus, the computed factor of safety depends on the location of the point we use to compute the moments.

Traditionally, engineers have taken these moments about the toe of the footing, as shown in Figure 24.9a. This implies that the ground below is infinitely strong, so the wall rotates about a knife-edge at the toe. Bruner et al. (1983) modified the traditional analysis by taking moments about the bottom midpoint of the footing, as shown in the same figure.

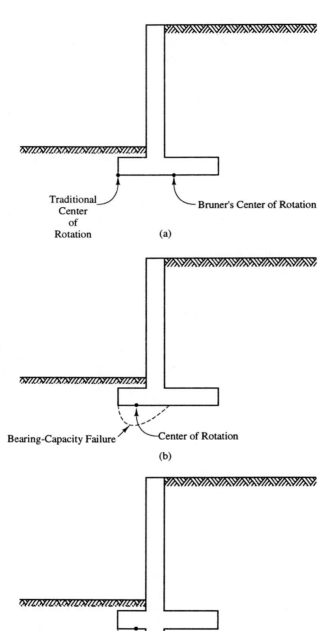

Figure 24.9 The overturning stability of a cantilever retaining wall: (a) traditional center of overturning and Bruner's center; (b) recommended center of overturning and design earth pressures when no key is present; and (c) recommended center of overturning and design earth pressures when a key is present.

However, it now is evident that neither of these points is correct. We must take the moments about the point of rotation to obtain the lowest, and therefore most critical, factor of safety. Observations of walls that have failed in overturning have shown that this point is somewhere between the toe and the midpoint, as shown in Figure 24.9b. Thus, as the wall overturns, a bearing capacity failure occurs in the soil below the toe of the footing. If the soil is strong, the ultimate bearing capacity is large, so the center of overturning is closer to the toe; if the soil is weak, it is closer to the midpoint.

Use trial-and-error to find the center of rotation by computing the factor of safety about different points along the bottom of the footing. The point that produces the lowest F is the most critical, and this is the global factor of safety against overturning. Usually, this point is between $0.1B$ and $0.3B$ behind the toe, as shown in Figure 24.6. Fortunately, the computer program RETWALL, included with this book, simplifies this trial-and-error process.

Because this is a limit equilibrium analysis, consider the forces that would act on the wall-soil unit if an overturning failure was occurring. The following forces generate driving moments:

- The horizontal component of the lateral earth pressures acting on the back of the wall-soil unit.
- Hydrostatic forces acting behind the wall-soil unit.
- The weight of the portions of the soil-wall unit between the toe of the footing and the point of rotation.

These moments are resisted by the following:

- The vertical component of the lateral earth pressures acting on the back of the wall-soil unit.
- The weight of the portions of the wall-soil between the point of rotation and the heel of the footing.
- Surcharge loads acting above the wall-soil unit.
- Hydrostatic pressure acting on the front of the footing.
- The bearing pressure acting on the bottom of the footing, forward of the point of rotation.

The footing may move slightly rearward if an overturning failure were to occur (i.e., the center of rotation may actually be slightly higher than the bottom of the footing). Therefore, the resistance offered by the passive pressure acting on the front of the footing is not reliable, and should be neglected in the overturning analysis.

The normal force between the bottom of the footing and the ground, P_f/b, requires special attention. In a limit equilibrium condition (i.e., if the wall was in the process of overturning), only the part of the footing forward of the point of rotation would be in contact with the soil, and the soil beneath this zone would be experiencing a bearing capacity failure. Therefore, for the overturning analysis, set P_f/b equal to the ultimate bearing capacity of a pseudo-footing with a width equal to the distance from the toe to the center of overturning and refer to it as $P_{f\text{-ult}}/b$.

The shear force, V_f/b, between the footing and the ground does not affect the overturning stability because it has no moment arm.

The required factor of safety is the same as for sliding: at least 1.5 for cohesionless soils, and 2.0 for cohesive soils.

Location of Normal Force Along Base of Footing

During in-service conditions (compared to the limit equilibrium condition used in the overturning analysis), the force P_f/b must be located within the middle third of the footing (i.e., the eccentricity, e, must be no greater than $B/6$) to maintain a compressive stress along the entire base of the footing. This avoids excessive rotation of the footing and wall, as discussed in Section 6.1 of Chapter 6. For walls supported on bedrock, the rotation will be small, so P_f/b should be located within the middle half of the footing (i.e. e must be no greater than $B/4$) (Duncan et al., 1990).

For the in-service conditions, the wall is in static equilibrium, so the sum of the moments about any point must be equal to zero. Use this fact to compute the location of P_f/b, as demonstrated in the next example.

Example 24.1 — Part D

5. Check the overturning stability of the wall described in Part B.

Using Figure 24.6 as a guide, try using a center of overturning 0.9 ft from the toe.

The overturning analysis considers the limit equilibrium condition (i.e., as if the wall were about to fail). Therefore, a bearing pressure is present only along the portion of the footing between the toe and the center of overturning. Therefore, compute the ultimate bearing capacity of a 0.9 ft wide footing.

For $\phi = 38°$: $N_c = 77.5$, $N_q = 61.5$, and $N_\gamma = 82.3$

$$q_u' = c N_c + \sigma_D'(N_q - 1) + 0.5\gamma B N_\gamma$$
$$= 0 + (1)(125)(61.5 - 1) + 0.5(125)(0.9)(82.3)$$
$$= 12,190 \text{ lb/ft}^2$$
$$P_{f\text{-ult}}/b = (12,190)(0.9) = 10,970 \text{ lb/ft}$$

Source	Force (lb)	Arm (ft)	M_D/b (ft-lb/ft)	M_R/b (ft-lb/ft)
P_a/b	4840	0.4 (15.18) = 6.07	29,379	
V_a/b	2055	7.75 - 0.9 = 6.85		14,077
Stem	1562	1.0 + 0.5 - 0.9 = 0.6		937
Footing (left)	168	0.9/2 = 0.45	76	
Footing (right)	1284	(7.75 - 0.9)/2 = 3.42		4,391
Soil	9272	1 + 1 + 5.75/2 - 0.9 = 3.98		36,902
$P_{f\text{-ult}}/b$	10970	0.9/2 = 0.45		4,936
			29,500	61,200

$$F = \frac{\Sigma(M_R/b)}{\Sigma(M_D/b)} = \frac{61200}{29500} = 2.08 > 1.5 \quad \text{OK}$$

Additional trials will demonstrate that this is the most critical center of overturning (i.e., the one that gives the lowest factor of safety).

6. Find the location of the resultant force acting on the footing.

This analysis differs from the overturning analysis in that it considers the in-service conditions. The objective is to find the location of the resultant force acting on the bottom of the footing and determine whether or not it is within the middle third. Because the wall is in static equilibrium, find the location of this force by taking the sum of moments about any point and setting it equal to zero. We already have computed moments about the center of overturning, so it is easiest to continue using this point.
    The magnitude of P_f/b is 14,342 lb/ft (from Part B).

Let x = horizontal distance from the center of overturning to the resultant force.

$$\sum M/b = 29500 + 14342x - (61200 - 4936) = 0$$

$$x = 1.87 \text{ ft}$$

Let e = eccentricity = distance from the center of footing to the resultant force.

$$e = \frac{B}{2} - 0.9 - x = \frac{7.75}{2} - 0.9 - 1.87 = 1.11 \text{ ft}$$

$$\frac{B}{6} = \frac{7.75}{6} = 1.29 \text{ ft}$$

$$e \leq B/6 \quad \therefore \text{ Resultant is within the middle third} \quad \textbf{OK}$$

If the overturning stability of the trial design is not satisfactory, or if the resultant is not in the middle third of the footing, consider any or all of the following modifications:

- **Extend the toe of the footing:** This is almost always the most cost-effective method because it increases the moment arms of all the resisting moments and increases the bearing capacity of the footing without increasing the driving moments or significantly increasing the volume of the construction excavation.
- **Extend the heel of the footing**, as shown in Figure 24.7: This method is much more expensive because it also increases the volume of the construction excavation. An engineer would most likely choose this approach only if a property line or some other restriction prevented extending the toe.
- **Use a stronger backfill soil**, as shown in Figure 24.7: This, too, would be expensive for the reasons described earlier.
- **Use tiedown or tieback anchors**, as shown in Figure 24.7: This is very effective, but may be expensive.

Example 24.1 — Part E

6. Try removing 0.25 ft from the toe extension.

Source	Force (lb)	Arm (ft)	M_D/b (ft-lb/ft)	M_R/b (ft-lb/ft)
P_a/b	4840	0.4 (15.18) = 6.07	29,379	
V_a/b	2055	7.50 - 0.9 = 6.60		13,563
Stem	1562	0.75 + 0.5 - 0.9 = 0.35		547
Footing (left)	168	0.9/2 = 0.45	76	
Footing (right)	1237	(7.50 - 0.9)/2 = 3.30		4,082
Soil	9272	0.75 + 1 + 5.75/2 - 0.9 = 3.73		34,585
$P_{f\text{-ult}}/b$	10970	0.9/2 = 0.45	————	4,936
			29,500	57,700

$$F = \frac{\Sigma(M_R/b)}{\Sigma(M_D/b)} = \frac{57700}{29500} = 1.96 > 1.5 \quad \text{OK}$$

Additional trials will demonstrate that this is the most critical center of overturning.

Check the location of P_f/b:

$$P_f/b = 14,342 - (0.25)(1.25)(150) = 14,295 \text{ lb/ft}$$

$$\sum M/b = 29,500 + 14,295x - (57,700 - 4,936) = 0$$

$$x = 1.63 \text{ ft}$$

$$e = \frac{B}{2} - 0.9 - x = \frac{7.50}{2} - 0.9 - 1.63 = 1.22 \text{ ft}$$

$$\frac{B}{6} = \frac{7.50}{6} = 1.25 \text{ ft}$$

$e \leq B/6$ ∴ Resultant is within the middle third   **OK**

Any further trimming of the toe extension would cause the resultant to be outside the middle third.

Bearing Capacity

The footing that supports the wall may be subject to a bearing capacity failure, as shown in Figure 24.4. Check this possibility using the techniques for footings with moment loads, as described in Section 6.9 of Chapter 6.

Example 24.1 — Part F

7. Check the bearing capacity.

$$B' = B - 2e = 7.50 - 2(1.22) = 5.06 \text{ ft}$$

$$q'_{equiv} = \frac{P_f/b}{B'} = \frac{14295}{5.06} = 2825 \text{ lb/ft}^2$$

Use Terzaghi's formula (Equation 6.14):

For $\phi = 38°$: $N_q = 61.5$, $N_\gamma = 82.3$

$$\sigma'_D = 1(125) = 125 \text{ lb/ft}^2$$

$$
\begin{aligned}
q'_u &= cN_c + \sigma'_D(N_q - 1) + 0.5\gamma B N_\gamma \\
&= 0 + (125)(61.5 - 1) + (0.5)(125)(5.06)(82.3) \\
&= 33,590 \text{ lb/ft}^2
\end{aligned}
$$

$$q'_a = \frac{q'_u}{F} = \frac{33,590}{3} = 11,000 \text{ lb/ft}^2$$

2825 << 11,000, therefore bearing capacity is OK

8. Compute the bearing pressures at heel and toe:

Using Equations 6.8 and 6.9 from Chapter 6:

$$q'_{heel} = \frac{P}{BL}\left(1 - \frac{6e}{B}\right) = \frac{14295}{(7.50)(1)}\left(1 - \frac{(6)(1.22)}{7.50}\right) = 46 \text{ lb/ft}^2$$

$$q'_{toe} = \frac{P}{BL}\left(1 + \frac{6e}{B}\right) = \frac{14295}{(7.50)(1)}\left(1 + \frac{(6)(1.22)}{7.50}\right) = 3766 \text{ lb/ft}^2$$

Walls that meet all of these criteria usually do not have problems with excessive settlement. Generally, a settlement analysis is necessary only if q'_{toe} exceeds the preconsolidation pressure, σ'_c.

9. Conclusion

The design shown in the following figure satisfies all stability criteria:

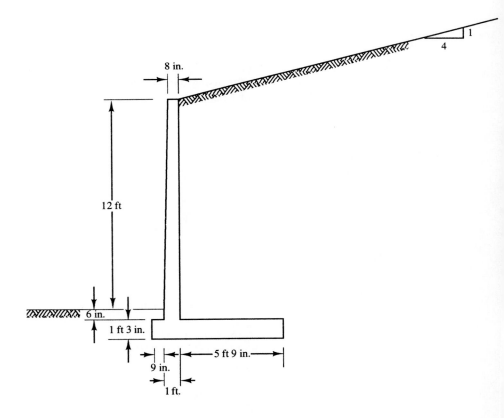

Deep-Seated Shear Failure

A deep-seated shear failure would be a catastrophic event that could be much larger in scope than any of the modes described earlier. Be concerned about this mode of failure if any of the following conditions is present:

- The soil is a soft or medium clay and is subjected to undrained conditions. Generally, this is a concern only if $\gamma H_w / s_u > 6$ where H_w is the height of the wall as defined in Figure 24.2, s_u is the undrained shear strength, and γ is the unit weight of the soil.
- Adversely oriented weak seams or bedding planes are present.
- Some or all of the soil is prone to liquefaction. This phenomenon occurs during earthquakes in loose, saturated sands and silty sands and causes the soil to lose most or all of its strength.

If so, evaluate the deep-seated shear stability using a slope stability analysis. Fortunately, the vast majority of walls do not satisfy any of these conditions, so we usually can conclude that a deep-seated failure is not a concern.

24.3 STRUCTURAL DESIGN

Once the wall has been sized to satisfy stability requirements, begin the structural design process. The footing is virtually always made of reinforced concrete, but the stem can be either reinforced concrete or reinforced masonry. Concrete stems have greater flexural strength, and greater flexibility in the design dimensions, and thus seem to be most cost-effective for tall walls. However, masonry stems do not require formwork, which is very expensive, and generally are preferred for shorter walls.

This chapter considers only walls with reinforced concrete stems. The design of masonry stems is very similar (see Schneider and Dickey, 1987).

Stem

It is convenient to begin by designing the stem.

Developing Shear and Moment Diagrams

Use ultimate strength analysis methods to develop the structural design. Therefore, it is necessary to multiply the design earth pressures from the stability analysis by a load factor. The ACI code uses a factor of 1.7 for lateral earth pressures. Then, use this factored load to develop shear and moment diagrams for the stem. Typical diagrams are shown in Figure 24.10.

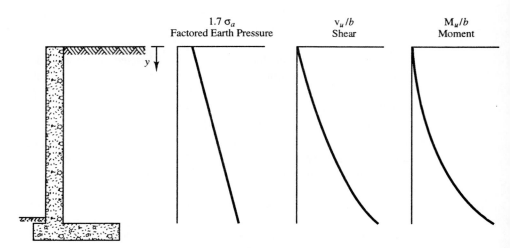

Figure 24.10 Earth pressure acting on the stem and typical shear and moment diagrams. The earth pressure diagram is that which varies linearly with depth and has a resultant that acts at 40% of the wall height.

Example 24.2 — Part A

Develop a structural design for the wall in Example 24.1.

1. Develop shear and moment equations for the stem.

   At bottom of the stem:

   $$P_a/b = \frac{\gamma H^2 K_a}{2} \cos\phi_w = \frac{(122)(12.5)^2(0.374)}{2} \cos 23° = 3280 \text{ lb/ft}$$

   $$P_u/b = 1.7\, P_a/b = (1.7)(3280) = 5580 \text{ lb/ft}$$

   $$M_u/b = (5580)(12.5)(0.4) = 27{,}900 \text{ ft–lb/ft}$$

   Find the slope, m, of the earth pressure vs. depth diagram by summing moments about the bottom of the stem:

   Let P_1/b = force due to the rectangular portion of the pressure diagram
   Let P_2/b = force due to the triangular portion of the pressure diagram

$$M_u/b = P_1/b\left(\frac{12.5}{2}\right) + P_2/b\left(\frac{12.5}{3}\right) = 27{,}900 \text{ ft–lb/ft}$$

$$P_1/b + P_2/b = P_u/b = 5580 \text{ lb/ft}$$

$$\therefore \quad P_1/b = 2230 \text{ lb/ft} \qquad P_2/b = 3350 \text{ lb/ft}$$

$$\frac{m(12.5)^2}{2} = 3350 \quad \rightarrow \quad m = 42.9$$

At any point on the stem:

$$V_u/b = 2230\left(\frac{y}{12.5}\right) + \frac{42.9\,y^2}{2} = 178\,y + 21.5\,y^2$$

$$M_u/b = 178\,y\left(\frac{y}{2}\right) + 21.5\,y^2\left(\frac{y}{3}\right) = 89.0\,y^2 + 7.17\,y^3$$

$$y = \text{distance from top of wall}$$

Determining Stem Thickness and Reinforcement

The thickness of the stem should be at least 8 in (200 mm) and expressed as a multiple of 1 in or 20 mm. For short walls, it may be possible to have a uniform thickness, but taller walls are usually tapered or stepped so the thickness at the bottom is greater than at the top. A greater thickness and more reinforcing steel are required at the bottom because the stresses there are larger. Typically, it is best to select a thickness such that the steel ratio, ρ, at the bottom of the wall is approximately 0.0200.

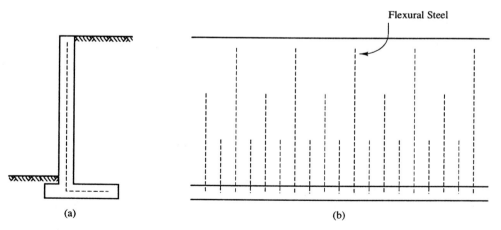

Figure 24.11 Different steel cutoff lengths are used to vary the steel ratio: (a) end view, and (b) side view.

Even if the wall is tapered or stepped, the bottom portion will require more steel than the top. Implement this by using different cutoff lengths for different bars, as shown in Figure 24.11. A technique for determining cutoff points is illustrated in Part C of Example 24.2.

The stem also should include some longitudinal steel. Although in theory there are no flexural stresses in that direction, such stresses may develop if the soil is not perfectly uniform or if isolated surcharge loads are present. This steel accommodates these loads as well as temperature and shrinkage stresses. Typically, use a longitudinal steel ratio, ρ, of about 0.0020. It is not necessary to conduct a longitudinal flexural analysis unless large surcharge point loads are present.

Example 24.2 — Part B

2. Determine the required wall thickness and flexural steel at the bottom of the stem.

Use $f_c' = 4000$ lb/in^2 and $f_y = 60$ k/in^2

$$
\begin{aligned}
M_u/b &= 89.0\,y^2 + 7.17\,y^3 \\
&= (89.0)(12.5)^2 + (7.17)(12.5)^3 \\
&= 27{,}900 \text{ ft--lb/ft} \\
&= 335{,}000 \text{ in--lb/ft}
\end{aligned}
$$

Using Equation 10.13:

$$
A_s/b = \left(\frac{f_c'\,b}{1.17\,f_y}\right)\left(d - \sqrt{d^2 - \frac{2.35\,M_u/b}{\phi\,f_c'\,b}}\right)
$$

$$
= \left(\frac{(4000)(12)}{(1.17)(60{,}000)}\right)\left(d - \sqrt{d^2 - \frac{(2.35)(335{,}000)}{(0.9)(4000)(12)}}\right)
$$

$$
= 0.684\left(d - \sqrt{d^2 - 18.2}\right)
$$

$$
\begin{aligned}
T &= d + 1/2 \text{ bar diameter} + 3 \text{ in} \\
&= d + 0.5 \text{ in} + 3 \text{ in}
\end{aligned}
$$

T (in)	d (in)	A_s/b (in^2/ft)	ρ
12	8.5	0.78	0.0077
10	6.5	1.09	0.0140
9	5.5	1.39	0.0210

$T = 9$ in gives $\rho \approx 0.0200$; use #8 bars at 6 in on center ($A_s = 1.57$ in^2/ft)

Example 24.2 — Part C

3. Design remainder of stem steel.

Although we could taper the stem to a thickness of 8 in at the top, we will choose to use a thickness of 9 in along the entire height.

For $d = 5.5$ in:

Steel	A_s/b (in^2/ft)	$\phi M_n/b$	
		(in-lb/ft)	(ft-lb/ft)
#8 at 6 in OC	1.57	330,000	27,500
#8 at 12 in OC	0.79	188,300	15,700
#8 at 18 in OC	0.52	128,100	10,700

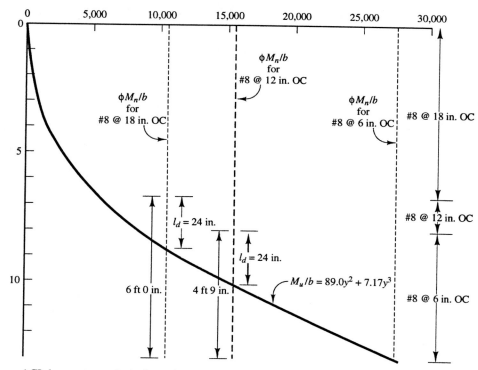

ACI does not permit steel spacings greater than 18 in on center.

Development length = 24 in (see MacGregor, 1992, for computation procedure).

The ultimate moments, nominal moment capacities, and required development lengths are shown in the figure on the previous page. If the stem were tapered, the nominal moment capacity lines in this figure would be inclined; if it were stepped, they too would be stepped.

Use this information to determine the steel cutoff points as follows:

Alternative A:

Distance from Bottom of Stem	Steel
0 - 4 ft 9 in	#8 at 6 in OC
4 ft 9 in - Top	#8 at 12 in OC

Alternative B:

Distance from Bottom of Stem	Steel
0 - 6 ft 0 in	#8 at 6 in OC
6 ft 0 in - Top	#8 at 18 in OC

Longitudinal steel:

$$A_s = \rho Bd = (0.0020)(12.5\,\text{ft} \times 12\,\text{in/ft})(5.5\,\text{in}) = 1.65\,\text{in}^2$$

Use 9 #4 bars ($A_s = 1.77\,\text{in}^2$).

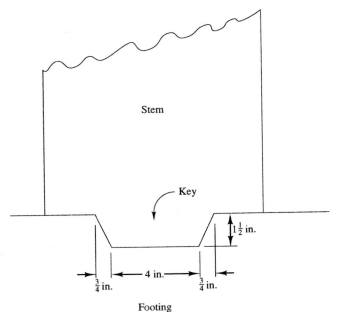

Footing

Figure 24.12 Use of a shear key between the stem and the footing.

Designing Connection Between Stem and Footing

The critical location for shear is the connection between the stem and the footing. Because they are cast separately, there is a weak plane at the interface. To prevent sliding along this plane, make a shear key in the footing, as shown in Figure 24.12.

The flexural steel must have sufficient embedment into the footing to develop its tensile load capacity.

Example 24.2 — Part D

4. Check shear at the bottom of the stem.

$$V_u/b = P_u/b = 5580 \text{ lb/ft}$$

Using Equation 10.19 from Chapter 10:

$$V_n = V_c = 0.54 \, b_w \, d \, \sigma_r \sqrt{f_c'/\sigma_r}$$

$$V_n/b = 0.54 \, d \, \sigma_r \sqrt{f_c'/\sigma_r}$$

$$= 0.54 \, (5.5) \, (14 \text{ lb/in}^2) \sqrt{\frac{4000 \text{ lb/in}^2}{14 \text{ lb/in}^2}}$$

$$= 703 \text{ lb/in}$$

$$= 8434 \text{ lb/ft}$$

$$\phi \, V_n/b = 0.85 \, (8434) = 7169 \text{ lb/ft}$$

$$5580 < 7169 \qquad \therefore \text{ Shear is OK}$$

5. Design the connection between the stem and the footing.

Development length (based on $f_c' = 2500$ lb/in^2 in footing) = 36 in.

$\therefore$ Extend steel 36 in into the footing

Footing

Design the heel extension (the part of the footing beneath the backfill) as a cantilever beam. It must support the weight of the backfill soil and the shear component of the lateral earth pressure. The weight of the backfill is a dead load, so multiply it by a load factor of 1.4. The earth pressure load requires a load factor of 1.7. Ignore the bearing pressure along the bottom of the heel extension, which is conservative.

The minimum required thickness of the footing is governed by one-way shear along a vertical plane immediately behind the stem. For reasons discussed in Chapter 10, do not use stirrups and determine the effective depth, d, from a shear analysis of the concrete. The total footing thickness, T, should be a multiple of 3 in or 0.1 m and never less than 12 in (300 mm).

Design the flexural steel in the heel extension to resist the moment at a vertical section immediately behind the stem. This steel will be near the top of the footing. Because this part of the footing is adjacent to soil, there must be at least 3 in (75 mm) of concrete cover over the steel.

The toe extension also is a cantilever beam. However, it bends in the opposite direction, so the flexural steel must be near the bottom.

The footing also should have longitudinal steel. A steel ratio of 0.0015 to 0.0020 is normally appropriate.

Example 24.2 — Part E

6. Design the footing.

Neglect sliding friction along the back of the stem (conservative), so the heel extension must support both the weight of the backfill soil and V_a/b.

Shear analysis - heel extension below the back of the stem:

$$W_s/b = 5.75\left(12.50 + \frac{5.75}{4\times2}\right)(122) = 9273 \text{ lb/ft}$$

$$V_a/b = \frac{(122)(12.5)^2(0.3741)}{2}(\sin 23) = 1393 \text{ lb/ft}$$

$$V_u/b = 1.4(9273) + 1.7(1393) = 15,400 \text{ lb/ft}$$

Try $T = 18$ in ($d = 14.5$ in):

$$V_n/b = 0.54 \, d \, \sigma_r \sqrt{f_c'/\sigma_r}$$

$$= (0.54)(14.5)(14)\sqrt{2500/14}$$

$$= 1460 \text{ lb/in}$$

$$= 17,500 \text{ lb/ft}$$

$$\phi V_n = (0.85)(17,500) = 14,900 < 15,300 \quad \text{NG}$$

Try $T = 21$ in ($d = 17.5$ in):

$$V_n/b = = 0.54 \, d \, \sigma_r \sqrt{f_c'/\sigma_r}$$

$$= (0.54)(17.5)(14)\sqrt{2500/14}$$

$$= 1770 \text{ lb/in}$$

$$= 21,200 \text{ lb/ft}$$

$$\phi V_n = (0.85)(21,200) = 18,000 > 15,300 \quad \text{OK}$$

$\therefore$ Use $T = 21$ in $(d = 17.5$ in$)$

Flexure analysis - heel extension:

$$M_u/b = (1.4)(9273)\left(\frac{5.75}{2}\right) + (1.7)(1393)(5.75)$$

$$= 51,000 \text{ ft–lb/ft}$$

$$= 612,000 \text{ in–lb/ft}$$

$$A_s/b = \left(\frac{f_c' b}{1.17 f_y}\right)\left(d - \sqrt{d^2 - \frac{2.35 M_u/b}{\phi f_c' b}}\right)$$

$$= \left(\frac{(2500)(12)}{(1.17)(60000)}\right)\left(14.5 - \sqrt{14.5^2 - \frac{(2.35)(612,000)}{(0.9)(2500)(12)}}\right)$$

$$= 0.84 \text{ in}^2/\text{ft}$$

$$\rho = \frac{0.84}{(14.5)(12)} = 0.0048 > \rho_{min}$$

Use #6 bars at 6 in OC $(A_s = 0.88 \text{ in}^2/\text{ft})$

Development length:

Need 26.5 in development length (see MacGregor, 1992, for details).   Since the toe extension is short, it will be necessary to hook the bars to obtain the required 26.5 in of development length.

Flexure analysis - toe extension:

Weight of wall and soil behind critical section:

W_s/b	= 9273 lb/ft
Stem = $(9/12)(12.5)(150 \text{ lb/ft}^3)$	= 1406 lb/ft
Footing = $(5.75 + 9/12)(21/12)(150 \text{ lb/ft}^3)$	= 1706 lb/ft
	12,385 lb/ft

Moment arm = $(5.75 + 9/12)/2 = 3.25$ ft

$$M_u/b = (M_u/b)_{stem} - (12385)(3.25)(1.4)$$
$$= 27,900 - 56,381$$
$$= -28,451 \text{ ft–lb/ft}$$
$$= -341,000 \text{ in–lb/ft}$$

In general, we would now design the steel based on this moment. However, because it is less than the moment in the heel and because the upper steel must be wrapped around the toe anyway, we can simply extend this steel along the bottom of the footing.

Longitudinal steel:

$$A_s = \rho dB = (0.0020)(14.5)(7.50 \times 12) = 2.61 \text{ in}^2$$

Use 6 #6 bars ($A_s = 2.65 \text{ in}^2$).

Final design:

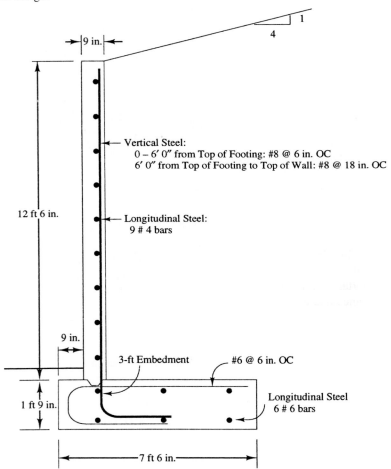

24.4 DRAINAGE AND WATERPROOFING

Example 23.4 in Chapter 23 demonstrated the impact of groundwater on lateral earth pressures. Because groundwater increases earth pressures so dramatically, engineers provide a means of drainage whenever possible to prevent the groundwater table from building up behind the wall. Two types of drains are commonly used: *weep holes* and *perforated pipe drains*. Both are shown in Figure 24.13. It is also helpful to include a means of intercepting water and bringing it to the drain, possibly using products such as that shown in Figure 24.14.

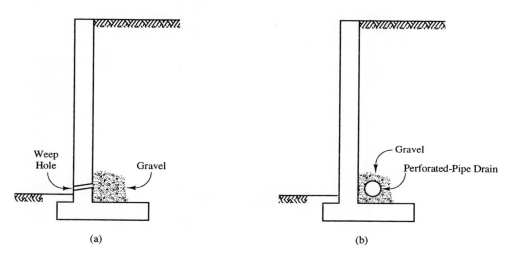

Figure 24.13 Methods of draining the soil behind retaining walls: (a) weep holes; and (b) perforated pipe drain.

In addition, walls are often covered with a waterproofing material to retard the possible migration of moisture from the soil through the wall. This is especially important on basement walls. Excessive moisture inside basements can be both a nuisance and a health hazard.

24.5 AVOIDANCE OF FROST HEAVE PROBLEMS

Concrete is a relatively poor insulator of heat, so retaining walls located in areas with frost heave problems[1] may be damaged as a result of the formation of ice lenses behind the wall. In some cases, ice lenses have caused retaining walls to move so far out of position that they became unusable.

Reduce frost heave problems by using all of the following preventive measures:

- Incorporate good drainage details in the design to avoid the buildup of free water behind the wall. Deep perforated drainage pipes are usually a better choice than

---

[1] See discussion of frost heave in Chapter 2.

weep holes because they are less likely to become blocked by frozen water.

- Use non frost-susceptible soil for the portion of the backfill immediately behind the wall. This zone should extend horizontally behind the wall for a distance equal to the depth of frost penetration in that locality.

Figure 24.14 A geosynthetic drainage panel, such as this, can be placed against a wall to capture water and direct it to weep holes or perforated pipe drains.

SUMMARY OF MAJOR POINTS

1. Cantilever retaining walls must satisfy two major requirements: They must have adequate *stability* and they must have sufficient *structural integrity*.

2. The active earth pressure may be computed as follows:
   - **With clayey soils in backfill or below footing:** Use Terzaghi and Peck's method.
   - **With sandy or gravelly soils in both the backfill and below the footing:** Use Coulomb's method.
   - **With sandy or gravelly backfill and footing on rock:** Use $K_0 = 0.45$ for compacted backfill and 0.55 for uncompacted backfill and $K_v = 0.1$.

3. For design purposes, place the resultant of the active earth pressure at a point 40% of the wall height above the bottom.

4. Use Rankine's method to compute passive pressures.

5. Use Table 23.4 in Chapter 24, or $\phi_f \approx 0.7\,\phi$ to assess the sliding friction resistance along the bottom of the footing.

6. The stability analysis consists of checking the following:
   - sliding stability
   - overturning stability
   - location of the resultant force acting on the bottom of the footing
   - possibility of a bearing capacity failure
   - possibility of a deep-seated shear failure

7. The wall dimensions, especially the heel and toe extensions and the key depth (if any) must be adjusted until all the stability criteria are satisfied.

8. The structural design begins with the stem. The bottom of the stem must have sufficient thickness to resist the shear forces, and sufficient reinforcement must be provided to resist flexure. Because the moment in the upper portion of the stem is much less than that below, the required steel area will also be less. Often, the wall is also tapered or stepped to give a smaller thickness at the top.

9. The footing is designed as two cantilever beams. The heel extension is designed to support the weight of the backfill, and the toe extension is designed to resist the soil bearing pressure.

10. Most walls include weep holes or perforated pipe drains to prevent large hydrostatic pressures from developing behind the wall. Sometimes waterproofing is also appropriate to keep moisture from seeping through the wall.

COMPUTER SOFTWARE – Program RETWALL

A computer program, RETWALL, is included with this book. It computes the sliding and overturning stability of cantilever retaining walls and checks the location of the normal force acting on the footing and the bearing capacity. The user can easily change the dimensions of the wall or footing, the soil properties, and methods of analysis, thus relieving much of the tedium associated with trial-and-error hand computations.

To use this program, install the software according to the instructions in Appendix C, and then type **RETWALL** at the DOS prompt. A brief introductory screen will appear, followed by the worksheet. Then, follow this procedure:

1. Select the units of measurement by pressing **ALT-U** and selecting ENGLISH or SI from the menu. English units are the default.

2. Select the method of computing the earth pressure behind the wall by pressing **ALT-P**. You may choose from the following:
   - Active pressure
   - At-rest pressure (Equations 23.2, 24.2 and 24.3)
   - Specified K_h and K_v values (Equations 24.1 and 24.2)
   - Specified G_h and G_v values (Equations 23.28 and 23.29)

3. Select the method of computing the earth pressure acting behind the wall (Coulomb or Rankine) by pressing **ALT-E** and selecting from the menu. The Coulomb method is the default. This information is used only if ACTIVE is selected in step 2. The program always uses Rankine's method to compute passive pressures.

4. Use the arrow keys to move the cursor and input the title, wall geometry, and soil properties. Replace all of the **XXXXX** indicators with appropriate data, as shown in Figure 24.15. You may wish to use the information in Figure 24.6 to obtain a first estimate of some of these dimensions. If no key is present, input zeros for key offset, key width, and key depth.

   You may wish to use a toe cover distance that is smaller than the actual cover. This accounts for possible disturbance of this soil and produces a smaller passive resistance.

5. Using a converging trial-and-error process, select a heel extension that produces a satisfactory factor of safety against sliding. Alternatively, select key dimensions to accomplish the same objective. Remember that these dimensions should be multiples of 3 in or 0.1 m.

6. Using a converging trial-and-error process, select a toe extension that produces a satisfactory factor of safety against overturning, a resultant force in the middle third and a satisfactory bearing pressure.

   The program computes the equivalent bearing pressure and the allowable bearing capacity using Terzaghi's method (Equation 6.14 in Chapter 6) with a reduced footing width, B' (Equation 6.48), and a factor of safety of 3. If the equivalent bearing pressure is less than or equal to the allowable bearing capacity, then the "Bearing OK?" question receives a "Yes" answer.

7. Using a converging trial-and-error process, select the center of overturning that produces the lowest overturning factor of safety.

8. Repeat steps 5 - 7 as necessary to fine-tune the design.

9. Check the V/P_f ratio. This is the portion of the normal load on the footing that is due to the shear force acting on the back of the wall/soil unit. It is generally good to keep this ratio below 0.33. If it is too high, reduce the wall friction angle, ϕ_w. The program will not allow the shear load on the stem to exceed ϕ_w times the normal load. Then, repeat steps 5 - 9.

10. If a printout is needed, press **ALT-M** to access the menu and select one of the following:

    PRINT - Prints the worksheet on a printer
    PRINT TO DISK - Prints the worksheet to a disk file

    When printing to disk, the program will prompt for a file name. You must select a new name; it will not write over an existing file.

11. If you wish to erase all the input data, select **REFRESH** from the menu.

12. To exit to DOS, select **QUIT** from the menu.

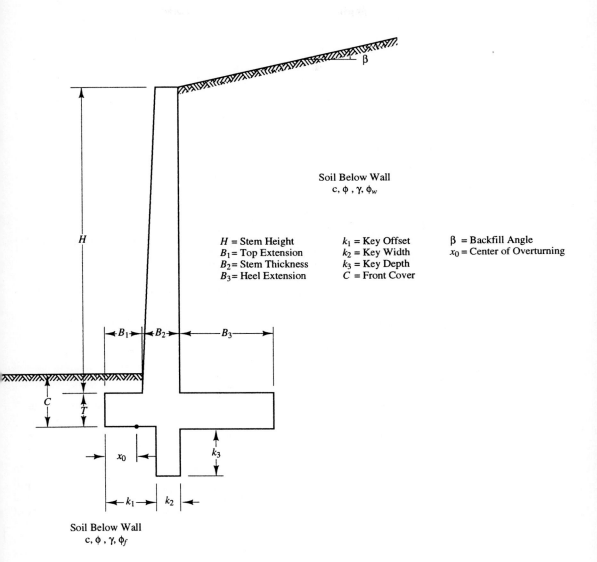

H = Stem Height
B_1 = Top Extension
B_2 = Stem Thickness
B_3 = Heel Extension

k_1 = Key Offset
k_2 = Key Width
k_3 = Key Depth
C = Front Cover

β = Backfill Angle
x_0 = Center of Overturning

Soil Below Wall
c, ϕ, γ, ϕ_w

Soil Below Wall
c, ϕ, γ, ϕ_f

Figure 24.15 Input data for RETWALL program.

Example 24.3

Use RETWALL to analyze the stability of the wall described at the end of Example 24.1.

Output:

```
            RETWALL - Version 1.0      (c) 1994 by Prentice Hall, Inc.
                    Stability of Cantilever Retaining Walls


    Title:     Example 24.3

    Date:      09-Jun-93                 Time:          08:13 PM


    Units of Measurement :       English
       (Press ALT-U to change)

    Pressure Behind Wall :       Active
       (Press ALT-P to change)

    Earth Pressure Method:       Coulomb
       (Press ALT-E to change)        *****************************************
                                      *                                       *
    INPUT DATA                        *               RESULTS                 *
                                      *                                       *
    Stem Height          =    12.50 ft  * Sliding FS           =   1.49      *
    Stem Thickness (top) =     0.67 ft  * Overturning FS       =   1.95      *
    Stem Thickness (bot) =     1.00 ft  * In Middle Third?           Yes     *
    Footing Thickness    =     1.25 ft  * Toe Bearing Pressure =   3774 psf  *
    Toe Extension        =     0.75 ft  * Heel Bearing Pressure =    39 psf  *
    Heel Extension       =     5.75 ft  * Equiv Bearing Pressure =  2831 psf *
    Key Offset           =     0.00 ft  * All Bearing Capacity  = 11182 psf  *
    Key Width            =     0.00 ft  * Bearing OK?                Yes     *
    Key Depth            =     0.00 ft  * Va/Pf                 =   0.14     *
    Front Cover          =     1.00 ft  * Eccentricity         =   1.22 ft  *
    Backfill Angle       =    14.00 deg *****************************************
    Center of Overturn   =     0.90 ft
    Soil Below Footing
       Cohesion          =     0.00 psf
       Friction Angle    =    38.00 deg
       Unit Weight       =   125.00 pcf
       Ftg Friction      =    26.00 deg
    Soil Behind Wall
       Cohesion          =     0.00 psf
       Friction Angle    =    35.00 deg
       Unit Weight       =   122.00 pcf
       Wall Friction     =    23.00 deg
```

QUESTIONS AND PRACTICE PROBLEMS

24.1 A 2.0 m tall cantilever retaining wall is to be backfilled with a compacted silty medium to coarse sand that has a unit weight of 20.0 kN/m^3. The ground surface above this wall will be level. Its footing will be supported on bedrock. Compute the horizontal and vertical components of the force between the backfill and the wall and express them in units of kN/m.

24.2 Determine the design pressure distribution that results from the forces computed in Problem 24.1 and show them in a sketch.

24.3 Why is it better to compute moments for an overturning analysis about a point between the toe and the midpoint of the footing instead of about the toe?

24.4 Using program RETWALL, compute the factors of safety against sliding and overturning for the wall shown in the following figure. Are they satisfactory?

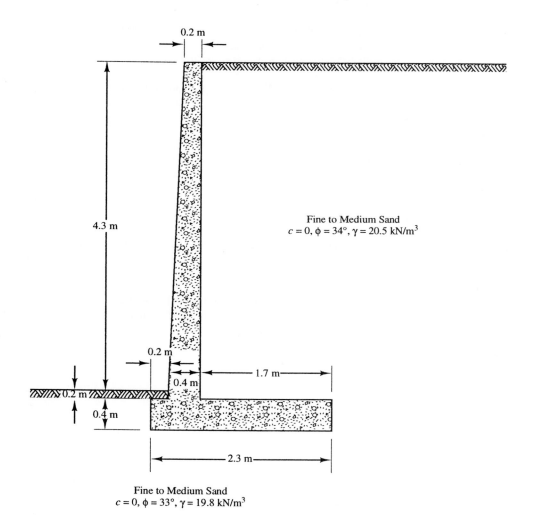

0.2 m

4.3 m

Fine to Medium Sand
$c = 0, \phi = 34°, \gamma = 20.5$ kN/m^3

0.2 m

1.7 m

0.4 m

0.2 m

0.4 m

2.3 m

Fine to Medium Sand
$c = 0, \phi = 33°, \gamma = 19.8$ kN/m^3

24.5 Check the computations from Figure 24.4 using hand computations.

24.6 Using hand computations, determine the location of the resultant normal force that acts on the bottom of the footing of the wall in Problem 24.4. Is this location satisfactory?

24.7 In the context of its stability, what changes, if any, would you suggest for the wall design in Problem 24.4?

24.8 Use program RETWALL to size the footings for the cantilever walls described below. Do not use a key in any of your designs, and assume proper drainage is provided to prevent accumulation of groundwater.

    a. Wall height = 8 ft with level backfill
       Backfill soils: Well graded sand, $c = 0$, $\phi = 37°$, and $\gamma = 128$ lb/ft^3
       Soils below foundation: Same as backfill soils

    b. Wall height = 2 m with backfill sloped upward at 3H:1V
       Backfill soils: Silty sand, $c = 0$, $\phi = 32°$, and $\gamma = 19.0$ kN/m^3
       Soils below foundation: Same as backfill soils

    c. Wall height = 16 ft with level backfill
       Backfill soils: Compacted gravelly sand, $c = 0$, $\phi = 39°$, and $\gamma = 132$ lb/ft^3
       Rock below foundation: Sandstone bedrock, $c = 6000$ lb/ft^2, $\phi = 46°$, and $\gamma = 145$ lb/ft^3

24.9 Use program RETWALL to redesign the wall in Problem 24.8b with a key combined with the shortest possible heel extension.

24.10 Develop a complete structural design for the wall in Problem 24.8a. Use $f_c' = 5000$ lb/in^2 in the stem, $f_c' = 3000$ lb/in^2 in the footing, and $f_y = 60$ k/in^2. Show your design in a sketch similar to that in Example 24.2, Part E.

24.11 Develop a complete structural design for the wall in Problem 24.8b. Use $f_c' = 35$ MPa in the stem, $f_c' = 20$ MPa in the footing, and $f_y = 400$ MPa. Show your design in a sketch similar to that in Example 24.2, Part E.

Appendix A

Recommended Resources for Further Study

General References

Books and Manuals

CGS (1985), *Canadian Foundation Engineering Manual*, Canadian Geotechnical Society, BiTech, Vancouver

U.S. Navy (1982a), *Soil Mechanics*, NAVFAC Design Manual 7.1, Naval Facilities Engineering Command, Alexandria, VA

U.S. Navy (1982b), *Foundations and Earth Structures*, NAVFAC Design Manual 7.2, Naval Facilities Engineering Command, Alexandria, VA

U.S. Navy (1982c), *Soil Dynamics, Deep Stabilization and Special Geotechnical Construction*, NAVFAC Design Manual 7.3, Naval Facilities Engineering Command, Alexandria, VA

Journals

Canadian Geotechnical Journal, National Research Council Canada, Ottawa

Géotechnique, Institution of Civil Engineers, London

Ground Engineering, Thomas Telford, Ltd, London

Journal of Geotechnical Engineering, American Society of Civil Engineers, New York

739

Chapter 2 – Performance Requirements

Johnston, G.H., ed. (1981), *Permafrost Engineering Design and Construction*, John Wiley, New York

> An extensive coverage of the theory and practice of civil engineering in cold climates, including a chapter on foundations.

LePatner, Barry B. and Johnson, Sidney M. (1982), *Structural and Foundation Failures: A Casebook for Architects, Engineers, and Lawyers*, McGraw-Hill, New York

> Case studies of several foundation failures from technical, managerial, and legal perspectives.

Peck, Ralph B. (1967), "Bearing Capacity and Settlement: Certainties and Uncertainties", in *Bearing Capacity and Settlement of Foundations*, Aleksandar S. Vesić, ed.; Department of Civil Engineering, Duke University, Durham, North Carolina

> Discusses uncertainties in foundation analysis and their impact on design.

Chapter 3 – Soil Mechanics

Holtz, Robert D. and Kovaks, William D. (1981), *An Introduction to Geotechnical Engineering*, Prentice Hall, Englewood Cliffs, NJ

> Includes much more detailed discussions of soil mechanics.

Chapter 4 – Site Characterization

Acker, W.L. (1974), *Basic Procedures for Soil Sampling and Core Drilling*, Acker Drill Co., Scranton, PA

> Discusses the "nuts and bolts" of drilling exploratory borings and recovering soil and rock samples.

ASCE (1972), "Subsurface Investigation for Design and Construction of Foundations of Buildings", *Journal of the Soil Mechanics and Foundations Division*, Vol. 98 No. SM5, SM6, SM7, and SM8

> Practical recommendations from a task committee.

Briaud, Jean-Louis and Miran, Jerome (1991), *The Cone Penetration Test*, Report No. FHWA-TA-91-004, Federal Highway Administration, McLean, VA

> Practical guidelines for the use and interpretation of CPT results.

Dunnicliff, John (1088), *Geotechnical Instrumentation for Monitoring Field Performance*, John Wiley & Sons, New York

> Discusses use of instruments to monitor pore water pressures, soil movements, and other useful data.

Hunt, Roy E. (1984), *Geotechnical Engineering Investigation Manual*, McGraw Hill, New York

   A very thorough discussion of various investigation tools and methods.

Hvorslev, M. Juul (1949), *Subsurface Exploration and Sampling of Soils for Civil Engineering Purposes*, ASCE

   The classic work on subsurface exploration.

Kulhawy, F.H. and Mayne, P.W. (1990), *Manual on Estimating Soil Properties for Foundation Design*, Report No. EL-6800, Electric Power Research Institute, Palo Alto, CA

   Discusses the interpretation of laboratory and in-situ tests to develop soil parameters for analysis and design.

Peck, Ralph B. (1975), *The Selection of Soil Parameters for the Design of Foundations*, Second Nabor Carrillo Lecture, Mexican Society for Soil Mechanics, Mexico City (reprinted in *Judgment in Geotechnical Engineering*, Dunnicliff, John and Deere, Don U., eds., John Wiley, 1984)

   The "big picture" of obtaining design parameters from field exploration; emphasizes the importance of geology, the performance of existing structures, and other factors.

Robertson, P.K. and Campanella, R.G. (1989), *Guidelines for Geotechnical Design Using the Cone Penetrometer Test and CPT with Pore Pressure Measurement*, 4th Ed., Hogentogler and Co., Columbia, MD

   Practical application of the CPT to engineering analysis and design.

Wray, Warren K. (1986), *Measuring Engineering Properties of Soil*, Prentice Hall, Englewood Cliffs, NJ

   Step-by-step procedures and techniques for conducting laboratory tests.

Chapter 5 – Shallow Foundations

Peck, Ralph B. (1948), *History of Building Foundations in Chicago*, University of Illinois Engineering Experiment Station, Bulletin Series No. 373, Urbana, IL

   Interesting technical and historical perspectives on foundation design, construction and performance during the late 1800s and early 1900s.

Chapter 6 – Spread Footings – Bearing Capacity

Meyerhof, G. G. (1982), *The Bearing Capacity and Settlement of Foundations*, Tech-Press, Technical University of Nova Scotia, Halifax

Vesić, Aleksandar S. (1973), "Analysis of Ultimate Loads of Shallow Foundations", *Journal of the Soil Mechanics and Foundations Division*, Vol. 99, No. SM1, ASCE

Chapter 7 – Spread Footings – Settlement

Simons, N. E. (1987), "Settlement", Chapter 14 in *Ground Engineer's Reference Book*, F. G. Bell, ed., Butterworths, London

Chapter 9 – Mat Foundations

ACI (1988), "Suggested Analysis and Design Procedures for Combined Footings and Mats", *ACI Structural Journal*, Vol. 85, No. 3, p. 304 - 324; Also in *ACI Manual of Concrete Practice*

> Practical recommendations for analysis and design of mat foundations. Developed by a committee of geotechnical and structural engineers.

Chapter 10 – Shallow Foundations – Structural Design

ACI (1989), *Building Code Requirements for Reinforced Concrete (ACI 318-89) and Commentary-ACI 318R-89*, (SI versions are identified by 318M-89 and 318RM-89), American Concrete Institute, Detroit, 1989

> The definitive reference for the ACI code. Includes references to research that supports the code requirements.

Dowrick, David J. (1987), *Earthquake Resistant Design: A Manual for Engineers and Architects*, 2nd Ed., John Wiley, New York

> Discusses structural design principles for seismic areas.

MacGregor, James G. (1992), *Reinforced Concrete: Mechanics and Design*, 2nd Ed., Prentice Hall, Englewood Cliffs, NJ

> A thorough coverage of ultimate strength design of reinforced concrete based on the ACI code.

Chapter 12 – Pile Foundations – Methods and Applications

ASCE (1984), *Practical Guidelines for the Selection, Design and Installation of Piles*

> Discussion of many practical issues other than methods of predicting load capacities.

Fleming, W.G.K.; Weltman, A.J.; Randolph, M.F. and Elson, W.K. (1985), *Piling Engineering*, John Wiley & Sons, New York

> Discusses practical considerations for pile design and construction.

Tomlinson, M. J. (1987), *Pile Design and Construction Practice*, 3rd Ed., Viewpoint, London

> Describes many aspects of the analysis, design, and construction of pile foundations with emphasis on British practice.

Chapter 13 – Pile Foundations – Axial Capacity

Crowther, Carroll L. (1988), *Load Testing of Deep Foundations*, John Wiley, New York

    Detailed discussions of the planning, execution, and interpretation of pile load tests.

Fellenius, Bengt H. (1990), *Guidelines for the Interpretation and Analysis of the Static Loading Test*, Deep Foundations Institute, Sparta, NJ

    Interpreting data from full-scale load tests.

Hannigan, Patrick J. (1990), *Dynamic Monitoring and Analysis of Pile Foundation Installations*, Deep Foundations Institute, Sparta, NJ

    Practical application of wave equation, Case method and CAPWAP analyses.

Poulos, H. G. and Davis, E. H. (1980), *Pile Foundation Analysis and Design*, John Wiley, New York

    Thorough discussions of the analysis and design of pile foundations.

Rausche, Frank; Goble, George G. and Likins, Garland E. (1985), "Dynamic Determination of Pile Capacity", *Journal of Geotechnical Engineering*, Vol. 111, No. 3, p. 367 - 383

    Describes the Case method and the use of the pile driving analyzer to predict static load capacities.

Vanikar, Suneel N. (1986), *Manual on Design and Construction of Driven Pile Foundations*, Revision 1, Report No. FHWA-DP-66-1, U.S. Department of Transportation, Federal Highway Administration, Washington, DC

    Analysis and design procedures, with emphasis on highway structures. Example problems.

Chapter 14 – Pile Foundations – Other Geotechnical Considerations

O'Neill, Michael W. (1983), "Group Action in Offshore Piles", in *Geotechnical Practice in Offshore Engineering*, Stephen G. Wright, ed., p. 25 - 64, ASCE

    A state-of-the-art paper on group effects for both onshore and offshore piles.

Chapter 15 – Drilled Shaft Foundations

ACI (1985), *Suggested Design and Construction Procedures for Pier Foundations*, ACI 336.3R-72, American Concrete Institute, Detroit (included in the *ACI Manual of Concrete Practice*)

ADSC (1989), *Drilled Shaft Inspector's Manual*, Association of Drilled Shaft Contractors, Dallas, TX

    Discusses construction methods and quality control techniques.

Greer, David M. and Gardner, William S. (1986), *Construction of Drilled Pier Foundations*, John Wiley, New York

> A very thorough coverage of construction equipment and methods along with information on specifications, inspections and contracts.

Kulhawy, F.H.; Trautmann, C.H.; Beech, J.F.; O'Rourke, T.D.; McGuire, W.; Wood, W.A.; and Capano, C. (1983), *Transmission Line Structure Foundations for Uplift-Compression Loading*, Report No. EL-2870, Electric Power Research Institute, Palo Alto, CA

> Methods of determining downward and uplift capacities of drilled shafts and other foundations with emphasis on foundations for electrical power transmission towers.

Reese, Lymon C. and O'Neill, Michael W. (1988), *Drilled Shafts: Construction Procedures and Design Methods*, Federal Highway Administration Publication No. FHWA-HI-88-042

> A comprehensive treatise on both the design and construction aspects of drilled shafts.

Reese, L. C. and O'Neill, M. W. (1989), "New Design Method for Drilled Shafts from Common Soil and Rock Tests", *Foundation Engineering: Current Principles and Practices*, F.H. Kulhawy, ed., ASCE

> Condensed discussion of the geotechnical design aspects of Reese and O'Neill (1988) with some revisions.

Stas, C.V. and Kulhawy, F.H. (1984), *Critical Evaluation of Design Methods for Foundations Under Axial Uplift and Compressive Loading*, Report No. EL-3771, Electric Power Research Institute, Palo Alto, CA

> A condensation and commentary on Kulhawy, et. al. (1983).

Chapter 16 – Hybrid Deep Foundations

Clemence, Samuel P., ed. (1985), *Uplift Behavior of Anchor Foundations in Soil*, ASCE

> Proceedings of a conference session on the behavior of anchors.

DFI (1990), *Auger Cast-in-Place Piles Manual*, Deep Foundations Institute, Sparta, NJ

> Construction methods and model specifications.

Franki Foundation Company *Technical Data Supplement*, Boston, MA (also included in Sweet's Catalog)

> Information on design and construction of PIFs.

Xanthakos, Petros P. (1991), *Ground Anchors and Anchored Structures*, John Wiley, New York

> Detailed discussion of grouted anchors.

Chapter 17 – Laterally Loaded Deep Foundations

Poulos, H. G. and Davis, E. H. (1980), *Pile Foundation Analysis and Design*, John Wiley, New York

Chapters 7 and 8 provide a thorough discussions of lateral load analyses.

Prakash, Shamsher, and Sharma, Hari D. (1990), *Pile Foundations in Engineering Practice*, John Wiley, New York

Discusses additional analytical techniques for evaluating lateral loads.

Reese, Lymon C. (1984), *Handbook on Design of Piles and Drilled Shafts Under Lateral Load*, Federal Highway Administration Report No. FHWA-IP-84-11

Discusses Broms' method and p - y analyses.

Reese, L.C. (1986), *Behavior of Piles and Pile Groups Under Lateral Load*, Federal Highway Administration Report No. FHWA/RD-85/106

Discusses the p-y method, the finite difference model, and methods for choosing design p-y curves.

Chapter 18 – Deep Foundations – Structural Design

ACI (1980), *Recommendations for Design, Manufacture and Installation of Concrete Piles*, ACI 543R-74, American Concrete Institute, Detroit

Code requirements and recommended practices.

ACI (1985), *Suggested Design and Construction Procedures for Pier Foundations*, ACI 336.3R-72, American Concrete Institute, Detroit (included in the *ACI Manual of Concrete Practice*)

Recommendations for drilled shaft design.

Davisson, M.T.; Manuel, F.S. and Armstrong, R.M. (1983), *Allowable Stresses in Piles*. Report No. FHWA/RD-83/059, Federal Highway Administration, Washington

Results of a study of allowable stresses. Includes design recommendations.

PCI (1993), "Recommended Practice for Design, Manufacture and Installation of Prestressed Concrete Piling", *PCI Journal*, Vol. 38, No. 2, p. 14 - 41

Practical recommendations from a committee of engineers.

Reese, Lymon C. and O'Neill, Michael W. (1988), *Drilled Shafts: Construction Procedures and Design Methods*, Publication No. FHWA-HI-88-042, Federal Highway Administration, McLean, VA

Chapter 19 – Foundations on Weak and Compressible Soils

Koerner, Robert M. (1990), *Designing With Geosynthetics*, 2nd Ed., Prentice Hall, Englewood Cliffs, NJ

> Thorough discussion of the use of geosynthetics in engineering design, including their application to foundation engineering.

Hausmann, Manfred R. (1990), *Engineering Principles of Ground Modification*, McGraw-Hill, New York

> Discusses different methods of soil improvement, many of which are applicable to foundation engineering.

Zeevaert, Leonardo (1983), *Foundation Engineering for Difficult Subsoil Conditions*, 2nd Ed., Van Nostrand Reinhold, New York

> Methods of foundation construction on difficult soils, with emphasis on techniques used in Mexico City.

Chapter 20 – Foundations on Expansive Soils

Chen, Fu Hua (1988), *Foundations on Expansive Soils*, 2nd Ed., Developments in Geotechnical Engineering, Vol. 54, Elsevier, Amsterdam

> A complete treatise on the subject from the perspective of a practicing engineer from Denver.

Greenfield, Steven J. and Shen, C.K. (1992), *Foundations in Problem Soils*, Prentice Hall, Englewood Cliffs, NJ

> Discusses state-of-practice and practical guidelines for construction on difficult soils, including expansive soils.

Gromko, Gerald J. (1974), "Review of Expansive Soils", *Journal of the Geotechnical Engineering Division*, Vol. 100, No. GT6, ASCE

> An excellent review of the identification, causes, and solutions to expansive soil problems.

Mitchell, James K. (1993), *Soil Behavior*, 2nd Ed. John Wiley, New York

> Includes discussions of clay mineralogy as it applies to geotechnical engineering.

Nelson, John D. and Miller, Debora J. (1992), *Expansive Soils: Problems and Practice in Foundation and Pavement Engineering*, John Wiley, New York

> Comprehensive discussion of expansive soil behavior and preventive design measures.

O'Neill, Michael W. and Poormoayed, Nader (1980), "Methodology for Foundations on Expansive Clays", *Journal of the Geotechnical Engineering Division*, Vol. 106, No. GT12, ASCE

> Discusses methods of evaluating expansive soils and the development of rational and semirational design procedures.

In addition, there have been seven international conferences devoted entirely to the subject of expansive soils:

> 1st International Research and Engineering Conference on Expansive Clay Soils, College Station, TX, 1965, proceedings published by Texas A&M Press

> 2nd International Research and Engineering Conference on Expansive Clay Soils, College Station, TX, 1969, proceedings published by Texas A&M Press

> 3rd International Conference on Expansive Soils, Haifa, Israel, 1973, proceedings published by Jerusalem Academic Press

> 4th International Conference on Expansive Soils, Denver, CO, 1980, proceedings published by ASCE

> 5th International Conference on Expansive Soils, Barton, Australia, 1984, proceedings published by the Institution of Engineers, Australia

> 6th International Conference on Expansive Soils, New Delhi, India, 1987, sponsored by the International Society for Soil Mechanics and Foundation Engineering, proceedings published by Balkema, Rotterdam, in 1988

> 7th International Conference on Expansive Soils, Dallas, TX, 1992, proceedings published by Texas Tech University

Chapter 21 – Foundations on Collapsible Soils

Clemence, Samuel P. and Finbarr, Albert O. (1981), "Design Considerations for Collapsible Soils", *Journal of the Geotechnical Engineering Division*, Vol. 107, No. GT3, p. 305 - 317, ASCE

> Discusses the identification, evaluation, and treatment of collapsible soil problems.

Dudley, John H. (1970), "Review of Collapsing Soils", *Journal of the Soil Mechanics and Foundations Division*, Vol. 96, No. SM3, p. 925 - 947, ASCE

> Discusses occurrence and consequences of collapsing soils along with methods of evaluating them. Extensive bibliography.

Houston, William N. and Houston, Sandra L. (1989), "State-of-the-Practice Mitigation Measures for Collapsible Soil Sites", *Foundation Engineering: Current Principles and Practices*, Vol. 1, p. 161 - 175, F.H. Kulhawy, ed., ASCE

> Discusses a wide range of mitigation measures and includes case histories. Extensive bibliography.

Chapter 22 – Earth Retaining Structures

Lambe, Philip C. and Hansen, Lawrence A. (1990), *Design and Performance of Earth Retaining Structures*, Geotechnical Special Publication No. 25, ASCE

> Proceedings of a conference on the design and construction of retaining structures.

O'Rourke, T.D. and Jones, C.J.F.P. (1990), "Overview of Earth Retention Systems: 1970-1990", *Design and Performance of Earth Retaining Structures*, Geotechnical Special Publication No. 25, p. 22 - 51, P.C. Lambe and L.A. Hansen, ed., ASCE

Chapter 23 – Lateral Earth Pressures

Huntington, Whitney (1957), *Earth Pressures and Retaining Walls*, John Wiley, New York

> A very thorough treatise on the various earth pressure theories and their application.

Schnabel, Harry (1982), *Tiebacks in Foundation Engineering and Construction*, McGraw-Hill, New York

> Includes information on estimating lateral earth pressures on flexible retaining structures.

U.S. Navy (1982b), *Foundations and Earth Structures*, NAVFAC Design Manual DM-7.2

> Includes several design charts for computing earth pressures for a variety of conditions, including various surcharge loads.

Chapter 24 – Cantilever Retaining Walls

Casagrande, Leo (1973), "Comments on the Conventional Design of Retaining Structures", ASCE *Journal of the Soil Mechanics and Foundations Division*, Vol. 99, No. SM2, P. 181 - 198

> Discusses the theory and practice of retaining structure design.

Duncan, J. Michael; Clough, G. Wayne; Ebeling, Robert M. (1990), "Behavior and Design of Gravity Earth Retaining Structures", *Design and Performance of Earth Retaining Structures*, Philip C. Lambe and Lawrence A. Hansen, ed., p. 251 - 277, ASCE

> Discusses performance history of retaining walls and gives guidelines for practical design.

Schneider, Robert R. and Dickey, Walter L. (1987), *Reinforced Masonry Design*, 2nd Ed., Prentice Hall, Englewood Cliffs, NJ

> Describes the structural design of masonry retaining walls.

Appendix B

Unit Conversion Factors

ENGLISH UNITS

Engineers in the United States usually use English units. The following units are commonly used in foundation engineering:

TABLE B1 COMMON ENGLISH UNITS

Unit	Measurement	Symbol
foot	distance	ft
inch	distance	in
pound	force or mass	lb
kip (kilopound)	force	k
ton	force or mass	t
second	time	s
pound per square foot	stress or pressure	lb/ft^2 or psf
pound per square inch	stress or pressure	lb/in^2 or psi
pound per cubic foot	unit weight	lb/ft^3 or pcf

SI AND METRIC UNITS

Engineers in most countries other than the United States use primarily metric or SI units. SI (Système International) is a subset of the metric systems that uses standardized units. The following units are commonly used in foundation engineering:

TABLE B2 COMMON SI UNITS

Unit	Measurement	Symbol
meter	distance	m
gram	mass	g
Newton	force	N
Pascal	stress or pressure	Pa
kilonewton per cubic meter	unit weight	kN/m^3
second	time	s

These units are often accompanied by the following prefixes:

TABLE B3 COMMON SI PREFIXES

Prefix	Symbol	Multiplier
milli	m	10^{-3}
centi	c	10^{-2}
kilo	k	10^3
mega	M	10^6

CONVERSION FACTORS

The conversion factors in Tables B4 - B7 are useful for converting measurements between English, metric and SI units. Most of these factors are rounded to four significant figures. Those in **bold type** are absolute conversion factors (for example 12 inches = 1 ft). When units of force are equated to units of mass, the acceleration (F = ma) is presumed to be 9.807 m/s^2 (32.17 ft/s^2), which is the acceleration due to gravity on the earth's surface.

There are at least three definitions for the word "ton": the 2000 lb short ton (commonly used in the United States and Canada), the 2240 lb long ton (used in Great Britain), and the 1000 kg (2205 lb) metric ton (also known as a tonne).

A useful approximate conversion factor: 1 short ton/ft^2 ≈ 1 kg/cm^2 ≈ 100 kPa ≈ 1 atmosphere. These are true to within 2 to 4%.

TABLE B4 UNITS OF DISTANCE

To Convert	To	Multiply by
ft	in	**12**
ft	m	0.3048
in	ft	**1/12**
in	mm	25.40
m	ft	3.281
mm	in	0.03937

TABLE B5 UNITS OF FORCE

To Convert	To	Multiply by
k	kN	4.448
k	lb	**1000**
kg	lb	2.205
kg	N	9.807
kg	ton (metric)	**1/1000**
kN	k	0.2248
lb	k	**0.001**
lb	kg	0.4536
lb	N	4.448
lb	ton (short)	**1/2000**
lb	ton (long)	**1/2240**
N	kg	0.1020
N	lb	0.2248
ton (short)	lb	**2000**
ton (long)	lb	**2240**
ton (metric)	kg	**1000**

TABLE B6 UNITS OF STRESS AND PRESSURE

To Convert	To	Multiply by
atmosphere	lb/ft^2	2117
atmosphere	kPa	101.3
bar	kPa	**0.01**
kg/cm^2	kPa	98.07
kPa	atmosphere	0.009869
kPa	bar	**100**
kPa	kg/cm^2	0.01020
kPa	lb/ft^2	20.89
kPa	lb/in^2	0.1451
kPa	metric ton/m^2	0.1020
lb/ft^2	atmosphere	4.725×10^{-4}
lb/ft^2	kPa	0.04787
lb/ft^2	lb/in^2	**1/144**
lb/in^2	kPa	145,000
lb/in^2	lb/ft^2	**144**
lb/in^2	MPa	145.0
metric ton/m^2	kPa	9.807
MPa	lb/in^2	0.006897

TABLE B7 UNITS OF UNIT WEIGHT

To Convert	To	Multiply by
kN/m^3	lb/ft^3	6.366
kN/m^3	metric ton/m^3	0.1020
kN/m^3	Mg/m^3	0.1020
lb/ft^3	kN/m^3	0.1571
metric ton/m^3	kN/m^3	9.807
Mg/m^3	kN/m^3	9.807

Appendix C

Computer Software

This book includes a 3 1/2 inch diskette with the following foundation analysis software:

FTGBC — Analyzes the bearing capacity of spread footings using Terzaghi's, Meyerhof's and Brinch Hansen's methods, as described in Chapter 6.

FTGSETT — Analyzes the settlement of spread footings on cohesive soils, as described in Chapter 7.

SCHMERT — Analyzes the settlement of spread footings on cohesionless soils using Schmertmann's method, as described in Chapter 7.

SHAFT — Analyzes the axial load capacity of drilled shafts, as described in Chapter 15.

RETWALL — Analyzes the stability of cantilever retaining walls, as described in Chapter 24.

Specific instructions for each program are included in the corresponding chapters. These programs are intended to be used as instructional tools in conjunction with this textbook. The purchaser of this book receives an individual license to use these programs.

Equipment Requirements

These programs are designed to be used on IBM or compatible personal computers that meet the following requirements:

- Operating system: MS-DOS® or PC-DOS™ version 2.10 or higher
- Disk drive: One 3 1/2 inch 720 kb or 1.4 Mb diskette drive
- Memory: Minimum 640 kb RAM
- Math coprocessor: Optional - will make programs run faster
- Printer: Optional - to directly obtain hard copy output (see discussion below). It must be connected to parallel port LPT1.

These programs may not be compatible with some memory-resident programs, such as network drivers and TSR (terminate and stay resident) programs. Try disabling them if you encounter problems (especially out-of-memory errors).

Printer

All of these programs are able to produce hard copy output on a printer. Any standard printer, including laser printers, should be compatible. However, PostScript printers must be operated in a mode that emulates a standard printer.

It also is possible to print the output to a disk file. When selecting this option, the program will ask for a file name. You must select a name that has not already been used. The program will not write over an existing file with the same name. Any standard DOS filename is acceptable; if no extension is given, .PRN will be used. The file is an ASCII file that can be printed later, or imported into a word processor.

Copying

The purchaser of this book is authorized to make a backup copy of this diskette or transfer it to a hard disk or another size diskette for personal use. However, copies may not be given or sold to others.

Limited Warranty

Although these programs have been carefully written and tested, neither the author nor Prentice Hall can state unequivocally that there are no mistakes. The programs are intended to be used as educational tools, and the author encourages the use of hand computations to confirm the computer analyses. Any use of these programs for actual design is at the user's risk. Any erroneous output encountered should be brought to the author's attention.

Technical Support

Unfortunately, neither the author nor Prentice Hall are able to provide technical support to users of these programs.

Appendix D

Glossary of Terms

absolute settlement — Same as *total settlement*.

active zone — The layer of expansive soil subjected to shrinking and swelling. This zone extends from the ground surface to the maximum depth of significant moisture fluctuation.

aeolian soil — A soil that was deposited by wind.

allowable bearing pressure — The maximum bearing pressure that satisfies both bearing capacity and settlement criteria.

allowable bearing value — The ultimate bearing value divided by an appropriate factor of safety. A footing designed using this bearing pressure will satisfy bearing capacity criteria, but may have an excessive settlement.

alluvial soil — A soil deposited by water.

angle of internal friction — See *friction angle*

Atterberg Limits — The results of three tests, the liquid limit test, the plastic limit test, and the shrinkage limit test (only the first two are commonly performed). These tests are a measure of the plasticity of a soil.

auger — A helically shaped drill used to bore holes into the ground. These holes may be made for the purpose of obtaining soil samples or it may be filled with concrete and form a *drilled shaft foundation*.

auger-cast pile foundation — A deep foundation made by drilling a hole with a hollow stem auger, and then pumping grout through the auger while it is being retracted.

backfill — Soil that has been placed back into a location where it had been excavated, such as behind a retaining wall.

batter pile — A pile driven at an angle from the vertical; normally used to resist combined vertical and lateral loads. Also known as a *raker pile* or a *spur pile*.

bearing capacity failure — A foundation failure that occurs when the shear stresses in the adjacent soil exceed the shear strength.

755

bearing wall — A wall that carries a structural load, such as that from upper floors or a roof.

bell — The enlarged portion at the bottom of a drilled shaft foundation. (Not all drilled shafts have bells). Also known as an *underream*.

bentonite slurry — A mixture of a highly plastic clay, known as bentonite, with water to form a material with a consistency similar to that of pea soup. Sometimes used during drilling of borings as a means of preventing the adjacent soil from caving into the boring.

blow count — (a) The number of hammer blows to drive a pile for a specified distance, such as 1 foot; (b) The number of hammer blows required to drive a soil sampler, especially an SPT sampler, for a specified distance.

bored pile foundation — See *drilled shaft foundation*.

boring — See *exploratory boring*.

brooming — The separation of wood fibers at the butt of a timber pile while it is being driven.

butt — The top of a pile.

caisson foundation — See *drilled shaft foundation*.

casing — A steel pipe that is temporarily inserted into a boring or drilled shaft in order to prevent the adjacent soil from caving.

cast-in-place pile foundation — See *drilled shaft foundation*.

cohesion — The parameter c in the Mohr-Coulomb strength formula. The component of shear strength in a soil that is independent of the normal stress.

cohesionless soil — A soil that has very little or no bonding between the individual particles and therefore obtains its shear strength only from interlocking and sliding friction between the particles. Sands and gravels are cohesionless soils. See *cohesive soil*.

cohesive soil — A soil in which the individual particles are bonded together. Such soils have strength even when unconfined. Clays are cohesive soils. See *cohesionless soil*.

colluvial soil — Soils transported downhill by gravity, such as by landslides or creep.

combined footing — A spread footing that supports more than one column.

compressibility — A measure of the normal stress-normal strain behavior of a soil. High compressibility means the soil has a large strain when subjected to a given stress. See Chapter 3 for mathematical definitions.

cone penetration test — An in-situ test for determining soil properties. It consists of slowly pressing a steel cone into the soil and measuring the penetration resistance.

dead load (DL) — Structural loads due to the weight of beams, columns, floors, roofs and other fixed members. Does not include nonstructural items such as furniture, snow, occupants, or inventory. See *live load*.

deep foundation — A foundation that partially or fully bypasses the upper soils and transmits most or all of the structural load to deeper soils. See *shallow foundation*.

desiccation — The drying of a soil. When soils are dried, either naturally or artificially, their engineering properties change. As the soil dries, the pore water is drawn to the particle contact points and a soil *suction* force develops. This causes the soil to become overconsolidated. Thus, simply rewetting the soil will not restore its original properties. Although this effect is most pronounced in fine-grained soils, it is also present in sandy soils.

dewatering — The process of removing water from a construction area, such as by pumping from sumps or from wells. Normally, a temporary measure.

differential settlement — The difference in settlement between two foundations or between two points on a single foundation. See *total settlement*.

downdrag - See *negative skin friction*.

drilled caisson foundation — see *drilled shaft foundation*.

drilled pier foundation — see *drilled shaft foundation*.

drilled shaft foundation - A deep foundation that is constructed by drilling a cylindrical shaft, with or without a bell, inserting a reinforcing steel cage (optional), and filling the shaft with concrete. Also known as *pier, drilled pier, bored pile, cast-in-place pile, caisson, and drilled caisson*. This category of deep foundation does not include certain other methods that involve cast-in-place concrete, such as *Franki piles, auger-cast piles, step-taper piles,* and *tie-back anchors*.

Dutch Cone — See *cone penetration test*.

excess pore water pressure — The portion of the pore water pressure that is due to transient effects such as consolidation. See *pore water pressure* and *hydrostatic pore water pressure*.

expansive soil — A clayey soil that expands when wetted and contracts when dried.

exploratory boring — A vertical hole drilled into the ground for the purpose of exploring the soil conditions.

footer — See *spread footing*.

footing — Same as *spread footing*. (Note: Although this book uses this term only as a synonym for *spread footing*, some engineers use it as a synonym for the word *foundation* and others use it to encompass both *spread footings* and *pile caps*.)

Franki pile foundation — A deep foundation made by ramming concrete into the ground. See Chapter 16.

freeze — An increase in the load capacity of a pile after it has been driven. Caused primarily by the dissipation of excess pore water pressures. Also known as *setup*.

friction angle — The parameter ϕ in the Mohr-Coulomb strength formula. Describes the frictional component of the shear strength of a soil.

frost heave — The heaving of the ground due to the formation of ice lenses.

frost susceptible soil — A soil that is capable of drawing in large quantities of water, usually by capillary action, and forming ice lenses. These lenses are the primary cause of *frost heave*.

hydraulic fill — A fill placed by transporting soils through a pipe using large quantities of water. These fills are generally loose because they had little or no mechanical compaction during construction.

hydrostatic pore water pressure — The portion of the pore water pressure that is due to the force of gravity acting on the water above. The pore water pressure that would be present when the soil has reached equilibrium. See *pore water pressure* and *excess pore water pressure*.

in situ — In its original place. For example, an in-situ soil test is one that is conducted in the ground as opposed to a test performed on a soil sample that has been brought out of the ground.

isotropic — An adjective used to describe materials with physical properties that are the same regardless of the direction of measurement. For example, the strength at a point in an isotropic soil is the same in the vertical, horizontal, or any other direction.

jetting — The use of a water jet to facilitate the installation of a pile.

kip — Kilopound; a unit of force equal to 1000 lb.

leads — The portion of a pile driving rig that guides the pile hammer.

liquefaction — A rapid loss in shear strength in a soil during cyclic loading, such as an earthquake. Occurs most commonly in loose, saturated sands.

live load (LL) — Structural loads due to nonstructural members, such as furniture, occupants, inventory and snow. See *dead load*.

mandrel — A steel tube used to install thin shell piles, such as a Raymond Step-Taper Pile. The mandrel is inserted into the shell and used to transmit the driving stresses, and then it is removed and the shell is filled with concrete.

mat foundation — A type of shallow foundation that consists of a reinforced concrete pad that is large enough to encompass the entire structure. Same as *raft foundation*.

necking — A reduction in cross-sectional area of a drilled shaft as a result of the inward movement of the adjacent soils.

negative skin friction — The means by which a consolidating soil imposes a downward load on a pile or drilled shaft.

permafrost — Soil or rock that remains at a temperature below 0°C for at least two consecutive winters and the intervening summer.

pier foundation — See *drilled shaft foundation*.

pore water pressure — The gage pressure in the water that fills the voids in a soil or rock. It may include both a hydrostatic pressure (due to the weight of the overlying water) and an excess pore water pressure (due to squeezing or expanding of the voids). Also see *hydrostatic pore water pressure* and *excess pore water pressure*.

prestressed concrete — Reinforced concrete made in such a way that a tensile force is applied to the reinforcing steel before the concrete is placed. After the concrete has cured, this external force is released and a compressive stress develops in the concrete. Thus, the steel is in tension and the concrete is in compression even before any external loads are applied.

quake — The elastic compression of the soil during pile driving.

raft foundation — See *mat foundation*.

raker pile — See *batter pile*.

Raymond step-taper pile foundation — See *step-taper pile foundation*.

refusal — The condition reached when a pile or a soil sampler being driven by a hammer has negligible penetration from each blow of the hammer. Occurs when very hard soil or rock is encountered. Also applies to cone penetration tests when the cone cannot advance further.

setup — See *freeze*.

shallow foundation — A foundation that transmits the structural loads to the near-surface soils. See *deep foundation*.

slump — A measure of the consistency of fresh concrete. Measured by means of a slump test and expressed in units of length. A high slump concrete is very fluid, whereas a low slump concrete is more stiff.

splice — The joining of two lengths of piling. Splicing is necessary when very long piles must be installed.

spread footing — A type of shallow foundation that transmits the load from a single column or wall (or in some cases, a couple of columns) to the near-surface soils. See *combined footing*. Sometimes referred to as a *footer*.

spur pile — See *batter pile*.

step-taper pile foundation — A type of deep foundation that consists of a thin steel shell driven into the ground with a mandrel, and then filled with concrete.

strain — A measure of the deformation of materials. *Normal strain* is a measure of compressive or tensile deformations, and is defined as the change in length divided by the initial length. *Shear strain* is a measure of shear deformations.

substructure — Foundations, retaining walls and appurtenances. The portion of a structure that supports the superstructure. See *superstructure*.

superstructure — The portion of a structure located above the foundation. Includes beams, columns, floors and other structural and architectural members. See *substructure*.

tie-back anchor — A drilled and grouted tensile foundation, often used with retaining walls.

total settlement — The absolute vertical movement of a foundation. See *differential settlement*.

underpinning — Piles or other types of foundations built to provide new support for an existing foundation. Often used as a remedial measure.

underream — See *bell*.

vane shear test (VST) — An in-situ test to determine the undrained shear strength of a cohesive soil. The test consists of inserting a four-bladed vane and rotating it about a vertical axis.

References

AASHTO (1981), *Standard Specifications for Highway Bridges*, American Association of State Highway and Transportation Officials, Washington, D.C.

ACI (1980), *Recommendations for Design, Manufacture and Installation of Concrete Piles*, ACI 543R-74, American Concrete Institute, Detroit

ACI (1985), *Suggested Design and Construction Procedures for Pier Foundations*, ACI 336.3R-72, American Concrete Institute, Detroit (included in the *ACI Manual of Concrete Practice*)

ACI (1988), "Suggested Analysis and Design Procedures for Combined Footings and Mats", *ACI Structural Journal*, Vol. 85, No. 3, p. 304 - 324; Also in the *ACI Manual of Concrete Practice*

ACI (1989), *Building Code Requirements for Reinforced Concrete (ACI 318-89) and Commentary-ACI 318R-89*, American Concrete Institute, Detroit, 1989

ACI-ASCE (1962), "Report of Committee on Shear and Diagonal Tension", *Proceedings, American Concrete Institute*, Vol. 59, No. 1

ACKER, W.L. (1974), *Basic Procedures for Soil Sampling and Core Drilling*, Acker Drill Co., Scranton, PA

ADSC (1989), *Drilled Shaft Inspector's Manual*, Association of Drilled Shaft Contractors, Dallas, TX

AIRHART, T.P.; COYLE, H.M.; HIRSH, T.J. AND BUCHANAN, S.J. (1969), "Pile-Soil System Response in a Cohesive Soil", *Performance of Deep Foundations*, STP 444, p. 264 - 294, ASTM, Philadelphia

AISC (1989), *Manual of Steel Construction, Allowable Stress Design*, 9th Ed., American Institute of Steel Construction, Chicago

AL-KHAFAJI, A.W.N. AND ANDERSLAND, O.R. (1992), "Equations for Compression Index Approximation", ASCE *Journal of Geotechnical Engineering*, Vol. 118, No. 1, p. 148 - 153

ANDERSON, J.N. AND LADE, P.V. (1981), "The Expansion Index Test", *ASTM Geotechnical Testing Journal*, Vol. 4, No. 2, June 1981, p. 58 - 67

API (1981), *Recommended Practice for Planning, Designing and Constructing Fixed Offshore Platforms*, American Petroleum Institute, Dallas

ARMSTRONG, R.M. (1978), "Structural Properties of Timber Piles", *Behavior of Deep Foundations*, ASTM STP 670, Raymond Lundgren, Ed., p. 118 - 152, ASTM, Philadelphia

ASCE (1972), "Subsurface Investigation for Design and Construction of Foundations of Buildings", ASCE *Journal of the Soil Mechanics and Foundations Division*, Vol. 98 No. SM5, SM6, SM7 and SM8

ASCE (1984), *Practical Guidelines for the Selection, Design and Installation of Piles*

ASTM (1991), *Annual Book of ASTM Standards*, Vol. 04.08 (Soil and Rock; Dimension Stone; Geosynthetics), ASTM, Philadelphia

760

AZZOUZ, AMR S.; BALIGH, MOSHEN M. AND WHITTLE, ANDREW J. (1990), "Shaft Resistance of Piles in Clay", ASCE *Journal of Geotechnical Engineering*, Vol. 116, No. 2, p. 205 - 221

AZZOUZ, A.S.; KRIZEK, R.J. AND COROTIS, R.B. (1976), "Regression Analysis of Soil Compressibility", *Soils and Foundations*, Vol. 16, No. 2, p. 19 - 29

BAGUELIN, F.; JÉZÉQUEL, J.F. AND SHIELDS, D.H. (1978), *The Pressuremeter and Foundation Engineering*, Trans Tech, Clausthal, Germany

BARDEN, L.; McGOWN, A. AND COLLINS, K. (1973), "The Collapse Mechanism in Partly Saturated Soil", *Engineering Geology*, Vol. 7, p. 49 - 60

BARTLETT, S.F. AND YOUD, T.L. (1992), *Empirical Analysis of Horizontal Ground Displacement Generated by Liquefaction-Induced Lateral Spreads*, Technical Report NCEER-92-0021, National Center for Earthquake Engineering Research, Buffalo, NY

BASMA, ADNAN A. AND TUNCER, ERDIL R. (1992), "Evaluation and Control of Collapsible Soils", ASCE *Journal of Geotechnical Engineering*, Vol. 118, No. 10, p. 1491 - 1504

BAUMANN, FREDERICK (1873), *The Art of Preparing Foundations, with Particular Illustration of the "Method of Isolated Piers" as Followed in Chicago*, Reprinted in Powell (1884)

BAZARAA, A. (1967), *Use of the Standard Penetration Test for Estimating Settlements of Shallow Foundations on Sand*, PhD Dissertation, Civil Engineering Dept., Univ. of Illinois

BELL, A.L. (1915), "The Lateral Pressure and Resistance of Clay, and the Supporting Power of Clay Foundations", p. 93 - 134 in *A Century of Soil Mechanics*, ICE, London

BETHLEHEM STEEL CORP. (1979), *Bethlehem Steel H-Piles*, Bethlehem Steel Corp., Bethlehem, PA

BHUSHAN, KUL (1982), Discussion of "New Design Correlations for Piles in Sand", ASCE *Journal of the Geotechnical Engineering Division*, Vol. 108, No. GT11, p. 1508 - 1510

BJERRUM, LAURITS (1963), "Allowable Settlement of Structures", *Proceedings, 3rd European Conference on Soil Mechanics and Foundation Engineering*, Vol. 2, p. 135 - 137, Weisbaden

BJERRUM, L.; JOHANNESSEN, I.J. AND EIDE, O. (1969), "Reduction of Negative Skin Friction on Steel Piles to Rock", *Proceedings, Seventh International Conference on Soil Mechanics and Foundation Engineering*, Mexico City, Vol. 2, p. 27 - 34

BLASER, HAROLD D. AND SCHERER, OSCAR J. (1969), "Expansion of Soils Containing Sodium Sulfate Caused by Drop in Ambient Temperatures", *Effects of Temperature and Heat on Engineering Behavior of Soils*, Special Report 103, Highway Research Board, Washington, DC

BOCA (1990), *The BOCA Basic/National Building Code*, Building Officials & Code Administrators, International, Country Club Hills, IL

BOGARD, DEWAINE AND MATLOCK, HUDSON (1983), "Procedures for Analysis of Laterally Loaded Pile Groups in Soft Clay", *Geotechnical Practice in Offshore Engineering*, Stephen G. Wright, ed., p. 499 - 535, ASCE

BOLENSKI, M. (1973), "Osiadania nowo wznoszonych budowli w zaleznosci od podloza gruntowego: Wyniki 20-letnich Badan w Instytucie Techniki Budowlanej" (Settlement of Constructions Newly Erected and Type of Subsoil: The Results of 20 Years Studies Carried Out in the Building Research Institute), Prace Instytutu Techniki Budowlanej, Warszawa (in Polish; partial results quoted in Burland and Burbidge, 1985)

BOLIN, H.W. (1941), "The Pile Efficiency Formula of the Uniform Building Code", *Building Standards Monthly*, Vol. 10, No. 1, p. 4 - 5

BOUSSINESQ, M.J. (1885), *Application Des Potentiels, à l'Étude de l'Équilibre et du Movvement Des Solides Elastiques*, Gauthier-Villars, Paris (in French)

BOZOZUK, M. (1972), "Downdrag Measurement on a 160-ft Floating Pipe Test Pile in Marine Clay", *Canadian Geotechnical Journal*, Vol. 9, No. 2, p. 127 - 136

BOZOZUK, M. (1978), "Bridge Foundations Move", *Transportation Research Record 678*, p. 17 - 21, Transportation Research Board, Washington, DC

BOZOZUK, M. (1981), "Bearing Capacity of Pile Preloaded by Downdrag", *Proceedings, Tenth International Conference on Soil Mechanics and Foundation Engineering*, Stockholm, Vol. 2, p. 631 - 636

BRANDON, THOMAS L.; DUNCAN, J. MICHAEL AND GARDNER, WILLIAM S. (1990), "Hydrocompression Settlement of Deep Fills", ASCE *Journal of Geotechnical Engineering*, Vol. 116, No. 10, p. 1536 - 1548

BRIAUD, J.L.; TUCKER, L.; LYTTON, R.L.; AND COYLE, H.M. (1985), *Behavior of Piles and Pile Groups in Cohesionless Soils*, Report No. FHWA/RD-83/038, Federal Highway Administration, Washington, DC

BRIAUD, JEAN-LOUIS AND LEPERT, PHILIPPE (1990), "WAK Test to Find Spread Footing Stiffness", ASCE *Journal of Geotechnical Engineering*, Vol. 116, No. 3, p. 415 - 431

BRIAUD, J.L.; TUCKER, L.M.; ANDERSON, J.S.; PERDOMO, D.; AND COYLE, H.M. (1986), *Development of an Improved Pile Design Procedure for Single Piles in Clays and Sands*, Report No. MSHD-RD-86-050-1, Texas A&M University, College Station, TX

BRIAUD, JEAN-LOUIS (1991), "Dynamic and Static Testing of Nine Drilled Shafts at Texas A&M University Geotechnical Research Sites", *Foundation Drilling*, Vol. 30, No. 7, p. 12 - 14

BRIAUD, JEAN-LOUIS AND MIRAN, JEROME (1991), *The Cone Penetration Test*, Report No. FHWA-TA-91-004, Federal Highway Administration, McLean, VA

BRIAUD, JEAN-LOUIS (1992), *The Pressuremeter*, A.A. Balkema, Rotterdam

BRINCH HANSEN, J. (1961a), "The Ultimate Resistance of Rigid Piles Against Transversal Forces", *Bulletin No. 12*, Danish Geotechnical Institute, Copenhagen

BRINCH HANSEN, J. (1961b), "A General Formula for Bearing Capacity", *Bulletin No. 11*, Danish Geotechnical Institute, Copenhagen

BRINCH HANSEN, J. (1970), "A Revised and Extended Formula for Bearing Capacity", *Bulletin No. 28*, Danish Geotechnical Institute, Copenhagen

BRINK, A.B.A. AND KANTEY, B.A. (1961), "Collapsible Grain Structure in Residual Granite Soils in South Africa", *Proceedings, Fifth International Conference on Soil Mechanics and Foundation Engineering*, p. 611 - 614

BROMS, BENGT B. (1964a), "Lateral Resistance of Piles in Cohesive Soils", ASCE *Journal of the Soil Mechanics and Foundations Division*, Vol. 90, No. SM2, p. 27 - 63

BROMS, BENGT B. (1964b), "Lateral Resistance of Piles in Cohesionless Soils", ASCE *Journal of the Soil Mechanics and Foundations Division*, Vol. 90, No. SM3, p. 123 - 156

BROMS, BENGT B. (1965), "Design of Laterally Loaded Piles", ASCE *Journal of the Soil Mechanics and Foundations Division*, Vol. 91, No. SM3, p. 79 - 99

BROMS, BENGT B. (1972), "Stability of Flexible Structures (Piles and Pile Groups)", *Proceedings, 5th European Conference on Soil Mechanics and Foundation Engineering*, Madrid, Vol. 2, p. 239 - 269

BROMS, BENGT B. (1981), *Precast Piling Practice*, Thomas Telford, London

BROMS, B.B. AND INGELSON, I. (1971), "Earth Pressure Against the Abutments of a Rigid Frame Bridge", *Géotechnique*, Vol. 21, No. 1, p. 15 - 28

BRONS, K.F. AND KOOL, A.F. (1988), "Methods to Improve the Quality of Auger Piles", *Deep Foundations on Bored and Auger Piles*, p. 269 - 272, W.F. Van Impe, Ed., Balkema, Rotterdam

BROWN, DAN A. AND SHIE, CHINE-FENG (1990), "Numerical Experiments into Group Effects on the Response of Piles to Lateral Loading", *Computers and Geotechnics*, Vol. 10, No. 3, p. 211 - 230

BROWN, RALPH E. AND GLENN, ANDREW J. (1976), "Vibroflotation and Terra-Probe Comparison", ASCE *Journal of the Geotechnical Engineering Division*, Vol. 102, No. GT10, p. 1059 - 1072

BRUNER, ROBERT F., COYLE, HARRY M. AND BARTOSKEWITZ, RICHARD E. (1983), *Cantilever Retaining Wall Design*, Research Report No. 236-2F, Texas Transportation Institute, Texas A&M University, College Station, TX

BUDGE, W.D., ET AL. (1964), *A Review of Literature on Swelling Soils*, Colorado Department of Highways, Denver, CO

BUHR, C.A.; HOUGHTON, L.E. AND LEONARD, R.J. (1982), "Design of Friction Piling in Mississippi Embayment Sands Using Cone Penetrometers", *Second European Symposium on Penetration Testing (ESOPT II)*, A. Verruijt, et al., Ed., Vol. 2, p. 487 - 492

BULLET, P. (1691), *L'Architecture Pratique*, Paris (in French)

BURLAND, JOHN (1973), "Shaft Friction of Piles in Clay - A Simple Fundamental Approach", *Ground Engineering*, Vol. 6, No. 3, p. 30 - 42

BURLAND, J.B. AND BURBIDGE, M.C. (1985), "Settlement of Foundations on Sand and Gravel", *Proceedings, Institution of Civil Engineers*, Part 1, Vol. 78, p. 1325 - 1381

BURLAND, J.B. AND WROTH, C.P. (1974), "Allowable and Differential Settlement of Structures, Including Damage and Soil-Structure Interaction", *Conference on Settlement of Structures*, p. 611 - 654, Pentech Press, Cambridge

BUSTAMANTE, M. AND GIANESELLI, L. (1982), "Pile Bearing Capacity Prediction by Means of Static Penetrometer CPT", *Second European Symposium on Penetration Testing (ESOPT II)*, A. Verruijt et al., Ed., Vol. 2, p. 493 - 500

BUTLER, F.G. (1975), "Heavily Over-Consolidated Clays", p. 531ff in *Settlement of Structures*, Halstead Press

BYRN, JAMES E. (1991), "Expansive Soils: The Effect of Changing Soil Moisture Content on Residential and Light Commercial Structures", *Journal of the National Academy of Forensic Engineers*, Vol. 8, No. 2, p. 67 - 84

CARTER, J.P. AND KULHAWY, F.H. (1988), *Analysis and Design of Drilled Shaft Foundations Socketed into Rock*, Report No. EL-5918, Electric Power Research Institute, Palo Alto, CA

CASAGRANDE, A. (1936), "The Determination of the Pre-Consolidation Load and Its Practical Significance", *Proceedings, First International Conference on Soil Mechanics and Foundation Engineering*, Vol. III, p. 60 - 64

CASAGRANDE, LEO (1973), "Comments on the Conventional Design of Retaining Structures", ASCE *Journal of the Soil Mechanics and Foundations Division*, Vol. 99, No. SM2, P. 181 - 198

CGS (1985), *Canadian Foundation Engineering Manual*, 2nd Ed., Canadian Geotechnical Society, BiTech, Vancouver

CHANDLER, RICHARD J. (1988), "The In-Situ Measurement of the Undrained Shear Strength of Clays Using the Field Vane", *Vane Shear Strength Testing in Soils: Field and Laboratory Studies*, ASTM STP 1014, p. 13 - 44, A.F. Richards, Ed., American Society for Testing and Materials, Philadelphia

CHANG, M.F. AND BROMS, B.B. (1991), "Design of Bored Piles in Residual Soils Based on Field-Performance Data", *Canadian Geotechnical Journal*, Vol. 28, p. 200 - 209

CHELLIS, ROBERT D. (1961), *Pile Foundations*, 2nd Ed., McGraw Hill, New York

CHELLIS, R.D. (1962), "Pile Foundations", p. 633 - 768 in *Foundation Engineering*, G.A. Leonards, editor, McGraw-Hill, New York

CHEN, FU HUA (1988), *Foundations on Expansive Soils*, 2nd Ed., Developments in Geotechnical Engineering Vol. 54, Elsevier, Amsterdam

CHRISTIAN, JOHN T. AND CARRIER, W. DAVID III (1978), "Janbu, Bjerrum and Kjaernsli's Chart Reinterpreted", *Canadian Geotechnical Journal*, Vol. 15, p. 123 - 128

CHRISTOPHER, BARRY R.; BAKER, CLYDE N. AND WELLINGTON, DENNIS L. (1989), "Geophysical and Nuclear Methods for Non-Destructive Evaluation of Caissons", *Foundation Engineering: Current Principles and Practices*, ASCE

CIBOR, JOSEPH M. (1983), "Geotechnical Considerations of Las Vegas Valley", *Special Publication on Geological Environment and Soil Properties*, ASCE Annual Convention, Houston, TX, p. 351 - 373

CLAYTON, C.R.I. (1990), "SPT Energy Transmission: Theory, Measurement and Significance", *Ground Engineering*, Vol. 23, No. 10, p. 35 - 43

CLEMENCE, SAMUEL P., ED. (1985), *Uplift Behavior of Anchor Foundations in Soil*, ASCE

CLEMENCE, SAMUEL P. AND FINBARR, ALBERT O. (1981), "Design Considerations for Collapsible Soils", ASCE *Journal of the Geotechnical Engineering Division*, Vol. 107, No. GT3, p. 305 - 317

COULOMB, C.A. (1776) "Essai sur une application des règles de maximis et minimis à quelques problèmes de statique relatifs à l'architecture", *Mémoires de mathématique et de physique présentés à l'Académie Royale des Sciences*, Paris, Vol. 7, p. 343 - 382 (in French)

COYLE, HARRY M. AND CASTELLO, RENO R. (1981), "New Design Correlations for Piles in Sand", ASCE *Journal of the Geotechnical Engineering Division*, Vol. 107, No. GT7, p. 965 - 986 (also see discussions, Vol. 108, No. GT11, p. 1508 - 1520)

CRAWFORD, CARL B. AND BURN, KENNETH N. (1968), "Building Damage from Expansive Steel Slag Backfill", *Placement and Improvement of Soil to Support Structures*, p. 235 - 261, ASCE

CRAWFORD, C.B. AND JOHNSON, G.H. (1971), "Construction on Permafrost", *Canadian Geotechnical Journal*, Vol. 8, No. 2, p. 236 - 251

CROCE, A., ET. AL. (1981), "The Tower of Pisa and the Surrounding Square: Recent Observations", *Proceedings, 10th International Conference on Soil Mechanics and Foundation Engineering*, Stockholm 1981, Vol. 3, p. 61 - 70, A. A. Balkema, Rotterdam

CRSI (1992), *CRSI Handbook*, Concrete Reinforcing Steel Institute, Schaumburg, IL

CROWTHER, CARROLL L. (1988), *Load Testing of Deep Foundations*, John Wiley and Sons, New York

CUMMINGS, A.E. (1940), "Dynamic Pile Driving Formulas", *Journal of the Boston Society of Civil Engineers*, Vol. 27, No. 1, p. 6 - 27

CURTIN, GEORGE (1973), "Collapsing Soil and Subsidence" in *Geology, Seismicity and Environmental Impact*, Special Publication, Association of Engineering Geologists, Douglas E. Moran et al., ed., University Publishers, Los Angeles

D'APPOLONIA, DAVID J., POULOS, HARRY G., AND LADD, CHARLES C. (1971), "Initial Settlement of Structures on Clay", ASCE *Journal of the Soil Mechanics and Foundations Division*, Vol. 97, No. SM10, p. 1359 - 1377

D'APPOLONIA, D.J., D'APPOLONIA, E.D. AND BRISSETTE, R.F. (1968), "Settlement of Spread Footings on Sand", ASCE *Journal of the Soil Mechanics and Foundations Division*, Vol. 94, No. SM3, p. 735359

DAVIS, E.H. AND POULOS, H.G. (1968), "The Use of Elastic Theory for Settlement Prediction Under Three-Dimensional Conditions", *Géotechnique*, Vol. 18, p. 67 - 91

DAVISSON, M.T. (1970), "Lateral Load Capacity of Piles", *Highway Research Record No. 333*, p. 104 - 112, Highway Research Board, Washington, DC

DAVISSON, M.T. (1973), "High Capacity Piles" in *Innovations in Foundation Construction*, proceedings of a lecture series, Illinois Section ASCE, Chicago

DAVISSON, M.T. (1979), "Stresses in Piles", *Behavior of Deep Foundations*, ASTM STP 670, Raymond Lundgren, Ed., p. 64 - 83, ASTM, Philadelphia

DAVISSON, M.T. (1989), "Driven Piles", *Foundations in Difficult Soils: State of Practice*, Foundations and Soil Mechanics Group, Metropolitan Section, ASCE, New York

DAVISSON, M.T.; MANUEL, F.S. AND ARMSTRONG, R.M. (1983), *Allowable Stresses in Piles*. Report No. FHWA/RD-83/059, Federal Highway Administration, Washington

DEBEER, E.E. AND LADANYI, B. (1961), "Experimental Study of the Bearing Capacity of Sand Under Circular Foundations Resting on the Surface", *Proceedings, 5th International Conference on Soil Mechanics and Foundation Engineering*, Vol. 1, p. 577 - 585, Paris

DeMELLO, V. (1971), "The Standard Penetration Test - A State-of-the-Art Report", *Fourth PanAmerican Conference on Soil Mechanics and Foundation Engineering*, Vol. 1, p. 1 - 86

DEMPSEY, J.P. AND LI, H. (1989), "A Rigid Rectangular Footing on an Elastic Layer", *Géotechnique*, Vol. 39, No. 1, p. 147 - 152

DENNIS, NORMAN D. AND OLSON, ROY E. (1983), "Axial Capacity of Steel Pipe Piles in Clay", ASCE *Geotechnical Practice in Offshore Engineering*, p. 370 - 388

DeRUITER, J. (1981), "Current Penetrometer Practice", *Cone Penetration Testing and Experience*, p. 1 - 48, ASCE

DeRUITER, J. AND BERINGEN, F.L. (1979), "Pile Foundations for Large North Sea Structures", *Marine Geotechnology*, Vol. 3, No. 3, p. 267 - 314

DFI (1990), *Auger Cast-in-Place Piles Manual*, Deep Foundations Institute, Sparta, NJ

DICKIN, E.A. AND LEUNG, C.F. (1990), "Performance of Piles with Enlarged Bases Subject to Uplift Forces", *Canadian Geotechnical Journal*, Vol. 27, p. 546 - 556

DiMILLIO, A.F.; NG, E.S.; BRIAUD, J.L.; O'NEILL, M.W., ET AL. (1987a), *Pile Group Prediction Symposium: Summary Volume I: Sandy Soil*, Report No. FHWA-TS-87-221, Federal Highway Administration, McLean, VA

DiMILLIO, A.F.; O'NEILL, M.W.; HAWKINS, R.A.; VGAZ, O.G. AND VESIC, A.S. (1987b), *Pile Group Prediction Symposium: Summary Volume II: Clay Soil*, Report No. FHWA-TS-87-222, Federal Highway Administration, McLean, VA

DISMUKE, THOMAS D.; COBURN, SEYMOUR K.; AND HIRSCH, CARL M. (1981), *Handbook of Corrosion Protection for Steel Pile Structures in Marine Environments*, American Iron and Steel Institute, Washington, DC

DOBRIN, MILTON B. (1988), *Introduction to Geophysical Prospecting*, 4th Ed., McGraw-Hill, New York

DONALDSON, G.W. (1973), "The Prediction of Differential Movement on Expansive Soils", *Proceedings of the Third International Conference on Expansive Soils*, Vol. 1, p. 289 - 293, Jerusalem Academic Press

DOWRICK, DAVID J. (1987), *Earthquake Resistant Design: A Manual for Engineers and Architects*, 2nd Ed., John Wiley, New York

DUDLEY, JOHN H. (1970), "Review of Collapsing Soils", ASCE *Journal of the Soil Mechanics and Foundations Division*, Vol. 96, No. SM3, p. 925 - 947

DUNCAN, J.M. AND BUCHIGNANI, A.L. (1976), *An Engineering Manual for Settlement Studies*, Department of Civil Engineering, University of California, Berkeley

DUNCAN, J. MICHAEL; CLOUGH, G. WAYNE; EBELING, ROBERT M. (1990), "Behavior and Design of Gravity Earth Retaining Structures", *Design and Performance of Earth Retaining Structures*, Philip C. Lambe and Lawrence A. Hansen, Ed., p. 251 - 277, ASCE

DUNCAN, J.M., WILLIAMS, G.W., SEHN, A.L., AND SEED, R.B. (1991), "Estimation Earth Pressures due to Compaction", ASCE *Journal of Geotechnical Engineering*, Vol. 117, No. 12, p. 1833 - 1847 (errata in Vol. 118, No. 3, p. 519)

EDWARDS, T.C. (1967), *Piling Analysis Wave Equation Computer Program Utilization Manual*, Texas Transportation Institute Research Report 33-11, Texas A&M University, College Station, TX

ENGEL, RICHARD L. (1988), "Discussion of Procedures for the Determination of Pile Capacity", *Transportation Research Record 1169*, p. 54 - 61, Transportation Research Board, Washington

EVANS, L.T. AND DUNCAN, J.M. (1982), *Simplified Analysis of Laterally Loaded Piles*, Report No. UCB/GT/82-04, Department of Civil Engineering, University of California, Berkeley

FARLEY, J. (1827), *A Treatise on the Steam Engine*, London (quoted in Golder, 1975)

FELLENIUS, BENGT H. (1980), "The Analysis of Results from Routine Pile Load Tests", *Ground Engineering*, Vol. 13, No. 6, p. 19 - 31

FELLENIUS, BENGT H. (1990), *Guidelines for the Interpretation and Analysis of the Static Loading Test*, Deep Foundations Institute, Sparta, NJ

FELLENIUS, BENGT H.; RIKER, RICHARD E.; O'BRIEN, ARTHUR J. AND TRACY, GERALD R. (1989), "Dynamic and Static Testing in Soil Exhibiting Set-Up", ASCE *Journal of Geotechnical Engineering*, Vol. 115, No. 7, p. 984 - 1001

FINNO, RICHARD J., ED. (1989), *Predicted and Observed Axial Behavior of Piles*, Geotechnical Special Publication No. 23, ASCE

FLEMING, W.G.K.; WELTMAN, A.J.; RANDOLPH, M.F. AND ELSON, W.K. (1985), *Piling Engineering*, John Wiley & Sons, New York

FOOTT, ROGER AND KOUTSOFTAS, DEMETRIOUS (1984), "Settlement of Natural Ground Under Static Loadings", p. 1 - 25 in *Ground Movements and Their Effects on Structures*, P.B. Attewell and R.K. Taylor, eds., Surrey University Press, Surrey, England

FOOTT, ROGER AND LADD, CHARLES C. (1981), "Undrained Settlement of Plastic and Organic Clays", ASCE *Journal of the Geotechnical Engineering Division*, Vol. 107, No. GT8, p. 1079 - 1094

GIBBS, HAROLD J. (1969), Discussion of Holtz, W. G. "The Engineering Problems of Expansive Clay Subsoils", *Proceedings of the Second International Research and Engineering Conference on Expansive Clay Soils*, p. 478 - 479, Texas A&M Press, College Station, TX

GIRAULT, PABLO D. (1987), "Analyses of Foundation Failures", *The Mexico Earthquakes-1985*, Michael A. Cassaro and Enrique Martinez Romero, Ed., p. 178 - 192, ASCE

GOBLE, G.G; LIKINS, GARLAND AND RAUSCHE, FRANK (1975), *Bearing Capacity of Piles from Dynamic Measurements*, FHWA Report No. OHIO-DOT-05-75, Federal Highway Administration

GOBLE, G.G. (1986), "Modern Procedures for the Design of Driven Pile Foundations", Short Course on Behavior of Deep Foundations Under Axial and Lateral Loading, University of Texas, Austin

GOLDER, H.Q. (1975), "Floating Foundations", Ch. 18 in *Foundation Engineering Handbook*, H.F. Winterkorn and H. Fang, Ed., Van Nostrand Reinhold, New York

GOULD, JAMES P. (1970), "Lateral Pressures on Rigid Permanent Structures" in *Lateral Stresses in the Ground and the Design of Earth-Retaining Structures*, p. 219 - 270, ASCE

GRAHAM, JAMES (1985), "Allowable Compressive Design Stresses for Pressure-Treated Round Timber Foundation Piling", American Wood Preservers Association Proceedings, Vol. 81

GRANT, R. ET AL. (1974), "Differential Settlement of Buildings", ASCE *Journal of the Geotechnical Engineering Division*, Vol. 100, No. GT9, p. 973 - 991

GREENFIELD, STEVEN J. AND SHEN, C.K. (1992), *Foundations in Problem Soils*, Prentice Hall, Englewood Cliffs, NJ

GREER, DAVID M. AND GARDNER, WILLIAM S. (1986), *Construction of Drilled Pier Foundations*, John Wiley, New York

GRIFFITHS, D. V. (1990), "Failure Criteria Interpretation Based on Mohr-Coulomb Friction", ASCE *Journal of Geotechnical Engineering*, Vol. 116, No. 6, p. 986 - 999

GROMKO, GERALD J. (1974), "Review of Expansive Soils", ASCE *Journal of the Geotechnical Engineering Division*, Vol. 100, No. GT6, p. 667 - 687

HADJIAN, ASADOUR H.; FALLGREN, RICHARD B. AND TUFENKJIAN, MARK R. (1992), "Dynamic Soil-Pile-Structure Interaction: The State of Practice", *Piles Under Dynamic Loads*, Shamsher Prakash, Ed., p. 1 - 26, ASCE

HAIN, STEPHEN J. AND LEE, IAN K. (1974), "Rational Analysis of Raft Foundation", ASCE *Journal of the Geotechnical Engineering Division*, Vol. 100, No. GT7, p. 843 - 860

HANDY, RICHARD L. (1980), "Realism in Site Exploration: Past, Present, Future, and Then Some — All Inclusive", *Site Exploration on Soft Ground Using In-Situ Techniques*, p. 239 - 248, Report No. FHWA-TS-80-202, Federal Highway Administration, Washington

HANNIGAN, PATRICK J. (1990), *Dynamic Monitoring and Analysis of Pile Foundation Installations*, Deep Foundations Institute, Sparta, NJ

HARDER, LESLIE F. AND SEED, H. BOLTON (1986), *Determination of Penetration Resistance for Coarse-Grained Soils Using the Becker Hammer Drill*, Report No. UCB/EERC-86/06, Earthquake Engineering Research Center, Richmond, CA

HART, STEPHEN S. (1974), *Potentially Swelling Soil and Rock in the Front Range Urban Corridor, Colorado*, Colorado Geological Survey publication EG-7, Denver

HAUSMANN, MANFRED R. (1990), *Engineering Principles of Ground Modification*, McGraw-Hill, New York

HEAD, K.H. (1982), *Manual of Soil Laboratory Testing*, Vol. 2, Pentech, London

HEARNE, THOMAS M., STOKOE, KENNETH H. AND REESE, LYMON C. (1981), "Drilled Shaft Integrity By Wave Propagation Method", ASCE *Journal of the Geotechnical Engineering Division*, Vol. 107, No. GT10, p. 1327 - 1344

HEIJNEN, W.J. (1974), "Penetration Testing in Netherlands", *European Symposium on Penetration Testing (ESOPT)*, Stockholm, Vol. 1, p. 79 - 83

HEIJNEN, W.J. AND JANSE, E. (1985), "Case Studies of the Second European Symposium on Penetration Testing (ESOPT II)", *LGM-Mededelingen*, Part XXII, No. 91, Delft Soil Mechanics Laboratory, Delft, The Netherlands

HEINZ, RONEY (1993), "Plastic Piling", *Civil Engineering*, Vol. 63, No. 4, p. 63 - 65, ASCE

HERTLEIN, BERNARD H. (1992), "Selecting an Effective Low-Strain Foundation Test", *Foundation Drilling*, Vol. 31, No. 2, p. 14 - 22

HETÉNYI, M. (1974), *Beams on Elastic Foundation*, University of Michigan Press, Ann Arbor

HEYMAN, JACQUES (1972), *Coulomb's Memoir on Statics: An Essay in the History of Civil Engineering*, Cambridge University Press, London

HODGKINSON, ALLAN (1986), *Foundation Design*, The Architectural Press, London

HOLDEN, J.C. (1988), "Integrity Control of Bored Piles Using SID", *Deep Foundations on Bored and Auger Piles* (Proceedings of the 1st International Geotechnical Seminar on Deep Foundations on Bored and Auger Piles), p. 91 - 95, W.F. VanImpe, ed., Balkema, Rotterdam

HOLLOWAY, D.M.; MORIWAKI, Y.; FINNO, R.J. AND GREEN, R.K. (1982), "Lateral Load Response of a Pile Group in Sand", *Second International Conference on Numerical Methods in Offshore Piling*, p. 441 456, ICE

HOLTZ, W.G. (1969), "Volume Change in Expansive Clay Soils and Control by Lime Treatment", *Proceedings of the Second International Research and Engineering Conference on Expansive Clay Soils*, p. 157 - 173, Texas A&M Press

HOLTZ, WESLEY G. AND GIBBS, HAROLD J. (1979), Discussion of "SPT and Relative Density in Coarse Sand", ASCE *Journal of the Geotechnical Engineering Division*, Vol. 105, No. GT3, p. 439 - 441

HOLTZ, WESLEY G. AND HART, STEPHEN S. (1978), *Home Construction on Shrinking and Swelling Soils*, Colorado Geological Survey publication SP-11, Denver

HORVATH, JOHN S. (1983), "New Subgrade Model Applied to Mat Foundations", ASCE *Journal of Geotechnical Engineering*, Vol. 109, No. 12, p. 1567 - 1587

HORVATH, JOHN S. (1992), "Lite Products Come of Age", ASTM *Standardization News*, Vol. 20, No. 9, p. 50 - 53

HOUGH, B.K. (1959), "Compressibility as the Basis for Soil Bearing Value", ASCE *Journal of the Soil Mechanics and Foundations Division*, Vol. 85, No. SM4, p. 11 - 39

HOUSEL, WILLIAM S. (1966), "Pile Load Capacity: Estimates and Test Results", ASCE *Journal of the Soil Mechanics and Foundations Division*, Vol. 92, No. SM4, p. 1 - 30

HOUSTON, SANDRA L. (1992), "Partial Wetting Collapse Predictions", *Proceedings, 7th International Conference on Expansive Soils*, Vol. 1, p. 302 - 306

HOUSTON, SANDRA L. AND EL-EHWANY, MOSTAFA (1991), "Sample Disturbance of Cemented Collapsible Soils", ASCE *Journal of Geotechnical Engineering*, Vol. 117, No. 5, p. 731 - 752. Also see discussions, Vol. 118, No. 11, p. 1854 - 1862

HOUSTON, SANDRA L.; HOUSTON, WILLIAM N. AND SPADOLA, DONALD J. (1988), "Prediction of Field Collapse of Soils Due To Wetting", ASCE *Journal of Geotechnical Engineering*, Vol. 114, No. 1, p. 40 - 58

HOUSTON, WILLIAM N. (1991), Personal communication

HOUSTON, WILLIAM N. AND HOUSTON, SANDRA L. (1989), "State-of-the-Practice Mitigation Measures for Collapsible Soil Sites", *Foundation Engineering: Current Principles and Practices*, Vol. 1, p. 161 175, F.H. Kulhawy, ed., ASCE

HUBER, FRANK (1991), "Update: Bridge Scour", ASCE *Civil Engineering*, Vol. 61, No. 9, p. 62 - 63

HUMMEL, CHARLES E. (1971), *Tyranny of the Urgent*, InterVarsity Press, Downers Grove, IL

HUNT, HAL W. (1987), "American Practice in Design and Installation of Driven Piles for Structure Support", *Second International Deep Foundations Conference*, p. 13 - 28, Deep Foundations Institute, Sparta, NJ

HUNT, ROY E. (1984), *Geotechnical Engineering Investigation Manual*, McGraw Hill, New York

HUNTINGTON, WHITNEY (1957), *Earth Pressures and Retaining Walls*, John Wiley, New York

ICBO (1991a), *Uniform Building Code*, International Conference of Building Officials, Whittier, CA

ICBO (1991b), *Uniform Building Code Standards*, International Conference of Building Officials, Whittier, CA

ICBO (1991c), *Handbook on the Uniform Building Code*, International Conference of Building Officials, Whittier, CA

INGOLD, TERRY (1979), "Retaining Wall Performance During Backfilling", ASCE *Journal of Geotechnical Engineering*, Vol. 105, No. GT5, p. 613 - 626

ISAACS, D.V. (1931), "Reinforced Concrete Pile Formula", *Transactions of the Institution of Australian Engineers*, Vol. 12, p. 312 - 323

ISMAEL, NABIL F. AND VESIĆ, ALEKSANDAR S. (1981), "Compressibility and Bearing Capacity", ASCE *Journal of the Geotechnical Engineering Division*, Vol. 107, No. GT12, p. 1677 - 1691

JAKY, J. (1944), "The Coefficient of Earth Pressure at Rest", *Journal for Society of Hungarian Architects and Engineers*, October 1944, p. 355 - 358 (in Hungarian); Also quoted in Jaky (1948); See derivation in Kézdi (1972)

JAKY, J. (1948), "Pressure in Silos", *Proceedings of the Second International Conference on Soil Mechanics and Foundation Engineering*, p. 103 - 107

JAMIOLKOWSKI, M.; GHIONNA, V.; LANCELLOTTA, R. AND PASQUALINI, E. (1988), "New Correlations of Penetration Tests for Design Practice", *First International Symposium on Penetration Testing* (ISOPT-1), Vol. 1, p. 263 - 296

JAMIOLKOWSKI, M.; LADD, C.C.; GERMAINE, J.T. AND LANCELLOTTA, R. (1985), "New Developments in Field and Laboratory Testing of Soils", *Eleventh International Conference on Soil Mechanics and Foundation Engineering*, San Francisco, Volume 1, p. 57 - 153, A.A. Balkema, Rotterdam

JANBU, N., BJERRUM, L AND KJAERNSLI, B. (1956), *Veiledning ved losning av fundamenteringsoppgaver*, Norwegian Geotechnical Institute Publication 16, p. 30 - 32, Oslo (in Norwegian)

JENNINGS, J.E. (1969), "The Engineering Problems of Expansive Soils", *Proceedings of the Second International Research and Engineering Conference on Expansive Clay Soils*, p. 11 - 17, Texas A&M Press, College Station, TX

JENNINGS, J.E. (1973) "The Engineering Significance of Constructions on Dry Subsoils", *Proceedings of the Third International Conference on Expansive Soils*, Vol. 1, p. 27 - 32, Jerusalem Academic Press, Israel

JENNINGS, J.E. AND KNIGHT, K. (1956), "Recent Experience with the Consolidation Test as a Means of Identifying Conditions of Heaving or Collapse of Foundations on Partially Saturated Soils", *Transactions, South African Institution of Civil Engineers*, Vol. 6, No. 8, p. 255 - 256

JENNINGS, J.E. AND KNIGHT, K. (1957), "The Additional Settlement of Foundations Due to a Collapse Structure of Sandy Subsoils on Wetting", *Proceedings, Fourth International Conference on Soil Mechanics and Foundation Engineering*, Section 3a/12, p. 316 - 319

JENNINGS, J.E. AND KNIGHT, K. (1975), "A Guide to Construction on or with Materials Exhibiting Additional Settlement Due to 'Collapse' of Grain Structure", *Sixth Regional Conference for Africa on Soil Mechanics and Foundation Engineering*, p. 99 - 105

JEYAPALAN, JEY K. AND BOEHM, ROLLAND (1986), "Procedures for Predicting Settlements in Sands", *Settlement of Shallow Foundations on Cohesionless Soils: Design and Performance*, Geotechnical Special Publication No. 5, p. 1 - 22, W.O. Martin, ed, ASCE

JIANG, DA HUA (1983), "Flexural Strength of Square Spread Footing", ASCE *Journal of Structural Engineering*, Vol. 109, No. 8, p. 1812 - 1819

JOHNSTON, G.H., ED. (1981), *Permafrost Engineering Design and Construction*, John Wiley, New York

JOHNSON, LAWRENCE D. AND STROMAN, WILLIAM R. (1976), *Analysis of Behavior of Expansive Soil Foundations*, Technical report S-76-8, U.S. Army Waterways Experiment Station, Vicksburg, Miss.

JONES, D. EARL AND HOLTZ, WESLEY G. (1973), "Expansive Soils - The Hidden Disaster", *Civil Engineering*, Vol. 43, No. 8, ASCE

JONES, D. EARL AND JONES, KAREN A. (1987), "Treating Expansive Soils", *Civil Engineering*, Vol. 57, No. 8, August 1987, ASCE

KANTEY, BASIL A. (1980), "Some Secrets to Building Structures on Expansive Soils", *Civil Engineering*, Vol. 50, No. 12, December 1980, ASCE

KERISEL, J. (1965), "Vertical and Horizontal Bearing Capacity of Deep Foundations in Clay", *Bearing Capacity and Settlement of Foundations*, Duke University, Raliegh, NC, p. 45 - 52

KERISEL, JEAN (1987), *Down to Earth, Foundations Past and Present: The Invisible Art of the Builder*, A.A. Balkema, Rotterdam

KÉZDI (1972), "Stability of Rigid Structures", *Fifth European Conference on Soil Mechanics and Foundation Engineering*, Madrid, Vol. II, p. 105 - 130

KNODEL, PAUL C. (1981), "Construction of Large Canal on Collapsing Soils", ASCE *Journal of the Geotechnical Engineering Division*, Vol. 107, No. GT1, p. 79 - 94

KOERNER, ROBERT M. (1990), *Designing with Geosynthetics*, 2nd Ed., Prentice Hall, Englewood Cliffs, NJ

KOSMATKA, STEVEN H. AND PANARESE, WILLIAM C. (1988), *Design and Control of Concrete Mixtures*, 13th Ed., Portland Cement Association, Skokie, Illinois

KOVACS, W.D.; SALOMONE, L.A. AND YOKEL, F.Y. (1981), *Energy Measurements in the Standard Penetration Test*, Building Science Series 135, National Bureau of Standards, Washington, DC

KRAFT, LELAND M. (1991), "Performance of Axially Loaded Pipe Piles in Sand", ASCE *Journal of Geotechnical Engineering*, Vol. 117, No. 2, p. 272 - 296

KRAFT, LELAND M.; FOCHT, JOHN A., JR.; AND AMERASINGHE, SRINATH F. (1981), "Friction Capacity of Piles Driven into Clay", ASCE *Journal of the Geotechnical Engineering Division*, Vol. 107, No. GT11, p. 1521 - 1541

KRAFT, LELAND M.; RAY, RICHARD P. AND KAGAWA, TAKAAKI (1981), "Theoretical *t-z* Curves", ASCE *Journal of the Geotechnical Engineering Division*, Vol. 107, No. GT11, p. 1543 - 1561

KRAMER, STEVEN L. (1991), *Behavior of Piles in Full-Scale, Field Lateral Loading Tests*, Report No. WA-RD 215.1, Federal Highway Administration, Washington, DC

KRINITZSKY, ELLIS L.; GOULD, JAMES P. AND EDINGER, PETER H. (1993), *Fundamentals of Earthquake Resistant Construction*, John Wiley, New York

KULHAWY, FRED H. (1984), "Limiting Tip and Side Resistance: Fact or Fallacy?", *Analysis and Design of Pile Foundations*, p. 80 - 98, ASCE

KULHAWY, FRED H. (1985), "Uplift Behavior of Shallow Soil Anchors - An Overview", *Uplift Behavior of Anchor Foundations in Soil*, Samuel P. Clemence, ed., ASCE

KULHAWY, FRED H. (1991), "Drilled Shaft Foundations", Chapter 14 in *Foundation Engineering Handbook*, 2nd Ed., Hsai-Yang Fang, ed., Van Nostrand Reinhold, New York

KULHAWY, FRED H. AND GOODMAN, RICHARD E. (1980), "Design of Foundations on Discontinuous Rock", *Structural Foundations on Rock*, P.J.N. Pells, ed., Vol. 1, p. 209 - 220, A.A. Balkema, Rotterdam

KULHAWY, FRED H. AND JACKSON, CHRISTINA STAS (1989), "Some Observations on Undrained Side Resistance of Drilled Shafts", *Foundation Engineering: Current Principles and Practices*, p. 1011 - 1025, ASCE

KULHAWY, F.H. AND MAYNE, P.W. (1990), *Manual on Estimating Soil Properties for Foundation Design*, Report No. EL-6800, Electric Power Research Institute, Palo Alto, CA

KULHAWY, F.H.; TRAUTMANN, C.H.; BEECH, J.F.; O'ROURKE, T.D.; MCGUIRE, W.; WOOD, W.A.; AND CAPANO, C. (1983), *Transmission Line Structure Foundations for Uplift-Compression Loading*, Report No. EL-2870, Electric Power Research Institute, Palo Alto, CA

KUMBHOJKAR, A.S. (1993), "Numerical Evaluation of Terzaghi's N_γ", ASCE *Journal of Geotechnical Engineering*, Vol. 119, No. 3, p. 598 - 607

KUWABARA, F. AND POULOS, H.G. (1989), "Downdrag Forces in a Group of Piles", ASCE *Journal of Geotechnical Engineering*, Vol. 115, No. 6, p. 806 - 818

LADD, C. C.; FOOTE, R.; ISHIHARA, K.; SCHLOSSER, F.; AND POULOS, H. G. (1977), "Stress-Deformation and Strength Characteristics", State-of-the-art report, *Proceedings, Ninth International Conference on Soil Mechanics and Foundation Engineering*, Vol. 2, p. 421 - 494, Tokyo

LADE, POUL V. AND LEE, KENNETH L. (1976), *Engineering Properties of Soils*, Report No. UCLA-ENG-7652, University of California, Los Angeles

LAM, IGNATIUS AND MARTIN, GEOFFREY R. (1986), *Seismic Design of Highway Bridge Foundations*, Report Nos. FHWA/RD-86/101, 102 and 103, Federal Highway Administration, McLean, VA

LAMBE, PHILIP C. AND HANSEN, LAWRENCE A. (1990), *Design and Performance of Earth Retaining Structures*, Geotechnical Special Publication No. 25, ASCE

LAMBE, T.W. AND WHITMAN, R.V. (1959), "The Role of Effective Stress in the Behavior of Expansive Soils", *Quarterly of the Colorado School of Mines*, Vol. 54, No. 4, p. 44 - 61

LAMBE, T. WILLIAM AND WHITMAN, ROBERT V. (1969), *Soil Mechanics*, John Wiley, New York

LANKFORD, WILLIAM T.; SAMWAYS, NORMAN L.; CRAVEN, ROBERT F. AND MCGANNON, HAROLD E. (1985), *The Making, Shaping and Treating of Steel*, 10th Ed., United States Steel Corp.

LAWTON, EVERT C.; FRAGASZY, RICHARD J. AND HARDCASTLE, JAMES H. (1989), "Collapse of Compacted Clayey Sand", ASCE *Journal of Geotechnical Engineering*, Vol. 115, No. 9, p. 1252 - 1267

LAWTON, EVERT C.; FRAGASZY, RICHARD J. AND HARDCASTLE, JAMES H. (1991), "Stress Ratio Effects on Collapse of Compacted Clayey Sand", ASCE *Journal of Geotechnical Engineering*, Vol. 117, No. 5, p. 714 - 730. Also see discussion, Vol. 118, No. 9, p. 1472 - 1474

LAWTON, EVERT C.; FRAGASZY, RICHARD J. AND HETHERINGTON, MARK D. (1992), Review of Wetting-Induced Collapse in Compacted Soil", ASCE *Journal of Geotechnical Engineering*, Vol. 118, No. 9, p. 1376 - 1394

LEE, A.R. (1974), *Blastfurnace and Steel Slag: Production, Properties and Uses*, John Wiley, New York

LEONARDS, GERALD A. (1979), Discussion of "Foundation Performance of Tower of Pisa", ASCE *Journal of the Geotechnical Engineering Division*, Vol. 105, No. GT1, p. 95 - 105

LEONARDS, GERALD A. (1982), "Investigation of Failures", The 16th Terzaghi Lecture, ASCE *Journal of the Geotechnical Engineering Division*, Vol. 108, No. GT2, p. 185 - 246

LEONARDS, G.A. AND FROST, J.D., (1988), "Settlement of Shallow Foundations on Granular Soils", ASCE *Journal of Geotechnical Engineering*, Vol. 114, No. GT7, p. 791 - 809, (also see discussions, Vol. 117, No. 1, p. 172 - 188)

LEPERT, P.; BRIAUD, J.-L.; AND MAXWELL, J. (1991), "A Dynamic Method to Assess the Stiffness of Soil Underlying Spread Foundations", Paper No. 910111, 70th Annual Meeting, Transportation Research Board, January 1991

LINDNER, ERNEST (1976), "Swelling Rock: A Review", *Rock Engineering for Foundations and Slopes*, Vol. 1, p. 141 - 181, ASCE

LITKE, SCOT (1986), "New Nondestructive Integrity Testing Yields More Bang for the Buck", *Foundation Drilling*, Sept/Oct, p. 8 - 11

LUSCHER, ULRICH; BLACK, WILLIAM T. AND NAIR, KESHAVAN (1975), "Geotechnical Aspects of Trans-Alaska Pipeline", ASCE *Transportation Engineering Journal*, Vol. 101, No. TE4, p. 669 - 680

LUTENEGGER, ALAN J. AND SABER, ROBERT T. (1988), "Determination of Collapse Potential of Soils", *Geotechnical Testing Journal*, Vol. 11, No. 3, p. 173 - 178

LYTTON, R.L. AND WATT, W.G. (1970), *Prediction of Swelling in Expansive Clays*, Research Report 118-4, Center for Highway Research, University of Texas, Austin

MACGREGOR, JAMES G. (1992), *Reinforced Concrete: Mechanics and Design*, 2nd Ed., Prentice Hall, Englewood Cliffs, NJ

MAHMOUD, HISHAM (1991), *The Development of an In-Situ Collapse Testing System for Collapsible Soils*, PhD Dissertation, Civil Engineering Department, Arizona State University

MARCHETTI, SILVANO (1980), "In Situ Tests by Flat Dilatometer", ASCE *Journal of the Geotechnical Engineering Division*, Vol. 106, No. GT3, p. 299 - 321 (also see discussions, Vol. 107, No. GT8, p. 831 - 837)

MASSARSCH, K. RAINER; TANCRÉ, ERIC AND BRIEKE, WERNER (1988), "Displacement Auger Piles with Compacted Base", *Deep Foundations on Bored and Auger Piles*, p. 333 - 342, W.F. Van Impe, Ed., Balkema, Rotterdam

MATLOCK, H. AND REESE, L.C. (1960), "Generalized Solution for Laterally Loaded Piles", ASCE *Journal of the Soil Mechanics and Foundations Division*, Vol. 86, No. SM5, p. 63-91

MAYNE, PAUL W. AND KULHAWY, FRED H. (1982), "K_0-OCR Relationships in Soil", ASCE *Journal of the Geotechnical Engineering Division*, Vol. 108, No. GT6, p. 851 - 872

MCCARTHY, DAVID F. (1988), *Essentials of Soil Mechanics and Foundations: Basic Geotechnics*, Prentice Hall, Englewood Cliffs, NJ

MCCLELLAND, BRAMLETTE AND FOCHT, JOHN A. JR. (1958), "Soil Modulus for Laterally Loaded Piles", *Transactions of ASCE*, Vol. 123, p. 1049 - 1086, Paper No. 2954

MCDOWELL, C. (1956), "Interrelationships of Load, Volume Change, and Layer Thicknesses of Soils to the Behavior of Engineering Structures", *Proceedings, Highway Research Board*, Vol. 35, p. 754 - 772

MEHTA, P.K. (1983), "Mechanism of Sulfate Attack on Portland Cement Concrete–Another Look", *Cement and Concrete Research*, Vol. 13, p. 401 - 406

MEIGH, A.C. (1987), *Cone Penetration Testing: Methods and Interpretation*, Butterworths, London

MESRI, G. (1973), "Coefficient of Secondary Compression", ASCE *Journal of the Soil Mechanics and Foundations Division*, Vol. 99, No. SM1, p. 123 - 137

MEYERHOF, G.G. (1953), "The Bearing Capacity of Foundations Under Eccentric and Inclined Loads", *Proceedings, 3rd International Conference on Soil Mechanics and Foundation Engineering*, Vol. 1, p. 440 - 445, Zurich (Reprinted in Meyerhof, 1982)

MEYERHOF, G.G. (1955), "Influence of Roughness of Base and Ground-Water Conditions on the Ultimate Bearing Capacity of Foundations", *Géotechnique*, Vol. 5, p. 227 - 242 (Reprinted in Meyerhof, 1982)

MEYERHOF, G.G. (1956), "Penetration Tests and Bearing Capacity of Cohesionless Soils", ASCE *Journal of the Soil Mechanics and Foundations Division*, Vol. 82, No. SM1, p. 1 - 19 (Reprinted in Meyerhof, 1982)

MEYERHOF, GEORGE GEOFFREY (1963), "Some Recent Research on the Bearing Capacity of Foundations", *Canadian Geotechnical Journal*, Vol. 1, No. 1, p. 16 - 26 (Reprinted in Meyerhof, 1982)

MEYERHOF, GEORGE G. (1965), "Shallow Foundations", ASCE *Journal of the Soil Mechanics and Foundations Division*, Vol. 91 No. SM2, p. 21 - 31 (Reprinted in Meyerhof, 1982)

MEYERHOF, GEORGE GEOFFREY (1976), "Bearing Capacity and Settlement of Pile Foundations", ASCE *Journal of the Geotechnical Engineering Division*, Vol. 102, No. GT3, p. 197 - 228

MEYERHOF, G.G. (1982), *The Bearing Capacity and Settlement of Foundations*, Tech-Press, Technical University of Nova Scotia, Halifax

MITCHELL, JAMES K. (1978), "In-Situ Techniques for Site Characterization", *Site Characterization and Exploration*, p. 107 - 129, C.H. Dowding, Ed., ASCE

MITCHELL, JAMES K. (1993), *Fundamentals of Soil Behavior*, 2nd Ed. John Wiley and Sons, New York

MITCHELL, JAMES K., ET AL. (1978), *Soil Improvement: History, Capabilities and Outlook*, ASCE

MITCHELL, JAMES K. AND GARDNER, WILLIAM S. (1975), "In Situ Measurement of Volume Change Characteristics", *In Situ Measurement of Soil Properties*, Vol. 2, p. 279 - 345, ASCE

MITCHELL, JAMES K. AND HUBER, TIMOTHY R. (1985), "Performance of a Stone Column Foundation", ASCE *Journal of Geotechnical Engineering*, Vol. 111, No. 2, p. 205 - 223

MITCHELL, JAMES K. AND KATTI, R.K. (1981), "Soil Improvement: State-of-the-Art", *Proceedings, 10th International Conference on Soil Mechanics and Foundation Engineering*, Stockholm

MITCHELL, JAMES K., VIVATRAT, VITOON, AND LAMBE, T. WILLIAM (1977), "Foundation Performance of Tower of Pisa", ASCE *Journal of the Geotechnical Engineering Division*, Vol. 103, No. GT3, p. 227 - 249 (discussions in Vol. 104 No. GT1 and GT2, Vol. 105 No. GT1 and GT11)

MITCHELL, P.W. AND AVALLE, D.L. (1984), "A Technique to Predict Expansive Soil Movements", *Proceedings, Fifth International Conference on Expansive Soils*, Institution of Engineers, Australia

MOE, JOHANNES (1961), *Shearing Strength of Reinforced Concrete Slabs and Footings Under Concentrated Loads*, Portland Cement Association Bulletin D47, Skokie, Illinois

MORGAN, M.H. (1914), Translation of Marcus Vitruvius *The Ten Books on Architecture*, Harvard University Press, Cambridge, MA

MORGENSTERN, N.R. AND EISENSTEIN, Z. (1970), "Methods of Estimating Lateral Loads and Deformations" in *Lateral Stresses in the Ground and the Design of Earth-Retaining Structures*, p. 51 - 102, ASCE

MURILLO, JUAN A. (1987), "The Scourge of Scour", *Civil Engineering*, Vol. 57, No. 7, p. 66 - 69, ASCE

MURPHY, E.C. (1908), "Changes in Bed and Discharge Capacity of the Colorado River at Yuma, Ariz.", *Engineering News*, Vol. 60, p. 344 - 345

NAGAO, YOSHIKAZU, ET AL. (1989), *Development of New Pavement Base Course Material Using High Proportion of Steelmaking Slag Properly Combined with Air-Cooled and Granulated Blast Furnace Slags*, Nippon Steel Technical Report No. 43, Oct 1989

NAHB (1988), *Frost-Protected Shallow Foundations for Houses and Other Heated Structures*, National Association of Home Builders, Upper Marlboro, MD

NAHB (1990, draft) *Frost-Protected Shallow Foundations for Uneated Structures*, National Association of Home Builders, Upper Marlboro, MD

NEATE, JAMES J. (1989), "Augered Cast-in-Place Piles", *Foundation Engineering: Current Principles and Practices*, Vol. 2, p. 970 - 978, F.H. Kulhawy, Ed., ASCE

NEELY, WILLIAM J. (1989), "Bearing Pressure - SPT Correlations for Expanded Base Piles in Sand", *Foundation Engineering: Current Principles and Practices*, Vol. 2, p. 979 - 990, F.H. Kulhawy, ed. ASCE

NEELY, WILLIAM J. (1990a), "Bearing Capacity of Expanded-Base Piles in Sand", ASCE *Journal of Geotechnical Engineering*, Vol. 116, No. 1, p. 73 - 87

NEELY, WILLIAM J. (1990b), "Bearing Capacity of Expanded-Base Piles with Compacted Concrete Shafts", ASCE *Journal of Geotechnical Engineering*, Vol. 116, No. 9, p. 1309 - 1324

NEELY, WILLIAM J. (1991), "Bearing Capacity of Auger-Cast Piles in Sand", ASCE *Journal of Geotechnical Engineering*, Vol. 117, No. 2, p. 331 - 345

NELSON, JOHN D. AND MILLER, DEBORA J. (1992), *Expansive Soils: Problems and Practice in Foundation and Pavement Engineering*, John Wiley, New York

NEWMARK, NATHAN M. (1935), *Simplified Computation of Vertical Pressures in Elastic Foundations*, Engineering Experiment Station Circular No. 24, University of Illinois, Urbana

NEWMARK, NATHAN M. (1942), *Influence Charts for Computation of Stresses in Elastic Foundations*, Engineering Experiment Station Bulletin 338, University of Illinois, Urbana

NIXON, IVAN K. (1982), "Standard Penetration Test State-of-the-Art Report", *Second European Symposium on Penetration Testing* (ESOPT II), Amsterdam, Vol. 1, p. 3 - 24, A. Verruijt, et. al., eds.

NORDLUND, REYMOND L. (1982), "Dynamic Formula for Pressure Injected Footings", ASCE *Journal of the Geotechnical Engineering Division*, Vol. 108, No. GT3, p. 419 - 437

NORDLUND, REYMOND L. AND DEERE, DON U. (1970), "Collapse of Fargo Grain Elevator", ASCE *Journal of the Soil Mechanics and Foundations Division*, Vol. 96, No. SM2, p. 585 - 607

NOTTINGHAM, LARRY C. AND SCHMERTMANN, JOHN H. (1975), *An Investigation of Pile Capacity Design Procedures*, Research Report D629, Dept. of Civil Engr., Univ. of Florida, Gainesville

NRCC (1990), *Canadian National Building Code*, National Research Council of Canada, Ottawa, Ontario

O'NEILL, MICHAEL W. (1983), "Group Action in Offshore Piles", *Geotechnical Practice in Offshore Engineering*, p. 25 - 64, Steven G. Wright, ed., ASCE

O'NEILL, MICHAEL (1987), "Use of Underreams in Drilled Shafts", *Drilled Foundation Design and Construction Short Course*, Association of Drilled Shaft Contractors, Dallas, TX

O'NEILL, MICHAEL W. AND POORMOAYED, NADER (1980), "Methodology for Foundations on Expansive Clays", ASCE *Journal of the Geotechnical Engineering Division*, Vol. 106, No. GT12, p. 1345 - 1368

O'ROURKE, T.D. AND JONES, C.J.F.P. (1990), "Overview of Earth Retention Systems: 1970-1990", *Design and Performance of Earth Retaining Structures*, Geotechnical Special Publication No. 25, p. 22 51, P.C. Lambe and L.A. Hansen, eds., ASCE

OLSON, LARRY D. AND WRIGHT, CLIFFORD C. (1989), "Nondestructive Testing of Deep Foundations With Sonic Methods", *Foundation Engineering: Current Principles and Practices*, ASCE

OLSON, ROY E. AND FLAATE, KARRE S. (1967), "Pile-Driving Formulas for Friction Piles in Sand", ASCE *Journal of the Soil Mechanics and Foundations Division*, Vol. 93, No. SM6, p. 279 - 296

OSTERBERG, JORJ O. (1984), "A New Simplified Method For Load Testing Drilled Shafts", *Foundation Drilling*, August, 1984, Association of Drilled Shaft Contractors, Dallas

OSTERBERG, JORJ (1992), "Rock Socket Friction in Drilled Shafts: It's Greater Than You Think", *Foundation Drilling*, Vol. 31, No. 1, p. 15 - 21, Association of Drilled Shaft Contractors, Dallas

PCA (1991), "Durability of Concrete in Sulfate-Rich Soils", *Concrete Technology Today*, Vol. 12, No. 3, p. 6 - 8, Portland Cement Association

PCI (1984), "Standard Prestressed Concrete Square and Octagonal Piles", Drawing No. STD-112-84, Prestressed Concrete Institute, Chicago

PCI (1993), "Recommended Practice for Design, Manufacture and Installation of Prestressed Concrete Piling", *PCI Journal*, Vol. 38, No. 2, p. 14 - 41

PECK, RALPH B. (1942), Discussion of "Pile Driving Formulas: Progress Report of the Committee on the Bearing Value of Pile Foundations", *ASCE Proceedings*, Vol. 68. No. 2, p. 323 - 324, Feb.

PECK, RALPH B. (1948), *History of Building Foundations in Chicago*, Bulletin Series No. 373, University of Illinois Engineering Experiment Station, Urbana, IL

PECK, RALPH B. (1958), *A Study of the Comparative Behavior of Friction Piles*, Highway Research Board Special Report 36, Washington, DC

PECK, RALPH B. (1967), "Bearing Capacity and Settlement: Certainties and Uncertainties", *Bearing Capacity and Settlement of Foundations*, Aleksandar S. Vesić, ed.; Department of Civil Engineering, Duke University, Durham, North Carolina

PECK, RALPH B. (1975), *The Selection of Soil Parameters for the Design of Foundations*, Second Nabor Carrillo Lecture, Mexican Society for Soil Mechanics, Mexico City (reprinted in *Judgment in Geotechnical Engineering*, Dunnicliff, John and Deere, Don U., eds, John Wiley, 1984)

PECK, RALPH B. (1976), "Rock Foundations for Structures", *Rock Engineering for Foundations and Slopes*, Vol. II, p. 1 - 21, ASCE

PECK, R. B. AND BRYANT, F. G. (1953), "The Bearing-Capacity Failure of the Transcona Elevator", *Géotechnique*, Vol. 3, p. 201 - 208

PECK, RALPH B.; HANSON, WALTER E. AND THORNBURN, THOMAS H. (1974), *Foundation Engineering*, 2nd Ed., John Wiley, New York

POLSHIN, D.E. AND TOKAR, R.A. (1957), "Maximum Allowable Non-Uniform Settlement of Structures", *Proceedings, 4th International Conference on Soil Mechanics and Foundation Engineering*, Vol. 1, p. 402 - 405

POTYONDY, J.G. (1961), "Skin Friction Between Various Soils and Construction Materials", *Géotechnique*, Vol. 11, No. 4, p. 339 - 353

POULOS, HARRY G. (1979), "Group Factors for Pile-Deflection Estimation", ASCE *Journal of the Geotechnical Engineering Division*, Vol. 105, No. GT12, p. 1489 - 1510

POULOS, H.G. AND DAVIS, E.H. (1974), *Elastic Solutions for Soil and Rock Mechanics*, John Wiley, New York

POULOS, H.G. AND DAVIS, E.H. (1980), *Pile Foundation Analysis and Design*, John Wiley, New York

POWELL, GEORGE T. (1884), *Foundations and Foundation Walls*, William T. Comstock, New York

PRAKASH, SHAMSHER AND SHARMA, HARI D. (1990), *Pile Foundations in Engineering Practice*, John Wiley, New York

PRANDTL, L. (1920), "Über die Härte plastischer Körper (On the Hardness of Plastic Bodies)", *Nachr. Kgl. Ges Wiss Göttingen, Math-Phys. Kl.*, p. 74 (in German)

PRICE, RICHARD; ROLLINS, KYLE M. AND KEANE, EDWARD (1992), "Comparison of Measured and Computed Drilled Shaft Capacities Based on Utah Load Tests", *Foundation Drilling*, Vol. 31, No. 4, p. 17-27, ADSC, Dallas, TX

PRYKE, J.F.S. (1987), "Pisa: More Inclination Than Time?", *Civil Engineering* (London), May 1987, p. 1045

PTI (1980), *Design and Construction of Post-Tensioned Slabs-on-Ground*, Post-Tensioning Institute, Phoenix, AZ

PUGSLEY, A. (1966), *The Safety of Structures*, Arnold, London

PYE, KENNETH (1987), *Aeolian Dust and Dust Deposits*, Academic Press, London

RANDOLPH, M.F. AND WROTH, C.P. (1982), "Recent Developments in Understanding the Axial Capacity of Piles in Clay", *Ground Engineering*, Vol. 15, No. 7, p. 17 - 32

RANKINE, W.J.M. (1857), "On the Stability of Loose Earth", *Philosophical Transactions of the Royal Society*, Vol. 147, London

RAO, KANAKAPURA S. SUBBA AND SINGH, SHASHIKANT (1987), "Lower-Bound Collapse Load of Square Footings", ASCE *Journal of Structural Engineering*, Vol. 113, No. 8, p. 1875 - 1879

RAUSCHE, F. AND GOBLE, G.G. (1978), "Determination of Pile Damage by Top Measuremsnts", *Behavior of Deep Foundations*, ASTM STP 670, p. 500 - 506, Raymond Lundgren, Ed., ASTM

RAUSCHE, FRANK; GOBLE, GEORGE G. AND LIKINS, GARLAND E. (1985), "Dynamic Determination of Pile Capacity", ASCE *Journal of Geotechnical Engineering*, Vol. 111, No. 3, p. 367 - 383

RAUSCHE, F., MOSES, F., AND GOBLE, G. (1972), "Soil Resistance Predictions from Pile Dynamics", ASCE *Journal of the Soil Mechanics and Foundations Division*, Vol. 98, No. SM9

REESE, LYMON C. (1984), *Handbook on Design of Piles and Drilled Shafts Under Lateral Load*, Report No. FHWA-IP-84-11, Federal Highway Administration

REESE, L.C. (1986), *Behavior of Piles and Pile Groups Under Lateral Load*, Report No. FHWA/RD-85/106, Federal Highway Administration

REESE, LYMON C. AND O'NEILL, MICHAEL W. (1988), *Drilled Shafts: Construction Procedures and Design Methods*, Report No. FHWA-HI-88-042, Federal Highway Administration

REESE, L.C. AND O'NEILL, M.W. (1989), "New Design Method for Drilled Shafts From Common Soil and Rock Tests", *Foundation Engineering: Current Principles and Practices*, p. 1026 - 1039, Fred H. Kulhawy, Ed., ASCE

REESE, LYMON C. AND WANG, SHIN TOWER (1986), "Method of Analysis of Piles Under Lateral Loading", *Marine Geotechnology and Nearshore/Offshore Structures*, ASTM STP 923, R.C. Chaney and H.Y. Fang, Ed., p. 199 - 211, ASTM

REMPE, D.M. (1979), "Building Code Requirements for Maximum Design Stresses in Piles", *Behavior of Deep Foundations*, p. 507 - 519, ASTM STP 670, Raymond Lundgren, Ed., ASTM, Philadelphia

RICHARDS, B.G. (1967), "Moisture Flow and Equilibria in Unsaturated Soils for Shallow Foundations", *Permeability and Capillarity of Soils*, p. 4 - 33, STP 417, ASTM

RICHARDSON, E.V.; HARRISON, LAWRENCE J. AND DAVIS, STANLEY R. (1991), *Evaluating Scour at Bridges*, Report No. FHWA-IP-90-017, Federal Highway Administration, McLean, VA

RICHART, FRANK E. (1948), "Reinforced Concrete Wall and Column Footings", *Journal of the American Concrete Institute*, Vol. 20, No. 2, p. 97 - 127, and Vol. 20, No. 3, p. 237 - 260

ROBERTSON, P.K. AND CAMPANELLA, R.G. (1983), "Interpretation of Cone Penetration Tests: Parts 1 and 2", *Canadian Geotechnical Journal* Vol. 20, p. 718 - 745

ROBERTSON, P.K.; CAMPANELLA, R.G.; AND BROWN, P.T. (1984), *The Application of CPT Data to Geotechnical Design: Worked Examples*, Department of Civil Engineering, University of British Columbia

ROBERTSON, P.K. AND CAMPANELLA, R.G. (1988), *Guidelines for Using the CPT, CPTU and Marchetti DMT for Geotechnical Design*, Volume II, Report No. FHWA-PA-87-023+84-24, Federal Highway Administration

ROBERTSON, P.K. AND CAMPANELLA, R.G. (1989), *Guidelines for Geotechnical Design Using the Cone Penetrometer Test and CPT With Pore Pressure Measurement*, 4th Ed., Hogentogler & Co., Columbia, MD

ROBERTSON, PETER K.; DAVIES, MICHAEL P. AND CAMPANELLA, RICHARD G. (1989), "Design of Laterally Loaded Driven Piles Using the Flat Dilatometer", ASTM *Geotechnical Testing Journal*, Vol. 12, No. 1, p. 30 - 38

ROMANOFF, MELVIN (1962), "Corrosion of Steel Pilings in Soils", *Journal of Research of the National Bureau of Standards*, Vol. 66C, No. 3; also in National Bureau of Standards Monograph 58, Washington, DC

ROMANOFF, MELVIN (1970), "Performance of Steel Pilings in Soils", *Proceedings, 25th Conference, National Association of Corrosion Engineers*, March 1969

SCHEIL, THOMAS J. (1979), "Long Term Monitoring of a Building Over the Deep Hackensack Meadowlands Varved Clays", presented at 1979 Converse Ward Davis Dixon, Inc. Technical Seminar, Pasadena, CA

SCHIFF, M.J. (1982), "What is Corrosive Soil?", *Proceedings, Western States Corrosion Seminar*, Western Region, National Association of Corrosion Engineers

SCHMERTMANN, JOHN H. (1955), "The Undisturbed Consolidation Behavior of Clay", *ASCE Transactions*, Vol. 120, p. 1201 - 1233

SCHMERTMANN, J.H. (1970), "Static Cone to Compute Settlement over Sand", ASCE *Journal of the Soil Mechanics and Foundations Division*, Vol. 96, No. SM3, p. 1011 - 1043

SCHMERTMANN, JOHN H. (1978), *Guidelines for Cone Penetration Test: Performance and Design*, Report FHWA-TS-78-209, Federal Highway Administration, Washington, DC

SCHMERTMANN, JOHN H. (1985), "Measure and Use of the Insitu Lateral Stress", *The Practice of Foundation Engineering*, p. 189 - 213, Department of Civil Engineering, Raymond J. Krizek, et. al. eds., Northwestern University, Evanston, IL

SCHMERTMANN, J.H. (1986a), "Suggested Method for Performing the Flat Dilatometer Test", *Geotechnical Testing Journal*, Vol. 9, No. 2, p. 93 - 101

SCHMERTMANN, JOHN H. (1986b), "Dilatometer to Compute Foundation Settlement", *Use of In-Situ Tests in Geotechnical Engineering*, Samuel P. Clemence, ed., ASCE, p. 303 - 319

SCHMERTMANN, JOHN H. (1988a), "Dilatometers Settle In", *Civil Engineering*, Vol. 58, No. 3, p. 68 - 70, March 1988

SCHMERTMANN, JOHN H. (1988b), *Guidelines for Using the CPT, CPTU and Marchetti DMT for Geotechnical Design*, Vol. I-IV, Federal Highway Administration

SCHMERTMANN, JOHN H.; HARTMAN, JOHN PAUL; AND BROWN, PHILLIP R. (1978), "Improved Strain Influence Factor Diagrams", ASCE *Journal of the Geotechnical Engineering Division*, Vol. 104, No. GT8, p. 1131 - 1135

SCHNABEL, HARRY (1982), *Tiebacks in Foundation Engineering and Construction*, McGraw-Hill, New York

SCHNEIDER, ROBERT R. AND DICKEY, WALTER L. (1987), *Reinforced Masonry Design*, 2nd Ed., Prentice Hall, Englewood Cliffs, NJ

SCHULTZE, EDGAR (1961), "Distribution of Stress Beneath a Rigid Foundation", *Proceedings, 5th International Conference on Soil Mechanics and Foundation Engineering*, p. 807 - 813, Paris

SCOTT, RONALD F. (1981), *Foundation Analysis*, Prentice-Hall, Englewood Cliffs, NJ

SEED, H. BOLTON (1970), "Soil Problems and Soil Behavior", Chapter 10 in *Earthquake Engineering*, Robert L. Wiegel, ed., Prentice Hall, Englewood Cliffs, NJ

SEED, H.B. AND CHAN, C.K. (1959), "Structure and Strength Characteristics of Compacted Clays", ASCE *Journal of the Soil Mechanics and Foundations Division*, Vol. 85, No. SM5

SEED, H.B., MITCHELL, J.K., AND CHAN, C.K. (1962), "Studies of Swell and Swell Pressure Characteristics of Compacted Clays", *Bulletin No. 313*, Highway Research Board, p. 12 - 39

SEED, H. BOLTON; TOKIMATSU, K.; HARDER, L.F. AND CHUNG, RILEY M. (1985), "Influence of SPT Procedures in Soil Liquefaction Resistance Evaluations", ASCE *Journal of Geotechnical Engineering*, Vol. 111, No. 12, p. 1425 - 1445

SEITZ, JÖRN M. AND RAUSCHE, FRANK (1987), "Dynamic Testing for Bored Pile Bearing Capacity", *Proceedings, 2nd International Deep Foundations Conference*, p. 179 - 200, Deep Foundations Institute, Sparta, NJ

SHEPHERD, R. AND DELOS-SANTOS, E.O. (1991), "An Experimental Investigation of Retrofitted Cripple Walls", *Bulletin of the Seismological Society of America*, Vol. 81, No. 5, p. 2111 - 2126

SHIELDS, DONALD; CHANDLER, NEIL; AND GARNIER, JACQUES (1990), "Bearing Capacity of Foundations on Slopes", ASCE *Journal of Geotechnical Engineering*, Vol. 116, No. 3, p. 528 - 537

SHUKLA, SHYAM N. (1984), "A Simplified Method for Design of Mats on Elastic Foundations", *Journal of the American Concrete Institute*, Vol. 81, No. 5, p. 469 - 475

SIMONS, KENNETH B. (1991), "Limitations of Residential Structures on Expansive Soils", ASCE *Journal of Performance of Constructed Facilities*, Vol. 5, No. 4, p. 258 - 270

SIMONS, N.E. (1987), "Settlement", Chapter 14 in *Ground Engineer's Reference Book*, F. G. Bell, editor, Butterworths, London

SIVAKUGAN, N. (1990), Discussion of "Constitutive Parameters Estimated by Plasticity Index", ASCE *Journal of Geotechnical Engineering*, Vol. 116, No. 10, p. 1594 - 1597

SKEMPTON, A.W. (1951), "The Bearing Capacity of Clays", *Proceedings, Building Research Congress*, Vol. 1, p. 180 - 189, London

SKEMPTON, A.W. (1986), "Standard Penetration Test Procedures and the Effects in Sands of Overburden Pressure, Relative Density, Particle Size, Aging and Overconsolidation", *Géotechnique*, Vol. 36, No. 3, p. 425 - 447

SKEMPTON, A.W. AND McDONALD, D.H. (1956), "Allowable Settlement of Buildings", *Proceedings, Institute of Civil Engineers*, Vol. 3, No. 5, p. 727 - 768

SKEMPTON, A.W. AND BJERRUM, L. (1957), "A Contribution to the Settlement Analysis of Foundations on Clay", *Géotechnique*, Vol. 7, p. 168 - 178

SLADEN, J.A. (1992), "The Adhesion Factor: Applications and Limitations", *Canadian Geotechnical Journal*, Vol. 29, No. 2, p. 322 - 326

SMITH, E.A.L. (1951), "Pile Driving Impact", *Proceedings, Industrial Computation Seminar*, Sept. 1950, International Business Machines Corp., New York

SMITH, E.A.L. (1960), "Pile Driving Analysis by the Wave Equation", ASCE *Journal of Soil Mechanics and Foundations*, Vol. 86, No. SM4, p. 35 - 61

SMITH, E.A.L. (1962), "Pile Driving Analysis by the Wave Equation", ASCE *Transactions*, Vol. 127, Part 1, p. 1145 - 1193

SMITH, TREVOR D. (1989), "Fact or Fiction: A Review of Soil Response to a Laterally Moving Pile", *Foundation Engineering: Current Principles and Practices*, Vol. 1, p. 588 - 598, Fred Kulhawy, Ed., ASCE

SNETHEN, D.R. (1980), "Characterization of Expansive Soils Using Soil Suction Data", *Proceedings of the Fourth International Conference on Expansive Soils*, ASCE

SNETHEN, D.R. (1984), "Evaluation of Expedient Methods for Identification and Classification of Potentially Expansive Soils", *Proceedings of the Fifth International Conference on Expansive Soils*, p. 22 - 26, National Conference Publication 84/3, The Institution of Engineers, Australia

SODERBERG, L. (1962), "Consolidation Theory Applied to Foundation Pile Time Effects", *Géotechnique*, Vol. 12, p. 217 - 225

SOWERS, GEORGE F. (1979), *Introductory Soil Mechanics and Foundations: Geotechnical Engineering*, 4th Ed., MacMillan, New York

SOWERS, G.F. AND KENNEDY, C.M. (1967), "High Volume Change Clays of the Southeastern Coastal Plain", *Proceedings of the Third Panamerican Conference on Soil Mechanics and Foundation Engineering*, Caracas

SPANOVICH, MILAN AND FEWELL, RICHARD (1968), Discussion of Crawford and Burn (1968), *Placement and Improvement of Soil to Support Structures*, p. 247 - 249, ASCE

STAMATOPOULOS, ARIS C. AND KOTZIAS, PANAGHIOTIS C. (1985), *Soil Improvement by Preloading*, John Wiley, New York

STARK, TIMOTHY D. AND DUNCAN, J. MICHAEL (1991), "Mechanisms of Strength Loss in Stiff Clays", ASCE *Journal of Geotechnical Engineering*, Vol. 117, No. 1, p. 139 - 154

STAS, C.V. AND KULHAWY, F.H. (1984), *Critical Evaluation of Design Methods for Foundations Under Axial Uplift and Compressive Loading*, Report No. EL-3771, Electric Power Research Institute, Palo Alto, CA

STEWART, J.P. AND KULHAWY, F.H. (1981), *Experimental Investigation of the Uplift Capacity of Drilled Shaft Foundations in Cohesionless Soil"*, Contract Report B49(6), Niagara Mohawk Power Corporation, Syracuse, NY

TALBOT, ARTHUR N. (1913), *Reinforced Concrete Wall Footings and Column Footings*, Bulletin No. 67, University of Illinois Engineering Experiment Station, Urbana

TAYLOR, BRIAN B. AND MATYAS, ELMER L. (1983), "Influence Factors for Settlement Estimates of Footings on Finite Layers", *Canadian Geotechnical Journal*, Vol. 20, p. 832 - 835

TAYLOR, DONALD W. (1948), *Fundamentals of Soil Mechanics*, John Wiley, New York

TENG, WAYNE C. (1962), *Foundation Design*, Prentice Hall, Englewood Cliffs, NJ

TERZAGHI, K. (1934a), "Die Ursachen der Schiefstellung des Turmes von Pisa", *Der Bauingenieur*, Vol. 15, No. 1/2, p. 1 - 4 (Reprinted in *From Theory to Practice in Soil Mechanics*, Bjerrum, L. et. al., eds., p. 198 - 201, John Wiley, New York, 1960) (in German)

TERZAGHI, KARL (1934b), "Large Retaining Wall Tests", a series of articles in *Engineering News-Record*, Vol. 112; 2/1/34, 2/22/34, 3/8/34, 3/29/34, 4/19/34, and 5/17/34

TERZAGHI, KARL (1936), Discussion of "Settlement of Structures", *Proceedings, First International Conference on Soil Mechanics and Foundation Engineering*, Cambridge, MA, Vol. 3, p. 79 - 87

TERZAGHI, KARL (1942), Discussion of "Pile Driving Formulas: Progress Report of the Committee on the Bearing Value of Pile Foundations", ASCE *Proceedings*, Vol. 68. No. 2, p. 311 - 323, Feb.

TERZAGHI, KARL (1943), *Theoretical Soil Mechanics*, John Wiley, New York

TERZAGHI, KARL (1955), "Evaluation of Coefficients of Subgrade Reaction", *Géotechnique*, Vol. 5, No. 4, p. 297 - 326

TERZAGHI, KARL AND PECK, RALPH B. (1967), *Soil Mechanics in Engineering Practice*, 2nd Ed., John Wiley, New York

THORNBURN, S. AND HUTCHISON, J.F. (1985), *Underpinning*, Surrey University Press

THORNTHWAITE, C.W. (1948), "An Approach Toward a Rational Classification of Climate", *The Geographical Review*, Vol. 38

THORSON, BRUCE M. AND BRAUN, J.S. (1975), "Frost Heaves–A Major Dilemma for Ice Arenas", ASCE *Civil Engineering*, Vol. 45, No. 3

TOMLINSON, M.J. (1957), "The Adhesion of Piles Driven into Clay Soils", *Proceedings, 4th International Conference on Soil Mechanics and Foundation Engineering*, Vol. 2, p. 66 - 71

TOMLINSON, M.J. (1971), "Some Effects of Pile Driving on Skin Friction", *Conference on Behavior of Piles*, p. 107 - 114, ICE, London

TOMLINSON, M.J. (1980), *Foundation Design and Construction*, Pitman, London

TOMLINSON, M.J. (1987), *Pile Design and Construction Practice*, 3rd Ed., Viewpoint, London

TOURTELOT, H.A. (1973), "Geologic Origin and Distribution of Swelling Clays", *Proceedings of Workshop on Expansive Clay and Shale in Highway Design and Construction*, Vol. 1

TRB (1984), *Second Bridge Engineering Conference*, Transportation Research Record 950, Vol. 2, Transportation Research Board, Washington, DC

TURNER, JOHN P. AND KULHAWY, FRED H. (1990), "Drained Uplift Capacity of Drilled Shafts Under Repeated Axial Loading", ASCE *Journal of Geotechnical Engineering*, Vol. 116, No. 3, p. 470 -491

USBR (1974), *Earth Manual*, 2nd Ed., Bureau of Reclamation, Department of the Interior, Washington

U.S. NAVY (1982a), *Soil Mechanics*, NAVFAC Design Manual 7.1, Naval Facilities Engineering Command, Arlington, VA

U.S. NAVY (1982b), *Foundations and Earth Structures*, NAVFAC Design Manual 7.2, Naval Facilities Engineering Command, Arlington, VA

U.S. NAVY (1982c), *Soil Dynamics, Deep Stabilization and Special Geotechnical Construction*, NAVFAC Design Manual 7.3, Naval Facilities Engineering Command, Alexandria, VA

UPPOT, JANARDANAN O. (1980), "Damage to a Building Founded on Expansive Slag Fill", *Proceedings, 7th African Regional Conference on Soil Mechanics and Foundation Engineering*, Vol. 1, p. 311 - 314

VANIKAR, SUNEEL N. (1986), *Manual on Design and Construction of Driven Pile Foundations*, Revision 1, Report No. FHWA-DP-66-1, U.S. Department of Transportation, Federal Highway Administration, Washington

VESIĆ, ALEKSANDAR S. (1961), "Bending of Beams Resting on Isotropic Elastic Solid", ASCE *Journal of the Engineering Mechanics Division*, Vol. 87, No. EM2, p. 35 - 53

VESIĆ, ALEKSANDAR (1963), "Bearing Capacity of Deep Foundations in Sand", *Highway Research Record*, No. 39, p. 112 - 153, Highway Research Board, National Academy of Sciences, Washington, DC

VESIĆ, ALEKSANDAR S. (1973), "Analysis of Ultimate Loads of Shallow Foundations", ASCE *Journal of the Soil Mechanics and Foundations Division*, Vol. 99, No. SM1, p. 45 - 73

VESIĆ, ALEKSANDAR S. (1975), "Bearing Capacity of Shallow Foundations", *Foundation Engineering Handbook*, 1st Ed., p. 121 - 147, Winterkorn, Hans F. and Fang, Hsai-Yang, eds., Van Nostrand Reinhold, New York

VESIĆ, ALEKSANDAR S. (1977), *Design of Pile Foundations*, National Cooperative Highway Research Program, Synthesis of Highway Practice #42, Transportation Research Board, National Research Council, Washington

VESIĆ, ALEKSANDAR S. (1980), "Pile Group Prediction Symposium, Summary of Prediction Results", *Proceedings, Pile Group Prediction Symposium*, Federal Highway Administration, Washington

VIJAYVERGIYA, V. N. AND FOCHT, J. A., JR. (1972), "A New Way to Predict the Capacity of Piles in Clay", *Proceedings, 4th Annual Offshore Technology Conference*, Vol. 2, p. 865 - 874

VIJAYVERGIYA, V.N. AND GHAZZALY, OSMAN I. (1973), "Prediction of Swelling Potential for Natural Clays", *Proceedings of the Third International Conference on Expansive Soils*, Vol. 1, p. 227 - 236, Jerusalem Academic Press, Israel

WAHLS, HARVEY E. (1981), "Tolerable Settlement of Buildings", ASCE *Journal of the Geotechnical Engineering Division*, Vol. 107, No. GT11, p. 1489 - 1504

WAHLS, H.E. (1985), "Comparisons of Predicted and Measured Settlements", in *The Practice of Foundation Engineering: A Volume Honoring Jorj O. Osterberg*, p. 309 - 318, Krizek, Raymond J. et. al., eds., Dept. of Civil Engr., Northwestern Univ., Evanston, IL

WALKINSHAW, JOHN L. (1975), *Reinforced Earth Construction*, Report No. FHWA-DP-18, Federal Highway Administration, Arlington, VA

WARRINGTON, DON C. (1992), "Vibratory and Impact-Vibration Pile Driving Equipment", *Pile Buck* newspaper, Second October Issue, Pile Buck, Inc., Jupiter, FL

WELLINGTON, ARTHUR M. (1887), *The Economic Theory of the Location of Railways*, 6th Ed., John Wiley, New York

WELLINGTON, A.M. (1888), "Formulas for Safe Loads of Bearing Piles", *Engineering News*, Dec. 29, 1888; Reprinted in Wellington (1893), p. 22 - 33

WELLINGTON, A.M. (1892), "Mr. Foster Crowell on Pile Driving Formulas", *Engineering News*, Oct. 27, 1892; Reprinted in Wellington (1893), p. 52 - 73

WELLINGTON, A.M. (1893) *Piles and Pile-Driving*, Engineering News Publishing Co., New York

WESTERGAARD, H.M. (1938), "A Problem of Elasticity Suggested by a Problem in Soil Mechanics: Soft Material Reinforced by Numerous Strong Horizontal Sheets", *Contributions to the Mechanics of Solids, Dedicated to Stephen Timoshenko*, p. 268 - 277, MacMillan, New York

WHITAKER, THOMAS (1976), *The Design of Piled Foundations*, 2nd Ed., Pergamon Press, Oxford

WHITE, EDWARD E. (1962), "Underpinning", Chapter 9 in *Foundation Engineering*, G.A. Leonards, ed., McGraw Hill, New York

WHITE, L. SCOTT (1953), "Transcona Elevator Failure: Eyewitness Account", *Géotechnique*, Vol. 3, p. 209 - 214

WHITE, LAZARUS (1936), Discussion No. H-10, *International Conference on Soil Mechanics and Foundation Engineering*, Volume III, p. H-9

WHITNEY, CHARLES S. (1957), "Ultimate Shear Strength of Reinforced Concrete Flat Slabs, Footings, Beams, and Frame Members Without Shear Reinforcement", *Journal of the American Concrete Institute*, Vol. 29, No. 4, p. 265 - 298

WINKLER, E. (1867), *Die Lehre von Elastizität und Festigkeit* (On Elasticity and Fixity), H. Dominicus, Prague (in German)

WOODWARD, R. AND BOITANO, J. (1961), "Pile Loading Tests in Stiff Clays", *Proceedings, 5th International Conference on Soil Mechanics and Foundation Engineering*, Vol. 2, p. 177

WRAY, WARREN K. (1986), *Measuring Engineering Properties of Soil*, Prentice Hall, Englewood Cliffs, NJ

XANTHAKOS, PETROS P. (1991), *Ground Anchors and Anchored Structures*, John Wiley, New York

YORK, DONALD L. AND SUROS, OSCAR (1989), "Performance of a Building Foundation Designed to Accommodate Large Settlements", *Foundation Engineering: Current Principles and Practices*, Vol. 2, p. 1406 - 1419, F.H. Kulhawy, Ed., ASCE

YOUD, T. LESLIE (1989), "Ground Failure Damage to Buildings During Earthquakes", *Foundation Engineering: Current Principles and Practices*, F.H. Kulhawy, Ed., Vol. 1, p. 758 - 770, ASCE

ZEEVAERT, LEONARDO (1957), "Foundation Design and Behavior of Tower Latino Americana in Mexico City", *Géotechnique*, Vol. 7, p. 115 - 133

ZEEVAERT, LEONARDO (1983), *Foundation Engineering for Difficult Subsoil Conditions*, 2nd Ed., Van Nostrand Reinhold, New York

Abbreviations for societies and organizations:

ACI	American Concrete Institute
ADSC	Association of Drilled Shaft Contractors
AISC	American Institute of Steel Construction
API	American Petroleum Institute
ASCE	American Society of Civil Engineers
ASTM	American Society for Testing and Materials
FHWA	Federal Highway Administration
PCA	Portland Cement Association
PCI	Precast/Prestressed Concrete Institute
PTI	Post-Tensioning Institute

Index